THE FORMATION AND DYNAMICS OF GALAXIES

Jean Arp: Constellation according to the laws of chance. *Circa* 1930, The Tate Gallery, London.
© by ADAGP, Paris, 1974. *(See p. 218.)*

INTERNATIONAL ASTRONOMICAL UNION
UNION ASTRONOMIQUE INTERNATIONALE

SYMPOSIUM No. 58

HELD IN CANBERRA, AUSTRALIA, AUGUST 12–15, 1973

THE FORMATION AND DYNAMICS OF GALAXIES

EDITED BY

JOHN R. SHAKESHAFT

Mullard Radio Astronomy Observatory, Cavendish Laboratory, Cambridge, U.K.

SPRINGER-SCIENCE+BUSINESS MEDIA, B.V.

1974

TABLE OF CONTENTS

SELECTED QUOTATIONS

"I took this highly conservative view of the matter." (H. C. ARP, not in text)

"I don't know how to make galaxies – that's my problem." (G. R. BURBIDGE, p. 107)

"Interacting or colliding systems form an interesting subject of study, but are not new types of galaxies. (After a collision a car is a wreck, not a new type of car!" (G. DE VAUCOULEURS, p. 6)

"History has shown that the Universe is complex." (W. C. SASLAW, p. 332)

"I can't possibly swear that some sort of fissioning of galaxies might not – even write ALAR TOOMRE in the sky." (ALAR TOOMRE, p. 363)

"I don't want to be considered lacking in photographic skill so I would point out that I too have published pictures where the connection (between NGC 4319 and Markarian 205) was not present." (H. C. ARP, not in text)

"What generates the enormous energies that pour from galactic nuclei? How do they evolve? And are new physical laws needed to understand them? Even though I've only asked three questions so far, already the ratio of questions to answers is infinite." (W. C. SASLAW, p. 305)

"The result you get is so improbable that I would try any way of escape first." (J. H. OORT, p. 241)

'Philosophical questions cannot be answered!" (A. TOOMRE, p. 242)

"It must be a general phenomenon, and theoreticians should explain it." (G. DE VAUCOULEURS, p. 340)

EDITORIAL NOTE

This volume contains all of the papers read at Canberra with the exception of that by W. L. W. Sargent, for which there is an extended abstract. The papers by Ozernoy and Pronik were read by title only.

The invited papers (indicated in the Table of Contents) are in the order presented, but the contributed papers have been interpolated as seemed most appropriate and are not in the order of presentation. I have also rearranged the discussion to relate more closely the remarks with the papers to which they refer. The edited discussion was derived mainly from the slips completed by participants but I have also had recourse to the tape recording of the proceedings, in an attempt to recapture some of the spirit of the discussion.

I thank Mrs I. M. Laing for her invaluable help with the typing.

SCIENTIFIC ORGANIZING COMMITTEE

E. M. Burbidge (Chairman), M. S. Roberts (Secretary), H. C. Arp, G. Contopoulos, K. C. Freeman, I. R. King, D. S. Mathewson, J. L. Sersic, A. Toomre, G. M. Tovmasyan, H. van Woerden.

LOCAL ORGANIZING COMMITTEE

K. C. Freeman (Chairman), Joanne Craft.

ACKNOWLEDGEMENTS

Our warm thanks are due to the following organizations and individuals for their part in making this Symposium so successful:

The Australian Academy of Science for the use of their fine auditorium;

The Australian National University for acting as host and for providing accommodation;

Professor D. N. F. Dunbar, Deputy Vice-Chancellor of the Australian National University, for opening the Symposium;

Ken Freeman and his team from Mount Stromlo for their very efficient local organization;

Morton Roberts for acting as Secretary to the Scientific Organizing Committee;

Alar Toomre for help in many ways.

In addition, my personal thanks are due to my husband, Geoffrey Burbidge, for his unstinting help with the planning.

E. MARGARET BURBIDGE

The Editor acknowledges with thanks permission to use copyright material from: *Astronomy and Astrophysics*, Copyright The European Southern Observatory; *Astrophysical Letters*, Copyright Gordon and Breach, Science Publishers Ltd.; *Monthly Notices of the Royal Astronomical Society*, Copyright The Royal Astronomical Society; *The Astrophysical Journal*, University of Chicago Press, Copyright The American Astronomical Society.

LIST OF PARTICIPANTS

G. O. Abell, U.C.L.A., California, U.S.A.

R. J. Allen, Kapteyn Astronomical Laboratory, Groningen, The Netherlands

H. C. Arp, Hale Observatories, Pasadena, California, U.S.A.

J. Ashbrook, *Sky and Telescope*

E. Athanassoula-Georgala, N.R.C., Athens, Greece

J. E. Baldwin, Mullard Radio Astronomy Observatory, Cambridge, U.K.

L. Bautz, N.S.F., Washington, D.C., U.S.A.

F. Bertola, Padova, Italy

F. Biraud, Observatoire de Paris, Meudon, France

A. Blaauw, European Southern Observatory

J. G. Bolton, C.S.I.R.O., Parkes, Australia

R. N. Bracewell, Stanford University, California, U.S.A.

A. Brahic, Observatoire de Paris, Meudon, France

J. C. Brandt, N.A.S.A., Greenbelt, Maryland, U.S.A.

E. M. Burbidge, R.G.O., Herstmonceux, U.K.

G. R. Burbidge, U.C.S.D., La Jolla, California, U.S.A.

E. Capriotti, Ohio State University, U.S.A.

G. Contopoulos, University of Thessaloniki, Greece

R. G. Conway, Jodrell Bank, University of Manchester, U.K.

A. D. Code, University of Wisconsin, U.S.A.

D. P. Cox, University of Wisconsin, U.S.A.

I. J. Danziger, European Southern Observatory

R. D. Davies, Jodrell Bank, University of Manchester, U.K.

A. de Vaucouleurs, University of Texas, Austin, U.S.A.

G. de Vaucouleurs, University of Texas, Austin, U.S.A.

M. J. Disney, Mount Stromlo Observatory, Canberra, Australia

R. D. Ekers, Kapteyn Astronomical Laboratory, Groningen, The Netherlands

M. W. Feast, Radcliffe Observatory, Pretoria, South Africa

K. C. Freeman, Mount Stromlo Observatory, Canberra, Australia

S. C. B. Gascoigne, Mount Stromlo Observatory, Canberra, Australia

R. J. Gott III, Hale Observatories, Pasadena, California, U.S.A.

J. Graham, Cerro Tololo, Chile

K. Gyldenkerne, Copenhagen, Denmark

F. Hartwick, University of Victoria, B.C., Canada

A. Hayli, Besançon, France

J. Heidmann, Observatoire de Paris, Meudon, France

C. Hunter, Florida State University, U.S.A.

K. A. Innanen, York University, Toronto, Canada

J. B. Irwin, Newark State College, N.J., U.S.A.

A. Kalnajs, Mount Stromlo Observatory, Canberra, Australia

K. I. Kellermann, N.R.A.O., Charlottesville, Virginia, U.S.A.

F. J. Kerr, University of Maryland, U.S.A.

R. Kirschner, Hale Observatories, Pasadena, California, U.S.A.

W. Kovach, Aerospace Research Labs., Ohio, U.S.A.

R. Larson, Yale University Observatory, New Haven, Connecticut, U.S.A.

G. Larsson-Leander, Lunds Universitet, Sweden

B. M. Lewis, Carter Observatory, Wellington, New Zealand

P. O. Lindblad, Stockholm Observatory, Sweden

J. L. Locke, N.R.C., Ottawa, Canada

M. S. Longair, Mullard Radio Astronomy Observatory, Cambridge, U.K.

J. W.-K. Mark, M.I.T., Cambridge, Massachusetts, U.S.A.

D. S. Mathewson, Mount Stromlo Observatory, Canberra, Australia

L. Mestel, University of Manchester, U.K.

G. K. Miley, Sterrewacht te Leiden, The Netherlands

R. H. Miller, University of Chicago, U.S.A.

B. Y. Mills, University of Sydney, Australia

D. C. Morton, Princeton University Observatory, N.J., U.S.A.

W. Morton, Princeton University Observatory, N.J., U.S.A.

K. Nandy, Royal Observatory, Edinburgh, U.K.

J. H. Oort, Sterrewacht te Leiden, The Netherlands

P. Osmer, Cerro Tololo, Chile

M. W. Ovenden, University of British Columbia, Canada

P. J. E. Peebles, Princeton University, N.J., U.S.A.

M. Peimbert, Instituto de Astronomia, U.N.A.M., Mexico

A. V. Peterson, A.N.U., Canberra, Australia

B. Peterson, Mount Stromlo Observatory, Canberra, Australia

P. Pishmish de Recillas, Instituto de Astronomia, U.N.A.M., Mexico

J. Rickard, European Southern Observatory

J. A. Roberts, C.S.I.R.O., Sydney, Australia

W. W. Roberts, University of Virginia, Charlottesville, U.S.A.

B. J. Robinson, C.S.I.R.O., Sydney, Australia

E. E. Salpeter, Cornell University, Ithaca, N.Y., U.S.A.

R. Sancisi, Kapteyn Astronomical Laboratory, Groningen, The Netherlands

W. L. W. Sargent, Hale Observatories, Pasadena, California, U.S.A.

W. C. Saslaw, University of Virginia, Charlottesville, U.S.A.

M. P. Savedoff, University of Rochester, N.Y., U.S.A.

R. E. Schild, Smithsonian Astrophysical Observatory, Cambridge, Mass., U.S.A.

G. A. Seielstad, California Institute of Technology, Pasadena, U.S.A.

J. R. Shakeshaft, Mullard Radio Astronomy Observatory, Cambridge, U.K.

U. W. Steinlin, University of Basel, Switzerland

W. C. Tifft, University of Arizona, Tucson, U.S.A.

A. Toomre, M.I.T., Cambridge, Mass., U.S.A.

A. J. Turtle, University of Sydney, Australia

M.-H. Ulrich, University of Texas, Austin, Texas, U.S.A.

T. S. van Albada, Kapteyn Astronomical Laboratory, Groningen, The Netherlands

S. van den Bergh, David Dunlap Observatory, Ontario, Canada

P. C. van der Kruit, Hale Observatories, Pasadena, California, U.S.A.

H. van Woerden, Kapteyn Astronomical Laboratory, Groningen, The Netherlands

K. Wellington, Sterrewacht te Leiden, The Netherlands

A. E. Whitford, Lick Observatory, Santa Cruz, California, U.S.A.

R. Wielen, Astronomisches Recheninstitut, Heidelberg, F.R.G.

D. Wills, University of Texas, Austin, Texas, U.S.A.

B. Wills, University of Texas, Austin, Texas, U.S.A.

B. Wilson, Canterbury, New Zealand

R. F. Wing, Perkins Observatory Ohio, U.S.A.

G. Wlérick, Observatoire de Paris, Meudon, France

A. E. Wright, C.S.I.R.O., Sydney, Australia

STRUCTURE, DYNAMICS AND STATISTICAL
PROPERTIES OF GALAXIES

G. DE VAUCOULEURS

The University of Texas at Austin, Tex., U.S.A.

Abstract. The following empirical or physical parameters describing individual and statistical properties of galaxies are reviewed: morphological type including new or revised types; intrinsic luminosity
distribution functions of two main components (spheroid and disk); true ellipticities of different types;
characteristic scale parameters (effective diameter and luminosity density) of spheroidal and flat
systems; masses and densities of spheroidal, disk and mixed systems from velocity dispersion and
rotation velocities; rotation periods, maximum rotation velocities, angular velocities and momenta of
different types; neutral hydrogen masses and densities; spectral energy distributions and colour indices; number of independent parameters from principal component (factor) analysis; luminosity
functions and selection effects; clustering and space distribution; maximum density-radius and
velocity dispersion-radius relations in systems of galaxies.

1. Introduction

Galaxies may be regarded, in one view, merely as a bewildering collection of different
individual objects each with its own peculiarities, but in another view, adopted here,
as forming a class of physical systems characterizable by a small set of empirical or
basic physical parameters (Table I). The average values and distribution functions of
these quantities once known may be used as inputs to theories of galactic structure and
evolution and, at least ideally, they should be predictable as outputs of realistic cosmological models.

In particular we need to define

(1) a characteristic length or scale factor L^*, for example the effective radius r_e^* of
the volume V_e within which half the total mass is located or within which half the

TABLE I
Dominant empirical and physical parameters

Parameters		Stars	Galaxies
Empirical {		Spectral type	Hubble type
		Luminosity	Luminosity } a
		Mass	Mass $\mathfrak{M}_T$, $\mathfrak{M}_H$ }
	Present	Radius	Effective radius
		Temperature	and Ellipticity
Physical {		Pressure	Kinetic energy
		Angular Momentum	Angular momentum
	Initial	Chemical composition	Chemical composition
		Mass function	Mass function
		(Model of Protogalaxy)	(Cosmological model)

a and radial density distribution functions.

*ohn R. Shakeshaft (ed.), The Formation and Dynamics of Galaxies. 1–53. All Rights Reserved.
Copyright © 1974 by the IAU.*

total power or bolometric luminosity is emitted. When this information is not yet accessible, we may use as substitutes some parameters of the optical luminosity distribution, for example the e-folding length Λ or scale height in an exponential distribution (Freeman, 1970), the structural length in a quasi-isothermal distribution (Zwicky, 1937, 1957) or some other *metric* scale factor[‡]

(2) A characteristic mass $\mathfrak{M}^*$, perhaps the *total* mass $\mathfrak{M}_T$, if it can be defined and evaluated, or at least some indicative mass value $\mathfrak{M}_i$ derived by a fixed rule from length and velocity data, for example $\mathfrak{M}_i = V_M^2 R_M / G\mu$, where $V_M = V(R_M)$ is the maximum rotational velocity in a flat rotating disk (Bottinelli *et al.*, 1968), or $\mathfrak{M}_i = \lambda \sigma_v^2 r_e^2 / G$, where σ_v is the central velocity dispersion in a spheroidal system of effective radius r_e (Poveda, 1958; Poveda *et al.*, 1960); λ and μ are empirical factors chosen to achieve approximate equality between $\mathfrak{M}_i$ and $\mathfrak{M}_T$, the latter derived from more realistic models.

If both $\mathfrak{M}_T$ and the mass distribution function $\mathfrak{M}(r)/\mathfrak{M}_T$ can be estimated, the ratio $\varrho_e = \mathfrak{M}_T / 2V_e$ is, at least in principle, a well-defined measure of the average (effective) density of the system.

(3) A characteristic time scale T^*, be it some typical rotation period, for example that corresponding to the maximum rotational velocity $V_M(R_M)$ in flat systems, or some typical crossing time, for example r_e^*/σ_v in spheroidal systems.

If both ϱ_e and σ_v can be evaluated, two scale factors of great dynamical interest may be derived: the Jeans length $L_e = (\sigma_v^2 / G\varrho_e)^{1/2}$ and the free fall time $T_e = (G\varrho_e)^{-1/2}$.

(4) The total energy output ε_T in the form of electromagnetic radiation of all wavelengths (bolometric luminosity) and its normalized spectral distribution $\varepsilon(\lambda)/\varepsilon_T$. When it cannot be evaluated, the total absolute magnitude M in one of the standard colour bands (B or V) and at least two colour indices, say $B-V$ and $U-B$, are useful though rather unsatisfactory substitutes.

An important derived quantity is the (bolometric) mass-luminosity ratio $f^* = \mathfrak{M}_T / \mathfrak{L}^*$ or at least its optical counterpart, say $f_v = \mathfrak{M}_T / \mathfrak{L}_v$, corrected for self-absorption if appropriate (both in solar units).

(5) The normalized laws of luminosity and mass density distributions in galaxies of different morphological types, if possible per unit volume $J(\mathbf{R})$ and $\mu(\mathbf{R})$ – or at least per unit area in the face-on projection $J(r)$ and $\mu(r)$ – from which the radial dependence of the mass-luminosity ratio $f(r)$ is derived.

(6) The normalized laws of density distribution of interstellar neutral hydrogen in galaxies of different types $H(\mathbf{R})$ and the total hydrogen mass $\mathfrak{M}_H$. Important derived quantities are the ratio of hydrogen mass to total mass $h = \mathfrak{M}_H / \mathfrak{M}_T$ and the distance-independent ratio of hydrogen to luminosity $g = \mathfrak{M}_H / \mathfrak{L}^*$ (both in solar units).

Some other derived quantities that are more or less directly accessible are also important, for example the total net angular momentum A of a galaxy (or the average

---

[‡] Isophotal diameters are, in general, less satisfactory size indicators since they depend on both luminosity density and space orientation as well as pure scale factors. Commonly quoted 'photographic' diameters are even less satisfactory since they depend additionally on the apparent luminosity gradient (de Vaucouleurs, 1959a; Heidmann *et al.*, 1971).

angular momentum per unit mass) may give clues to initial conditions since it should be conserved in the evolution of an isolated galaxy (although galaxies can hardly be regarded as 'isolated' on a time scale of 10^{10} yr).

Finally, a property of great physical and dynamical significance, but which does not easily yield to quantitative expression or measurement, is the structural or morphological type T, e.g. in the revised Hubble scheme (de Vaucouleurs 1956a, 1959a, 1962, 1964; Sandage, 1961, 1973). In particular the stage t along the Hubble sequence, from ellipticals to Magellanic irregulars through lenticulars and spirals, correlates well with a number of measurable parameters, but cannot yet be precisely derived from such measurements. In the present state of the art, subjective classifications by experienced observers agree better than objective estimates from quantitative correlations (de Vaucouleurs, 1961; Corwin, 1968, 1970; Heidmann *et al.*, 1971; Brosche, 1973).

The present review will deal mainly with the 'normal' galaxies which are still by far the most common type of extragalactic objects and, as far as we know, are the major contributors to the mean density of space.

1.1. REMARKS ON SELECTION EFFECTS

The discovery of galaxies is severely limited by observational selection of surface brightness and apparent diameters. Figure 1, adapted from Arp (1965), illustrates this point. Objects having an average surface luminosity less than $\mu_B \simeq 27$ mag (arc sec^{-2}) ($\simeq 1 \, \mathfrak{L}_\odot$ pc^{-2}) are not optically detectable by present techniques. Objects having an apparent diameter less than $\sim 1''$ are not readily distinguishable from stars with current instruments. It is probably not by accident that the average surface brightness of the so-called 'normal' galaxies is only slightly above that of the night sky (de Vaucouleurs, 1957a).

Further selection effects arise in the formation of catalogues. No galaxy catalogue is complete to any given total magnitude or even 50% complete at some fixed magnitude. The effective limit is mainly a function of surface brightness b for large objects, of total luminosity bD^2 for smaller objects, and of apparent diameter D only for very small objects which necessarily must have high b. It is secondarily a function of shape and light concentration (i.e. morphological type) and subject to external factors such as star field density and interstellar extinction. There is very little quantitative information on the completeness factors involved and the relative space densities of different galaxy types other than 'normal' remain highly uncertain. All the distribution functions describing the statistical properties of galaxies are subject to this fundamental bias of unknown magnitude.

2. Morphology and Classification

The revised Hubble system is applicable to 95% or more of the galaxy population; its general validity and basic significance is demonstrated by the fact that the *stage T* along the Hubble sequence from E to I (or on a convenient numerical scale from

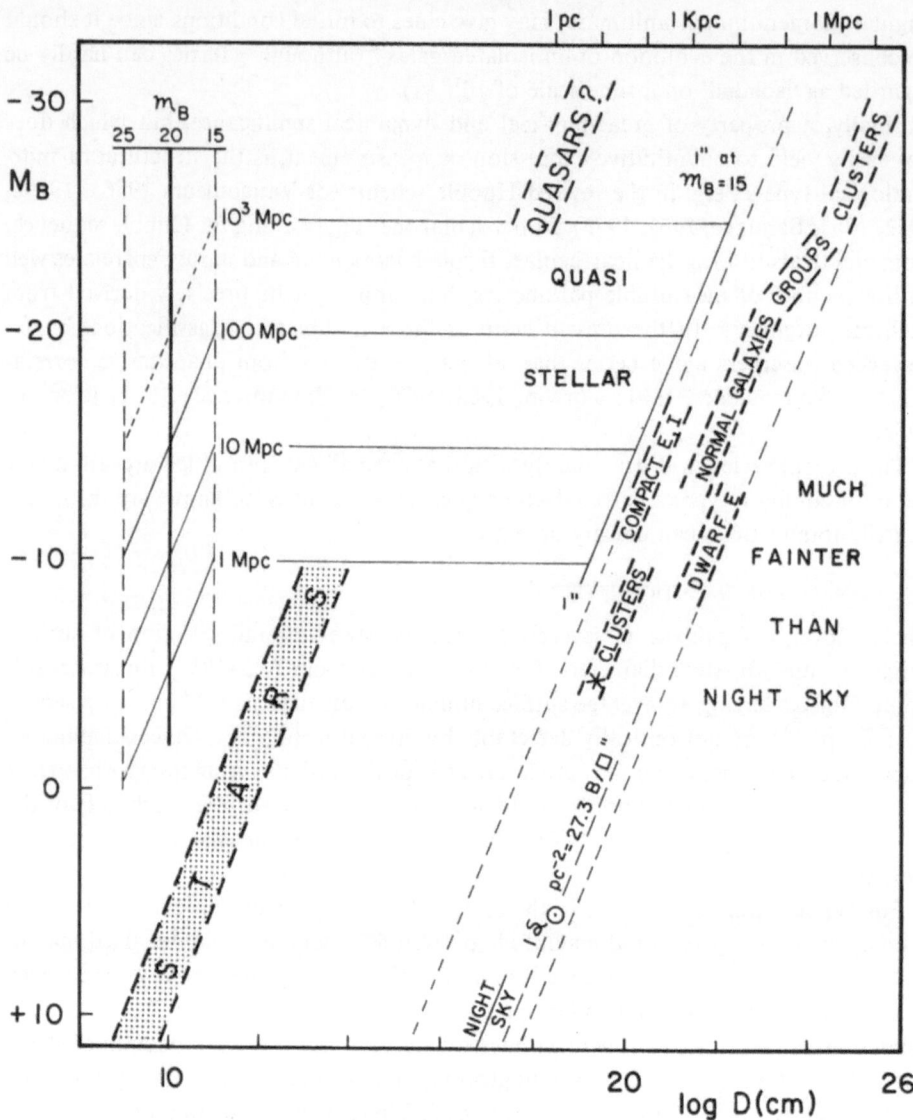

Fig. 1. Luminosity-diameter relation for stars and galaxies, adapted from Arp (1965). Normal galaxies and star clusters are restricted to a narrow range of surface brightness close to that of the night sky. Optical detection of galaxies smaller than ~ 1″ or fainter than ~ 1$\mathfrak{L}_\odot$ pc^{-2} is not possible by current techniques. Relation between apparent magnitude m_B, absolute magnitude M_B, linear diameter D and distance in Mpc for a 1″ apparent diameter is illustrated.

$t=-5$ to $+10$) correlates well with several fundamental parameters such as photometric structure, colour index, hydrogen content, etc. [‡] On the other hand the structural differences between *families* (A, AB, B) and *varieties* (r, rs, s) of lenticulars and spirals, that is the presence or absence of a bar or a ring, seem to reflect relatively minor differences of detailed dynamics rather than basic differences in physical properties and composition. In brief, the long axis of the 3-dimensional classification volume relates to the basic physics of galaxies, while the cross-section displays dynamical details.

The quantitative studies of recent years suggest some minor corrections and additions to the sequence of stages. For example Morgan's cD galaxies fit best at stage $t=-4 (E^+)$; compact ellipticals may be added as an extension (E^-) of the sequence to $t=-6$. Similarly, the dwarf compact Magellanic irregulars recently described as isolated extragalactic H II regions (Sargent and Searle, 1970) may be included in the revised Hubble sequence as an additional step (Im^+), $t=11$, beyond stage Im $(t=10)$. Low density irregulars of the IC 2574 and W-L-M nebula type are already recognized, being designated as dIm.

Irregular galaxies of the rare, non-Magellanic type, originally described by Hubble (1926) as merely 'chaotic', may be designated as 'type II' (Holmberg, 1950) or preferably I0 (de Vaucouleurs, 1963) because, except for their dust lanes, their structure and physical properties resemble most closely those of normal galaxies near the transition stage S0/a, $t=0$, between lenticulars and spirals (Heidmann *et al.*, 1971).

The revised classification sequence appears in Table II. A number of recently reported 'new' galaxy types deserve some discussion:

(a) *Gamma* $(\gamma; \gamma\text{-}R)$ *types:* as described by Vorontsov-Velyaminov (1962) γ-types

TABLE II
Revised Hubble sequence

Stage	t	-6	-5	-4	-3	-2	-1						
Type	T	E^- [a]	E [b]	E^+ [c]	L^-	L^0	L^+						

Stage	0	1	2	3	4	5	6	7	8	9	10	11
Type	S0/a [d]	Sa	Sab	Sb	Sbc	Sc	Scd	Sd	Sdm	Sm	Im [e]	Im+ [f]

[a] compact E.
[b] and dE.
[c] Morgan's cD.
[d] also I0.
[e] and dIm.
[f] compact Im.

[‡] The precision with which t may be estimated by a trained observer on plates having adequate resolution ($\sim 10^4$ picture elements in area) is indicated by the standard error $\sigma_t = 0.75$ (or 5% of the 15-step range) objectively derived through factor analysis by Brosche (1973) (Section 11). This is consistent with earlier estimates ($\sigma_t \simeq 1$) from intercomparison of independent classifications by several observers (de Vaucouleurs, 1962; Corwin, 1970).

are mainly late-type barred spirals, especially of the Magellanic type SBm. The Large Cloud itself, photographed at low resolution and with a short exposure, is a good example (Figure 2). The arms B_1, B_2 form the branches of the γ, the bar A is the stem. When on longer exposures the circular outer loop C is visible, the γ-R type is produced. In some cases γ-R may describe other galaxy types with a cardioid outer ring, e.g. NGC 3368 (Sandage, 1961). Many small Magellanic barred spirals appear γ-shaped on the Palomar Sky Survey prints.

(b) *Ring galaxies.* The ring-type galaxies of which Mayall's nebula was the first example (Smith, 1941; Baade and Minkowski, 1954; Arp, 1966), and the chaotic irregulars rich in emission regions, of which NGC 2444-45 and 6438 are two well studied examples (Sandage, 1963; Sersic, 1966, 1968a; Burbidge and Burbidge, 1972), appear to result from the interaction of a spiral or Magellanic-type system with a heavier and denser elliptical or lenticular galaxy (Sandage, 1963), or they may be produced by collisions between spirals and intergalactic hydrogen clouds‡ (Freeman and de Vaucouleurs, 1974).

(c) *Outer ring structures.* Although spiral patterns have attracted more attention, ring structures are perhaps equally common both in the central regions of lenticulars and spirals of the (r) variety (Randers, 1940; de Vaucouleurs, 1959b; Sandage, 1961) and around the main body of galaxies of various types. This outer (R) structure is often less conspicuous because of its lower surface brightness but it is not rare; it is especially frequent in late lenticulars and early type spirals, where it is observed in about 10 to 20% of these systems. Often a pseudo ring is formed by the faint extensions of the outer arms which are, therefore, not at all branches of 'spirals' but rather arcs of circles or ellipses. Plausible mechanisms for the trapping of particles in near-circular quasi-periodic orbits around a barred stellar system have been discussed by Danby (1965) and especially by Freeman‡‡ (de Vaucouleurs and Freeman, 1972).

3. Intrinsic Laws of Luminosity Distribution in Normal Galaxies

Detailed photographic or photoelectric surface photometry in one or two colours is now available for some 100 bright galaxies. Recent studies of this material, supplemented by integral photoelectric photometry of about 1000 systems in two or three colours (de Vaucouleurs, 1961; de Vaucouleurs and de Vaucouleurs, 1972), confirm the subjective impression from photographs (Sandage, 1961) that the luminosity distribution in normal galaxies may be resolved into two major components:

‡ Interacting or colliding systems form an interesting subject of study (Vorontsov-Velyaminov, 1959; Arp, 1966), but are not new types of galaxies. (After a collision a car is a wreck, not a new type of car!) Computer simulations (Toomre and Toomre, 1972) demonstrate that tidal distortions can produce many of the curious forms observed in interacting galaxies.

‡‡ The frequency of outer ring structures observed among Seyfert galaxies (three cases: NGC 1068, 4151 and 7469 out of a dozen Seyfert objects in the BGC) does not significantly exceed that in the general population in the same range of types. It seems unlikely that such ring structures, which are resolvable into stars in the nearer examples (Sandage, 1961), might be somehow induced by the transient nuclear activity.

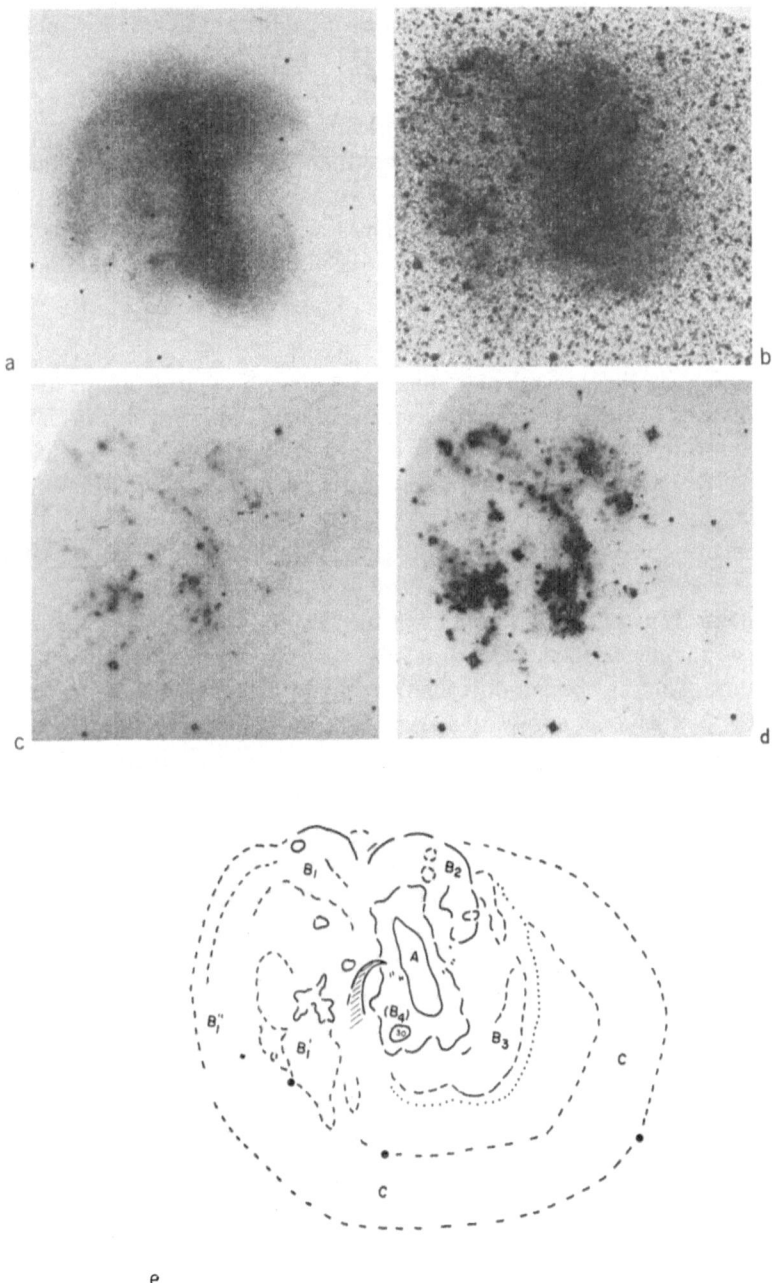

Fig. 2.   The Large Magellanic Cloud illustrates the γ-shaped appearance of SBm galaxies: (a) naked-eye view, after J. Herschel (1847); (b) visible light photograph from Mount Stromlo, 1953; (c, d) ultraviolet photographs by C. Young from the Moon, April 22, 1972 (courtesy Th. Page, NASA Apollo 16 mission), (c) $\lambda\lambda$ 1050–1550 (including Lα), exposure 1 min; (d) $\lambda\lambda$ 1230–1550 (excluding Lα), exposure 10 min; (e) key to Cloud structure (de Vaucouleurs and Freeman, 1972).

(I) a spheroidal component characteristic of elliptical galaxies (Hubble, 1930; de Vaucouleurs, 1948, 1953, 1962; Fish, 1964; King 1962, 1966) and of the bulge of early type spirals (de Vaucouleurs, 1958a; van Houten, 1961);

(II) a flat or exponential component characteristic of the disk of late-type spirals (de Vaucouleurs, 1958a, 1959b; Freeman, 1970) and of Magellanic irregulars (Ables, 1971; de Vaucouleurs and Freeman, 1972).

In normalized units the radial surface brightness (specific intensity) distribution in face-on systems is closely approximated by the following reduced luminosity laws

(I) $\log J_1(\varrho^*) = -3.33(\varrho^{*1/4} - 1)$,   (1)

(II) $\log J_2(\varrho^*) = -0.729(\varrho^* - 1)$,   (2)

where $\varrho^* = r^*/r_e^*$ is the equivalent radius $r_e^* = (A/\pi)^{1/2}$ of an isophote of area A, expressed in units of the (equivalent) effective radius r_e^* within which is emitted half the total luminosity; $J = I/I_e$ is the specific intensity I expressed in units of its value at $r^* = r_e^*$. Note that Equations (1) and (2) are intrinsic and dimensionless; there is no free, arbitrary fitting parameter.

Ellipticals obey very closely Equation (1), Magellanic irregulars Equation (2); intermediate stages of the Hubble sequence are well represented by linear combinations of (1) and (2); that is, all galaxies of the same morphological type follow the same reduced luminosity law, except for relatively minor effects of structural detail and inclination. This may be demonstrated directly by averaging the reduced luminosity curves $J(\varrho^*)$ of galaxies at several stages of the Hubble sequence and by matching the average luminosity curves for each stage t by model calculations with two free parameters, being the ratios of the scale factors (I_e or r_e^*) of the two components (I) and (II) which vary more or less continuously along the Hubble sequence. It can also be verified indirectly on the *integrated* luminosity functions $\Delta m(\varrho^*)$ (Figure 3) normalized to total luminosity L_T and effective radius, that is, in magnitudes

$$\Delta m(\varrho^*) = m(\varrho^*) - m_T = -2.5 \log \frac{L(\varrho^*)}{L_T},\qquad(3)$$

where

$$L(\varrho^*) = 2\pi \int_0^{\varrho^*} I(\varrho^*)\,\varrho^* \cdot d\varrho^* \quad \text{and} \quad L_T = L(\infty).$$

The curves for successive stages of the Hubble sequence display a smooth transition from type I (E) to type II (Im) and, in first approximation at least, form a one-parameter family depending only on the Hubble stage.

(a) *Sersic's formula*. A simple one-parameter expression of the overall luminosity distribution laws in galaxies of different types has been given by Sersic (1968b):

$$S(m) = k(m - m'')^n,\qquad(4)$$

where $m'' = m_0 - 1.086\,n$, and $m = -2.5 \log[0.921 \int_0^m I \cdot S(m)\,dm]$ is the integrated magnitude within the isophote of surface area S; m_0 is the surface magnitude at the centre, perhaps excluding a semi-stellar nucleus. The fitting parameter n varies smoothly with morphological type from $n = 8$ at type E to $n = 2$ at Im. Sersic's formula

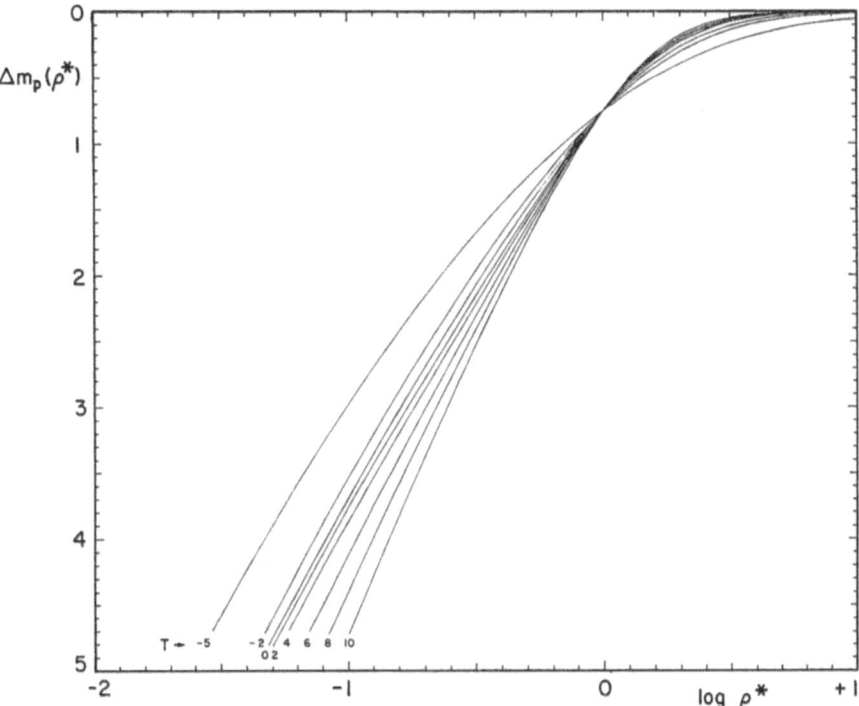

Fig. 3.   Normalized integrated luminosity distributions in galaxies as a function of stage t along Hubble sequence. The reduced integrated magnitude curves $\Delta m(\varrho^*)$, (normalized to $L_T=1$), vs reduced equivalent radius (normalized to $r^*_e = 1$) form a one-parameter family depending on stage t.

is more general than either Equation (1) or (2) and includes them as special cases for $n=8$ and 2, respectively. However, it does not define the characteristic scale factors as directly as (1) and (2) (For further comparisons between the Mt. Stromlo and Cordoba photometric parameters see de Vaucouleurs and Agüero (1973)).

(b) *Concentration indices.* Perhaps the simplest quantitative measure of the gross structural properties of a stellar system is a light concentration index. Such indices may be derived from the relative integrated luminosity curve $k(\varrho^*)=L(\varrho^*)/L_T$ giving the fraction of the total luminosity emitted within a given radius. In practice two indices have been used (de Vaucouleurs, 1962; Fraser, 1972; de Vaucouleurs and Agüero, 1973), namely those corresponding to the quartiles $k(\varrho^*_1)=\frac{1}{4}$ and $k(\varrho^*_3)=\frac{3}{4}$ of the integrated luminosity distribution function, that is

$$C_{21} = r^*_e/r^*_1 \quad \text{and} \quad C_{32} = r^*_3/r^*_e. \tag{5}$$

Note that $k(r^*_e)=\frac{1}{2}$ by definition.

Plots of C_{21} and C_{32} vs morphological type T (Figure 4) display a smooth transition from early to late galaxy types and are consistent with expected values if the luminosity distribution in galaxies can be resolved into two components, one following the $r^{1/4}$ law (1), and the other obeying the exponential law (2). For a pure $r^{1/4}$ distribution the

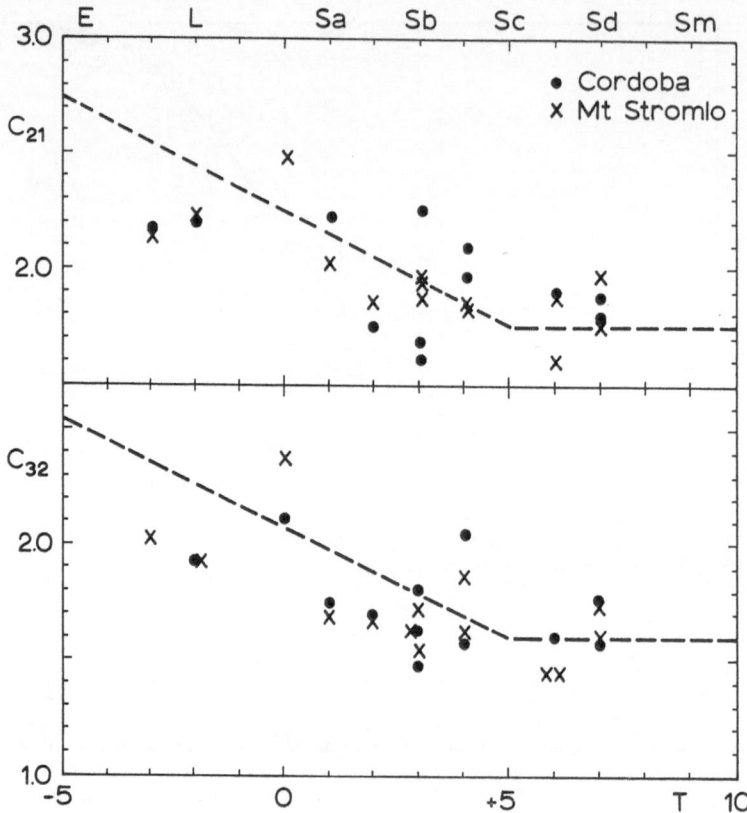

Fig. 4. Concentration indices versus stage along Hubble sequence. The trend reflects transition from a pure spheroidal distribution in ellipticals ($t = -5$) to a pure exponential distribution in late type spirals ($t \geqslant 6$) (de Vaucouleurs and Aguero, 1973).

calculated values are $C_{21} = 2.74$, $C_{32} = 2.55$ and for a pure exponential distribution $C_{21} = 1.75$, $C_{32} = 1.61$. It is clear, however, from Figure 4 that the observational scatter is rather large and, consequently, that concentration indices are not good indicators of morphological types nor acceptable substitutes for the Hubble type.

(c) *Tidally truncated dwarf ellipticals: King's formula.* Elliptical galaxies in gravitational interaction with a nearby system, in particular dwarf satellites of giant galaxies, appear to be tidally truncated, a situation similar to that of globular clusters in our Galaxy (von Hoerner, 1957; King, 1962). For such systems King (1962, 1966) has shown that the projected star density (and by extension the luminosity) distribution is well represented by the semi-empirical two-parameter expression

$$f = k \left[\frac{1}{\sqrt{1 + (r/r_c)^2}} - \frac{1}{\sqrt{1 + (r_t/r_c)^2}} \right]^2, \tag{6}$$

where r_c is a 'core' radius and r_t a 'tidal' radius or, in other words, a scale factor (r_c) and a free fitting parameter (r_t/r_c). This expression gives *a fortiori* a good representation of the luminosity distribution in isolated ellipticals for a sufficiently large value of r_t/r_c, when Equation (6) and (1) are numerically in close agreement (King, 1966). King's formula has been extensively used in analyses of the low density Sculptor-type satellites of our Galaxy which do not obey Equation (1) at all (Hodge, 1961; de Vaucouleurs and Ables, 1968; Hodge and Michie, 1969; Karachentseva, 1972).

A second type of truncated dwarf more recently identified is the high-density, compact type, for example NGC 4486B, the compact companion of M87 in the Virgo cluster (Rood, 1965; Faber, 1973) and NGC 5846A, the satellite of NGC 5846 in the NGC 5850 group (de Vaucouleurs, 1960; King and Kiser, 1973). Unlike the low-density dwarfs these high density systems obey Equation (1) quite well up to the centre, but suffer a sudden drop-off near the tidal limit as per Equation (6).[‡]

(d) *D galaxies*. As defined by Morgan (Mathews *et al.*, 1964) these are supergiant galaxies with an elliptical-like core in an extensive lenticular-like envelope. Variants or transition types are denoted as ED, DE and cD. Many coincide with core-halo radio sources in rich clusters, e.g. NGC 6166, classified cD. Others are isolated or in small groups, e.g. NGC 1316 (D) and 5128 (DE). Detailed photometry of several typical objects (Sersic, 1957; de Vaucouleurs, 1969, G. and A. de Vaucouleurs, 1970; Arp and Bertola, 1969, 1971) confirms Morgan's description and demonstrates quantitatively that the luminosity profiles can be resolved neatly into a spheroidal core obeying Equation (1) and an exponential corona obeying Equation (2), the latter with an unusually large scale factor (~ 100 kpc) and an abnormally low brightness factor.

Typically, for instance in M87, one-third of the total luminosity arises from the central spheroidal component and two-thirds from the corona. Such coronas contribute to – or perhaps are the sources of – the intergalactic luminosity observed in several rich galaxy clusters (de Vaucouleurs and de Vaucouleurs, 1970; Welsh and Sastry, 1971; Oemler, 1973). These diffuse optical sources are apparently co-extensive with the large radio and X-ray sources that have been detected in some clusters (Gursky *et al.*, 1971; Kellogg *et al.*, 1971; Owen, 1973) on a scale of order ~ 1 Mpc, but there is no evidence in the colour data that the optical source is of non-thermal origin; in particular the metallicity index Q agrees well with that of metal-poor globular clusters and Sculptor-type dwarf ellipticals.

4. True and Maximum Ellipticities of Galaxies

In principle the frequency function of true ellipticities ($e = 1 - c/a = 1 - q_0$) of spheroidal galaxies can be derived easily from the observed frequency function of apparent ellipticities ($\varepsilon = 1 - b/a = 1 - q$) under the assumption of random orientation of the

---

[‡] Because of the very small angular scale of these compact objects ($r*_e = 5\farcs0$ for NGC 4486B, de Vaucouleurs and Fraser, unpubl.) two-dimensional deconvolution of the raw photometry is essential to derive correctly the true luminosity distribution. Some early statements on the structure of NGC 4486B (Minkowski, 1962; Rood, 1965) are incorrect because of insufficient allowance for seeing effects.

spin axes. This problem has been often treated, mainly with respect to elliptical and lenticular galaxies and most recently by Sandage *et al.* (1970) from statistics of the BGC data. (See de Vaucouleurs, 1959b for a review of earlier work and references.)

Recently we have critically re-analyzed this problem on the basis of new and better data on the axis ratios of *isophotal* diameters of over 2000 galaxies. The data, if not quite complete for a given volume of space, were restricted to a limit of corrected face-on apparent diameters down to which the catalogue is $\sim 50\%$ complete.

Diameters refer now to a definite, photometrically calibrated isophote ($\mu_B = 25.0 \text{ mag} (\text{arc sec})^{-2}$) and, in comparisons between observed and calculated distributions, proper allowance was made for the effects of accidental measuring errors in b/a. Results are illustrated in Figures 5 and 6. The most significant conclusions are:

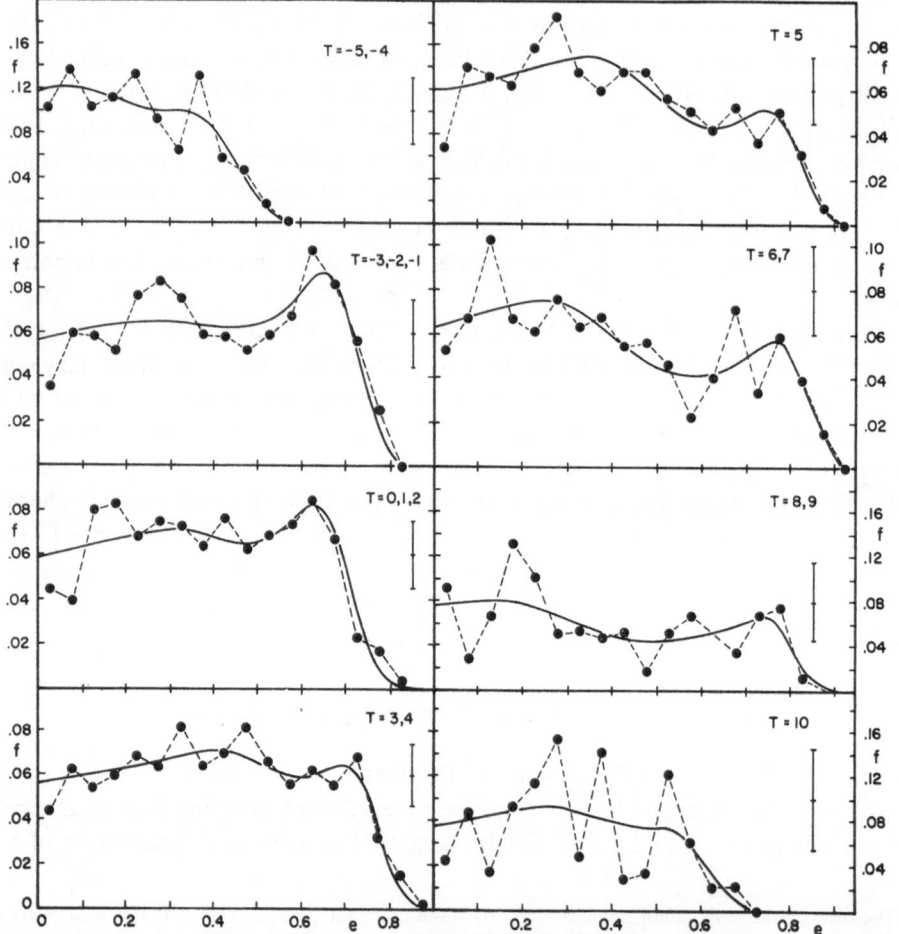

Fig. 5.   Relative frequencies of apparent ellipticities among galaxies having face-on isophotal diameters $D(0)_{25} \gtrsim 2'$. Curves are calculated frequencies for 2-component Gaussian models. Error bars (2σ) are expected random fluctuations for average count per 0.05 interval in e.

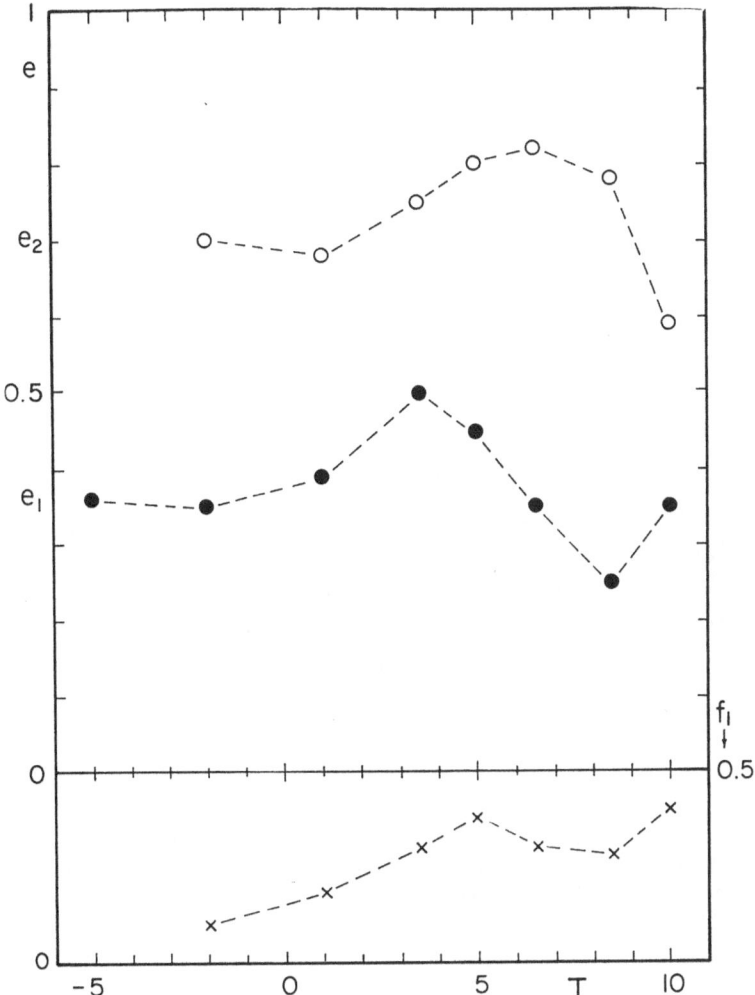

Fig. 6. Mean ellipticities of Gaussian component models of ellipticity distribution functions vs. stage along Hubble sequence. Ellipticity of flatter population component is maximum near type Sd ($t = 7$). At bottom: relative frequency f_1 of less flat component whose $\langle e_1 \rangle$ is near that of ellipticals ($t = -5$).

(a) the distribution of true ellipticities among E galaxies is definitely not uniform up to $\sim \frac{2}{3}$ ($q_0 \simeq \frac{1}{3}$), in agreement with the conclusions of Sandage $et\ al.$ (1970). The best fitting single gaussian model has $\langle e \rangle \simeq 0.36$ (i.e. E 6.5) with small cosmic dispersion $\sigma_e \simeq 0.1$. Spherical galaxies are rare or absent.

(b) Lenticular galaxies, as is well known, tend to be more flattened than ellipticals, but the analysis suggests that two components are present: a major group ($\sim 90\%$ of sample) with $\langle e \rangle = 0.65$ and $\sigma_e \simeq 0.1$, and a smaller group (10%) with $\langle e \rangle \simeq 0.35$ and $\sigma_e \simeq 0.05$.

(c) Spiral galaxies from S0/a to Sm have ellipticity functions very similar to the lenticulars' with a major group ($\sim 70\%$ of sample) with $\langle e \rangle = 0.7$ to 0.8 and, again, a minor one ($\sim 30\%$) with $\langle e \rangle \simeq 0.4$. These two components may be tentatively identified with the two classes of spirals having small bulges and large bulges respectively.

(d) Magellanic irregulars Im may also form two groups, but more nearly balanced with $\langle e \rangle \simeq 0.6$ (60%) and $\langle e \rangle \simeq 0.35$ (40%).

Figure 6 confirms the results of an earlier study of the Reference Catalogue data (Heidmann *et al.*, 1971) showing that $q_0 = c/a$ decreases steadily and smoothly along the Hubble sequence from E ($t = -5$) to Sd ($t = 7$) where it reaches a minimum. The flattest galaxies are observed at this stage, for example IC 2233 (Figure 7) which has an isophotal axis ratio of almost 10 to 1[‡]. This property is consistent with our previous

Fig. 7.   One of the flattest galaxies, IC 2233, an Sd system seen almost exactly edge-on, has no perceptible spheroidal component, only an extremely thin exponential disk. (McDonald Observatory, Struve 205-cm reflector, March 20, 1969; IIa-0, 61 min).

[‡] Similar and even more extreme examples are given by Vorontsov-Velyaminov (1967), but the axis ratios quoted (1/25 for MCG 7-30-11, 1/35 for MCG 4-9-60) are probably exaggerated by the well-known measuring errors first discussed by Holmberg (1946) and corrected in the BCG (de Vaucouleurs and de Vaucouleurs, 1964).

conclusion from photometry that the ratio of the two components (spheroidal and flat) of the luminosity distribution varies smoothly as a function of stage t and it suggests that the Hubble sequence from Sa to Sd is basically an angular momentum sequence (Section 8). Beyond stage Sd the ellipticity decreases rapidly so that Magellanic spirals (Sm) have typical axis ratios $q_0 \simeq 0.2$, and at the chaotic stage (Im) $q_0 \simeq 0.4$; nevertheless, contrary to early ideas, irregulars are by no means spherical (de Vaucouleurs, 1970a; Hodge and Hitchcock, 1966); most, perhaps all, are clearly rotating although more slowly than earlier types (Section 8). Examples of edge-on Magellanic systems such as NGC 55, 1507, 2188, etc. are given and illustrated by de Vaucouleurs and Freeman (1972).

5. Characteristic Parameters of Spheroidal Systems

There are not yet enough reliable photometric data to derive the mean values and distribution functions of the characteristic linear scale and density factors (r_e, I_e) of spheroidal systems or of the spheroidal component of mixed systems. Only isolated examples are known for some typical giant, intermediate and dwarf systems (Table III).

TABLE III

Effective parameters of spheroidal systems

Galaxy	m-M	Δ (Mpc)	$2r^*_e$ (arc min)	D_e (kpc)	$m_T(B)$	$M_T(B)$	$\mu'_e(B)$ (arc sec)$^{-2}$	L'_e $\odot$ pc^{-2}	Notes
N221	24.7	0.7	1.5	0.30	8.96	−15.74	19.22	1600	
N4486B	30.5:[b]	12.5?	0.16$_5$	0.60	14.28	−16.22	19.72	1000	
N4494	29.75	8.0	1.8	4.2	10.64	−19.11	21.29	230	
N3379	29.50	7.2	2.0	4.1	10.25	−19.25	21.13	270	
N224	24.7	0.7	27.	5.5	4.91	−19.79	21.45	200	a
N4649	30.5:	12.5?	2.5	9.1	9.73	−20.77	21.09	280	
N4486	30.5:	12.5?	3.5	12.8	9.50	−21.00	21.59	180	a
Fornax	22.0:	0.2:	23.	1.35	9.04	−13.0:	24.8	9	

[a] for spheroidal ($r^{1/4}$) component only.
[b] the colons indicate uncertain values.

(a) *High density systems* have effective diameters ranging from ~ 0.5 kpc in dwarf, compact E galaxies such as M32 and NGC 4486B, to ~ 10 kpc in giant systems such as NGC 4649 and the spheroidal components of M31 and NGC 4486, with a ratio of 20 to 1.

The specific intensity I_e at r_e (μ_e in mag per unit solid angle) is related to the average surface brightness μ'_e *within* r_e; for an $r^{1/4}$ distribution $\mu'_e = \mu_e - 1.40$. Values of μ'_e in B mag (arc sec)$^{-2}$ and in solar units per square parsec are listed in Table III. The range for these dense systems is from ~ 19.5 to ~ 21.5 mag (arc sec)$^{-2}$ (~ 800 to ~ 200 suns pc^{-2}), with a ratio of only 4 to 1.

In the denser systems the $r^{1/4}$ law apparently applies right up to the centre where, according to Equation (1), the surface brightness should be $2.5 \times 3.33 = 8.3$ mag brighter than at r_e; that is $\mu_0 = \mu_e - 8.3 = \mu_e' - 6.9$. For example, in the nucleus of M31 the observed luminosity peak, after correction for instrumental smoothing, is $\mu_0 \simeq 15.0$ mag (arc sec)$^{-2}$ (Johnson 1961), or ~ 6.5 mag brighter than μ_e', in close agreement with the $r^{1/4}$ formula. The situation is about the same for M32, where, as in M31, the central luminosity density approaches or possibly exceeds $10^5 \, \mathfrak{L}_\odot \, pc^{-3}$ (de Vaucouleurs, 1953; Kinman, 1965).

(b) *Low density dwarf spheroidal systems* which do not obey the $r^{1/4}$ law may be characterized by their core and tidal radii. Examples have been collected by van den Bergh (1968) and Karachentseva (1972) mainly from star counts by Hodge (1961–65). Typical values are in the range 0.2–0.3 kpc for r_c and 0.8–1.2 kpc for r_t, at least for the dwarf satellites of our Galaxy. Photometric effective radii are not yet known for these objects, except for Fornax where $r_e^* = 11'.5 = 0.7$ kpc (de Vaucouleurs and Ables, 1968); for comparison $r_c = 0.9$ kpc and $r_t = 3.1$ kpc from star counts (Hodge, 1961).

The light concentration in low density dwarfs is slight and the central area with radius $r \simeq r_e^*/2$ has a nearly constant surface brightness; for example in Fornax the maximum reaches only $\mu_0 = 24.6$ mag (arc sec)$^{-2} = 11 \mathfrak{L}_\odot \, pc^{-2}$, while at $r = r_e^* = 10' = 0.65$ kpc, $\mu_e = 25.2$ mag (arc sec)$^{-2}$ and the average between $r = 0$ and $r = r_e^*$ is $\mu_e' = 24.8$ mag (arc sec)$^{-2} = 9 \mathfrak{L}_\odot \, pc^{-2}$ (de Vaucouleurs and Ables, 1968; Hodge, 1973) or 1% of the corresponding light density in M32. Preliminary photometry of Sculptor and other nearby dwarfs indicates even lower densities, $\mu_0 > 25$ mag (arc sec)$^{-2}$ and down to the current detection threshold for such objects ($\mu_0 \simeq 27$ mag (arc sec)$^{-2} \simeq$ $\simeq 1 \mathfrak{L}_\odot \, pc^{-2}$) which is 10^{-4} to 10^{-5} of the central density of M32. Large numbers of dwarfs with such densities or lower ones might remain undetectable by current techniques.

6. Characteristic Parameters of Flat Systems

Although photometric information is still very incomplete, statistical studies of available data (Freeman, 1970; Heidmann *et al.*, 1971) lead to rather remarkable general conclusions. As noted in Section 3 the flat (disk) components of lenticulars and spirals follow closely the exponential luminosity law (2) which may also be written as

$$I(r) = I_0 \exp(-\alpha r), \tag{2'}$$

where α is the inverse of the scale length Λ; it is measured by the photometric gradient $G(r) = d(\log I)/dr$ since $\Lambda = -0.4343/G(r)$. For a pure exponential distribution the relation between effective radius r_e and scale length Λ is $r_e = 1.6785\Lambda$, and the relation between central and effective intensities is $I_0 = 0.729 \, I_e$ or, in magnitudes, $\mu(0) = = \mu_e - 1.82$. Similarly, the average surface brightness within r_e is $\mu_e' = \mu_e - 0.70 =$. $= \mu(0) + 1.12$.

Freeman (1970) has collected values of $\mu(0)$ and Λ for 36 disk galaxies having well-defined exponential components and small enough inclination angles that the corresponding values of $\mu(0)_c$, corrected to $i = 0$ and to zero galactic absorption, could be reasonably calculated. A number of correlations with morphological types emerge from this sample; in particular

(1) Λ has a range of about 10 to 1, from 0.5 kpc to $\sim$5 kpc for types earlier than Sc;

(2) the *maximum* value of Λ decreases from $\sim$5 kpc at Sc ($t=5$) to $\sim$1 kpc at Im ($t=10$), confirming earlier results on the dependence of galaxy 'diameters' on Hubble type (cf. Holmberg, 1950; de Vaucouleurs, 1959b, p. 314–5);

(3) for some three-quarters of the sample (28 out of 36), $\mu(0)_c$ varies little from $\langle\mu(0)_c\rangle=21.65$ mag (arc sec)$^{-2}$ ($\sigma\simeq0.3$ mag) and is independent of morphological type from L^- ($t=-3$) to Im ($t=10$) (Figure 8). However, some individual values depart from this mean by up to $+2$ mag (IC 1613) and -3 mag (NGC 5236).

Since the total luminosity of the exponential disk is $L_T=2\pi I_0\Lambda^2$, the result that $I_0\simeq$ constant implies that a close correlation must also exist between the total absolute magnitude of the disk and the scale length (Figure 9). This conclusion is in general agreement with, and accounts for the results of, earlier studies demonstrating a rather close luminosity-diameter correlation for spiral galaxies (Holmberg, 1958, 1964; de Vaucouleurs, 1959b; Heidmann, 1967, 1969; Heidmann *et al.*, 1971).

Nevertheless, two questions present themselves:

(1) is $\langle\mu(0)\rangle$ really independent of galaxy type or is it an effect of bias in the selec-

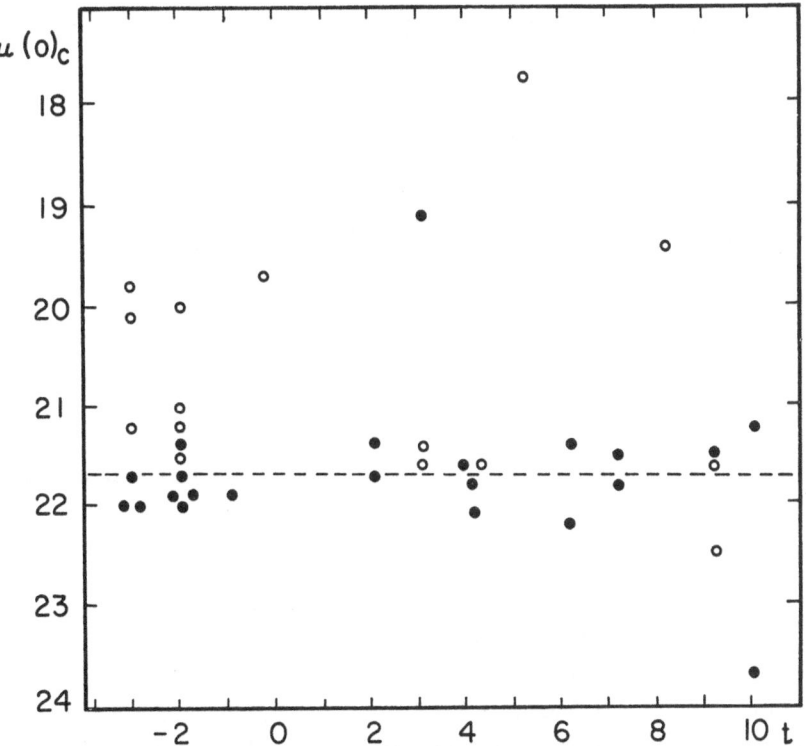

Fig. 8. Corrected face-on central luminosity $\mu(0)_c$ for exponential disks of 35 galaxies vs morphological type, after Freeman (1970). Note that $\langle\mu(0)_c\rangle$ is independent of Hubble type, but some individual galaxies are aberrant.

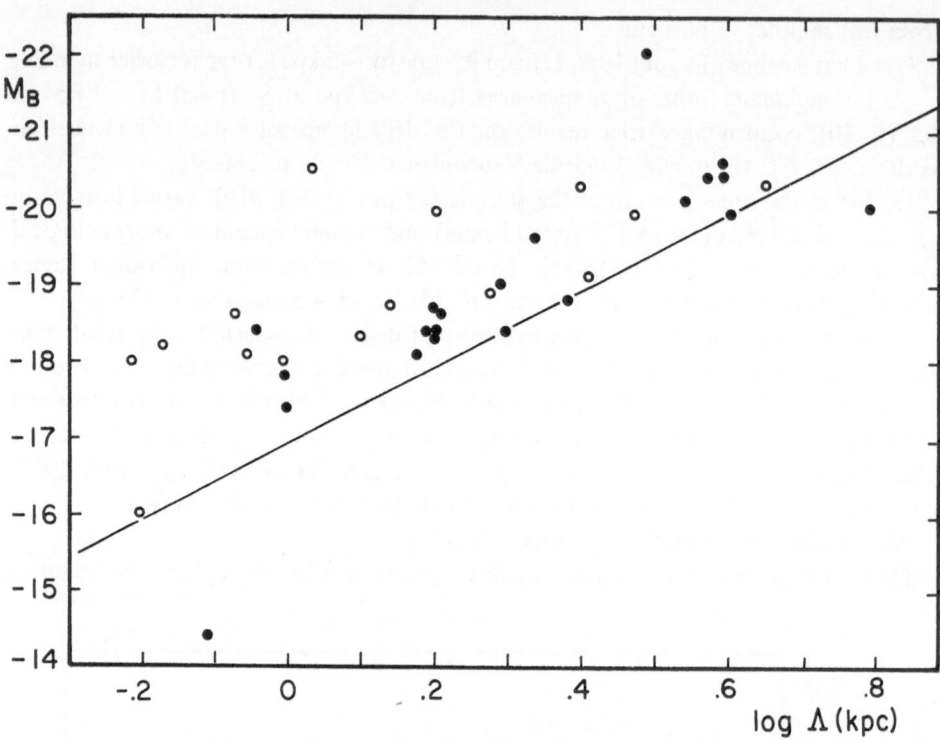

Fig. 9.   Correlation between absolute magnitude M_B and length scale $\alpha^{-1} = \Lambda$(kpc). Straight line
is theoretical relation for exponential disk with $B(0)_e = 21.65$ mag (arc sec)$^{-2}$.

tion of objects for detailed photometry (there is very little photometry of low-surface-
brightness objects)?

(2) is the dispersion about the mean really as small as 0.30 mag or, again, is it a
result of selection effects in a small sample (8 objects out of 36 were rejected in the
analysis)?

Some recent results suggest that both questions must be answered affirmatively at
least for the early and late-type systems:

(1) the discovery of large and very faint coronas around some early-type cD galaxies
(Section 3) demonstrates the existence of at least *some* exponential components as
faint as $\mu(0)_c \simeq 24.5$ mag (arc sec)$^{-2}$ (de Vaucouleurs, 1969); the elongated shape of
the isophotes (Arp and Bertola, 1969, 1971) suggests that the coronas are flattened;

(2) comparison of several late-type spirals and irregulars with pure exponential
distributions (de Vaucouleurs and Freeman, 1972, Fig. 47) indicates among them a
range of at least 3.0 mag in $\mu(0)$ from ~ 20.5 to ~ 23.5 mag (arc sec)$^{-2}$. As to other
galaxy types, some lenticulars and spirals with very faint disks are known to exist,
e.g. NGC 5365 (de Vaucouleurs, 1956a), but no photometry is available for them.

7. Characteristic Masses and Densities

Too little information is available yet to estimate reliably the total mass and density distribution functions of galaxies either in general or of particular types. However, fragmentary results on some typical objects give useful orders of magnitudes.

7.1. SPHEROIDAL SYSTEMS: VELOCITY DISPERSION

The classical method assumes that a spheroidal system may be approximated by a spherical distribution of point masses in stable, statistical, dynamical equilibrium with a constant velocity dispersion σ_v. Then, with the further hypothesis that the mass-density distribution in space may be derived from the projected luminosity-density distribution, i.e. $f = \mathfrak{M}/\mathfrak{L} = $ constant, and that the latter obeys the $r^{1/4}$ law (Poveda, 1958, 1960) – or some equivalent distribution – the total mass is given by

$$\mathfrak{M}_T \simeq 3r_e \sigma_v^2 / G, \tag{7}$$

where σ_v is the central velocity dispersion. It is derived from the observed line-of-sight component by $\sigma_v = \lambda \sigma_r$ where λ is a factor allowing for the shape of the velocity ellipsoid within the range $1 \leqslant \lambda \leqslant 3$ from linear (radial oscillations) to spherical (isotropic).

Many assumptions and approximations are involved here:

(1) the hypothesis $f = \mathfrak{M}/\mathfrak{L} = $ constant, independent of radius, conflicts with the observation that the colour index is not strictly constant but decreases outwards from the centre;

(2) the proper value of λ is uncertain; most authors argue that $2 < \lambda < 3$ at least near the centre (Poveda et al., 1960; Burbidge et al., 1961);

(3) the basic assumption of sphericity is contradicted by the recent analyses of distribution of ellipticity (Section 4). As shown by King (1961), even a small intrinsic ellipticity $e = 1 - c/a$ implies a rotational kinetic energy T of the same order as that in random motions W, and in a first approximation (homogeneous spheroid) $T/(W+T) = = (8/5) e$. Thus for the average true ellipticity of elliptical galaxies $e = 0.36$ (Section 4), $T/(W+T) \simeq 0.6$ or $T \simeq 1.5 W$, implying a total mass 2.5 times that calculated for a globular non-rotating model (King, 1962).

(4) the assumption $\sigma_v = $ constant, independent of r, cannot be exact since the only exact equilibrium solution is the isothermal gas sphere which has infinite mass and radius; σ_v must decrease outwards (Woolley and Robertson, 1956; Mitchie, 1963; King, 1966). Up to now technical difficulties have prevented direct measurement of $\sigma_r(r)$ except at $r \simeq 0$. However, a slight compensation arises through the fact that the line width measured at $r = 0$ is a luminosity-weighted average along the line of sight.

(5) the derivation of σ_r from the observed line broadening in the galaxy spectrum compared with a stellar spectrum of 'matching' spectral type (Burbidge et al. 1961; Minkowski, 1962; Morton and Chevalier, 1972) is subject to systematic errors; in particular the composite nature of galaxy spectra introduces a spurious broadening which, until recently, had not been evaluated.

(6) Finally, the standard simplifying assumption of a unique stellar mass value

in the theoretical model needs re-examination even if equipartition is not as pronounced as was once presumed (de Vaucouleurs, 1953; Woolley, 1954).

It is apparent that values commonly quoted for the masses of spheroidal systems (Poveda, 1961; Fish, 1964) are subject to large uncertainties. Some authors have concluded that the only quantities that can be safely inferred are the central density $\varrho(0)$ and mass-luminosity ratio (King and Minkowski, 1972). Recent determinations by photoelectric scanning or television techniques (Morton and Chevalier 1972, 1973; de Vaucouleurs, 1973) using more rigorous model fitting procedures suggest, however, that the older photographic estimates of $\sigma_r(0)$ are often too large for various reasons, in particular the neglect of compositeness in galaxy spectra. Some old and new values of $\sigma_r(0)$ are compared in Table IV. The net effect is a significant reduction of previous mass estimates, but more observations by modern techniques are needed to replace all older data.

For the present, provisional conclusions are:

(1) total masses of spheroidal (mainly E) galaxies cover an extremely wide range $(10^5 < \mathfrak{M}/\mathfrak{M}_\odot < 10^{13})$ from the Local Group 'pygmy' systems to supergiant cD galaxies (Table V).

TABLE IV

Central radial velocity dispersion $\sigma_r(0)$ in spheroidal systems

NGC	Type	Photographic			Photoelectric	
		[1]	[2]	[3]	[4]	[5]
221	−6	100	96	–	60	60
224	+2	225	–	–	150	120
3115	−3	205	–	–	230[a]	215
3379	−5	–	187	–	125	–
4486	−4	490:	–	550	–	–
4486B	−6	370:	–	–	160:	–
4494	−5	–	–	–	–	160
7332	−2	–	–	–	–	160

[1] Minkowski 1962.
[2] Burbidge et al. (1961).
[3] Brandt and Roosen (1969).
[4] scanner: de Vaucouleurs (1973).
[5] SEC Vidicon: Morton and Chevalier (1972, 1973).
[a] may be affected by rotation of nucleus.

TABLE V

Masses and mass/luminosity ratios of spheroidal systems

Galaxy	Examples	Mass range[a]	$\mathfrak{M}/\mathfrak{L}^*_B$
Supergiant E and cD	N4486, N4889	$> 10^{12}$	> 30
Giant E, bulges Sa-Sb	N224, N3379	$10^{10}-10^{12}$	10–30
High-density dwarfs	N221, N4486B	10^8-10^{10}	3–10
Low-density dwarfs	For, Scl	10^6-10^8	1–3
Extreme dwarfs	Dra, UMi	$< 10^6$	$\sim 1?$

[a] solar units.

(2) the average mass-luminosity ratio has a much smaller range of variation from ~ 2 in low-density dwarfs whose stellar population is similar to that of metal-poor globular clusters (McClure and van den Bergh, 1968; de Vaucouleurs and Ables, 1968) to perhaps ~ 50 in high density giants whose spectra and colours have super-metal rich characteristics (Spinrad *et al.*, 1971; van den Bergh, 1972a). The relationship is illustrated in Figure 10, after Einasto and Kaasik (1973). This correlation between $\mathfrak{M}/\mathfrak{L}$ and $\mathfrak{M}$ gives some hope of deriving the mass function from the luminosity function (Section 12).

(3) in giant systems the mass-luminosity ratio decreases outwards from high values, $f_B \sim 50$, in the core to low values, $f \sim 3$, in the 'halo' (or corona) which, as judged by its colour, has a metal-poor population similar to that of dwarf ellipticals (Figure 11). Models of spheroidal systems based on this concept agree well with observations (Einasto, 1972a).

7.2. Flat systems: rotation

In flat systems the z component of the velocity dispersion is small compared with the rotational velocity and contributes little to the total kinetic energy. The mass may be derived from thin disk or flat spheroid approximations by a number of well-known standard methods (Perek, 1950; Kuzmin, 1952; de Vaucouleurs, 1959b; Brandt, 1960; Burbidge *et al.*, 1960, 1963; Rubin *et al.*, 1964).

A large number of mass estimates of spiral galaxies have been derived during the past 20 yr from optical and radio (21-cm) observations of rotational velocities. Results have been collected and analyzed in several recent review papers (Vorontsov-Velyaminov 1970a, b; Kogure and Toya, 1970). However, many such mass estimates are at best lower limits only; in particular, mass values derived from polynomial fits of rotation curves, by disk models having an equatorial radius equal to the maximum range of optically measured velocities, refer to some unknown fractions of the total masses, since in all cases galaxies can be detected photometrically far beyond the range of spectroscopic data. Radio observations suffer much less from this limitation since 21-cm emission is frequently measurable beyond the range of optical detection; however, radio observations are often hampered by insufficient resolution, except in the largest galaxies. Recently, realistic models consistent with both spectroscopic and photometric data have been developed for a few nearby galaxies (Einasto, 1968–1972a).

Late-type and Magellanic barred spirals SBm with little or no spheroidal component have typical masses of the order of $10^{10} \mathfrak{M}_\odot$ and $\mathfrak{M}/\mathfrak{L} \simeq 3$ (de Vaucouleurs and Freeman, 1972); the latter value may be typical of the flat component of spirals. Flat systems of slightly earlier types Sc-Sd, for example M33, have masses of the order of $3 \times 10^{10} \mathfrak{M}_\odot$ with $\mathfrak{M}/\mathfrak{L} \simeq 5$ (Gordon, 1971; Huchtmeier, 1973; Warner *et al.*, 1973).

Freeman (1970) has pointed out that if $\mathfrak{M}/\mathfrak{L} = $ constant in the disk, the exponential distribution of luminosity (Section 3) implies that the surface-density distribution of mass is also exponential $\varrho(r) = \varrho_0 \exp(-\alpha r)$ and the total mass of the disk is

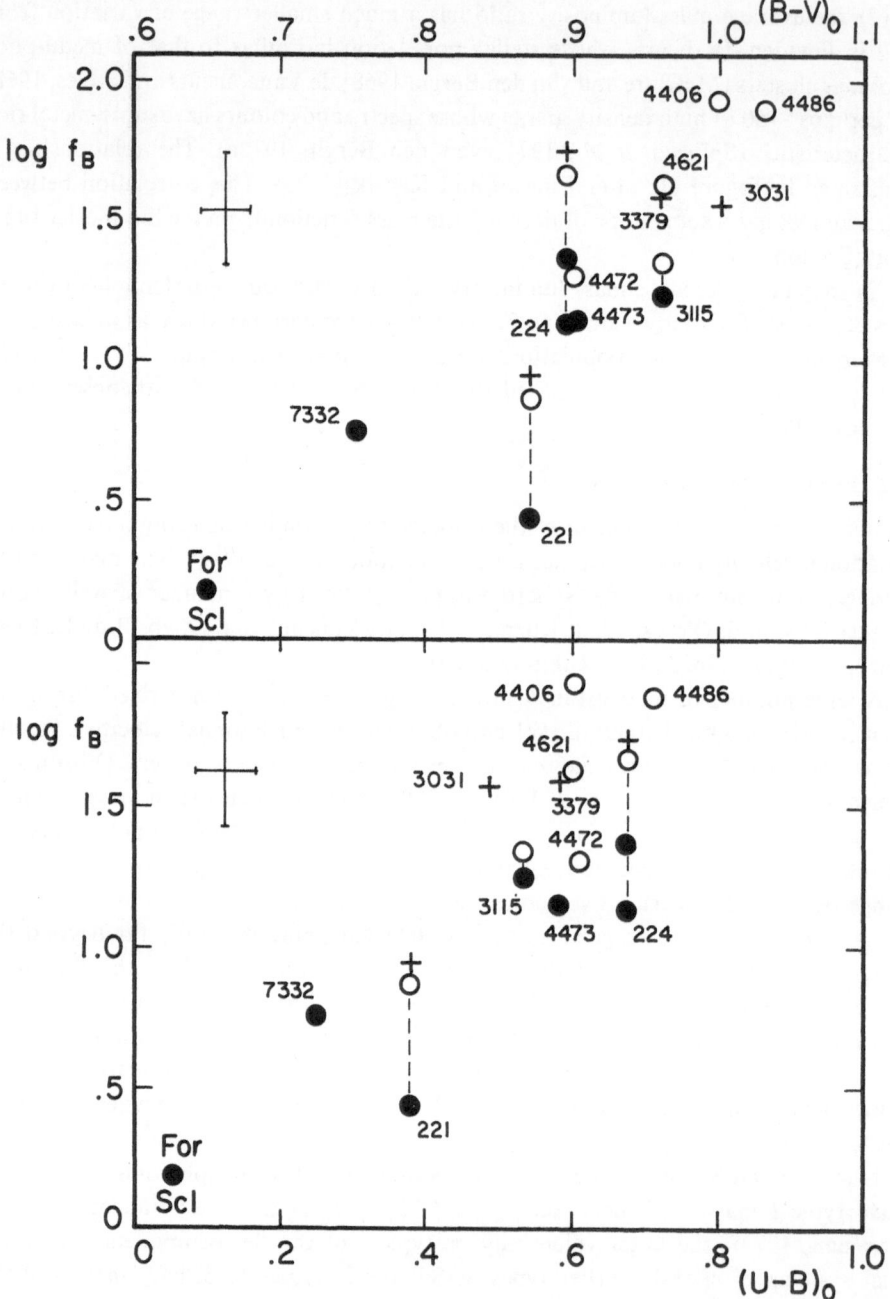

Fig. 10. Correlation between mass/luminosity ratio f_B and colour indices $(B - V)_0$, $(U - B)_0$ for spheroidal systems, after Einasto and Kaasik (1973). Recent photoelectric estimates of $\sigma_r(0)$ tend to give lower masses (dots) than earlier photographic data (circles). Estimates of f_B from spectro-photometric models of galaxy populations (crosses) are also too high.

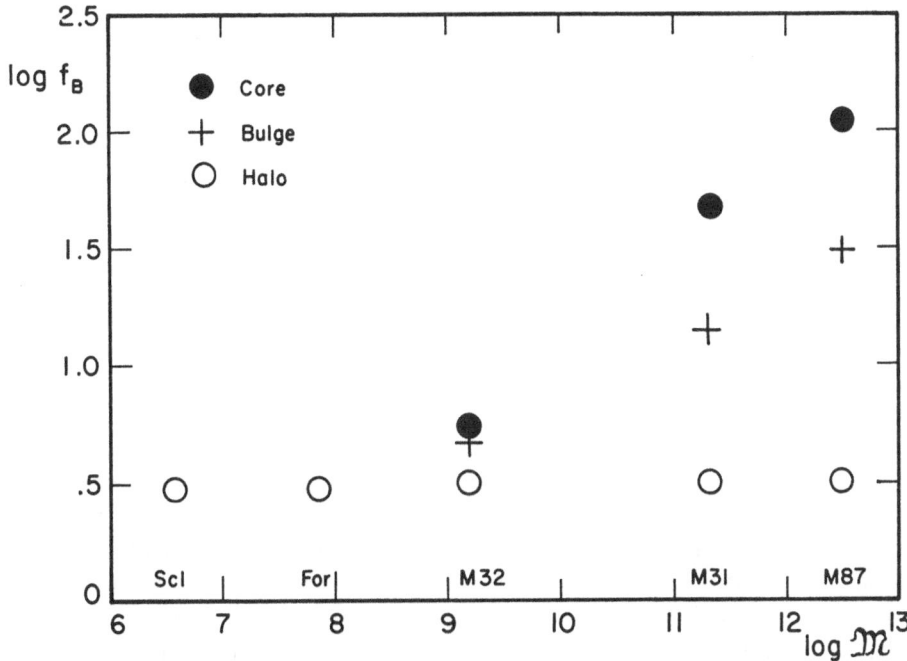

Fig. 11. Mass/luminosity ratio f_B in the core, bulge, and halo of spheroidal systems as a function of total mass $\mathfrak{M}_T$, after Einasto (1972). The halo component is the only population present in low-density dwarf ellipticals (Scl, For).

$\mathfrak{M}_D = 2\pi\varrho_0/\alpha^2$. The calculated rotation curve of an exponential disk agrees well with observations of some flat systems.

7.3. MIXED SYSTEMS: INDICATIVE MASSES

In galaxies of intermediate types from L to Sb or Sbc, both spheroidal and flat components contribute significantly to the total mass and potential energy of the system. Random motions and rotation make comparable contributions to the total kinetic energy and both must be taken into account (Oort, 1940, 1965). Realistic mass models of such galaxies are hampered by insufficient data on the velocity dispersion, and relatively detailed results are available only for a few systems including NGC 3115 (Oort, 1940), NGC 4111 (van Houten, 1961) and especially M31 (Einasto, 1972b).

For statistical purposes the best that can be done at present is to compute an *indicative* mass $\mathfrak{M}_i = V_M R_M^2/G\mu$, where the maximum rotational velocity $V_M = V(R_M)$ is derived from (a) optical and radio rotation curves for the larger galaxies, and (b) the total width W of the 21-cm profile (corrected for band-width and projection factors) for unresolved objects.

In case (b) R_M is not directly observable, but it may be estimated from an optical diameter with which it is correlated by case (a) data. Then the scale factor μ is chosen empirically so that on the average $\mathfrak{M}_i \simeq \mathfrak{M}_T$ for the galaxies with total masses derived by more detailed model fitting.

For example, the Nançay group (Bottinelli *et al.*, 1968; Heidmann, 1969) has adopted

$$\mathfrak{M}_i = 3 \times 10^4 \, aW^2 \quad \text{(solar units)}, \tag{8}$$

where the isophotal *diameter* a (in kpc) is on Holmberg's system (1958) and W is the corrected total line width (in km s^{-1}). The coefficient gives $\langle \mathfrak{M}_T/\mathfrak{M}_i \rangle \simeq 1$ within a factor 3 for galaxies which have been best observed at 21 cm. This precision is good enough compared with the mass range of hydrogen-rich galaxies which covers almost 3 orders of magnitudes.

An important result of this approach is a striking confirmation of the conclusion that $\mathfrak{M}/\mathfrak{L} = $ constant, independent of galaxy type for $t > 0$[‡] (Figure 15).

8. Rotation Periods, Maximum Velocities, Angular Velocities and Angular Momenta

Early analyses of line inclinations in galaxy spectra (Mayall, 1948, 1960; Mayall and Lindblad, 1970) revealed a loose correlation between rotation period P and Hubble type, in the sense that early types rotate faster than later types; a range from $P < 10^7$ yr for $t < 0$ to $P > 10^8$ yr for $t > 5$ was indicated. The correlation was rather loose because the angular velocity $\omega = 2\pi/P$ decreases rapidly from the centre outwards in most galaxies and the radius R_l of the 'linear' inner region where $\omega \simeq$ constant varies with galaxy type (Figure 12) and also with apparent diameter (i.e. angular resolution). Nevertheless, the basic trend has been confirmed by more recent studies (e.g. Vorontsov-Velyaminov, 1970c; Table VI) and may be restated as a correlation between maximum (linear) rotational velocity V_M and stage t along the Hubble sequence (Brosche 1971).[‡‡]

For spirals a definite correlation ($\varrho \simeq -0.8$) (Figure 13) is present and in the mean

$$V_M = 290 - 24 \, t \quad (2 \leqslant t \leqslant 10). \tag{9}$$

For $t < 2$ the velocity dispersion σ_v becomes comparable to ωr and a distinction must be made between circular V_c and rotational V_r velocities (Oort, 1940, 1965; van Houten, 1961), but available data, especially on σ_v, are insufficient to determine whether the linear relationship continues for earlier types, including lenticulars.

TABLE VI

Mean rotation periods of galactic disks[a]

Type	L, SO/a	Sa, ab	Sb, bc	Sc, cd	Sd, m
$\langle \log P \rangle$	7.77	7.80	8.16	8.29	8.45
n	3	11	25	30	30

[a] in years for linear branch of velocity curve, after Vorontsov-Velyaminov (1970).

[‡] Evidence for this remarkable conclusion first appeared in 1968–69 (Ables, 1971; de Vaucouleurs *et al.*, 1969a, b; Roberts, 1969; Vorontsov-Velyaminov, 1970b); it is in sharp contrast with ideas prevailing only 10 yr ago when the meagre data available suggested parallel decreases of $\mathfrak{M}/\mathfrak{L}$ and colour index as a function of stage along the Hubble sequence (cf. de Vaucouleurs, 1959b; Holmberg, 1964).

[‡‡] A search has also been made for possible correlations between ω and R_l, or between R_l and spectral indices for a given type Sb or Sc (Saslaw, 1970, 1971a, b, 1972) but no significant relationship has been found.

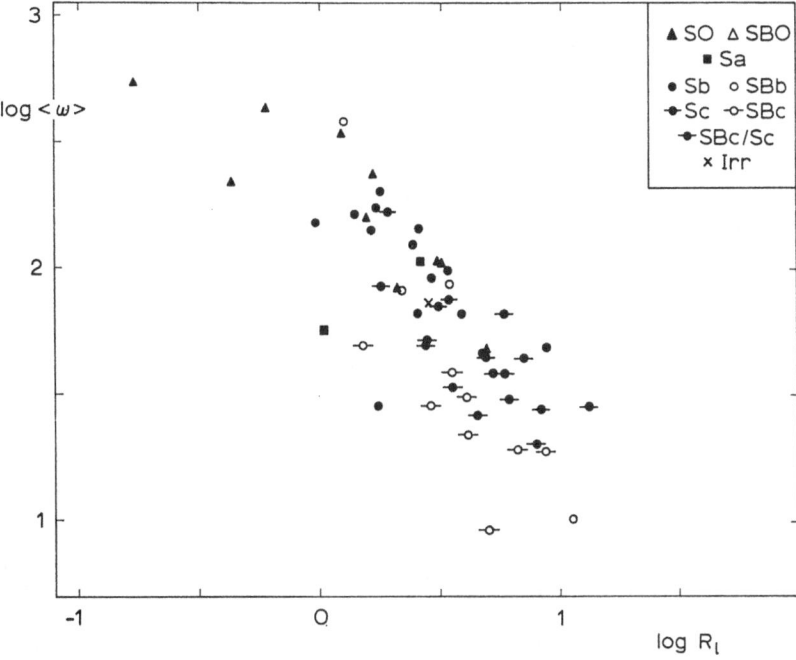

Fig. 12.   Correlation between mean angular velocity $\langle\omega\rangle$(km s^{-1} kpc^{-1}) and radius R_l(kpc) of linear part of rotation curve, after Mayall and Lindblad (1970).

There is also a correlation between R_M and other scale factors, such as the isophotal diameters in the BGC or Holmberg systems. In Equation (8) a mean value $\langle 2R_M/a\rangle = 0.2$ was assumed, independent of type. There is good evidence, however, that this ratio is not a constant but varies with stage t along the Hubble sequence as shown in Figure 14 and suggesting a possible improvement in estimates of indicative masses.

Angular momenta of galaxies (total A_T or per unit mass A^*) have attracted much interest in recent years (Brosche, 1963; Ozernoy, 1967; Takase, 1967; Takase and Kinoshita, 1967; Heidmann, N., 1969; Freeman, 1970) because it may be supposed that both mass and angular momentum are conserved as a primordial cloud contracts into a galaxy (Mestel, 1963).

For the larger galaxies with detailed velocity data, a mass distribution model $\mathfrak{M}(r) = 2\pi \int_0^r \mu(r)\, r\, dr$, derived from the rotation curve $V(r)$, is combined with it to estimate the angular momentum distribution

$$A(r) = \int_0^r rV(r)\, d\mathfrak{M}(r) = 2\pi \int_0^r r^2 V(r)\, \mu(r)\, dr \tag{10}$$

and the total angular momentum $A_T = A(\infty)$ (μ=projected surface density).

For small galaxies, especially when V_M only is given by the 21-cm line width, an *indicative* specific angular momentum (per unit mass)

$$A_i^* = 0.10 a V_M \tag{11}$$

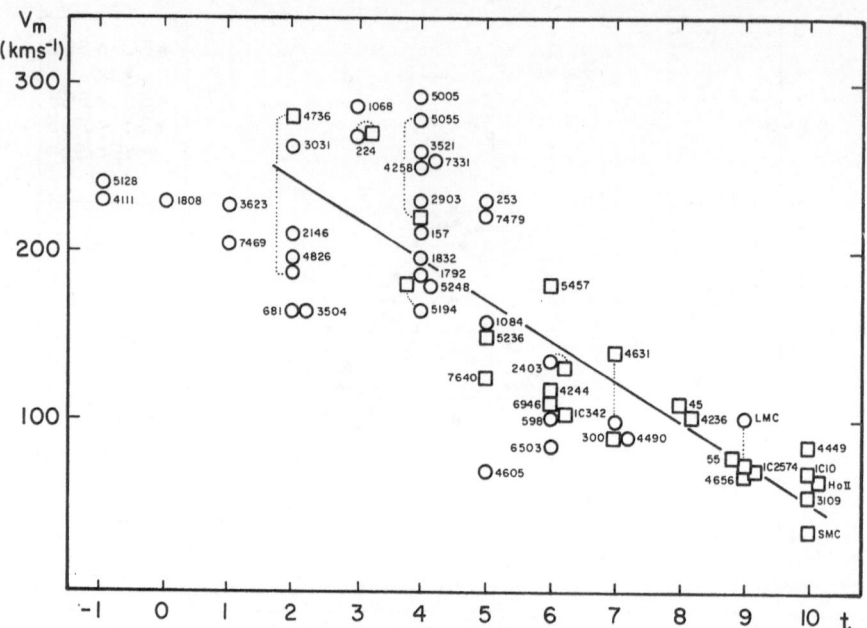

Fig. 13. Correlation between maximum rotation velocity V_M and revised Hubble
type, after Brosche (1971).

may be calculated (Heidmann, N., 1969). The numerical coefficient is chosen so that
A_i would be the actual angular momentum of a spinning solid disk of diameter $a/5$.

Remarkably tight linear correlations between $\log A$ and $\log \mathfrak{M}$ have been found in
the empirical data, i.e. $A \propto \mathfrak{M}_T^\theta$; for example $A_T \propto \mathfrak{M}_T^2$ (Brosche, 1963) or, more
precisely, $A_T \propto \mathfrak{M}_T^{7/4}$ (Takase and Kinoshita, 1967) or, again, $\mathfrak{M}_i \propto A_i^{*3/2}$ (Heidmann,
N., 1969). However, Freeman (1970) has pointed out that for an exponential disk in
centrifugal equilibrium

$$A_D = 1.109 \, (G\mathfrak{M}_D^3 \Lambda)^{1/2} \tag{12}$$

and since $\mathfrak{M}_D = 2\pi\mu_0\Lambda^2$ (Section 7.3) it follows necessarily that $A_D \propto \mathfrak{M}_D^{7/4}$. Freeman
shows, further, that if $\mu_0 =$ constant and $\mathfrak{M}_D$ depends only on the scale parameter Λ,
a relation of the form $\log A = \theta \log \mathfrak{M}_D + C$, with $\theta \simeq 7/4$, will always obtain when A
and $\mathfrak{M}$ are derived from a two-parameter representation of the rotation curve (e.g. by
a Brandt function), because of a fortuitous compensation of numerical coefficients.
Conversely, if $A_T \propto \mathfrak{M}_T^{7/4}$ is established in an early phase of evolution of the proto-
galaxy (and if A and $\mathfrak{M}_T$ are conserved in the subsequent collapse to a disk), the
constancy of μ_0 would be a consequence.

9. Interstellar Gas Content

A small, but significant fraction of the mass of galaxies is in the interstellar gas, mainly
neutral and ionized hydrogen, with a small admixture ($\sim 10\%$) of helium, heavy ele-

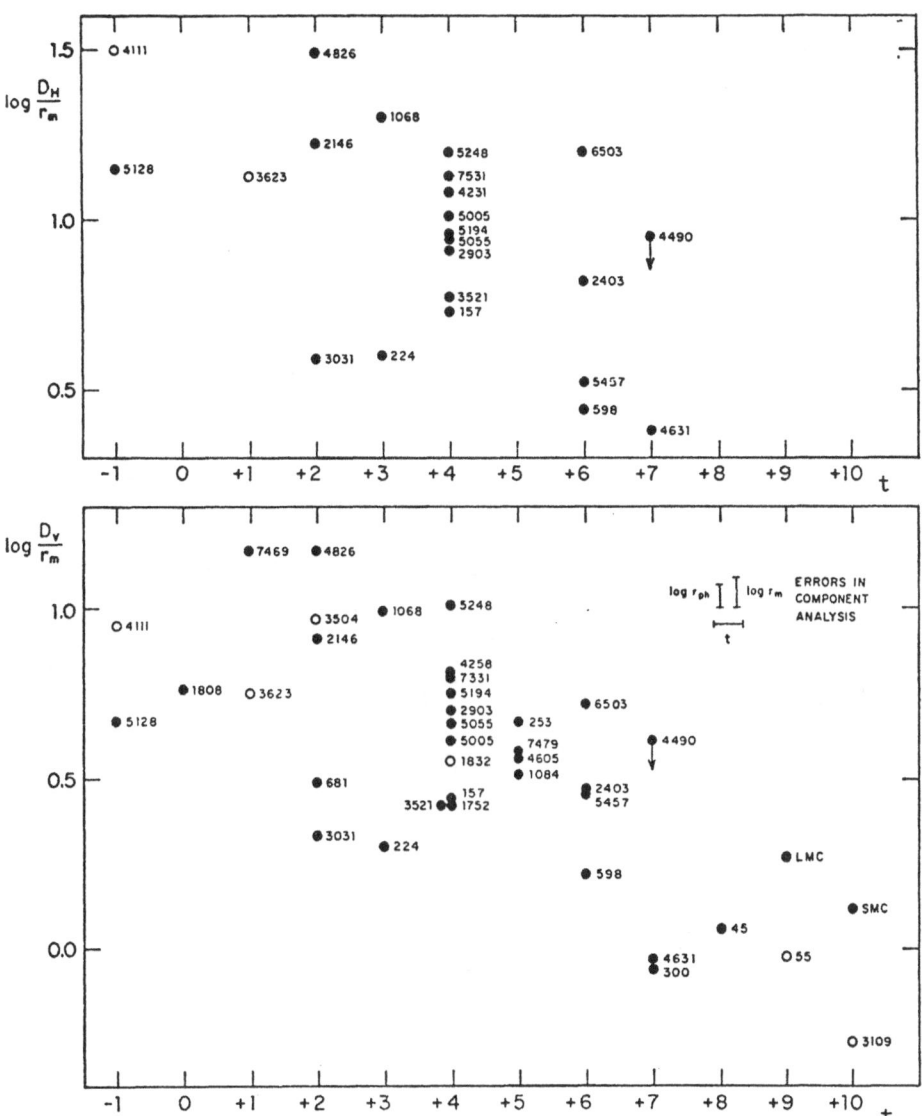

Fig. 14.   Correlation between ratio of optical diameter D to radius $R_M(V_M)$ at maximum rotation velocity with stage t along revised Hubble sequence, after Brosche (1973). D_H is in Holmberg's system, D_V in BGC system. The ratio is clearly not constant, indicating that the numerical coefficient in Equation (8) should be a function of t.

ments and molecules. Whether this gas is a primordial remnant or a by-product of stellar evolution its presence is a dominant factor in the present make-up of galaxies. Although a variable and in some cases significant fraction of this gas is ionized, it is believed that with allowance for self-absorption (Heidmann et al., 1971) its mass can be estimated with little error by observation of the 21-cm emission line.

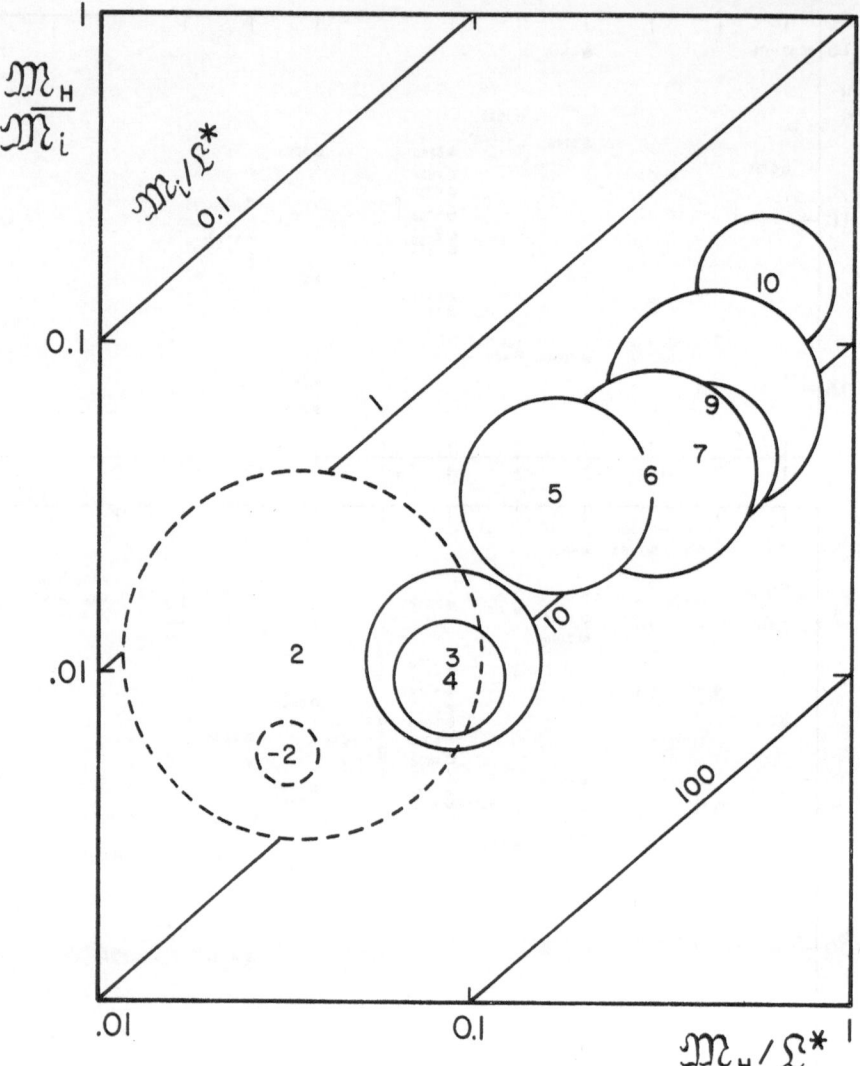

Fig. 15. Mass-luminosity ratio as a function of Hubble type and neutral hydrogen content, after N. Heidmann (1969). $\mathfrak{M}_i/\mathfrak{L}^*$ is constant and independent of stage t along spiral sequence ($t > 0$).

Some general conclusions may be drawn from the mass of data collected during the past decade, in particular with the large reflectors at Nançay, Parkes and Green Bank:[‡]

(1) H I emission has now been measured for nearly 200 galaxies, mainly spirals and irregulars ($t \geqslant 2$); very few lenticulars and early spirals are detectable. The total hydrogen mass $\mathfrak{M}_H$ is strongly correlated with Hubble type t and absolute luminosity $\mathfrak{L}^*$ or indicative mass $\mathfrak{M}_i$.

[‡] For reviews of results prior to 1969–70 see Roberts (1969, 1972).

For example Figure 16, after N. Heidmann (1969), displays the relationships between the three ratios of hydrogen mass $\mathfrak{M}_{\mathrm{H}}$, indicative mass $\mathfrak{M}_i$ and absorption-free blue luminosity $\mathfrak{L}^*$ taken 2 by 2 as a function of Hubble stage t; we see that $\mathfrak{M}_{\mathrm{H}}/\mathfrak{M}_i$ and $\mathfrak{M}_{\mathrm{H}}/\mathfrak{L}^*$ increase by nearly 2 orders of magnitude from early spirals to Magellanic irregulars, while $\mathfrak{M}_i/\mathfrak{L}^*$ remains nearly constant. In particular, the hydrogen-to-total mass ratio increases from $<1\%$ in L types $(t<0)$ to $\sim 10\%$ in Sm-Im types $(t=10)$. $\mathfrak{M}_i$ and $\mathfrak{M}_{\mathrm{H}}$ are plotted vs. $\mathfrak{L}^*$ and t in Figure 16a, b; the largest total masses $(\mathfrak{M}_{\mathrm{T}}>3\times 10^{11}\,\mathfrak{M}_\odot)$ are found among Sa to Sc supergiants, while the largest H I masses $(>10^{10}\,\mathfrak{M}_\odot)$ occur in high-luminosity Sc and Sd galaxies (Balkowski, 1972).

(2) The density distributions of interstellar hydrogen in galaxies are known in some detail for only a few nearby objects having large apparent diameters and types $t \geqslant 3$. This small sample strongly suggests that the distribution of 21-cm emission is closely correlated with that of luminosity in the flat (exponential) component, but with a lower density gradient. For example, near the centre of M31 where the spheroidal component is dominant, H I emission is weak or absent (Roberts, 1966) so that the H I distribution appears as a flat annulus more or less coinciding with the spiral structure in the disk; a similar situation was first noted in our Galaxy (Oort, 1965). In Magellanic spirals the H I surface density follows closely that of luminosity and star density, in particular that of the extreme population I component (de Vaucouleurs 1955, 1957b). More precisely, the effective diameters of the optical and 21-cm distributions are very nearly equal (6° in LMC, 12' in IC 1613; de Vaucouleurs and Freeman, 1972). Some minor departures from this general correlation are significant, especially an absence of concentration of H I in the bar of the Large Cloud where an older population component is dominant (de Vaucouleurs 1956b; McGee, 1964; McGee and Milton, 1966a, b; Walker *et al.*, 1969)*.

In intermediate systems Sc-Sd, such as M33, the H I distribution is also intermediate. It could be represented by a "large double ring structure overlaid on a uniform disk with a central density depression of 35%" (Huchtmeier, 1973) and it is much less concentrated as well as more extensive than the optical distribution.

(3) Smaller galaxies observed at lower relative resolution (Bottinelli, 1970, 1971) provide additional information. In particular, statistical comparisons of ring vs. gaussian models for 33 galaxies suggest that the ring model may be appropriate in only 1/3 of the cases. The H I diameter, defined as the half-intensity width a_{H}, is loosely correlated with the isophotal diameter in the Holmberg system $(\varrho \simeq 0.5)$ and Hubble type, with

$$\langle \log a_{\mathrm{H}}/a \rangle = -0.45 + 0.046t \quad (t \geqslant 3); \tag{13}$$

---

* Walker *et al.* refer to the bar population on their IR photographs of the Large Cloud as 'Population II'; this identification is questionable because neither the stellar luminosity function (de Vaucouleurs, 1956b), nor the integrated spectrum (de Vaucouleurs and de Vaucouleurs, 1959a, b) match that of a globular cluster or of a Sculptor type dwarf elliptical (Population II by definition), but are appropriate for an old (or ageing) Population I. This population was labelled type Ic to differentiate it from the 'zero age' type Ia and possible intermediate age type Ib (de Vaucouleurs, 1956b; Figure 2). Star counts in the core of the Small Cloud (Schilt *et al.*, 1955) lead to similar conclusions.

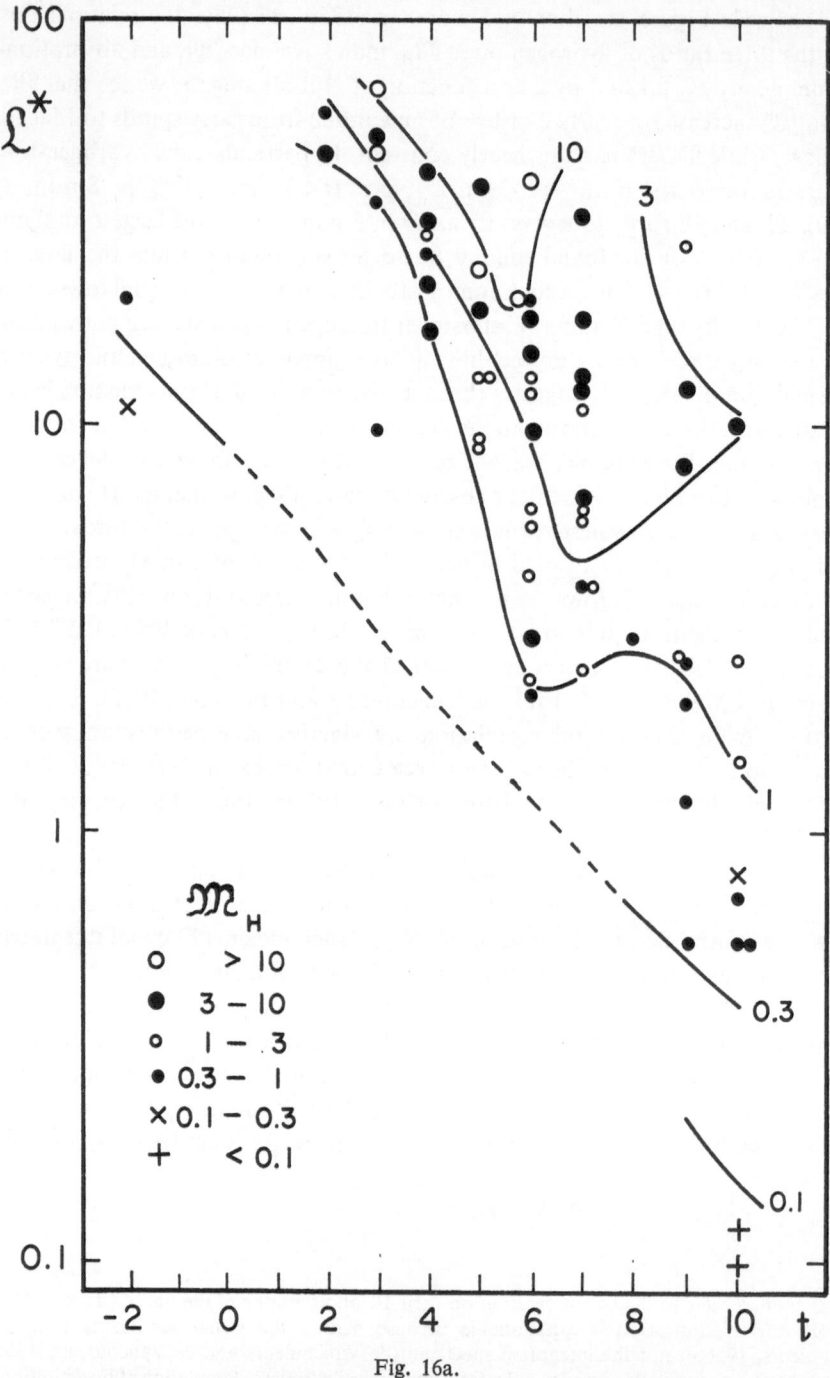

Fig. 16a.

Fig. 16.   Correlations between hydrogen mass $\mathfrak{M}_H$, indicative mass $\mathfrak{M}_i$, absolute luminosity $\mathfrak{L}*$ and revised type t, after N. Heidmann (1969). (a) $\mathfrak{M}_H$ (in $10^9\ \mathfrak{M}_\odot$) as a function of $\mathfrak{L}*$ and t. (b) $\mathfrak{M}_i$ (in $10^9\ \mathfrak{M}_\odot$) as a function of $\mathfrak{L}*$ and t.

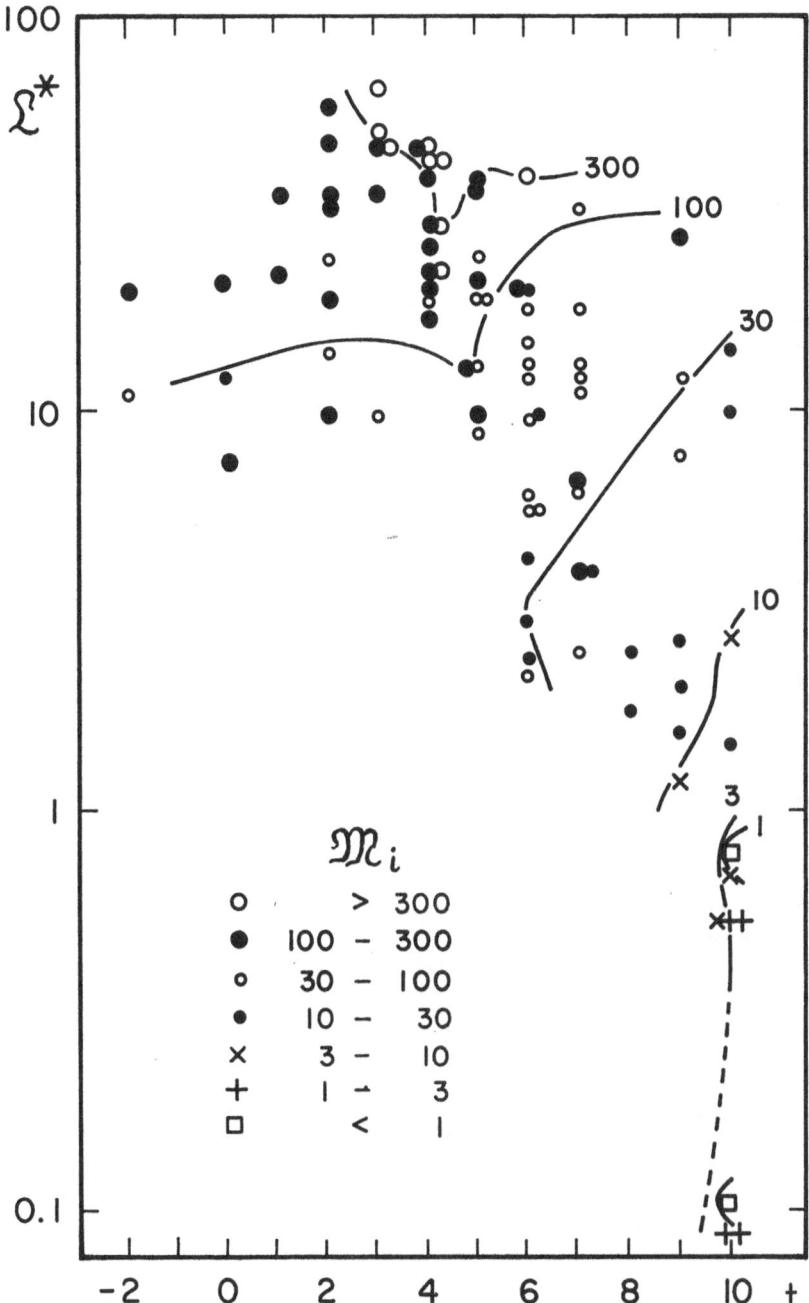

Fig. 16b.

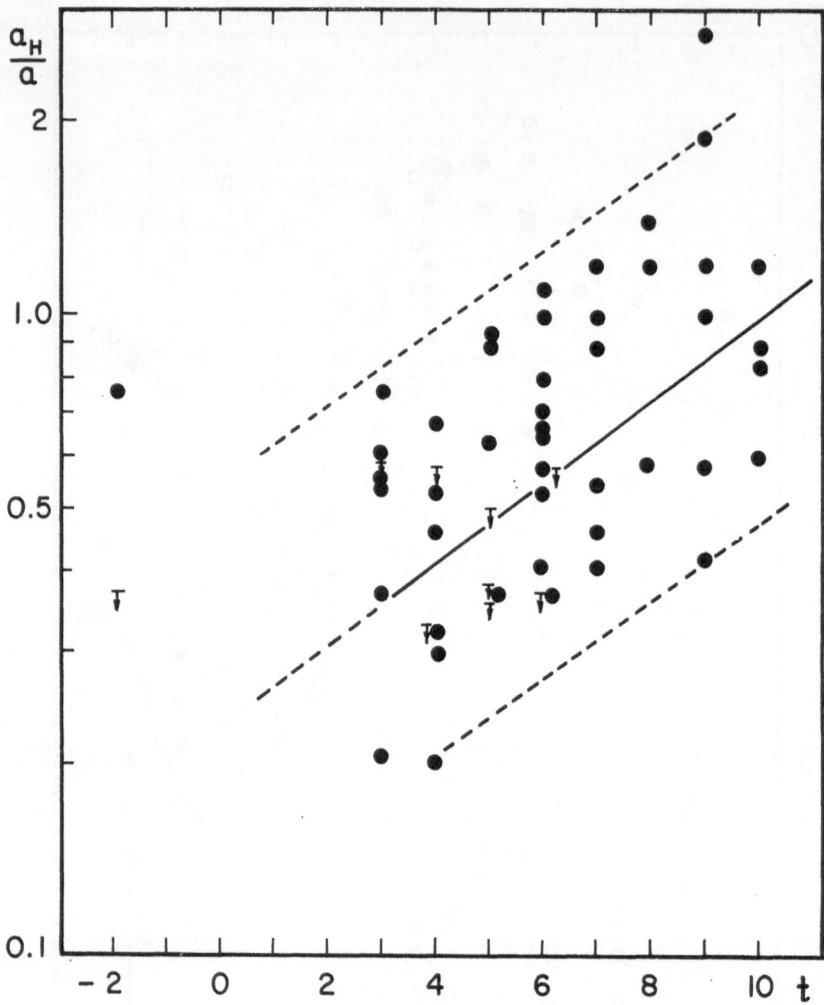

Fig. 17.   Correlation between ratio of hydrogen to isophotal diameters a_H/a and revised type t, after Bottinelli (1970). a_H is the half--intensity H I diameter, a is isophotal diameter in Holmberg's system.

thus $a_H \simeq a$ in the Magellanic irregulars ($t=10$), decreasing to $a_H \simeq a/2$ in the early spirals (Figure 17). It follows that the relation between mean hydrogen surface density σ_H^* and type depends on how the diameter is defined; using optical diameters it was initially concluded that σ_H^* increases toward later types (Gouguenheim, 1969), a result that was regarded as consistent with the parallel increase in the rate of star formation, abundance of extreme Population I supergiants, etc. However, when H I diameters are used, σ_H^* appears to be very nearly independent of morphological type, with an average $\langle \sigma_H^* \rangle \simeq 1$ mg cm^{-2} for $t>0$ (Figure 18).

(4) Recently, special efforts were made to detect 21-cm emission from early type galaxies ($t<0$); weak emission could be detected from lenticulars ($-3<t<-1$)

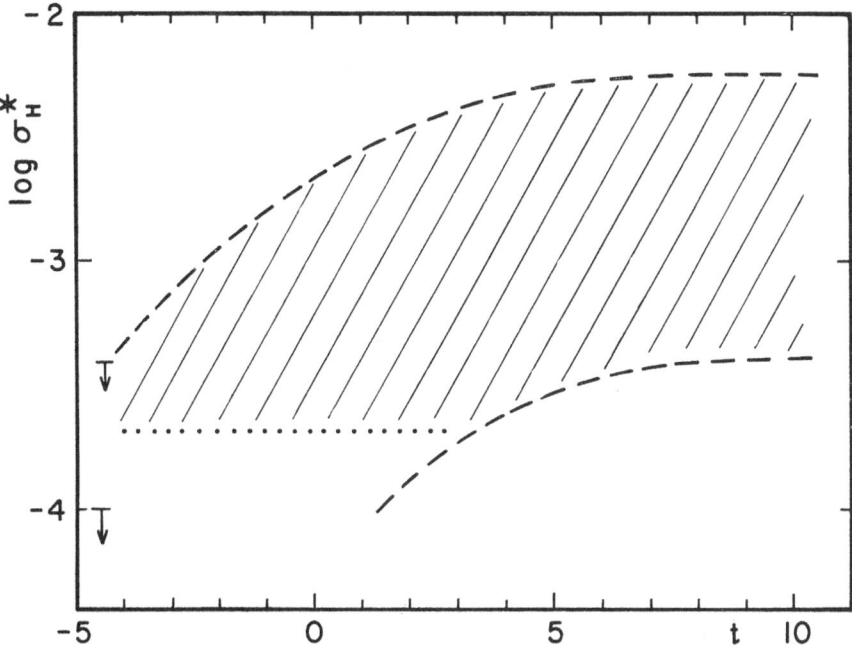

Fig. 18.   Average surface density of neutral hydrogen $\sigma^{*}{}_{H}$ (g cm^{-2}) as a function of revised type t, adapted from Balkowski (1972). The average density $\sigma^{*}{}_{H} \simeq 1$ mg cm^{-2} is nearly independent of $t > 0$ when calculated for half-intensity diameter a_{H} of H I distribution.

(Balkowski *et al.*, 1972), but none from normal ellipticals (Gallagher, 1972; Bottinelli *et al.*, 1973; Kerr, 1973). The upper limits for the latter appear to be much lower than could have been expected from an extrapolation to $t < -3$ of the trend of $\mathfrak{M}_{H}/\mathfrak{M}_{i}$ and σ_{H}^{*} vs type (Figure 19); this result suggests that a discontinuity in physical properties exists between E and L galaxies. A similar conclusion was reached independently from analyses of optical properties (luminosity distribution laws, ellipticities) (Sections 3, 4).

Optical line emission from interstellar gas in the centres of ~ 10 to 20% of the elliptical galaxies has been known for a long time (Mayall, 1939), sometimes with a remarkably high velocity dispersion ($\sigma_{v} \simeq 500$–1000 km s^{-1}); the best known examples are NGC 4278 and 4486 (Minkowski and Osterbrock, 1959; Osterbrock, 1960). This gas is believed to come from evolving stars and may be abnormally rich in heavy elements because, if the ratio $(Y + Z)/X$ of helium + heavy elements to hydrogen has the 'normal' value ~ 0.1, there is an apparent conflict between current estimates of

(a) the rate of mass loss by corpuscular radiation from evolving stars, $\sim 1 \mathfrak{M}_{\odot}$ yr^{-1} (Gallagher, 1972; van den Bergh, 1972b),

(b) the mass of interstellar gas responsible for the observed line emission, $\sim 10^{5} \mathfrak{M}_{\odot}$ in NGC 4278 (Osterbrock, 1960), and

(c) the upper limit to the mass of neutral hydrogen, which is probably less than $10^{8} \mathfrak{M}_{\odot}$ in NGC 4472 (Bottinelli *et al.*, 1973; Kerr, 1973).

Taking the numbers at face value, (c) is 10^{-2} of what (a) would produce in 10^{10} yr;

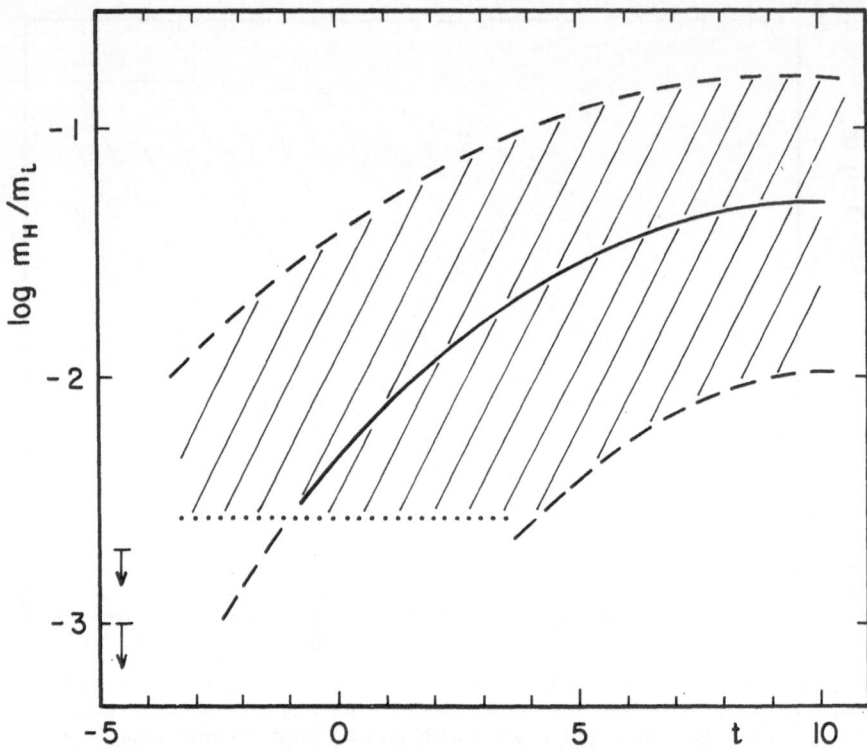

Fig. 19. Ratio of neutral hydrogen to indicative mass $\mathfrak{M}_H/\mathfrak{M}_i$ as a function of revised type t,
adapted from Balkowski (1972).

this suggests that intermittent bursts of star formation at intervals $\sim 10^8$ yr efficiently remove the accumulated gas. (A few ellipticals, e.g. NGC 185 and 205, are known to have blue stars near their centres). It is well to remember, however, that (a), (b) and (c) are all uncertain by ± 1 order of magnitude, that a large fraction of the gas might be ionized, that the $(Y+Z)/X$ ratio may be higher in the centres of E galaxies than in the disks of spirals where its 'normal' value is defined (Roberts, 1972), that a fraction of the gas might be ejected (e.g. in supernova exposions) at velocities exceeding the escape velocity and, if ϱ_H^* is low enough, may actually escape from the galaxy. In this case, also, the 21-cm line profile might be so broad and the emitting region so large that the line could easily have escaped detection (Bottinelli *et al.*, 1973), and if so then (c) would be greatly underestimated.

10. Spectral Energy Distributions and Colour Indices

Information on the integrated spectral energy distribution functions of galaxies of different types is still fragmentary. Because of the basic differences in the compositions of the spheroidal and flat components, spectral data on the nuclear region of a galaxy are not in general applicable to the rest of the galaxy. During the past few years reliable data became available on the integrated spectra of giant ellipticals (Whitford, 1971; Schild and Oke, 1971) which form a remarkably homogeneous group (Lasker, 1970; Sandage, 1972), but it was not until recently that similar information was secured for other major types spanning the whole Hubble sequence from E to Im (Wells, 1972; 1973) (Figure 20).

These studies at intermediate resolution (~ 50 Å) provide a link between high resolution (~ 1–5 Å) work on galactic nuclei for detailed population analyses from line intensities (de Vaucouleurs and de Vaucouleurs, 1959a, b; Spinrad *et al.*, 1971), and medium resolution (~ 100–200 Å) multicolour surveys (Wood, 1966; Tifft, 1961–1969;

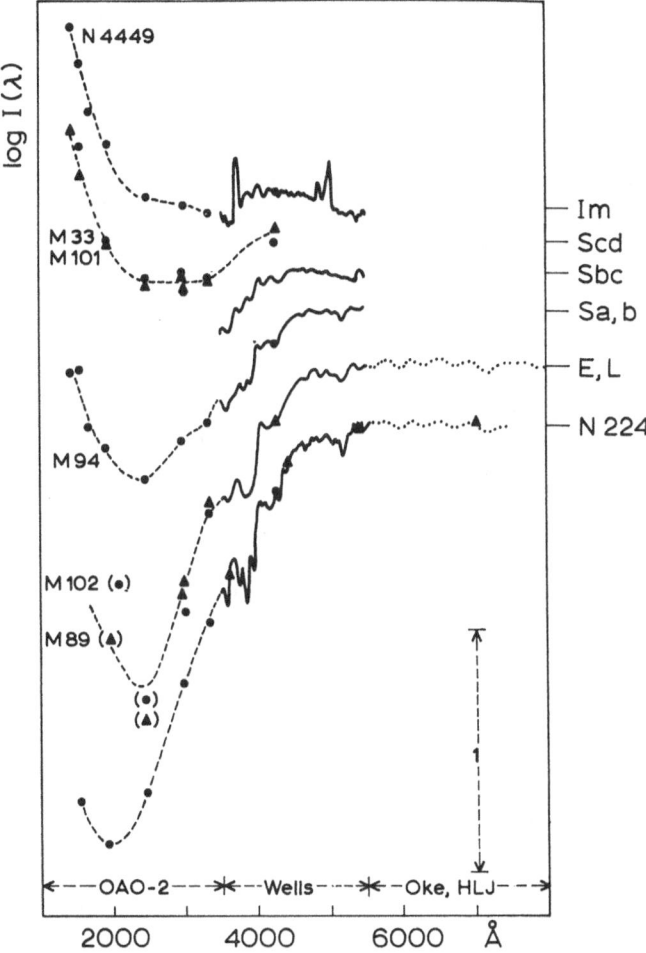

Fig. 20. Spectral energy distribution in galaxies of different morphological types along Hubble sequence from OAO-2 and ground-based data (identified at bottom). Galaxies in OAO set are identified at left and matched at $\lambda 4400$ Å to continuous curves from McDonald scanner observations (Wells, 1972) for corresponding galaxy types identified at right. Spectrum of N224 is for central region only.

Westerlund and Wall, 1969). Fortunately, principal component analysis (Martin and Bingham, 1970) demonstrates that only 3 *independent* parameters, of which 2 are dominant, can be defined by a narrow-band 12-colour system (Wood, 1966). Therefore, a broad-band 3-colour system generating 2-colour indices $U-B$, $B-V$ already provides much spectral information. *UBV* data are now available for over 1200 galaxies (de Vaucouleurs and de Vaucouleurs, 1964, 1972, where older references are given). New analyses of this material confirm and refine earlier conclusions:

(1) $U-B$ and $B-V$ colour indices, corrected for aperture, redshift, inclination and line emission effects, are closely correlated with each other (Figure 21) and with stage along the Hubble sequence (Figure 22). There is a smooth colour transition from reddish pure spheroidal systems $(t<-3)$ with intrinsic colours $U-B=+0.45$, $B-V=+0.90$ to blueish flat systems $(t>5)$ $(-0.2, +0.45)$; in the range $-3<t<+6$ the relation is very nearly linear: $(U-V)_0 = +0.98-0.12t$.

(2) At a given stage t, galaxies in the different families (A or B) and varieties (r or s)

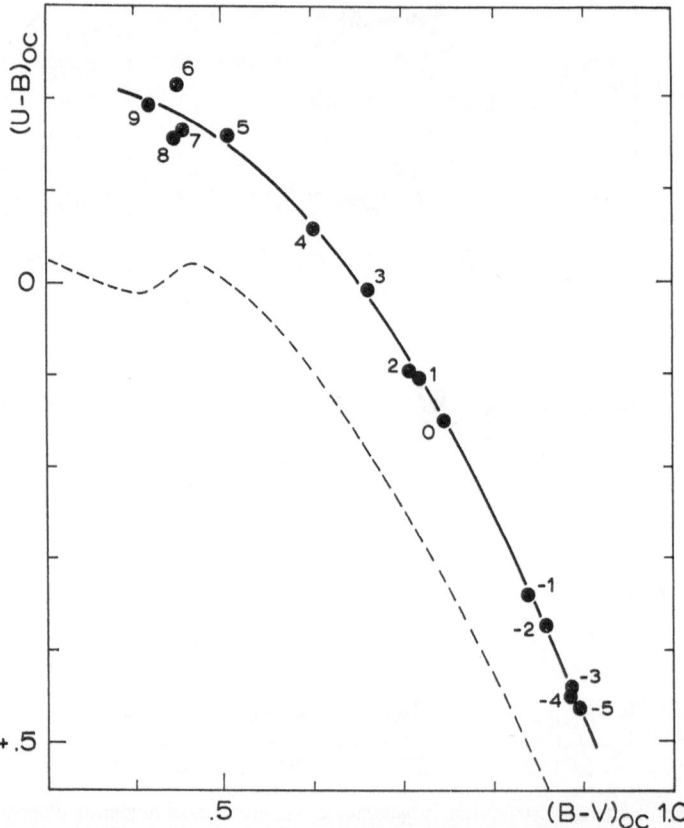

Fig. 21.   Mean colour-colour relation for normal galaxies as a function of stage t (G. & A. de Vaucouleurs, 1972). Colours are corrected for galactic absorption, redshift and inclination effects. Displacement from stellar main sequence (dashed) results from compositeness of galaxy spectra.

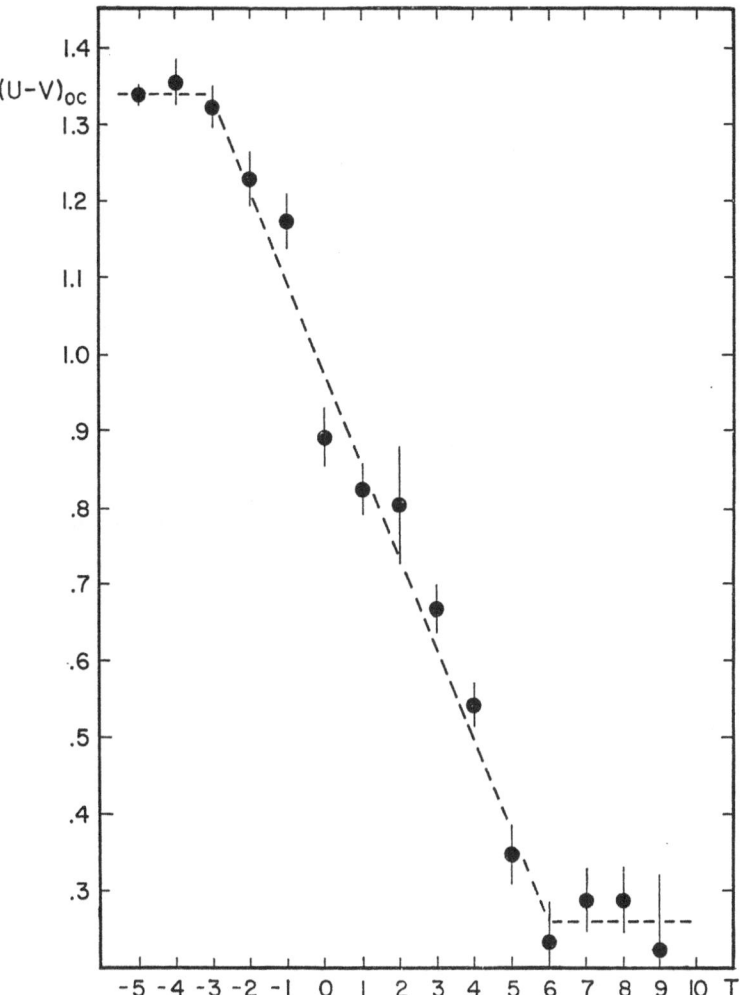

Fig. 22. Relation between mean corrected $U - V$ colour and stage t (G. & A. de Vaucouleurs, 1972). The relation is linear within the statistical errors in the range $-3 \leqslant t \leqslant 6$.

have the same colours; this suggests that structural details reflect relatively minor dynamical differences, not basic physical properties. Photometric and 21-cm studies lead to the same conclusions (Sections 3 and 9).

(3) The absorption-free or metallicity index $Q_0 = (B-V)_0 - 0.72(U-B)_0$ varies little with Hubble type for normal giant galaxies, increasing from 0.56 at E to 0.65 at Sb $(-5 \leqslant t \leqslant 3)$, and decreasing past Sb back to 0.55 at Sm $(3 \leqslant t \leqslant 9)$. This is merely a way of saying that the colour-colour relation is very nearly linear and parallel to the reddening line.

(4) Normal galaxies of types earlier than Sc – and possibly of all types – are redder toward the centre, except for line emission effects. The radial colour gradient is very

small in ellipticals and Magellanic irregulars (some of which are perhaps bluer in the centre), but large in early-type spirals where it is associated with strong radial gradients in various 'line strength' indices (McClure, 1969) and other spectral characteristics (McClure and van den Bergh, 1968) indicative of rapid changes in the stellar population mix or in the chemical composition, in particular with respect to metal abundances.

The presence of 'super-metal-rich' stars has been postulated to account for the strength of absorption lines from neutral metals and molecules (MgH, TiO) in the nuclear region of M31 and other giant galaxies of early types (Spinrad *et al.*, 1971).

(5) There is a significant correlation between integrated colours and absolute luminosities of elliptical and lenticular galaxies first noted 15 yr ago (Baum, 1959; de Vaucouleurs, 1961) and confirmed many times since (McClure and van den Bergh, 1968; Faber, 1971; Sandage, 1972; de Vaucouleurs and de Vaucouleurs 1972) (Figure 23). Low density dwarf ellipticals fainter than $M \simeq -15$ have the colour characteristics of metal-poor globular clusters ($Q_0 \simeq -0.45$); high density dwarfs, however, have normal or nearly normal colours (de Vaucouleurs, 1961; Faber, 1971).

(6) Ultraviolet multicolour photometry ($1200 < \lambda < 3300$ Å) with the OAO-2 satellite (Code *et al.*, 1972) correlates well with *UBV* data and follows similar colour-colour and colour-type relationships at wavelengths $\lambda > 2400$ Å. At shorter wavelengths an unexpectedly steep rise is observed in the spectral energy curve (per unit wavelength) of each galaxy type (Figure 20); it is even steeper than the λ^{-4} function for a black body

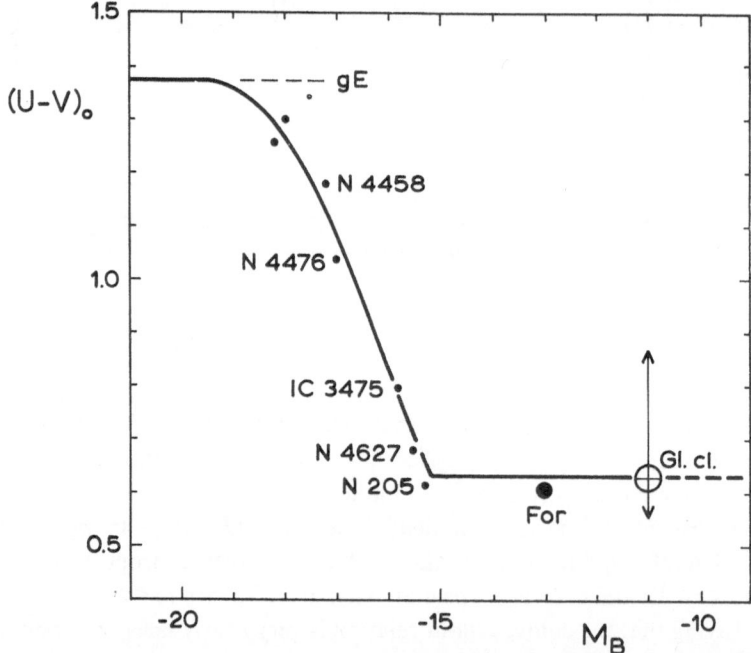

Fig. 23.   Luminosity-colour relation for spheroidal systems (de Vaucouleurs and Ables, 1968). The intrinsic $U - V$ colour of elliptical and lenticular galaxies is closely correlated with absolute magnitude $M_B < -15$. A few typical examples are illustrated.

at infinite temperature. The resulting deep minimum of the energy curve near $\lambda = 2400$ Å stands out most clearly in early type galaxies whose thermal continua are very weak in the UV as could be expected from the optical spectra, but it is still marked in the Magellanic irregulars whose UV spectra are dominated by the strong continua from their OB supergiants (Figure 20).

The favoured interpretation of the minimum invokes absorption and scattering by interstellar or possibly circumstellar grains (Gilra, 1971, 1972) with an absorption maximum near $\lambda \simeq 2200$ Å and strong forward scattering with high albedo at $\lambda < 2200$ Å. The implication of a significant dust component in E galaxies is interesting but, as the OAO experimenters point out, the detection of early type galaxies was marginal and needs confirmation*.

(7) Infra-red data are still very limited (Johnson 1966), except for a small and strongly biased sample of peculiar sources (Kleinmann and Low, 1970 a, b; Rieke and Low, 1972) in particular Seyfert galaxies, which have abnormally high IR fluxes, apparently correlated with radio continuum emission. Such galaxies emit 10^{22}–10^{24} W Hz^{-1} at 10μ compared with 10^{21}–10^{24} W Hz^{-1} at 21 cm. The IR luminosity in the 7.9–13.3μ band is in the range 10^{35} to 10^{38} W. Normal galaxies are much weaker IR emitters (10^{32} to 10^{38} W) and in most or all cases the IR source is localized in the nucleus or the nuclear region.

11. Independent Parameters

A large number of measurable properties of galaxies (m) are mutually correlated; the number of possible combinations of m parameters 2 by 2 can be very large and the results confusing. It is important to discover: (1) how many (say, p) of these m parameters are really independent, and, if possible, (2) which of those p parameters have the most basic physical significance. The empirical correlations between observables could then be re-interpreted in terms of a small number of fundamental physical parameters such as mass, length and time scales, or energy and angular momentum, etc.

In the past few years the methods of principal component analysis (Deeming, 1968) have been applied to multi-colour photometry and other observables. An analysis of Wood's (1966) 12-colour photometry by Martin and Bingham (1970) suggested that only 2 or possibly 3 independent variables are sufficient to describe colour properties. From a more complete analysis of $n=31$ galaxies with respect to $m=6$ variables (type t, radius R_M and maximum velocity V_M, face-on photometric radius $r_p = a/2$, luminosity L and colour index C) Brosche (1973) concludes that just 2 independent variables contribute 82% of the total variance. The residual variance is mainly due to noise in the data (measuring errors, approximations in the face-on corrections, departures from linearity). Except for this noise, the original variables x_{ij} (normalized to $\sigma_j = 1$) are given by $x_j = \eta_{j1}\xi_1 + \eta_{j2}\xi_2$, where ξ_1, ξ_2 are uncorrelated, rectangular

---

* Code et al. (1972) lump together M102 (NGC 5866) and M89 (NGC 4552) as examples of 'ellipticals'; it should be noted that M102 is a late lenticular (S0+, $t = -1$) seen edge-on, with a well-known dust belt (see e.g. Sandage, 1961; Burbidge and Burbidge, 1960).

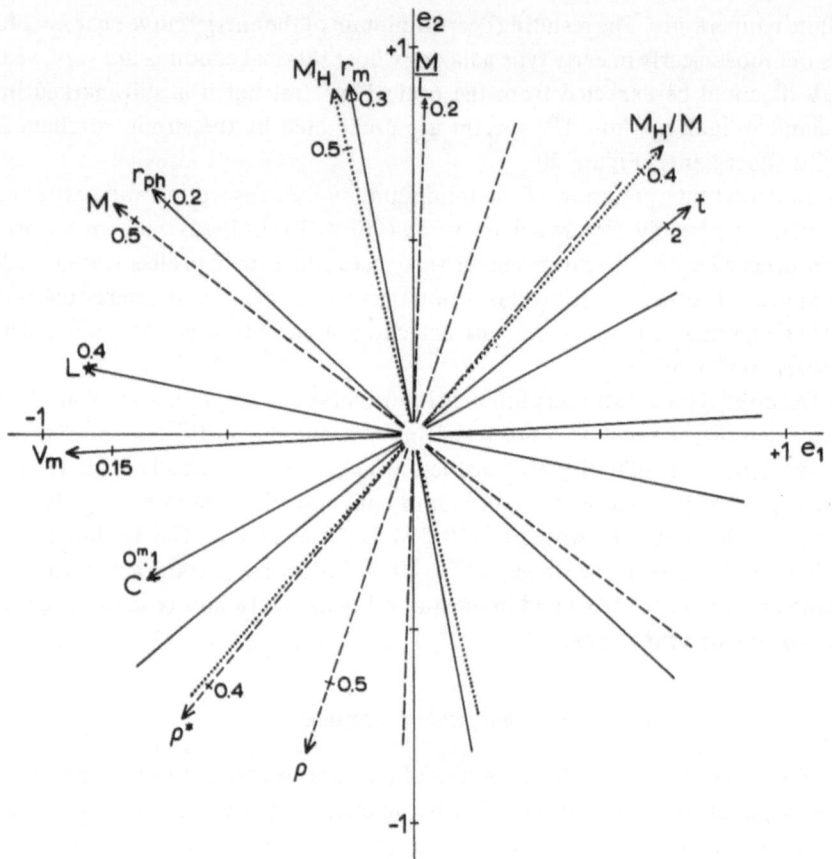

Fig. 24. Gradients of variables (observable and derived parameters t, r_m, V_m, r_p, $\mathfrak{L}^*$, C, M_H, M, ϱ) in eigenvector plane, after Brosche (1973). Gradients of uncorrelated variables are mutually rectangular.

coordinates. In this $\xi_1 \xi_2$ plane the lines $x_j =$ constant are parallel to a vector $\mathbf{J}(\eta_{1j}, \eta_{2j})$ with slope η_{2j}/η_{1j} (Figure 24). Highly correlated variables are represented by $\mathbf{J}$ vectors forming a small angle, uncorrelated variables by orthogonal vectors. From the former, the (linearized) relations between correlated variables (or linear functions of same, e.g. space density ϱ^*, mass-luminosity ratio f, etc.) may be obtained; for example $\log f = 0.009\,\xi_1 + 0.233\,\xi_2 + 0.257$. From the latter, the location of a galaxy in the manifold may be represented by its projection onto the ξ_1, ξ_2 plane (Figure 25).

Thus factor analysis tends to support the conclusions of the Nançay group (from a large number of 2- and 3-dimensional correlograms) that optical and 21-cm properties depend on two dominant and more or less independent parameters. The two most significant of the independent empirical parameters are the stage t along the Hubble sequence and the absolute luminosity $\mathfrak{L}^*$, or some photometric radius r_p with which $\mathfrak{L}^*$ is strongly correlated (cf. Holmberg, 1969; Heidmann et al., 1971). The corresponding basic physical parameters may tentatively be identified with total mass and

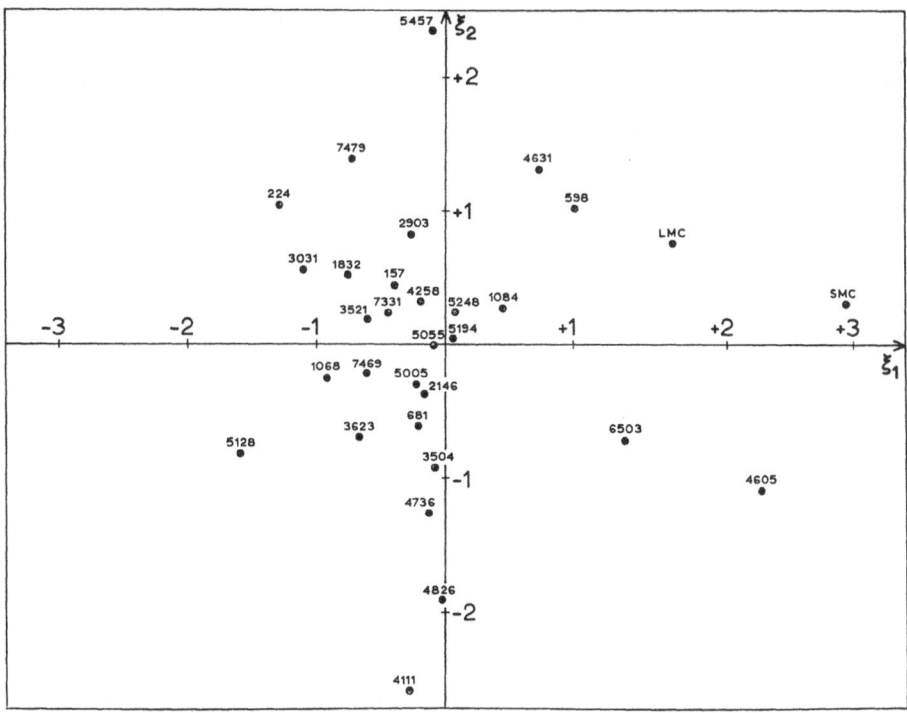

Fig. 25. Projections of galaxy representative points in x_j space onto the eigenvector plane, after Brosche (1973). Stage t increases from lower left to upper right; total masses increase from lower right to upper left, parallel to the gradient vectors of Figure 24.

total angular momentum (Brosche, 1971), although it is by no means proven that these two quantities suffice to determine all the others.

12. Luminosity Functions

Spectral data over a sufficiently large range of wavelengths are only beginning to appear and are still too few, especially in the infra-red, to allow meaningful estimates of the bolometric luminosity functions (LF) of normal galaxies. For the present, attempts to derive luminosity functions must still rely mainly on various catalogues and photo-graphic magnitudes affected by ill-defined errors and selection effects. Until a few years ago most of the discussions were based on the subset of Shapley-Ames galaxies having radial velocities (from Humason *et al.*, 1956) with corrections for incompleteness based on the questionable assumptions of spatial homogeneity and selection by apparent magnitude only (Kiang, 1961; van den Bergh, 1961). More recent studies of groups and clusters (Abell, 1962; Holmberg, 1969) using new magnitude data may be more directly applicable to unit volumes of space but, because of the small volumes and mode of selection of the data, the validity of the derived LF needs to be verified by independent analysis of residuals in the $(m, \log z)$ relation.

Since we are not immediately concerned here with the absolute value of the average space density of matter or luminosity, the general agreement between the shapes of the functions derived by the different approaches is encouraging (Figure 26). It is interesting to remark that the general LF of galaxies turns out to be very nearly the sum of the gaussian and exponential functions originally proposed by Hubble (1936) and by Zwicky (1957).

More specifically the gaussian component with mean $\bar{M}_{pg} \simeq -18.5$ and dispersion $\sigma_M \simeq 0.8$ mag (cf. de Vaucouleurs, 1958b) appears to describe mainly the giant spirals $(0 \leqslant t \leqslant 6)$, while the exponential component with slope 0.2 applies to earlier and later types. In other words, there are no dwarf spirals Sa to Sc, a fact first noticed by Shapley. However, the gaussian component includes also some giant ellipticals and lenticulars, at least in clusters (Figure 27). The corresponding integrated luminosity function $N(M_i > M)$ can be roughly approximated in the observable range by two straight segments, as noted by Abell (1962, 1972).

13. Clustering and Space Distribution

Statistical studies of nearby groups and of the space distribution of galaxies (de Vaucouleurs, 1965, 1971) bring out a number of general properties of galaxies that may be relevant to the problem of their formation and evolution:

(1) The majority of galaxies are not isolated and randomly distributed in space but are members of small groups similar to the Local Group (Karachentsev, 1967; Zonn, 1968); less than 10 to 20% of the largest or nearest galaxies are not clearly members of identified groups (de Vaucouleurs, 1965; Corwin, 1967). Apparently isolated galaxies such as NGC 1313, 2903, 6744 and 6946 do not seem to differ in any of their properties from galaxies of the same type and luminosity in groups.

(2) The galaxy population of small, loose groups and clouds is dominated by spirals and Magellanic irregulars often to the point of an almost total absence of giant ellipticals and lenticulars. The Local Group and nearest groups illustrate this point (Figure 28). Examples of almost 'pure' clouds of spirals are the Grus Cloud (de Vaucouleurs, 1956a), the UMaI cloud (Morgan, 1958a) and the M94 group (van den Bergh, 1960).

(3) Dwarf ellipticals and irregulars are strongly concentrated in the same areas as spirals and appear as *satellites* of the giant spirals and not as an independent, dominant population of space (Reeves, 1956; van den Bergh, 1959; Vorontsov-Velyaminov and Noskova, 1971; Karachentseva, 1972).

It follows that the faint branch of the luminosity function (Figures 26, 27) derived from counts in groups or clusters may tend to overestimate the abundance of dwarfs in space. It is very difficult at present to make allowance for this bias.

According to Holmberg (1969) the space distribution of these companions with respect to their primaries is anisotropic, with a deficiency near the plane, and an excess near the minor axis of the primaries (Figure 29). Ejection from the nucleus of the primary has been suggested as a possible explanation of this strange

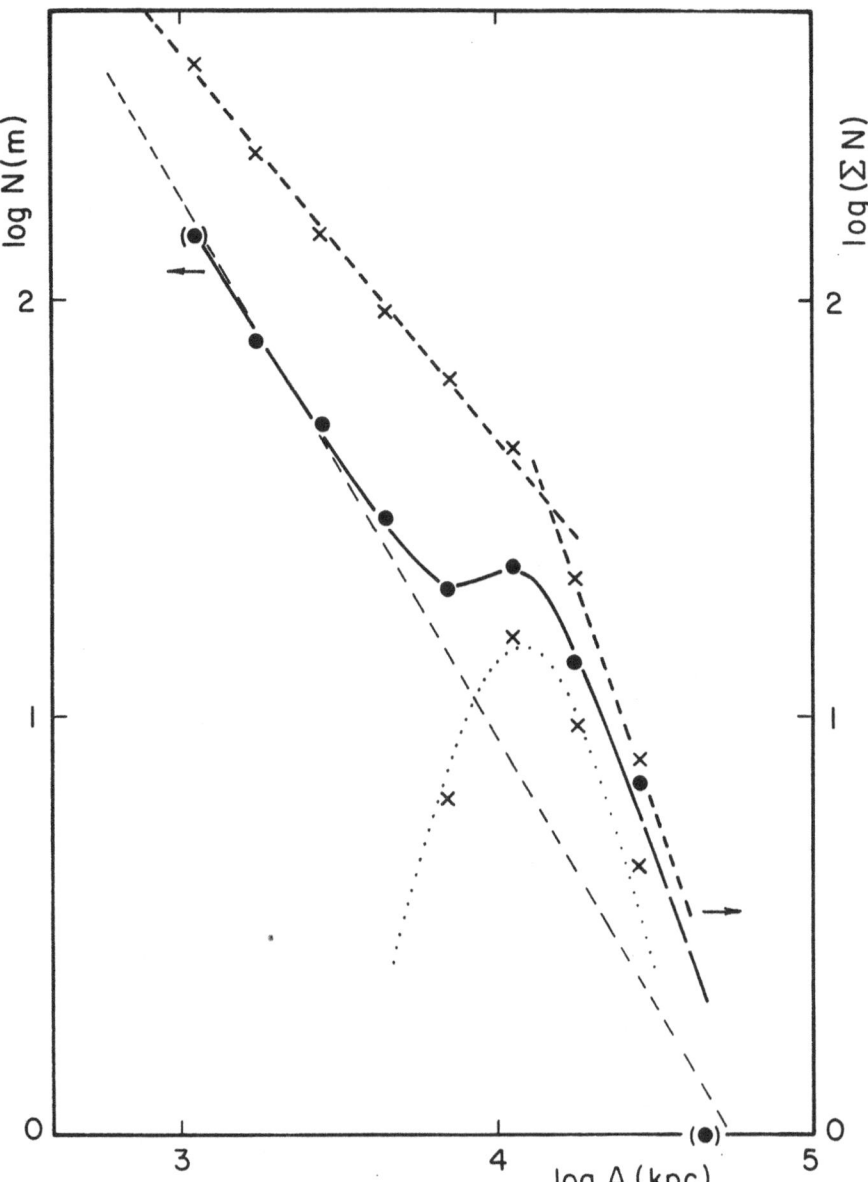

Fig. 26. Frequency function (dots and full line, scale at left, and cumulative frequency function (crosses and dashed lines, scale at right) of linear diameters of galaxies and their companions, after Holmberg (1969). The observed differential function can be resolved into two components, one obeying Hubble's gaussian function (dotted parabola), the other following Zwicky's exponential function (thin dashes). The integrated function can be approximated by two straight segments (heavy dashes) as observed by Abell in clusters (Figure 27).

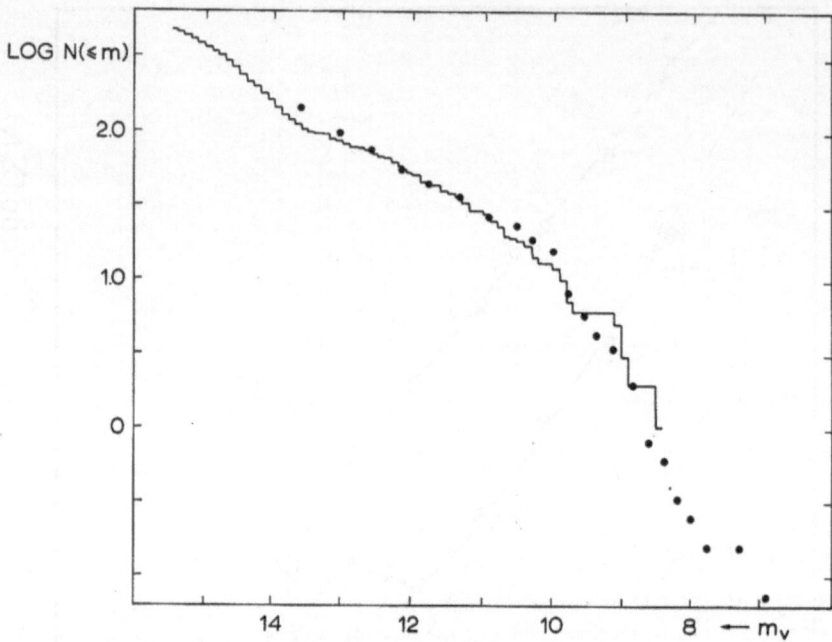

Fig. 27.　Integrated luminosity functions of Coma (dots) and Virgo clusters, after Abell (1972). Scales apply to Virgo cluster.

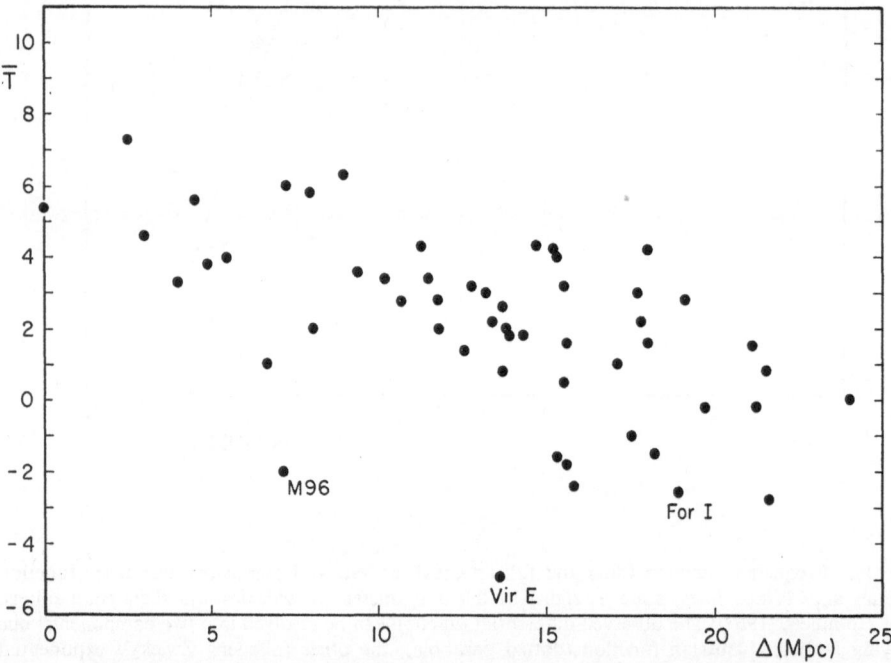

Fig. 28.　Average morphological type *t* of five brightest members of nearby groups as a function of distance $\varDelta$ (de Vaucouleurs 1965; revised 1973). The Local Group, at $\varDelta = 0$, is typical of nearby groups dominated by late-type spirals.

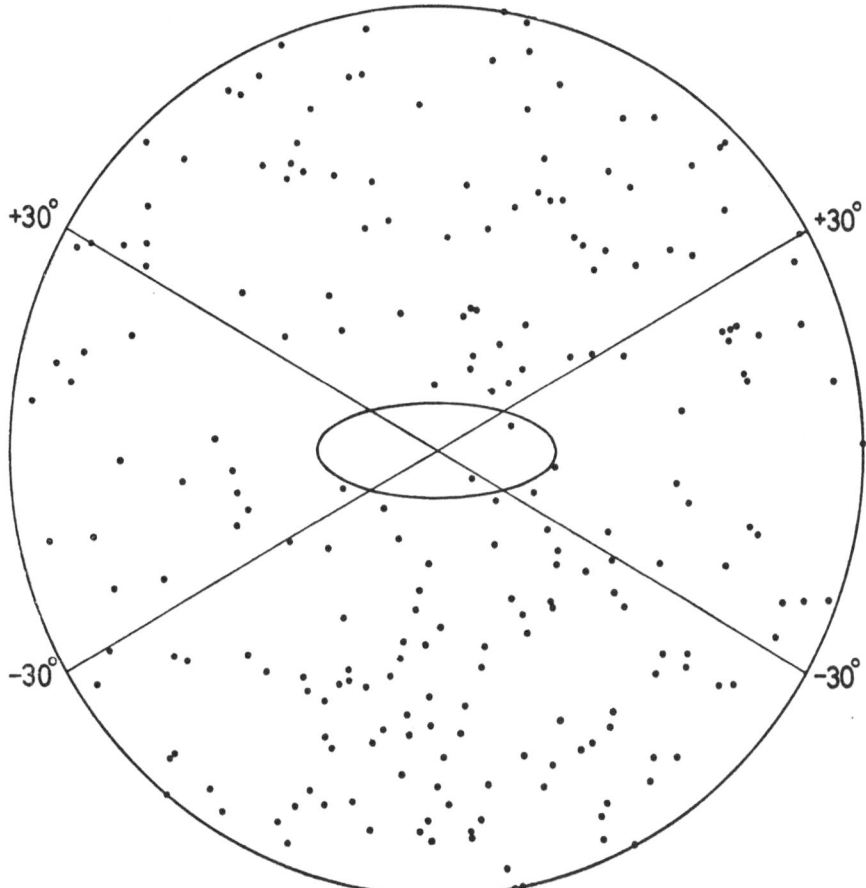

Fig. 29. Distribution of 218 companion galaxies in the vicinity of 58 edge-on systems, after Holmberg (1969). Ellipse shows average size of central system; note excess density within ± 60° from minor axis of primary.

distribution. The 'lines' of galaxies (and quasars) discussed by Arp (1968) are also suggestive of symmetric ejection of condensed matter by giant galaxies, but the physical cause and possible mechanism of the postulated phenomenon elude us.

(4) Giant ellipticals and lenticulars are dominant in *dense* clusters of galaxies, whether small as the Fornax I cluster or large as Coma I; very few spirals, if any, are members of such clusters. The rare exceptions may be interlopers (captures, encounters or optical coincidences). It is possible that apparently mixed clusters, such as Virgo I, are optical superpositions of a cluster of E, L types and a cloud of S types (de Vaucouleurs, 1961; G. and A. de Vaucouleurs, 1973), but even if the two systems were in fact concentric and co-extensive (Kowal, 1969; Tammann, 1972) the drastically different radial distributions of galaxies of different morphological types ($t<0$ vs $t>0$) remain unexplained. This strong segregation of galaxies of different morphological

types in groups and clusters of different structural forms has been known and discussed for more than a decade (de Vaucouleurs, 1956a, 1962, 1965; Morgan, 1958b) but it has not yet received the attention it deserves from the point of view of galaxy formation and/or evolution.

14. Maximum Space Density in Galaxies and Clusters

Selection effects discriminate mainly against small *and* faint galaxies; large and/or bright objects dominate our catalogues. Among objects having a given diameter – the effective diameter, say, for definiteness – a range of masses and densities exist. While the lower limits may be set more by selection factors than by physical conditions, the upper limits should not be so biased.

From such data (Table VII) we may, perhaps, answer two important questions:

(1) Is there an upper limit to the mass (or density) present in a given volume of space?

(2) What is the mass-radius relation for the densest objects or regions of space?

It is remarkable that the densest known galaxies, groups, clusters and superclusters

TABLE VII

Mass-radius-density data[a]

Class of Objects	Examples	$\log \mathfrak{M}_R$ (g)	$\log R$ [b] (cm)	$\log \varrho$ (g cm^{-3})
Centres of dense nuclei	M31, Centre ($R=1$ pc)	–	18.5	-16.5:
Dense nuclei	M31, nucleus	40.3	19.3	-18.2
of galaxies	M32, core	41.0	19.5:	-18.1
Compact dwarf	M32	42.5	20.65	-20.0
ellipticals	N4486-B	43.2	21.1	-20.7
Normal giant	M33	43.5	21.8	-22.5
spirals	M31	44.6	22.3	-22.9
Giant ellipticals	N3378	44.3	22.0	-22.35
	N4486	45.1	22.5	-22.95
Compact groups of spirals	Seyfert, Stephan	45.5	22.6:	-23.1
Small dense clusters of ellipticals	Virgo E, core Fornax I	46.5	23.7	-25.2
Small loose groups of spirals	Nearby groups	46.5	24.1	-26.4
Small clouds of spirals	Virgo S, Ursa Major	47.0	24.3	-26.5
Small clusters of ellipticals	Virgo E	47.2	24.3	-26.3
Large clusters of ellipticals	Coma I	47.9	24.5	-26.2
Superclusters	Local	48.7:	25.5	-28.4:

[a] adapted and revised from de Vaucouleurs (1961, 1970b, 1971).
[b] except for first 3 entries, data refer to effective radii and masses.

of galaxies of increasing radii define a rather tight correlation over the large range of radii $10^{18} < R < 10^{26}$ cm (de Vaucouleurs, 1970b, 1971) (Figure 30). In this range the maximum space density observed in various volumes centred on galaxies and systems of galaxies is given (in cgs units) by

$$\log \varrho_R = -21.7 - 1.7(\log R - 21.7). \tag{14}$$

This maximum density is from 4 to 6 orders of magnitude lower than the Schwarzschild

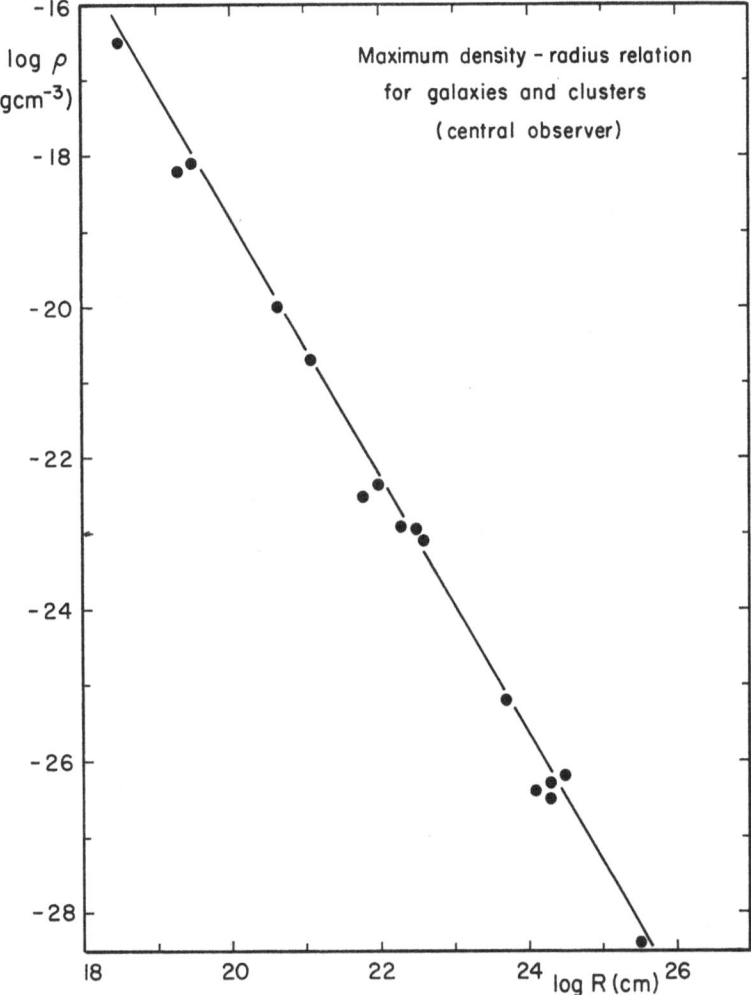

Fig. 30. Correlation between maximum density and radius of spheres centred on galaxies and systems of galaxies (de Vaucouleurs, 1971). The maximum density observable in a given volume of space by a central observer has an upper bound outlined here by some examples in the range ($18 < \log R < 26$).

limit, $\log \varrho_s = 27.2 - 2 \log R_s$. Whether the relation extends to systems larger than super-clusters ($\log R > 26$) or not is still in doubt.*

Since ϱ and R are related through the virial theorem, another expression of this result is a relation between velocity dispersion σ_v and cluster radius for groups and clusters of galaxies; according to Karachentsev (1967) and Ozernoy (1969) the relation

$$\log \sigma_v = 0.40 \log R + \text{constant} \quad \text{(cgs)} \tag{15}$$

applies in the range $23 < \log R < 26$ for 143 systems of galaxies (pairs to supercluster). Still another expression of this correlation is the much discussed apparent dependence of the 'missing mass' ratio upon the multiplicity of galaxy systems (Page, 1965; Rood *et al.*, 1970). Reference is made to a recent review article on galaxy clustering for further details (de Vaucouleurs, 1971).

References

Abell, G. O.: 1962, in G. C. McVittie (ed.), 'Problems of Extragalactic Research', *IAU Symp.* **15**, 213.
Abell, G. O.: 1972, in D. S. Evans (ed.), 'External Galaxies and Quasi-Stellar Objects', *IAU Symp.* **44**, 341.
Ables, H. D.: 1971, *Publ. U.S. Naval Obs.* **XX**, Part IV.
Arp, H. C.: 1965, *Astrophys. J.* **142**, 402.
Arp, H. C.: 1966, *Atlas of Peculiar Galaxies*, Calif. Inst. of Technology, Pasadena.
Arp, H. C.: 1968, *Publ. Astron. Soc. Pacific* **80**, 129.
Arp, H. C. and Bertola, F.: 1969, *Astrophys. Letters* **4**, 23.
Arp, H. C. and Bertola, F.: 1971, *Astrophys. J.* **163**, 195.
Baade, W. and Minkowski, R.: 1954, *Astrophys. J.* **119**, 225.
Balkowski, C.: 1972, *Etude statistique des propriétés intégrales des galaxies mesurées en raie 21 cm*, Thesis, Univ. of Paris.
Balkowski, C., Bottinelli, L., Gouguenheim, L., and Heidmann, J.: 1972, *Astron. Astrophys.* **21**, 203.
Baum, W.: 1959, *Publ. Astron. Soc. Pacific* **71**, 106.
Bonnor, W. B.: 1972, *Monthly Notices Roy. Astron. Soc.* **159**, 261.
Bottinelli, L.: 1970, *La distribution spatiale à grande échelle de l'hydrogéne neutre dans les galaxies*, Thesis, Univ. of Paris.
Bottinelli, L.: 1971, *Astron. Astrophys.* **10**, 437.
Bottinelli, L., Gouguenheim, L., Heidmann, J., and Heidmann, N.: 1968, *Ann. Astrophys.* **31**, 205.
Bottinelli, L., Gouguenheim, L., and Heidmann, J.: 1973, *Astron. Astrophys.* **25**, 451.
Brandt, J. C.: 1960, *Astrophys. J.* **131**, 293.
Brandt, J. C. and Roosen, R. G.: 1969 *Astrophys. J. Letters* **156**, L59.
Brosche, P.: 1963, *Z. Astrophys.* **57**, 143.
Brosche, P.: 1971, *Astron. Astrophys.* **13**, 293.
Brosche, P.: 1973, *Astron. Astrophys.* **23**, 259.
Burbidge, E. M. and Burbidge, G. R.: 1960, *Astrophys. J.* **131**, 224.
Burbidge, E. M. and Burbidge, G. R.: 1972, *Astrophys. J.* **171**, 253.
Burbidge, E. M., Burbidge, G. R., and Prendergast, K. H.: 1960, *Astrophys. J.* **132**, 654.
Burbidge, E. M., Burbidge, G. R., and Fish, R. A.: 1961, *Astrophys. J.* **134**, 251.
Burbidge, E. M., Burbidge, G. R., and Prendergast, K. H.: 1963, *Astrophys. J.* **137**, 376.

---

* This density-radius relation has often been grossly misunderstood and incorrectly applied (Haggerty and Wertz, 1972; Sandage *et al.*, 1972; Bonnor, 1972). Equation (14) applies only to *centrally* located observers in the *densest* known systems of each size; it must be understood as describing an *upper bound* to the space densities seen by central observers. It should be obvious that it is not an expression for the average density observed from an arbitrary location in space.

Code, A. D., Welch, G. A., and Page, T. L.: 1972, in *Scientific Results from the Orbiting Astronomical Observatory (OAO-2)*, NASA SP-310, p. 559.
Corwin, H. G.: 1967, *Galaxy Groups*, M. A. Thesis, Univ. of Kansas.
Corwin, H. G.: 1968, *Publ. Dept. Astron. Univ. of Texas*, Austin, II, No. 12.
Corwin, H. G.: 1970, *Publ. Dept. Astron. Univ. of Texas*, Austin, III, No. 5.
Danby, J. M. A.: 1965, *Astron. J.* 70, 501.
Deeming, T. J.: 1968, *Vistas in Astronomy* 10, 125.
de Vaucouleurs, G.: 1948, *Compt. Rend. Acad. Sci. Paris* 227, 548.
de Vaucouleurs, G.: 1953, *Monthly Notices Roy. Astron. Soc.* 113, 134.
de Vaucouleurs, G.: 1955, *Astron. J.* 60, 126, 319.
de Vaucouleurs, G.: 1956a, *Mem. Comm. Obs. Mt. Stromlo*, III, No. 13.
de Vaucouleurs, G.: 1956b, *Irish Astron. J.* 4, 13.
de Vaucouleurs, G.: 1957a, *Ann. Obs. Le Houga* II, fasc. 1.
de Vaucouleurs, G.: 1957b, *Astron. J.* 62, 69.
de Vaucouleurs, G.: 1958a, *Astrophys. J.* 128, 465.
de Vaucouleurs, G.: 1958b, *Astron. J.* 63, 253.
de Vaucouleurs, G.: 1959a, *Astron. J.* 64, 397.
de Vaucouleurs, G.: 1959b, in *Handbuch der Physik* 53, Springer-Verlag, Berlin, Göttingen, p. 275, 311.
de Vaucouleurs, G.: 1960, *Astrophys. J.* 131, 585.
de Vaucouleurs, G.: 1961, *Astrophys. J. Suppl.* 5, No. 48, 233.
de Vaucouleurs, G.: 1962, in G. C. McVittie (ed.), 'Classification of Galaxies by Form, Luminosity and Color', *IAU Symp.* 15, 3.
de Vaucouleurs, G.: 1963, *Astrophys. J. Suppl.* 8, No. 74, 31.
de Vaucouleurs, G.: 1964, *Astron. J.* 69, 737.
de Vaucouleurs, G.: 1965, 'Nearby Groups of Galaxies', *Stars and Stellar Systems*, 9, Chapter 17 (in press).
de Vaucouleurs, G.: 1969, *Astrophys. Letters* 4, 17.
de Vaucouleurs, G.: 1970a, *Bull. Am. Astron. Soc.* 2, 308.
de Vaucouleurs, G.: 1970b, *Science* 167, 1203.
de Vaucouleurs, G.: 1971, *Publ. Astron. Soc. Pacific* 83, 113.
de Vaucouleurs, G.: 1973, Report to Commission 30, IAU meeting, Sydney.
de Vaucouleurs, G. and Ables, H.: 1968, *Astrophys. J.* 151, 105.
de Vaucouleurs, G. and Agüero, E.: 1973, *Publ. Astron. Soc. Pacific* 85, 150.
de Vaucouleurs, G. and Freeman, K. C.: 1972, *Vistas in Astronomy* 14, 163.
de Vaucouleurs, G. and de Vaucouleurs, A.: 1959a, *Lowell Obs. Bull.* IV, No. 92, 58.
de Vaucouleurs, G. and de Vaucouleurs, A.: 1959b, *Publ. Astron. Soc. Pacific* 71, 83.
de Vaucouleurs, G. and de Vaucouleurs, A.: 1964, *Reference Catalogue of Bright Galaxies*, The Univ. of Texas Press, Austin.
de Vaucouleurs, G. and de Vaucouleurs, A.: 1970, *Astrophys. Letters* 5, 219.
de Vaucouleurs, G. and de Vaucouleurs, A.: 1972, *Mem. Roy. Astron. Soc.* 77, Part I, 1.
de Vaucouleurs, G. and de Vaucouleurs, A.: 1973, *Astron. Astrophys.* 28, 109.
de Vaucouleurs, G., de Vaucouleurs, A., and De Cesare, M.: 1969a, *Bull. Am. Astron. Soc.* 1, 186.
de Vaucouleurs, G., de Vaucouleurs, A., and De Cesare, M.: 1969b, *Contr. McDonald Obs.* II, No. 26 (see also *Sky Telesc.* 37, No. 3)
Einasto, J.: 1968, *Publ. Tartu Astron. Obs.* 36, 396, 414.
Einasto, J.: 1969, *Astrofizika* 5, 137.
Einasto, J.: 1970, *Astrofizika* 6, 149, 241.
Einasto, J.: 1972a, *Tartu Astron. Obs.*, Preprint No. 40.
Einasto, J.: 1972b, in D. S. Evans (ed.), 'External Galaxies and Quasi-Stellar Objects', *IAU Symp.* 44, 37.
Einasto, J. and Kaasik, A.: 1973, Private communication for presentation to this Symposium.
Faber, S. M.: 1971, Unpublished Ph.D. Thesis, Harvard Univ.
Faber, S. M.: 1973, *Astrophys. J.* 179, 423.
Fish, R. A.: 1964, *Astrophys. J.* 139, 284.
Fraser, C. W.: 1972, *Observatory* 92, 51.
Freeman, K. C.: 1970, *Astrophys. J.* 160, 811.

Freeman, K. C. and de Vaucouleurs, G.,: 1974, in preparation.
Gallagher III, J. S.: 1972, *Astron. J.* **77**, 568.
Gilra, D. P.: 1971, *Nature* **229**, 237.
Gilra, D. P.: 1972, in Scientific Results from the Orbiting Astronomical Observatory (OAO-2), NASA SP-310, p. 295.
Gordon, K. J.: 1971, *Astrophys. J.* **169**, 235.
Gouguenheim, L.: 1969, *Astron. Astrophys.* **3**, 281.
Gursky, H., Kellogg, E., Murray, S., Leong, C., Tananbaum, H., and Giacconi, R.: 1971, *Astrophys. J. Letters* **165**, L49.
Haggerty, M. J. and Wertz, J. R.: 1972, *Monthly Notices Roy. Astron. Soc.* **155**, 495.
Heidmann, J.: 1967, *Compt. Rend. Acad. Sci. Paris* **263B**, 1186.
Heidmann, J.: 1969, *Astrophys. Letters* **3**, 19.
Heidmann, J., Heidmann, N., and de Vaucouleurs, G.: 1971, *Mem. Roy. Astron. Soc.* **75**, Parts 4–6, 85, 105, 121.
Heidmann, N.: 1969, *Astrophys. Letters* **3**, 153, and Thesis, Univ. of Paris.
Herschel, J.: 1847, in *Results of Astronomical Observations Made During the Years 1834, 5, 6, 7, 8 at the Cape of Good Hope*, Smith & Co., London.
Hodge, P. W.: 1961, *Astron. J.* **66**, 249, 384.
Hodge, P. W.: 1963, *Astron. J.* **68**, 470.
Hodge, P. W.: 1964, *Astron. J.* **69**, 438.
Hodge, P. W.: 1965, *Astrophys. J.* **142**, 1390.
Hodge, P. W.: 1973, *Astrophys. J.* **182**, 671.
Hodge, P. W. and Hitchcock, J. L.: 1966, *Publ. Astron. Soc. Pacific* **78**, 79.
Hodge, P. W. and Michie, R. W.: 1969, *Astron. J.* **74**, 587.
Holmberg, E.: 1946, *Medd. Lund Obs.* Ser II, No. 117.
Holmberg, E.: 1950, *Medd. Lund Obs.* Ser. II, No. 128.
Holmberg, E.: 1958, *Medd. Lund Obs.* Ser. II, No. 136.
Holmberg, E.: 1964, *Arkiv Astron.* **3**, 387 (= *Medd. Uppsala Obs.* No. 148).
Holmberg, E.: 1969, *Arkiv Astron.* **5**, 305 (= *Medd. Uppsala Obs.* No. 166).
Hubble, E.: 1926, *Astrophys. J.* **64**, 321.
Hubble, E.: 1930, *Astrophys. J.* **71**, 231.
Hubble, E.: 1936, *Astrophys. J.* **84**, 158.
Huchtmeier, W.: 1973, *Astron. Astrophys.* **22**, 91.
Humason, M. L., Mayall, N. U., and Sandage, A. R.: 1956, *Astron. J.* **61**, 97.
Johnson, H. L.: 1966, *Astrophys. J.* **143**, 187.
Johnson, H. M.: 1961, *Astrophys. J.* **133**, 314.
Karachentsev, I. D.: 1967, *Commun. Byurakan Obs.* **39**, 96.
Karachentseva, V. E.: 1967, *Astrofizika* **3**, 535.
Karachentseva, V. E.: 1972, in Problems of Cosmical Physics, *Kiev Univ. Publ.* No. 7, p. 150.
Kellogg, E., Gursky, H., Leong, C., Schreier, E., Tananbaum, H., and Giacconi, R.: 1971, *Astrophys. J. Letters* **165**, L49.
Kerr, F. J.: 1973, Report to Commission 40, IAU meeting, Sydney.
Kiang, T.: 1961, *Monthly Notices Roy. Astron. Soc.* **122**, 263.
King, I. R.: 1961, *Astron. J.* **66**, 68.
King, I. R.: 1962, *Astron. J.* **67**, 471.
King, I. R.: 1966, *Astron. J.* **71**, 64, 276.
King, I. R. and Kiser, J.: 1973, *Astrophys. J.* **187**, 27.
King, I. R. and Minkowski, R.: 1972, in D. S. Evans (ed.), 'External Galaxies and Quasi-Stellar Objects', *IAU Symp.* **44**, 87.
Kinman, T. D.: 1965, *Astrophys. J.* **142**, 1376.
Kleinmann, D. E. and Low, F. J.: 1970a, *Astrophys. J. Letters* **159**, L165.
Kleinmann, D. E. and Low, F. J.: 1970b, *Astrophys. J. Letters* **161**, L203.
Kogure, T. and Toya, N.: 1970, *Mem. Fac. Sci. Kyoto Univ.* **XXXIII**, No. 2 (*Dept. Astron. Kyoto Univ. Rep.* No. 41).
Kowal, C. T.: 1969, *Publ. Astron. Soc. Pacific* **81**, 608.
Kuzmin, G. G.: 1952, *Publ. Astron. Obs. Tartu* **32**, 211.
Lasker, B. M.: 1970, *Astron. J.* **75**, 21.

Martin, W. L. and Bingham, R. G.: 1970, *Observatory* **90**, 13.
Mathews, T. A., Morgan, W. W., and Schmidt, M.: 1964, *Astrophys. J.* **140**, 35.
Mayall, N. U.: 1939, *Lick Obs. Bull.* **19**, No. 497, 33.
Mayall, N. U.: 1948, *Sky Telesc.* **8**, 3.
Mayall, N. U.: 1960, *Ann. Astrophys.* **23**, 344 (= *Lick Obs. Bull.* No. 566).
Mayall, N. U. and Lindblad, P. O.: 1970, *Astron. Astrophys.* **8**, 364.
McClure, R. D.: 1969, *Astron. J.* **74**, 50.
McClure, R. D. and van den Bergh, S.: 1968, *Astron. J.* **73**, 313, 1008.
McGee, R. X.: 1964, *Australian J. Phys.* **17**, 515.
McGee, R. X. and Milton, J. A.: 1966a, *Australian J. Phys.* **19**, 433.
McGee, R. X. and Milton, J. A.: 1966b, *Australian J. Phys., Astrophys. Suppl.* No. 2.
Mestel, L.: 1963, *Monthly Notices Roy. Astron. Soc.* **126**, 553.
Minkowski, R.: 1962, in G. C. McVittie (ed.), 'Problems of Extragalactic Research', *IAU Symp.* **15**, 112.
Minkowski, R. and Osterbrock, D.: 1959, *Astrophys. J.* **129**, 583.
Mitchie, R. W.: 1963, *Monthly Notices Roy. Astron. Soc.* **125**, 127.
Morgan, W. W.: 1958a, *Publ. Astron. Soc. Pacific* **70**, 372.
Morgan, W. W.: 1958b, in R. Stoops (ed.), *La Structure et l'Evolution de l'Univers*, Solvay Conference Rep., Bruxelles, p. 297.
Morton, D. C. and Chevalier, R. A.: 1972, *Astrophys. J.* **174**, 489.
Morton, D. C. and Chevalier, R. A.: 1973, *Astrophys. J.* **179**, 55.
Oemler, A.: 1973, *Astrophys. J.* **180**, 11.
Oort, J.: 1940, *Astrophys. J.* **91**, 273.
Oort, J.: 1965, *Trans. IAU* **XIIA**, 789.
Osterbrock, D.: 1960, *Astrophys. J.* **132**, 325.
Owen, F. N.: 1973, *Radio Sources in Clusters of Galaxies*, Ph.D. Dissert., Univ. of Texas at Austin.
Ozernoy, L. M.: 1967, *Astron. Tsirk. USSR*, No. 407.
Ozernoy, L. M.: 1969, *Zh. Eksperim. Teor. Fiz. (Letters)* **10**, 394 (= *JETP Letters* **10**, 251)
Page, T. L.: 1965, *Smithsonian Astrophys. Obs. Special Rep.* No. 195.
Perek, L.: 1950, *Bull. Astron. Inst. Czech.* **2**, 75.
Poveda, A.: 1958, *Bol. Obs. Tonantzintla Tacubaya* No. 17.
Poveda, A.: 1961, *Astrophys. J.* **134**, 910.
Poveda, A., Iturriaga, R., and Orozco, I.: 1960, *Bol. Obs. Tonantzintla Tacubaya* No. 20, 3.
Randers, G.: 1940, *Astrophys. J.* **92**, 235.
Reeves, G.: 1956, *Astron. J.* **61**, 69.
Rieke, G. H. and Low, F. J.: 1972, *Astrophys. J. Letters* **176**, L95.
Roberts, M. S.: 1966, *Astrophys. J.* **144**, 639.
Roberts, M. S.: 1969, *Astron. J.* **74**, 859.
Roberts, M. S.: 1972, in D. S. Evans (ed.), 'External Galaxies and Quasi-Stellar Objects', *IAU Symp.* **44**, 12.
Rood, H. J.: 1965, *Astron. J.* **70**, 689.
Rood, H. J., Rothman, V. C. A., and Turnrose, B. E.: 1970, *Astrophys. J.* **162**, 411.
Rubin, V. C., Burbidge, E. M., Burbidge, G. R., and Prendergast, K. H.: 1964, *Astrophys. J.* **140**, 80.
Sandage, A.: 1961, *The Hubble Atlas of Galaxies*, Publ. 618, Carnegie Inst. of Washington, D. C.
Sandage, A.: 1963, *Astrophys. J.* **138**, 863.
Sandage, A.: 1972, *Astrophys. J.* **176**, 21.
Sandage, A.: 1973, *Stars and Stellar Systems* **9**, in press.
Sandage, A., Freeman, K. C., and Stokes, N. R.: 1970, *Astrophys. J.* **160**, 831.
Sandage, A., Tammann, G. A., and Hardy, E.: 1972, *Astrophys. J.* **172**, 253.
Sargent, W. L. W.: 1970, *Astrophys. J.* **160**, 405.
Sargent, W. L. W. and Searle, L.: 1970, *Astrophys. J. Letters* **162**, L155.
Saslaw, W. C.: 1970, *Astrophys. J.* **160**, 11.
Saslaw, W. C.: 1971a, *Astrophys. J.* **163**, 249.
Saslaw, W. C.: 1971b, *Monthly Notices Roy. Astron. Soc.* **152**, 351.
Saslaw, W. C.: 1972, in D. S. Evans (ed.), 'External Galaxies and Quasi-Stellar Objects', *IAU Symp.* **44**, 93.
Schild, R. and Oke, J. B.: 1971, *Astrophys. J.* **169**, 209.

Schilt, J., Epstein, I., and Hill, S. J.: 1955, *Astron. J.* **60**, 341.
Sersic, J. L.: 1957, *Revista Astron.* **XXIX**–**II**, 68 (see also *Observatory* **78**, 24, 1958).
Sersic, J. L.: 1966, *Z. Astrophys.* **64**, 202.
Sersic, J. L.: 1968a, in 'Non-Stable Phenomena in Galaxies', *IAU Symp.* **29**, Acad. Sci. Armen. SSR, Yerevan, p. 403.
Sersic, J. L.: 1968b, *Atlas de Galaxias Australes*, Obs. Astron., Cordoba, Argentina.
Smith, R. T.: 1941, *Publ. Astron. Soc. Pacific* **53**, 187.
Spinrad, H.: 1966, *Publ. Astron. Soc. Pacific* **78**, 370.
Spinrad, H., Gunn, J. E., Taylor, B. J., McClure, R. D., and Young, J. W.: 1971, *Astrophys. J.* **164**, 11.
Takase, B.: 1967, *Publ. Astron. Soc. Japan* **19**, 427.
Takase, B. and Kinoshita, H.: 1967, *Publ. Astron. Soc. Japan* **19**, 209.
Tammann, G. A.: 1972, *Astron. Astrophys.* **21**, 355.
Tifft, W. G.: 1961, *Astron. J.* **66**, 390.
Tifft, W. G.: 1963, *Astron. J.* **68**, 302.
Tifft, W. G.: 1969, *Astron. J.* **74**, 354.
Toomre, A. and Toomre, J.: 1972, *Astrophys. J.* **178**, 623.
van den Bergh, S.: 1959, *Publ. David Dunlap Obs.* **II**, No. 5, 147.
van den Bergh, S.: 1960, *Astrophys. J.* **131**, 558.
van den Bergh, S.: 1961, *Z. Astrophys.* **53**, 219.
van den Bergh, S.: 1968, *J. Roy. Astron. Soc. Canada* **62**, No. 2, 3; *Commun. David Dunlap Obs.* No. 195.
van den Bergh, S.: 1972a, in D. S. Evans (ed.), 'External Galaxies and Quasi-Stellar Objects', *IAU Symp.* **44**, 1.
van den Bergh, S.: 1972b, *J. Roy. Astron. Soc. Canada* **66**, 237.
van Houten, C. J.: 1961, *Bull. Astron. Inst. Neth.* **16**, 1.
von Hoerner, S.: 1957, *Astrophys. J.* **125**, 451.
Vorontsov-Velyaminov, B. A.: 1959, *Atlas and Catalogue of Interacting Galaxies*, Sternberg Astron. Inst., Moscow.
Vorontsov-Velyaminov, B. A.: 1962, *Morphological Catalogue of Galaxies* **I**, Moscow.
Vorontsov-Velyaminov, B. A.: 1967, in *Modern Astrophysics*, Gauthier-Villars, Paris, p. 347.
Vorontsov-Velyaminov, B. A.: 1970a, *Comm. Sternberg Astron. Inst. Moscow* No. 166, p. 3.
Vorontsov-Velyaminov, B. A.: 1970b, *Astron. Zh.* **47**, 271 (= *Soviet Astron.* **14**, 222).
Vorontsov-Velyaminov, B. A.: 1970c, *Astron. Zh.* **47**, 16 (= *Soviet Astron.* **14**, 11).
Vorontsov-Velyaminov, B. A. and Noskova, R. I.: 1971, *Astron. Zh.* **48**, 513 (= *Soviet Astron.* **15**, 402).
Walker, M. F., Blanco, V. M., and Kunkel, W. E.: 1969, *Astron. J.* **74**, 44, 123.
Warner, P. J., Wright, M. C. H., and Baldwin, J. E.: 1973, *Monthly Notices Roy. Astron. Soc.* **163**, 163.
Wells, D. C.: 1972, *Integrated Spectral Energy Distributions of Galaxies*, Ph.D. Dissert. Univ. of Texas, Austin.
Wells, D. C.: 1973, *Bull. Am. Astron. Soc.* **5**, 26.
Welsh, G. A. and Sastry, G. N.: 1971, *Astrophys. J. Letters* **169**, L3.
Westerlund, B. E. and Wall, J. V.: 1969, *Astron. J.* **74**, 335.
Whitford, A. E.: 1971, *Astrophys. J.* **169**, 215.
Wood, D. B.: 1966, *Astrophys. J.* **145**, 36.
Woolley, R. v. d. R.: 1954, *Monthly Notices Roy. Astron. Soc.* **114**, 191.
Woolley, R. v. d. R. and Robertson, D. A.: 1956, *Monthly Notices Roy. Astron. Soc.* **116**, 288.
Zonn, W.: 1968, *Acta Astron.* **18**, 338.
Zwicky, F.: 1937, *Astrophys. J.* **86**, 217.
Zwicky, F.: 1957, *Morphological Astronomy*, Springer-Verlag, Berlin.

DISCUSSION

Schild: Your slides of the concentration index and other luminosity distribution-related parameters vs Hubble type showed poor correlation, and you regretted being unable to derive quantitative Hubble types. Is not the integrated spectral type, or what may be more quantitatively measured, the integrated energy distribution, a very much better indicator of Hubble type, especially if the absolute magnitudes are approximately known from the magnitude and redshift?

G. de Vaucouleurs: No, colour indices depend on too many extraneous variables (inclination, redshift, galactic extinction, line emission, etc.), and are not usually measured precisely enough to yield a good equivalent Hubble type (see e.g. G. & A. de Vaucouleurs: 1972, *Mem. Roy. Astron. Soc.* **77**, Part 1). Neutral hydrogen mass, or density, or M_H/M_T, etc. likewise give only poor correlations at present. From principal component analysis of T variables, Brosche (*Astron. Astrophys.* **23**, 259, 1973) has demonstrated that the standard error of the Reference Catalogue revised Hubble types is $\sigma_T = 0.75$ or only 5% of the range of the T variable.

van den Bergh: If E0 galaxies are indeed flattened objects seen pole-on, then how is one to interpret the geometry of an E0 galaxy like M87 in which the inner isophotes are circular and the outer isophotes are elongated?

G. de Vaucouleurs: I have no good answer to this good question. Note, however, that the statistics of ellipticities refer to $\mu_B = 25$ mag (arc sec)$^{-2}$ isophotes, not to the very low luminosity outer coronas which may be subject to tidal interactions (or mixing) in dense regions such as the core of the Virgo cluster. Also it may no longer be justifiable to classify M87 as E0; it may be just the globular nucleus of a larger lenticular object, i.e. a cD galaxy, with a faint flat component (cf. *Astrophys. Letters* **4**, 17, 1969).

Arp: When you classified the compacts off the end of the ellipticals in the Hubble sequence, did you distinguish between red and blue compacts?

G. de Vaucouleurs: The red compacts such as M32, NGC 4486B are at $T = -6$. The emission line – H II regions – compacts were placed beyond the Im type at $T = +11$. I have no quantitative information on early-type or continuum-blue compacts such as NGC 1510, which may also be placed at $T = -6$. The Hubble sequence is basically one of morphology, not colour.

Arp: That is what I understood, but I have taken spectra of Zwicky-type blue compacts which showed no emission lines whatsoever – only early-type absorption lines.

PROPERTIES AND DISTRIBUTION OF GALAXIES
IN RELATION TO THEIR FORMATION AND
EVOLUTION: THEORETICAL SITUATION

P. J. E. PEEBLES

Dept. of Physics, University of California, Berkeley, Calif., and
Joseph Henry Laboratories, Princeton University, Princeton, N.J., U.S.A.

Abstract. This report is a review of several proposed pictures for how matter came to be concentrated in lumps like galaxies and clusters of galaxies. The proposals cover a broad range of unrelated possibilities. I attempt here to subject these proposals to critical discussion, to discover the major problems for each case and the more promising lines for further study.

1. Introduction

The search for some theoretical handle on the significance of galaxies in cosmology centres on three questions:

(1) According to accepted or conjectured laws of physics, what are the processes by which structure on the scale of galaxies can form and evolve? This has been the subject of several recent reviews,* and for the most part I will not attempt to repeat any of the computational details here.

(2) What are the phenomena that bear clearest witness to processes of evolution? This question seems fair enough, but we recognize that it is loaded. The great trick is to pick out from the welter of peripheral effects those phenomena that have a direct bearing on the central issues of origin and evolution. The great danger is that our decisions about what is direct and what is peripheral depend on our prejudices about what is the best candidate for the physical process.

(3) Which of the processes listed under question (1) fit into the astrophysical situation, as delineated under (2), without logical contradiction or undue forcing? Again, we should approach this question with some care. One can find in the literature many different proposals for how galaxies may have formed, and in many cases the author shows how, if all works out as conjectured, the picture will account for a broad variety of phenomena. This is hardly surprising. There is a strong systematic bias against publishing models that do not give the 'right' answers. Also, by the nature of the problem, the models are often so complicated that it is hard to make definite statements about how they really would behave. One must introduce conjectures, and it is a natural human tendency to choose the optimistic possibility at each step along

---

* For different aspects of the problem see the preprint by G. B. Field (intended for the fabled Volume 9 of *Stars and Stellar Systems*), the review papers in Supplement 49 of *Progress of Theoretical Physics* (1971), the review by Rees (1971), and the textbooks *Physical Cosmology* (Peebles 1971a; hereafter referred to as PC), Chapter 7, and *Gravitation and Cosmology* (Weinberg, 1972; hereafter called GC), Chapters 8–10.

John R. Shakeshaft (ed.), The Formation and Dynamics of Galaxies. 55–74. All Rights Reserved.
Copyright © 1974 by the IAU.

the way. The result is that the 'big picture' in this subject is uncertain ground indeed. Under the circumstance it may be best to narrow the line of attack, to seek out the spots of firmer ground here and there where some aspect of the situation appears clean and simple, and we might see how one or another of the proposed schemes may be put to serious and objective test. It would be unrealistic to hope that this approach will soon lay to rest any of the more popular schemes for galaxy formation, but it should show the way to less cluttered ground for the debate.

In thinking about how to attack this third question, we may find it helpful to recall the role of *uniformitarianism* in the origins of modern geology (in the 18th century). One argued that observed geologic features might be understood as the result of observed processes (like erosion and sedimentation) acting over a sufficiently long time. A most beneficial feature of this concept was (and is) that it allowed people to approach the problem in a rational and scientific way *without* having to address the enormously complicated ultimate issue of the origin of the Earth. I suspect that our attempts to find a theoretical account of galaxies in cosmology has been unduly influenced by the fact that modern 'establishment cosmology' came into being in a peculiar way, through the inspired guesses of a few people. Even if these guesses prove accurate we have no reason to think that we can repeat the coup. But even if we lack inspired guesses we can fall back on the uniformitarian method. I would state the lesson as follows: (1) form an opinion of the relevant 'observed' phenomena; (2) draw up a list of presumed processes and/or laws of physics; (3) *ask whether one can trace back in time from the phenomena according to the processes without running into a situation that appears contrived or contradictory.*

Of course this is a slow and uncertain business. The singular big-bang surely does violence to uniformitarianism – so do we blame the assumed laws of physics – general relativity in this instance; the assumed situation – a uniformly expanding Universe; or the notion of uniformitarianism? Still, it offers a clear line of attack, which is to be treasured.

In the present review I consider mainly structure on scales larger than galaxies – groups, clusters and superclusters. This is somewhat out of line with the main subject of the conference, the nature of galaxies, but the phenomenon surely is closely related, and the great attraction is that the physics appear to be a good deal simpler. We might even hope that this large-scale structure is a reasonably undisturbed fossil of what was happening in the past, and at the very least the apparent simplicity of the phenomenon strongly limits the theoretical speculation. In contrast, the galaxies seem like fossils that have been pulverized by the complex processes of evolution.

Not included in this review is the search for young galaxies – perhaps newly-formed objects in our neighbourhood of the Universe, perhaps objects at high redshift. Its importance can hardly be overstated, however. Once we have a genuine young galaxy in our hands, so to speak, it will settle a lot of arguments.

I base the discussion on the standard big-bang cosmological model. This is an assumption, and not one that all would adopt, for the big-bang cosmology is not estab-lished beyond reasonable doubt. On the other hand, my impression of the observa-

tional situation is that this cosmology has strong credentials as a working hypothesis.*
In an effort at ecumenicity I have sought where possible to frame arguments that are
more broadly based than the big-bang picture. I have also restricted the discussion
to redshifts $z \lesssim 1000$. This eliminates an active field of research, the possible role of
the primeval fireball radiation in shaping and controlling the evolution of irregularities. This subject may prove to be of decisive importance to the problem of galaxy
formation, but it is still quite complex and highly uncertain.

2. Large-Scale Irregularities in the Matter Distribution

The nature of the large-scale distribution of the observed matter, as galaxies, has been
reviewed by de Vaucouleurs (1971). At this conference G. Burbidge (p. 93) has
discussed a second component of possibly decisive importance, the intergalactic
phenomena. I only wish to describe here a particular aspect of the observational
situation that seems to me to be of considerable interest – the clouds of galaxies in
the Shane-Wirtanen Catalog (1967) and the superclusters found in the distribution
of rich clusters cataloged by Abell (1958). Shane and Wirtanen counted galaxies down
to limiting brightness $m_{pg} \lesssim 19$, which means that the galaxies they counted are typically
at distances in the range of $200\,h^{-1}$ Mpc to $400\,h^{-1}$ Mpc.** The published data
give the counts in $1° \times 1°$ cells. These counts are correlated over angular distances
as large as $5°$, which corresponds to a spatial coherence or correlation in the galaxy
distribution over distances as large as $\sim 30\,h^{-1}$ Mpc.‡ Following Zwicky and Shane,
this phenomenon is conveniently described as the 'cloudy' nature of the large-scale
distribution of galaxies. The clouds are considerably bigger than the rich clusters
cataloged by Abell, but we know that the rich clusters must be in the clouds of galaxies
because the Shane-Wirtanen galaxy counts show a strong cross-correlation (over
angular scales $\lesssim 5°$, depending on cluster distance class) with the positions of the
Abell cluster centres (Peebles, 1974). What is more, there is convincing evidence of
correlations amongst the positions of the Abell clusters.† The observed correlation

* For a less optimistic view see Burbidge (1971). For a recent review of the current observational
situation in cosmology see PC. The only significant change since PC is the status of the microwave
background shortward of 1 mm wavelength. All observers now appear to be agreed that there is no
evidence for or against a black-body spectrum shortward of 1 mm wavelength (Houck *et al.*, 1972;
Williamson *et al.*, 1973; Muehlner and Weiss, 1973).
** All lengths, times, etc. are based on a nominal value for Hubble's constant, $H = 100$ km s⁻¹ Mpc⁻¹,
$H^{-1} \cong 1 \times 10^{10}$ yr. The dimensionless number h is the 'true' value of Hubble's constant measured in
units of 100 km s⁻¹ Mpc⁻¹. This very convenient practice is becoming common, but it was too much
to hope that people would settle on a standard nominal value – in different papers one can find the
same symbol h used to represent Hubble's constant in units of 50, 75, and 100 km s⁻¹ Mpc⁻¹.
‡ The Lick counts were analyzed by Limber (1954) and by Neyman *et al.* (1953). For a recent
detailed rediscussion see Peebles and Hauser (1974).
† The first demonstration of superclusters that I found truly clean and convincing was that of Bogart
and Wagoner (1973). Hauser and I had independently attacked the problem, following the method
used earlier to set an upper limit on superclustering (Yu and Peebles, 1969), and we independently
concluded that superclusters do indeed exist, at about the earlier upper limit. The numbers quoted
here are from Hauser and Peebles (1973).

has roughly the same linear scale as that of the clouds, and there are about two clusters per supercluster, in the sense that there is on the average ~ 1 cluster in excess of random near a randomly chosen cluster. The superclustering or correlations among cluster centres again extends to lengths $\sim 30\ h^{-1}$ Mpc.

The question of structure on scales larger than 20 to 30 h^{-1} Mpc has been reviewed by de Vaucouleurs (1971). The observational problems are very difficult, so I have chosen not to consider such effects here. As for the clouds (superclusters), many details of the galaxy organization and motion remain obscure. However, the simple fact that we can detect structure on scales as large as 30 h^{-1} Mpc gives some interesting constraints on models for evolution, as is described below.

3. The Evolution of Irregularities

3.1. WHITE HOLES

Phenomena associated with galactic nuclei have occasioned in a number of different contexts the thought that galaxies might have issued out of primeval nuclei or 'white holes.'* This is a spectacular conception, as befits spectacular and mysterious phenomena, and for that very reason it seems clear that its role will be understood only when we have a much clearer understanding of the physics of galaxies. We can observe, however, that if each galaxy issues out of its own primeval nucleus then we are left with the problem of accounting for the large-scale clustering of primeval nuclei, in the clusters and clouds of galaxies. It has been suggested that only the dominant galaxy in a group or cluster is associated with a primeval nucleus, the lesser members having formed by some sort of calving process (e.g. Hoyle, 1965, pp. 19, 20). But how does one account for the clouds of galaxies? Here is a true hierarchical structure, groups and clusters within clouds. In the present scheme, we must assume either that the primeval nuclei are distributed in a hierarchical fashion, or else that the primeval nuclei occur one to a cloud, the ejected matter fragmenting down the hierarchy into clusters, groups and galaxies. Either answer seems to lead us back to the original problem.

A related difficulty is that of the peculiar velocities of galaxies. If the galaxies in a cloud issued from one source, then to spread over 30 h^{-1} Mpc in one expansion time $\sim 10^{10}\ h^{-1}$ yr, the matter would have to have a dispersal velocity ~ 3000 km s^{-1}. If the dispersal velocity were close to this limit the expansion of the cloud would roughly mock the general expansion of the Universe in the big-bang cosmology. If the dispersal velocity were much less, the galaxies would have appeared tightly

---

* Early references to this concept are found in Jeans (1928) and Milne (1948). In cosmological discussions, McCrea (1964) considered 'embryos' of galaxies ejected from old galaxies, and Hoyle and Narlikar (1966) considered 'pockets of creation' of matter. The subject was first vigorously pursued from the point of view of phenomenology by Ambartsumian (1958, 1965). Other studies of phenomena that may point to (or test) the concept are described by Arp at this conference (p. 199, cf. also Arp 1971 and earlier references therein), Burbidge *et al.* (1963), Hoyle (1965), Holmberg (1969), Bahcall and Joss (1972) and van der Kruit *et al.* (1972). Novikov (1964) and Ne'eman (1965) discussed a similar picture for the nature of quasars.

concentrated around the sources, which is not observed. If the velocity were much greater, say 6000 km s^{-1}, twice the critical value, galaxies should now have peculiar velocities on the order of 3000 km s^{-1}. This may well be possible for the occasional object, but it is quite unreasonable for the general field. This picture thus requires a very special ejection velocity, which would seem to require some explanation.

These arguments have no direct bearing on the role of primeval nuclei in internal dynamics of galaxies. They may suggest that something else is required to account for the long-range ordering of galaxies.

3.2. THERMAL INSTABILITY

The whole panoply of fluid dynamics – thermal phenomena, shocks, turbulence, magnetohydrodynamic effects – may be expected to have played a role in galaxy formation, and to have added to the complexity of these systems. The situation appears much easier for the physics of the cloud structure. This and the following sections deal with several such effects by which people have proposed that irregularities on the scale of galaxies might have evolved. I begin with thermal instability.*

One could suppose that thermal instability acts as the primary agent causing matter which was initially more uniformly distributed to collect in structures like galaxies and groups and clouds of galaxies, or it might be a secondary agent that helps fashion evolution once proto-systems have formed. We are interested here in the former role. This role can be significant if two order-of-magnitude conditions are satisfied – that the velocity of sound in the pre-galactic medium is large enough that a pressure wave can traverse the desired length scale in an expansion time H^{-1}, and that the cooling time for the medium is less than or comparable to the expansion time. The first condition applied in the present state of the universe implies that the velocity of sound is $\gtrsim 3000$ km s^{-1}, or that the matter temperature $\gtrsim 10^9$ K, in order to make irregularities on the scale of $30\,h^{-1}$ Mpc. But if the gas is this hot and at the mean cosmological density (or, what is equivalent for this point, at the mean density of the matter within $\sim 300\,h^{-1}$ Mpc), the cooling time is at least three orders of magnitude larger than the expansion time. It does not help to imagine that the thermal instability occurred in the early Universe, at high redshift, for although the matter density is higher at larger redshift, so is the matter temperature needed to get irregularities on a scale of $\sim 30\,h^{-1}$ Mpc now, the temperature varying as $(1+z)$ in the Einstein-de Sitter model. The bremsstrahlung radiation from this dense relativistic plasma would make for severe problems with the X-ray background. Thus, while it may be that the thermal instability effect played an important role in the evolution of structure on a sub-galactic or even galactic scale, it could not have been a significant factor in the physics of clouds of galaxies.

3.3. MAGNETOHYDRODYNAMICS

The origin of the interstellar magnetic field is obscure. One possibility is that the

---

* Thermal instability in cosmology is discussed by Hoyle (1958), Gold and Hoyle (1959), Field (1965), Kondo *et al.* (1971) and Arons (1972).

field existed before the Galaxy. If so, it would be natural to ask whether the magnetic field played a role in the formation of the Galaxy and, by extension, the formation of systems of galaxies.* The dynamic effect of the field on the expanding matter distribution is important if the Alfvén velocity is comparable to the size of the structure of interest divided by the expansion time-scale. For the present values of expansion time-scale, mean mass density, and the size of clouds of galaxies, this criterion yields **

$$B \gtrsim 5 \times 10^{-6}\, \Omega^{1/2}\, h\, G.$$

In the interstellar medium the field is $\sim 10^{-6}$ G, but the matter density is some 5 or 6 orders of magnitude higher than the cosmological mean. If the interstellar medium were isotropically expanded down to the cosmological mean density, conserving flux, it would reduce the interstellar field to $\sim 10^{-10}$ G, some four orders of magnitude below the critical value. It might be noted that this factor 10^4 is nearly independent of time in the expanding Universe.‡ Thus it appears that, unless the Galaxy is composed of matter which contained usually small flux, magnetic field acting along could not have played a significant role in the behaviour of structure on the scale of clouds of galaxies.

A more interesting question is the possible role of magnetic stresses acting in concert with the gravitational instability effect discussed in Section 3.5 below. A primeval magnetic field would make an anisotropic (and presumably inhomogeneous) contribution to the stress-energy tensor. We might expect to find that, as we trace the expansion of the Universe back in time toward the singularity, this anisotropic stress forces deviations from a homogeneous isotropic expansion (Thorne, 1967). Turning the question around, does this mean that the assumption of a primeval magnetic field would require the assumption of highly special initial conditions to assure that the Universe ends up looking no more irregular than it does? That is, does the assumption of a primeval magnetic field agree with uniformitarianism?

3.4. PRIMEVAL TURBULENCE

The picture of cosmic turbulence has figured in one form or another in many different

---

* The possible role of magnetic fields in galaxy formation has not received the detailed attention it perhaps deserves. For some discussion see Zel'dovich (1969), Rees (1971), Harrison (1970a), Peebles (1969a), and Rees and Reinhardt (1972).
** Ω is the ratio of the mean mass density of the Universe (or of the part within a few hundred Mpc distance) to the density in the Einstein-de Sitter cosmological model. Thus the matter density is $\varrho = 2 \times 10^{-29}\, \Omega\, h^2$ g cm^{-3}. If the galaxies are the main contribution to ϱ and if galaxy masses are computed from observed angular sizes and velocity dispersions of galaxies then Ω is independent of h.
‡ The diameter of a chosen system expanding with the Universe varies in proportion to $(1+z)^{-1}$, where z is the redshift. In the Einstein-de Sitter cosmology, $q_0 = \frac{1}{2}$, the expansion time varies as $(1+z)^{-3/2}$, so the critical velocity varies as $(1+z)^{1/2}$. Since the Alfvén speed $B \propto \varrho^{-1/2}$, and $\varrho \propto (1+z)^3$, the critical field is $B \propto (1+z)^2$, which is also the law for the time variation of B under uniform isotropic expansion and flux conservation. The expansion parameter might be as low as $q_0 \cong 0.02$. Here, the critical field varies as $B \propto (1+z)^{3/2}$ back to $z \cong 50$. At higher redshift the model behaves like the Einstein-de Sitter case.

discussions of galaxy formation.* The attractions are evident. It is a common observation that fluid flow tends to go turbulent – whether it is smoke rising from a cigarette, crude oil flowing in a pipeline, or clouds moving in the interstellar medium. As Gamow (1952) and Oort (1958) have pointed out, the distribution and appearance of galaxies seem like the fossil remnants of turbulent eddies. And for the theorist, turbulence is a lovely point of departure for computation. However, there are some problems.

Let us consider first the possible role of turbulence in the present state of the Universe. We will suppose that matter moves about more or less after the fashion of developed turbulence, in random currents with typical velocity $v_T(\lambda)$ for eddies of size λ. These currents can appreciably alter the distribution of matter on the scale of λ if $v_T(\lambda) \gtrsim \lambda/t_0$, where $t_0 = H^{-1}$ fixes the expansion time-scale. For the clouds of galaxies this gives the condition, in the present Universe,

$$v_T \gtrsim 3000 \text{ km s}^{-1}.$$

But one would expect the peculiar velocities of galaxies and clusters of galaxies to be comparable with the velocities of the large-scale random currents, and it seems that such high peculiar velocities can be ruled out. Of course the situation is much easier if one only wants to make individual galaxies, which would require $v_T \sim 300$ km s^{-1}. But another problem arises when one attempts to fit these random currents into the big-bang cosmology.

It will be assumed that the pressure of matter and radiation may be neglected compared with ϱc^2, where ϱ is the mean mass density. This may be valid back to redshift $z \sim 4 \times 10^4 \, \Omega \, h^2$, at which epoch the Primeval Fireball radiation pressure becomes comparable to ϱc^2. Now at any epoch the presumed state of turbulence of matter (the matter imagined fairly smoothly distributed in some pre-galactic condition) may be placed in one of two convenient categories, *weak turbulence* if turbulence velocities $v_T(\lambda)$ on scales λ satisfy $v_T(\lambda) \, t/\lambda < 1$, where t is the expansion time-scale, and *strong turbulence* if the inequality goes the other way.** As was remarked, if the turbulence is weak nothing much happens in one expansion time because the eddies do not have time to turn over. What is more, one finds that as the Universe expands the number $v_T(\lambda) \, t/\lambda$, computed for a given eddy mass (that is, computed for a fixed comoving scale of length) decreases: *weak turbulence grows weaker as the Universe expands*. Thus, as long as the weakly turbulent eddy expands with the

---

* For a review see Jones and Peebles (1972). For somewhat less pessimistic views of turbulence see for example Oort (1970), Ozernoy (1971), Stecker and Puget (1972).

** For various parts of the argument given here see Peebles (1971b) and Jones (1973). Apparently the process of turbulent dissipation was first discussed in this context by Tomita *et al.* (1970). An elegant and considerably more rigorous discussion of some aspects of how strong and weak turbulence evolve is given by Olsen and Sachs (1973). Silk and Lea (1973) have discussed the evolution of primordial random motions of galaxies. The problem is similar to that of turbulent dissipation, but since galaxies are compact the collision rate would be lower than for a more nearly uniformly distributed pre-galactic medium, and the direct collision of two galaxies would be less catastrophic than the collision of two eddies in a gaseous fluid.

general expansion of the Universe, the eddy never will have time to turn over, or to alter appreciably the matter distribution. As in the discussion of primeval magnetic fields, we can note that a more interesting situation is obtained when we consider the combined effects of weak turbulence and gravitational instability – the weak matter currents feed growing modes of density irregularity (Peebles, 1971b; Ozernoy, 1971). However, it seems difficult to distinguish this situation from the straight assumption that there are small density irregularities in the early Universe.

If the turbulence is strong, eddies can turn over. According to conventional understanding the eddies in this case are strongly unstable, so that in one eddy turn-over time the energy of each eddy is dumped into eddies of smaller scale and/or shocks, depending on how v_T compares with the velocity of sound. The former effect is a result of inertia, not viscosity, and is thought to lead to the Kolmogorov spectrum for fully developed turbulence. Of course, it may be that the primeval turbulence eddies are stabilized by some unexpected effect, as is enjoyed by tornadoes (cf. the 'spinning cores' concept of Harrison, 1970b). The point is important, for in the absence of some stabilization effect, we must expect that the energy in strong primeval turbulence would rapidly cascade down to smaller and smaller eddies until it is dissipated by viscosity. If so, we are in a bind. *If we assume that at some epoch t primeval turbulence was playing a significant role in the evolution of irregularities, then we expect that at epochs earlier than t the turbulence was even stronger and the role even more significant.*

There are three ways to avoid the blow-up:

(1) *Spontaneous Generation of Turbulence.* The general expansion of the Universe is a fluid motion, with which one can associate a Reynolds number that is enormous. It is natural to ask whether this general expansion might break up into turbulent motion. The linear perturbation calculation says that this does *not* happen if the expansion is nearly homogeneous and isotropic – for only compressional (acoustic) perturbations grow, and these only as a power of time. The situation is more difficult if the expansion is highly anisotropic.* However, the precise isotropy of the microwave background shows that the expansion in fact must have been nearly isotropic at least back to a redshift $z \sim 1000$.

(2) *Driven turbulence.* Turbulent motion in intergalactic (or pre-galactic) matter might be a secondary phenomenon driven by local disturbances like quasars, young galaxies, regions of matter-antimatter annihilation, or primeval nuclei.** This merits further careful study, for it can neatly avoid the bind and open the way for turbulence phenomena at modest redshift, when galaxies might be expected to have formed.

---

* For conflicting indications on the generation of turbulence in anisotropic cosmological models, see Silk (1973) and Perko *et al.* (1972).

** Stecker and Puget (1972) discuss turbulence driven by matter-antimatter annihilation. In the picture discussed by Doroshkevich *et al.* (1967), turbulence (or mass motion sufficient to initiate galaxy cluster formation) is driven by heating by massive protostars. Silk and Solinger (1973) discuss turbulence driven by active radio galaxies. One might include in this category Sciama's (1955) picture of matter motion driven by the motion of galaxies through the intergalactic medium.

(3) *Turbulence in the radiation-dominated phase.* One considers here turbulence at high redshift, $z \gtrsim 4 \times 10^4 \, \Omega \, h^2$. There are two new effects – the Primeval Fireball radiation would have been coupled to the matter through Thomson scattering, and the mass density of the radiation would have dominated that of the matter. Under these conditions, one finds that the expansion of the Universe causes the number $v_T(\lambda) \, t/\lambda$ to *increase* with time (while subsequent to $z \sim 4 \times 10^4 \, \Omega \, h^2$ the expansion of the Universe causes the number to *decrease*). Thus *weak turbulence can evolve into strong turbulence, so one can assume that turbulence is primordial.** The very attractive feature of this picture is that the spectrum of density irregularities may be determined up to only a few adjustable parameters by the theory of fully developed turbulence. The main difficulty is the question of what happens subsequent to the epoch at redshift $z \sim 1000$ when, in the absence of any source of excitation, the matter is expected to recombine and decouple from the radiation. Any peculiar matter velocities left over from the turbulence are liable to initiate formation of bound systems sooner than we want them. There are many uncertainties here, but the dimensions of the problem perhaps are indicated by the following outline of the simple-minded arguments.

The turbulence can be considered to be *fully developed* (that is, in the regime of of strong turbulence) for some range of eddy sizes. Rough estimates for upper and lower bounds of this range are shown as lines (1) and (2) in Figure 1. The existence of an upper bound λ_m on the eddy size follows from the assumption, which will be adopted here, that the turbulence velocities are non-relativistic, $v_T < c$. This assures that the turbulence (at $z \sim 1000$) can be considered a small perturbation to a Friedman-Lemaître cosmological model (Jones, 1973). But then λ_m satisfies $\lambda_m/t \equiv v_T(\lambda_m) < c$, so λ_m is smaller than the horizon ct. Since there is an upper bound on the peculiar velocity, *the kinetic energy (per unit mass) of the turbulence is bounded.* Also, the matter currents in eddies larger than λ_m may be a large reservoir of kinetic energy, but the linear perturbation calculation says that this energy, in weakly turbulent eddies, is *not* fed to smaller-scale eddies. Now this limited fund of kinetic energy of the developed turbulence is dissipated in two ways. First is the general expansion of the Universe: in effect the expansion adiabatically 'cools' the random currents. Second is the turbulent dissipation process mentioned above, the cascade of energy from large eddies to smaller ones, the energy ending up in eddies small enough to be dissipated by viscosity or whatever. Figure 1 shows the possible course of evolution of λ_m at epochs near redshift $z = 1000$. The vertical axis is a co-moving length scale, that is, the length is adjusted to take account of the general expansion from the epoch z to the present. The right-hand scale is the mean mass (in solar masses) contained within a sphere with diameter equal to the length plotted on the left-hand

---

* This result was first exploited by Ozernoy and Chernin (1967). For recent discussions see Ozernoy (1971), Silk and Ames (1972), Dallaporta and Lucchin (1972), Jones (1973), Stein (1974). Zel'dovich and Novikov (1970) have remarked that even taking account of the fireball radiation the primeval turbulence picture has the unfortunate property that, when traced back to the singularity in the big-bang model, it gives rise to divergent irregularities in curvature.

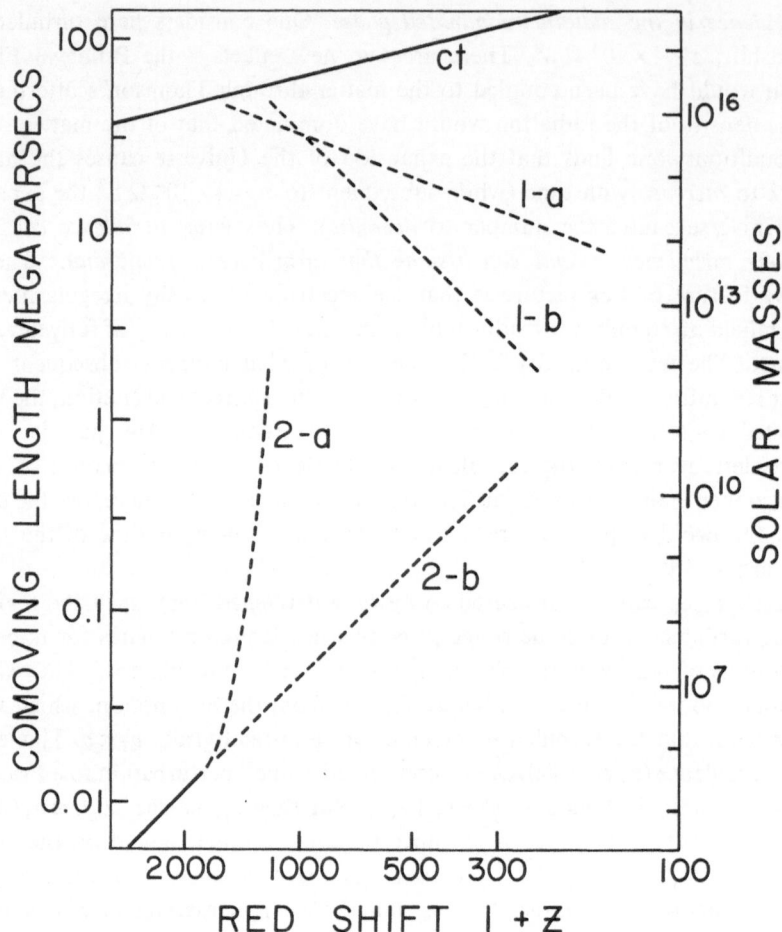

Fig. 1.   Limits of primeval turbulence in the early Universe.

scale. In plotting this figure, I have assumed $\Omega = h = 1$. Other values for these parameters change the numbers but not the general idea. The bounding wavelength λ_m has
been adjusted to pass through 30 Mpc/$(1+z)$ at $z = 1000$. This is about as large as it
could be, consistent with the assumption $v_T < c$. Line 1-a in the figure shows how λ_m
varies due to the adiabatic 'cooling' of the turbulence alone. It is assumed here that
the turbulence velocity has the Kolmogorov spectrum $v_T(\lambda) \propto \lambda^{1/3}$. The actual
attenuation of the turbulence, taking account of the turbulence dissipation process,
is presumably much more rapid than is indicated by line 1-a. Line 1-b is a free-hand
estimate of what both processes together might do.

Lines 2 in the figure are estimates of the photon mean free path as a function of
redshift. Line 2-a is computed on the assumption that the recombination of the plasma
is unaffected by the turbulence. For line 2-b it is assumed that the matter remains
fully ionized.

One can imagine that there is fully developed turbulence on scales extending from line 1-b to somewhere above line 2-b. In this range, matter and radiation move roughly like a single fluid, the radiation providing the pressure and making the turbulence sub-sonic. Now the question is, what happens when lines 1 and 2 cross? We have the following possibilities:

(a) The plasma recombines rapidly, following line 2-a. Matter and radiation sharply decouple, leaving turbulent currents that are supersonic, and can slap together because $v_T(\lambda)\, t/\lambda > 1$. I presume the matter would then pile up in shocks, forming lumps much denser than the mean, and that these lumps would be gravitationally bound systems. This would not necessarily be a good thing, because the lumps would be much denser than galaxies, but conceivably the lumps could form the nuclei of galaxies, or quasars, or some sort of undiscovered object.

(b) Dissipation of the turbulence keeps the plasma ionized. Then the turbulence 'dies' at $z \gtrsim 300$, and the matter recombines. But this leaves residual matter currents with velocity $v_T(\lambda) \sim \lambda/t$ at the point of intersection of the lines 1 and 2, for the dissipation process ceases once the turbulence becomes weak. These residual currents can be treated by perturbation theory (Peebles, 1971b). One finds that the currents feed matter density irregularities, and that, if $v_T \sim \lambda/t$ when the strong turbulence dies, gravitationally bound systems form shortly thereafter. Again, this is not necessarily a good thing, for the redshift is still high.

(c) The turbulence dies before the matter recombines. Then the radiation drag can sharply reduce the residual matter currents, avoiding the difficulty with premature formation of bound systems. Here the main point of interest is the residual acoustic 'noise' generated by the turbulence. This 'noise' may grow under the influence of gravity, fragmenting in the course of time into bound systems like galaxies. As was remarked, the attractive feature of this game is that the noise spectrum may be determined by the physics of fully developed turbulence. This is a subject that is just coming into lively discussion, and may well turn out to be profitable.*

In all the preceding discussion it has been imagined that the primeval turbulence can be treated as a perturbation to the Friedman-Lemaître cosmological model ($v_T < c$). Rees (1972) has proposed dropping even this assumption, and introducing the picture of large initial fluctuations in space curvature on the scale of galaxies (although the Universe would have to be homogeneous and isotropic when averaged over scales $\sim 3000\, h^{-1}$ Mpc, to account for the large-scale isotropy of the microwave background). A serious worry here, as in the milder versions of primeval turbulence, is that dense bound systems (including black holes) may form too soon in too great numbers.

One last aspect of primeval turbulence might be mentioned. It is often remarked that the rotation of the Galaxy might find a natural explanation as a residuum of primeval turbulence currents.** There is a simple problem with this idea if one assumes

---

* Acoustic noise generated by turbulence is discussed by Silk (1973), Jones (1973), and Stein (1974).
** cf. e.g. Silk (1973), Tomita (1972), Oort (1970).

that the turbulence is left over from the radiation-dominated epoch. Let us concentrate attention on the 'main bulk' of our own Galaxy, the part within 10 kpc radius of the centre. The mass contained within this radius is $\sim 10^{11} M_\odot$. If this mass were uniformly distributed within the sphere of 10 kpc radius, the density would be ~ 1 proton cm^{-3}. Now let us trace the expansion of the Universe back to the time when the mean cosmological density was equal to this mean value within the Galaxy, and let us suppose that, at this epoch, the material destined to end up in the Galaxy occupied a more or less spherical volume, radius ~ 10 kpc. This happens at redshift $z \sim 50$ to 300, depending on Ω and h. Finally, let us suppose that this roughly spherical proto-galaxy has angular momentum about its centre of mass equal to the angular momentum of the Galaxy. Since the Galaxy derives at least a significant fraction of its support from rotation, centrifugal force and self-gravitation are in a rough balance in the proto-galaxy at the epoch we are considering. Now what happens if we try to trace the evolution back still further in time? Can we assume that the proto-galaxy remains as a coherent lump with its final angular momentum? It is hard to see how this could work, for compression ought to increase the ratio of centrifugal to self-gravitation forces, forcing apart the material. This would be a reasonable situation in fully developed turbulence, where there is a steady transfer of angular momentum among eddies. However, when our proto-galaxy reaches 10 kpc radius *we cannot invoke turbulence*. The maximum optical depth of the proto-galaxy, if fully ionized, is 0.03. *Thus radiation can act only as a drag, not as a means of transferring angular momentum.* Since the velocity of sound in the matter is much smaller than the assumed matter currents, 100 km s^{-1}, collisions among eddies would be expected to produce shocks, not angular momentum transfer.

As is so often the case, this argument is hardly conclusive. It could be, for example, that primeval turbulence eddies are stabilized by some undiscovered process. The argument does show that there is considerable room for future work.

3.5. GRAVITATIONAL INSTABILITY

The gravitational instability picture also has been subject to doubt and uncertainty.* This effect figured in early discussions of structure in the big–bang cosmology, but then was criticized by later authors who felt that irregularities would grow too slowly to account for the origin of galaxies from 'reasonable' initial perturbations. The trend of some recent discussion has been in two directions – on the one hand, to reconsider what might be called a 'reasonable' initial density irregularity, and on the other to ignore the question of initial conditions as beyond our grasp for now, and to

---

* The role of gravitational instability in the evolution of structure in the expanding Universe was first considered by Lemaître (1933) and by Gamow and Teller (1939). For a review written at a time when the instability picture was less popular than it is now, see Layzer (1964). For reviews written after the renaissance, see the references in the first footnote of this report.

concentrate instead on the 'uniformitarian' method.* The physics has not changed since the early criticisms, and the instability picture still is quite incapable of accounting for galaxies *ab initio*. On the other hand, the uniformitarian method does lead to some limited but possibly significant results.

We have the picture that the matter within some $300 \, h^{-1}$ Mpc (the depth of the Shane-Wirtanen survey) is distributed fairly smoothly overall and expanding in a roughly homogeneous and isotropic way according to Hubble's law, $v = Hr$. Within this distance, matter is noticeably clumpy on a scale of $30 \, h^{-1}$ Mpc, but still expanding roughly in accordance with Hubble's law (for otherwise, peculiar velocities would be too high). The expected course of evolution of these $30 \, h^{-1}$ Mpc irregularities is very simply described. Let $\varrho(\mathbf{r})$ be the mass density smoothed over a running average of scale $30 \, h^{-1}$ Mpc, and write

$$\varrho(\mathbf{r}) = \varrho_0(1 + \delta(\mathbf{r})),$$

where ϱ_0 is the mean and δ the fractional departure from the mean. Apparently $\delta \lesssim 1$. Then the method of linear perturbation theory says that if non-gravitational forces may be neglected the time variation of δ is a linear combination of growing and decaying modes. The decaying mode generally varies as t^{-1}. If the Universe is close to the Einstein-de Sitter model, where

$$\varrho_0 \cong \varrho_c = 3H^2/8\pi \, G,$$

so that $\Omega \cong 1$, and pressure may be neglected, then the growing mode varies as $t^{2/3}$. If $\Omega \ll 1$, the 'growing' mode is very nearly constant. But in this case, as we trace the expansion of the Universe back in time, we find that the value of Ω approaches unity (because ϱ varies as $(1+z)^3$, H^2 as $(1+z)^2$). Since the local value of Ω is not less than ~ 0.02, we conclude that, unless some other effect intervenes, the growing mode would have been growing like $t^{2/3}$ at $z \gtrsim 50$.

These results do not depend very much on what the Universe is like at distances much greater than $300 \, h^{-1}$ Mpc. These distant parts can only affect the local situation through the intrusion of matter, which seems unlikely in view of the large peculiar velocities that would be required, or through the generation of tidal gravitational fields, which do not affect the time rate of change of δ in the linear approximation. Also, the very complicated and non-linear processes that obtain on scales much smaller than $30 \, h^{-1}$ Mpc (rich clusters, galaxies) appear to be unimportant. The bulk motion of matter as measured by the coarse-grained average $\varrho(\mathbf{r})$ is not influenced by the complex details of the interaction of neighbouring galaxies, for these stresses cancel in pairs.

---

* The lack of enthusiasm for the instability picture mainly stems from the fact that the growing mode (or modes) grows as a power of time, not as an exponential. However, the argument that is occasionally advanced (cf. Layzer, 1964), that galaxies *cannot* grow out of 'reasonable' irregularities like thermal fluctuations, seems insecure because we have no theory to fix the time at which the initial fluctuations are set, and clearly the smaller this time the greater the growth factor. For discussion of irregularities in the early Universe, see Peebles (1968), Kundt (1971), Zel'dovich (1972), Harrison (1973). For discussion of 'uniformitarianism' in this context see Peebles (1967, 1972).

Now let us consider the relative sizes of growing and decaying modes in δ. We would be surprised to find that the decaying mode is very much larger than the growing mode, for that would require a cunning balance of irregularities in the matter and velocity fields. But we can imagine applying the same argument in the Universe as it was at a redshift $z \sim 10$, or 100, or however large it might be without violating the basic picture and assumptions that have been evoked. Assuming this maximum redshift is large, we conclude that *the decaying mode now must be much smaller than the growing mode*.

How can one avoid this conclusion? The most direct way is to deny that the observed system of galaxies really is expanding, or, if expanding, has expanded from a dense initial state. In the model of Klein and Alfvén (cf. Alfvén, 1971), the observed system of galaxies is finite and has expanded from a point of maximum compression that perhaps is not very much denser than the present state. If during the course of expansion the effective value of Ω stays much less than unity, the conclusion does not apply. Perhaps the most difficult point for this picture is the isotropy of the microwave background. Since the system of galaxies would have an observable boundary, the microwave background would have to be an external sea of radiation through which most galaxies are moving at high speed, and it would be surprising that we do not observe this as a 24-h anistropy in the radiation intensity. Another possibility is to go to the Steady-State picture. Apparently the major problem here is the spectrum of the microwave background, which agrees with black-body in some detail (PC, pp. 129–142). In the framework of the big-bang cosmology, the argument is vitiated if non-gravitational phenomena are important. The difficulty of finding anything other than gravity that would materially affect the course of evolution of irregularities on a scale of $30 \, h^{-1}$ Mpc, at least back to $z \sim 1000$, helps explain why there has been so little controversy over the conclusion, but of course that is hardly a proof. The question is, to what extent can we account for the observed pattern of distribution and motion of matter as originating from small, growing irregularities in an expanding Universe? The rest of this section is devoted to a list of possible tests.

The available information on irregularities on a scale of $30 \, h^{-1}$ Mpc is so scanty that it is hard to see how one could directly estimate the relative amplitude of growing and decaying modes. An interesting (and difficult) task will be to learn whether the details of the character of the large-scale distribution of galaxies, like the distribution function of the masses of systems of galaxies, could have evolved without contrivance from growing density irregularities. The work of de Vaucouleurs and Sandage and others to map out the local velocity field, within $\sim 30 \, h^{-1}$ Mpc, may be of decisive importance. For example, it might be hard to account for the rotation of the Local Supercluster, as proposed by de Vaucouleurs (cf. de Vaucouleurs and Peters, 1968) within the framework of the gravitational instability picture.

Under the instability picture, groups and clusters of galaxies fragment out as bound systems because the potential energy of interaction among separating mass elements exceeds the kinetic energy of expansion. One does not expect to find that associations of galaxies are freely expanding, with time-scales $\ll 10^{10} \, h^{-1}$ yr, unless there has been some violent mass loss. If the free expansion of groups and clusters can be truly established, it will have been demonstrated that the instability picture lacks some very essential element.

The rich compact clusters of galaxies appear to be simple systems, the major part of the mass being contained in a few hundred objects, the brightest galaxies. If these galaxies separated out as bound systems before the cluster formed, then the cluster formation process ought to be just as simple, and reasonably well approximated by a numerical N-body model.* It has been found that such models are quite successful in reproducing observed features of the Coma Cluster. Typically, the numerical computation commences when the proto-cluster is at the presumed point of maximum expansion, having just separated out from the general expansion as a distinct system. The N-body computation then follows the collapse and subsequent evolution. One finds that 10^{10} yr is time enough for a proto-cluster to separate out from the general expansion, to collapse, and then relax to a state that matches the Coma Cluster fairly well – including the spatial distribution of galaxies and the variation of line-of-sight velocity dispersion with projected distance from the cluster centre. This is an important result because it shows that the instability picture can give a reasonable phenomenological account of at least some aspects of the large-scale structure. Of course one such example is hardly conclusive, because one certainly can think of other models for evolution of the clusters that could also fit the data.

In the gravitational instability picture in its most simple-minded form, a galaxy would originate as a collapsing cloud of gas (or, perhaps, of smaller clouds). This agrees with the picture Eggen *et al.* (1962) arrived at from a study of galactic structure, which may be significant, although we recognize that the connection is a tenuous one.

A simple property of galaxies that has been the subject of some controversy is their rotation. In the instability picture as outlined here, the initial growing density irregularities have negligible angular momentum and negligible circulation ($\nabla \times \mathbf{v} \cong 0$). A proto-galaxy can pick up angular momentum, for it is an irregular lump in the tidal field of neighbouring developing proto-galaxies.** Two questions have been discussed:

(1) Does the proto-galaxy pick up enough angular momentum? Although there has been some discussion of this point, my impression is that the order of magnitude is about in the right range, considering the substantial uncertainties.

(2) What is the origin of the circulation of matter in galaxies? The matter is supposed to commence with negligible circulation, $\oint \mathbf{v} \cdot d\mathbf{r} \cong 0$, for $\nabla \times \mathbf{v} \cong 0$. Under

---

* A first attempt at this was described by van Albada (1961). By using fast numerical computers I was able to make a more direct model computation, but still under the simplifying assumption that all galaxies have the same mass (Peebles 1970). The effect of a more realistic mass function has been discussed by Gunn and Aarseth (1972).

** This process was proposed by Hoyle (1949) and used by Sciama (1955) in his theory of the origin of galaxies in the Steady-State model. For a computation of the effect in the big-bang model, cf. Peebles (1969b) Oort (1970), Harrison (1971), Peebles (1971c). For the uncertainty in the angular momentum of the Galaxy, see Ostriker and Peebles (1973). For discussions of the circulation theorem, see Hunter (1970), Sunyaev and Zel'dovich (1972), Tomita (1973), Peebles (1973).

the assumption that non-gravitational forces may be ignored (or more generally that the force per unit mass can be derived from a potential), Kelvin's circulation theorem says that $\nabla \times v$ remains negligibly small. Does this contradict the observed circulation of matter in galaxies? It is my impression that here is a situation where the complexities of non-gravitational processes may quite becloud the theoretical position. Processes like turbulence, shocks, and fine-scale mixing all can violate the assumptions of the circulation theorem (cf. Peebles, 1973). As I understand the situation, therefore, the rotation of galaxies is not yet an embarrassment for the instability picture.

The available calculations of angular momentum transfer are at best estimates of 'typical' values. Before the angular momentum test can be tightened we will have to have not only more accurate calculations of the mean but also calculations of the expected dispersion about the mean. We might then hope to compare this dispersion with the frequency distribution of spirals and S0 galaxies relative to ellipticals of diverse eccentricities. Another possible test is based on an expected scaling law: as long as the expansion of the Universe approximates to the Einstein-de Sitter model ($\Omega \sim 1$), the mean square angular momentum transfer satisfies

$$L^2 E/G^2 M^5 = \text{constant},$$

where M is the mass and $-E$ is the energy of the proto-system (kinetic plus potential, and neglecting dissipation). This says that, in the absence of dissipation, the eccentricity due to centrifugal flattening is independent of mass. The situation is complicated for proto-galaxies, because the collapse may be strongly dissipative, substantially increasing E. Proto-clusters of galaxies may provide a useful test, as these systems might be expected to conserve E. The Coma Cluster of galaxies shows no significant rotation, the limit being below the value expected from the scaling law. This is not yet a significant test, for we do not know the expected dispersion in the angular momentum, or the projection effects in the Coma Cluster. It will be interesting to have comparable redshift data on several more of the rich clusters. On the theoretical side, it will be interesting to study how the addition of angular momentum to the proto-cluster affects the evolution in the N-body models. Does a compact core still form? What is the rotation curve?

If galaxies formed before clusters, as seems reasonable (although certainly not demanded) in the instability picture, then how can one account for the special nature of the population of galaxies in rich clusters (more S0 galaxies, few spirals)? It still seems quite possible that some variant of the Baade-Spitzer collision process might account for superficial differences like gas content. It will be of considerable interest to learn whether the differences go deeper than that, whether the mass function of galaxies in clusters differs from the mass function for galaxies elsewhere, for example, or whether the angular momentum functions differ.

My final example is more indirect, involving even more assumptions, than the preceding ones. This is the proposal that globular star clusters might be the direct descendents of processes operating in the early Universe. One invokes here the big-bang cosmology and the Primeval Fireball. As has been remarked, in this cosmology at redshift $z \gtrsim 1000$ the Fireball radiation is hot enough to ionize the plasma, and Thomson scattering by the free electrons causes matter and radiation to act more or less like a single fluid. The pressure of the radiation is large, making the critical Jeans length for gravitational instability large, encompassing a mass comparable to that of clouds of galaxies. At $z \cong 1000$ the plasma is expected to recombine and decouple from the radiation. Now the critical Jeans length drops to a much smaller value. Under a fairly broad range of possible assumptions about the spectrum of density irregularities, the first systems to separate out from the general expansion would have mass fixed by this critical Jeans length.* The interesting coincidence is that this critical mass agrees with the typical masses of globular star clusters, systems which, by and large, are thought to be old, and which appear in what otherwise might be thought to be strange circumstances – around galaxies large and small, and even in intergalactic space. All this is highly suggestive, and I think therefore merits further careful attention. The picture must explain the differences among globular star clusters – the systematic variation of heavy element abundance with position in the Galaxy and the variations of globular star cluster abundances around otherwise similar-looking galaxies. At this point I have no idea whether these phenomena will prove embarrassing to the theory.

Acknowledgement

Research supported in part at Princeton by the National Science Foundation.

References

Abell, G. O.: 1958, *Astrophys. J. Suppl.* **3**, 211.
Alfvén, H.: 1971, *Phys. Today* **24**, 28.
Ambartsumian, V. A.: 1958, in R. Stoops (ed.), 11th Solvay Conference on *The Structure and Evolution of the Universe*, Brussels, p. 241.
Ambartsumian, V. A.: 1965, in 13th Solvay Conference on *The Structure and Evolution of Galaxies*, Interscience Publishers, London, New York and Sydney, p. 1.
Arons, J.: 1972, *Astrophys. J.* **172**, 553.
Arp, H.: 1971, *Science* **174**, 1189.
Bahcall, J. N. and Joss, P. C.: 1972, *Comments Astrophys. Space Phys.* **4**, 95.
Bogart, R. S. and Wagoner, R. V.: 1973, *Astrophys. J.* **181**, 609.
Burbidge, G. R.: 1971, *Nature* **233**, 36.
Burbidge, G. R., Burbidge, E. M., and Sandage, A.: 1963, *Rev. Mod. Phys.* **35**, 947.
Dallaporta, N. and Lucchin, F.: 1972, *Astron. Astrophys.* **19**, 123.
de Vaucouleurs, G.: 1971, *Publ. Astron. Soc. Pacific* **83**, 113.
de Vaucouleurs, G. and Peters, W. L.: 1968, *Nature* **220**, 868.
Doroshkevich, A. G., Zel'dovich, Ya. B., and Novikov, I. D.: 1967, *Astron. Zh.* **44**, 295; *Soviet Astron.* **11**, 233, 1967.

---

* For a discussion of the Jeans length in an expanding cosmological model see PC, pp. 217–220. Gamow first considered in detail the role of the Jeans length in cosmology (cf. Gamow and Teller, 1939). Recognizing that the Jeans length would be smaller than a typical large galaxy for the entropy per baryon that he and Alpher were contemplating (and that agrees well with modern observations), Gamow (1952) proposed that primeval turbulence might have increased the effective matter pressure, increasing the effective Jeans limit. Peebles and Dicke (1968) proposed that the naïve estimate of the Jeans mass be accepted, but interpreted as fixing globular cluster masses rather than galaxy masses.

Eggen, O. J., Lynden-Bell, D., and Sandage, A.: 1962, *Astrophys. J.* **136**, 748.
Field, G. B.: 1965, *Astrophys. J.* **142**, 531.
Gamow, G.: 1952, *Phys. Rev.* **86**, 251.
Gamow, G. and Teller, E.: 1939, *Phys. Rev.* **55**, 654.
Gold, T. and Hoyle, F.: 1959, in R. N. Bracewell (ed.), 'Paris Symposium on Radio Astronomy', *IAU Symp.* **9**, 583.
Gunn, J. E. and Aarseth, S.: 1972, preprint.
Harrison, E. R.: 1970a, *Monthly Notices Roy. Astron. Soc.* **147**, 279.
Harrison, E. R.: 1970b, *Monthly Notices Roy. Astron. Soc.* **148**, 119.
Harrison, E. R.: 1971, *Monthly Notices Roy. Astron. Soc.* **154**, 167.
Harrison, E. R.: 1973, *Ann. Rev. Astron. Astrophys.* **11**, 155.
Hauser, M. G. and Peebles, P. J. E.: 1973, *Astrophys. J.* **185**, 757.
Holmberg, E.: 1969, *Arkiv Astron.* **5**, 305.
Houck, J. R., Soifer, B. T., Harwit, M., and Pipher, J. L.: 1972, *Astrophys. J.* **178**, L29.
Hoyle, F.: 1949, *Problems of Cosmical Aerodynamics*, Symp. of Int. Un. Theor. Appl. Math. and IAU, p. 195.
Hoyle, F.: 1958, in Proc. 11th Solvay Conf. on *The Structure and Evolution of the Universe*, p. 53.
Hoyle, F.: 1965, *Galaxies, Nuclei and Quasars*, Harper and Row, New York, pp. 18–20.
Hoyle, F. and Narlikar, J. V.: 1966, *Proc. Roy. Soc. London* **A290**, 143 and 185.
Hunter, C.: 1970, *Astrophys. J.* **162**, 445.
Jeans, J.: 1928, *Astronomy and Cosmogony*, Cambridge University Press, p. 352.
Jones, B. J. T.: 1973, *Astrophys. J.* **181**, 269.
Jones, B. J. T. and Peebles, P. J. E.: 1972, *Comments Astrophys. Space Phys.* **4**, 121.
Kondo, M., Sofue, Y., and Unno, W.: 1971, *Prog. Theor. Phys. Kyoto, Suppl.* **49**, 120.
Kundt, W.: 1971, *Springer Tracts Mod. Phys.* **58**, 1.
Layzer, D.: 1964, *Ann. Rev. Astron. Astrophys.* **2**, 341.
Lemaître, G.: 1933, *Compt. Rend. Acad. Sci. Paris* **196**, 1085; *Ann. Soc. Sci. Brussels* **A53**, 51.
Limber, D. N.: 1954, *Astrophys. J.* **119**, 655.
McCrea, W. H.: 1964, *Monthly Notices Roy. Astron. Soc.* **128**, 335.
Milne, E. A.: 1948, *Kinematic Relativity*, Oxford, pp. 167–169.
Muehlner, D. J. and Weiss, R.: 1973, *Phys. Rev. Letters* **30**, 757.
Ne'eman, Y.: 1965, *Astrophys. J.* **141**, 1303.
Neyman, J., Scott, E. L., and Shane, C. D.: 1953, *Astrophys. J.* **117**, 92.
Novikov, I. D.: 1964, *Astron. Zh.* **41**, 1075; *Soviet Astron.* **8**, 857, 1965.
Olsen, D. W. and Sachs, R. K.: 1973, *Astrophys. J.* **185**, 91.
Oort, J. H.: 1958, in 11th Solvay Conference on *The Structure and Evolution of the Universe*, p. 163.
Oort, J. H.: 1970, *Astron. Astrophys.* **7**, 381.
Ostriker, J. P. and Peebles, P. J. E.: 1973, *Astrophys. J.* **186**, 467.
Ozernoy, L. M.: 1971, *Astron. Zh.* **48**, 1160; *Soviet Astron.* **15**, 923, 1972.
Ozernoy, L. M. and Chernin, A. D.: 1967, *Astron. Zh.* **44**, 1131; *Soviet Astron.* **11**, 907, 1968.
Peebles, P. J. E.: 1967, *Astrophys. J.* **147**, 859.
Peebles, P. J. E.: 1968, *Nature* **220**, 237.
Peebles, P. J. E.: 1969a, *J. Roy. Astron. Soc. Can.* **63**, 27–28.
Peebles, P. J. E.: 1969b, *Astrophys. J.* **155**, 393.
Peebles, P. J. E.: 1970, *Astron. J.* **75**, 13.
Peebles, P. J. E.: 1971a, *Physical Cosmology*, Princeton University Press.
Peebles, P. J. E.: 1971b, *Astrophys. Space Sci.* **11**, 443.
Peebles, P. J. E.: 1971c, *Astron. Astrophys.* **11**, 377.
Peebles, P. J. E.: 1972, *Comments Astrophys. Space Phys.* **4**, 53.
Peebles, P. J. E.: 1973, *Publ. Astron. Soc. Japan* **25**, 291.
Peebles, P. J. E.: 1974, *Astrophys. J. Suppl.* No. 252.
Peebles, P. J. E. and Dicke, R. H.: 1968, *Astrophys. J.* **154**, 891.
Peebles, P. J. E. and Hauser, M. G.: 1974, *Astrophys. J. Suppl.* No. 252.
Perko, T. E., Matzner, R. A., and Shepley, L. C.: 1972, *Phys. Rev. D.* **6**, 969.
Rees, M. J.: 1971, in R. Sachs (ed.), *Proc. Int. School of Phys. Enrico Fermi; Course XLVII*, Academic Press, New York and London, p. 315.
Rees, M. J.: 1972, *Phys. Rev. Letters* **28**, 1669.

Rees, M. J. and Reinhardt, M.: 1972, *Astron. Astrophys.* **19**, 189.

Sciama, D. W.: 1955, *Monthly Notices Roy. Astron. Soc.* **115**, 3.

Shane, C. D. and Wirtanen, C. A.: 1967, *Publ. Lick Obs.* **22**, part 1.

Silk, J.: 1973, *Comments Astrophys. Space Phys.* **5**, 9.

Silk, J. and Ames, S.: 1972, *Astrophys. J.* **178**, 77.

Silk, J. and Lea, S.: 1973, *Astrophys. J.* **180**, 669.

Silk, J. and Solinger, A. B.: 1973, *Nature Phys. Sci.* **244**, 101.

Stecker, F. W. and Puget, J. L.: 1972, *Astrophys. J.* **178**, 57.

Stein, R.: 1974, to be published.

Sunyaev, R. A. and Zel'dovich, Ya. B.: 1972, *Astron. Astrophys.* **20**, 189.

Thorne, K. S.: 1967, *Astrophys. J.* **148**, 51.

Tomita, K.: 1972, *Prog. Theor. Phys. Kyoto* **48**, 1503.

Tomita, K.: 1973, *Publ. Astron. Soc. Japan* **25**, 287.

Tomita, K., Nariai, H., Sato, H., Matsuda, T., and Takeda, H.: 1970, *Prog. Theor. Phys. Kyoto* **43**, 1511.

van Albada, G. B.: 1961, *Astron. J.* **66**, 590.

van der Kruit, P. C., Oort, J. H., and Mathewson, D. S.: 1972, *Astron. Astrophys.* **21**, 169.

Weinberg, S.: 1972, *Gravitation and Cosmology*, Wiley, New York.

Williamson, K. D., Blair, A. G., Catlin, L. L., Hiebut, R. D., Loyd, E. G., and Romero, H. V.: 1973, *Nature Phys. Sci.* **241**, 79.

Yu, J. T. and Peebles, P. J. E.: 1969, *Astrophys. J.* **158**, 103.

Zel'dovich, Ya. B.: 1969, *Astron. Zh.* **46**, 775; *Soviet Astron.* **13**, 608, 1970.

Zel'dovich, Ya. B.: 1972, *Monthly Notices Roy. Astron. Soc.* **160**, 1P.

Zel'dovich, Ya. B. and Novikov, I. D.: 1970, *Astrofizica* **6**, 379.

DISCUSSION

Tifft: Why has rotation not been seen in clusters? Coma is well out of 'round' but shows no velocity gradient.

Peebles: In the gravitational instability picture, one could expect that clusters of galaxies may be out of round and have non-zero angular momentum. Because the two-body relaxation rate is slow, the oblateness need not be aligned along the angular momentum. It will be of considerable interest to learn whether the lack of a velocity gradient in the Coma cluster is found in other rich clusters.

G. de Vaucouleurs: Yes, it is true and surprising that no clear indication of general rotation has been detected in clusters or clouds of galaxies, even in ellipsoidal clouds of spirals, such as the S-cloud of the Virgo cluster (cf. *Astrophys. J. Suppl.* **5**, 233, 1961).

Mestel: I am not clear as to where you agree and where you disagree with Lifshitz – is it that you assume *large* initial perturbations?

Peebles: The mathematics has not changed, only the interpretation. We *observe* that the Universe is now irregular. The perturbation analysis leads one to believe that the irregularities will increase with time and would have been less prominent in the past.

Mestel: Then perhaps you will be able to work backwards and learn something about the primeval turbulence?

Peebles: The great goal would be to determine whether the present nature of the matter distribution can be fitted to some combination of growing and decaying modes of irregularities, and, if so, to interpret the results in terms of the possible situation in the earlier Universe. Of course we are a long way from being able to do this.

Wright: Would you tell us how circulation might grow during the collapse of an originally non-circulating cloud of gas?

Peebles: Kelvin's law of conservation of circulation assumes ideal fluid flow with a single-valued relation between pressure and density. The collapse of a proto-galaxy seems likely to involve violent collisions among different parts, giving rise to turbulent dissipation and shock waves, both of which violate the assumptions of the circulation theorem. For details see the article in *Publ. Astron. Soc. Japan* **25**, 291, 1973.

E. M. Burbidge: You said that the virial theorem takes hold in the condensations, and the film projected during your talk showed the persistence of these condensations. Does this mean we must find the 'missing mass' sufficient to balance the kinetic energy and stabilize the clusters?

Peebles: Yes. If there is a large difference between the kinetic energy and the magnitude of the potential energy for typical groups or clusters, it will show that the gravitational instability picture is in serious trouble.

Ekers: In your statistical analysis showing the 30 Mpc scale length, can you distinguish between superclusters of clusters and single clusters with a large halo of galaxies?

Peebles: I suspect both effects are involved. One observes that positions of cluster centres are correlated, and that there is a general halo of galaxies (or groups or small clusters of galaxies) around the average rich cluster.

THE DISTRIBUTION OF GALAXIES IN RELATION TO THEIR FORMATION AND EVOLUTION

BRUCE A. PETERSON

Mount Stromlo and Siding Spring Observatory, The Australian National University, Canberra, Australia

Abstract. The distribution of galaxies on the sky is not random, but has structure on a scale of ~ 30 Mpc. Similar structure in the cosmic background radiation is not present. Another expected source of structure in the background radiation is the finite size of the light horizon in the early Universe. The lack of observable structure in the background radiation implies extreme initial isotropy. Quantum effects may be responsible for this isotropy, and for the initial perturbations from which galaxies and clusters of galaxies formed.

The isotropic 2.7 K cosmic background radiation (Penzias and Wilson, 1965) and the recession of the extragalactic nebulae (Hubble, 1929) are strong evidence that the Universe expanded from a hot, dense fireball. The expansion rate $H_0 = 50$ km s^{-1} Mpc^{-1} (Abell, 1972) implies, in the context of the isotropic solutions of the Einstein equations (with no cosmological constant), that the expansion began $t_0 \lesssim 20 \times 10^9$ yr ago.

The early Universe was hot, and the matter and radiation were in equilibrium. Fluctuations in density involving less than the critical Jeans' mass could not grow, and dissipated. As the Universe expanded and cooled, this critical mass increased as $t^{3/2}$ to reach a limiting value of $M_J \sim 10^{18} \mathfrak{M}_\odot$ shortly before the Universe had cooled enough for the matter to recombine. As the matter recombined, the critical mass dropped to $M_J \sim 10^6 \mathfrak{M}_\odot$. Fluctuations involving more than $\sim 10^6 \mathfrak{M}_\odot$, that survived the period of dissipation before recombination, were then gravitationally unstable and grew as $\delta\varrho/\varrho \sim t^{2/3}$ (for $\delta\varrho/\varrho < 1$). Calculations of the transfer function for fluctuations through the period of dissipation (Peebles and Yu, 1970; Weinberg, 1971) show that fluctuations involving more than $\sim 10^{11} \mathfrak{M}_\odot$ could survive. These fluctuations in the density of the early Universe are seen today in the irregularities of the matter distribution.

The nearby galaxies are shown on an equal-area projection in Figure 1. They are plotted from the catalogue of Shapley and Ames (1932) which lists the galaxies with $m_{pg} \leqslant 13$. For comparison, a random space distribution viewed through a plane-parallel layer of absorbing material (with half-thickness $a = 0.5$ mag) aligned along the galactic plane has been simulated in Figure 2 using the same number of points as in Figure 1. The great circle of the galactic plane is traced out by the zone containing no galaxies in both Figures 1 and 2. The distribution of the random galaxies is more homogeneous than that of the real galaxies. The real galaxies are concentrated along the supergalactic equator, a band, nearly perpendicular to the galactic plane, running from the left of the top to the lower right of the centre in Figure 1. The prominent concentration to the upper right of centre in Figure 1 contains the Virgo cluster. The local supercluster (de Vaucouleurs, 1955) consists of the galaxies concentrated along

John R. Shakeshaft (ed.), The Formation and Dynamics of Galaxies · 75–84. All Rights Reserved.

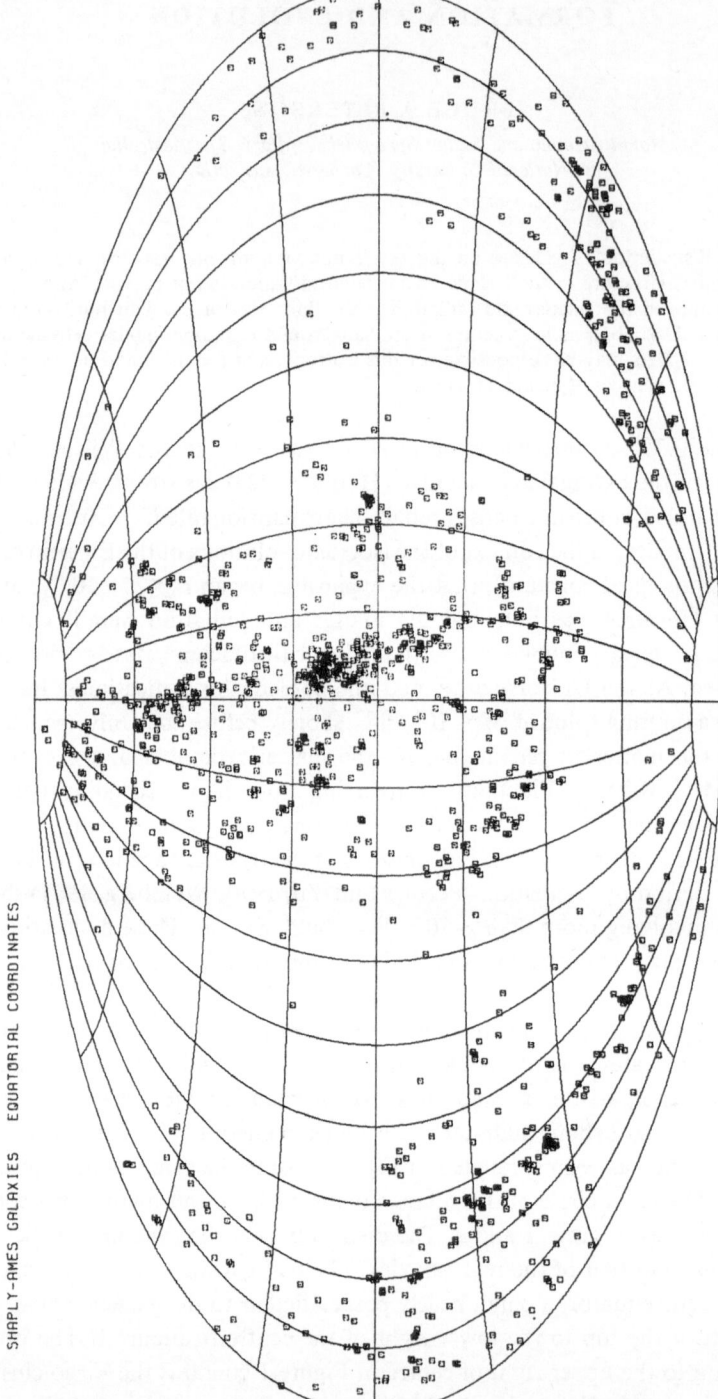

SHAPLY-AMES GALAXIES   EQUATORIAL COORDINATES

Fig. 1.   Galaxies from the catalogue of Shapley and Ames plotted on an equal-area projection.

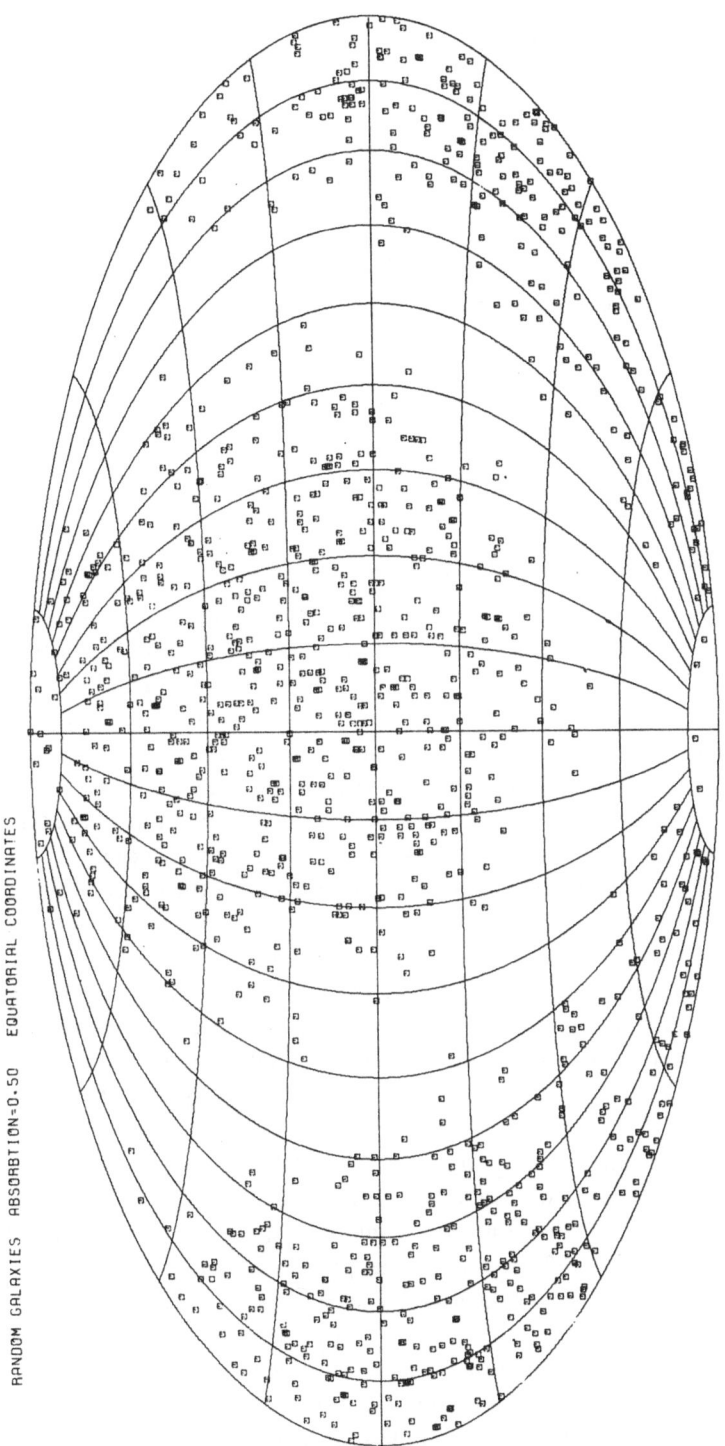

Fig. 2. Simulated, random, space distribution of galaxies seen through a plane parallel layer of absorbing material of half-thickness $a = 0.5$ mag. This figure contains the same number of points as Figure 1.

the supergalactic equator, including the Virgo cluster. It has a scale of ~ 30 Mpc and contains $\sim 10^{14}\,\mathfrak{M}_\odot$.

In Figures 3–6 are the galaxies in the catalogues of Zwicky and his co-workers (Zwicky *et al.*, 1961–68). Figure 3 reaches to $m_{pg} \leqslant 13$, the same as Figure 1. More distant galaxies dominate the successively fainter magnitude intervals shown in Figures 4–6. The galaxy distribution in Figure 6 has a mottled appearance on a scale of $\sim 20°$ on the sky. This corresponds to ~ 30 Mpc at the distance of the galaxies in Figure 6. Structure in the matter distribution on a similar scale has been found out to greater distances in the Abell clusters (Hauser and Peebles, 1973) and in the galaxy counts of Shane and Wirtanen (Peebles and Hauser, 1974).

The fluctuations in the early Universe which produce structure in the matter distribution at the present day would also produce anisotropy in the temperature of the cosmic background radiation. If we assume that the density of the intergalactic medium is low, then the background radiation comes to us from $z \sim 1500$ when it was last scattered during the recombination era. The angular scale of the temperature fluctuations expected from galaxies and superclusters is

$$\theta \sim 2q_0^{2/3}\,(MGH_0/c^3)^{1/3} \sim 1'.$$

The amplitude of the expected fluctuations is $\Delta T/T \sim \tfrac{1}{3}(\delta\varrho_0/\varrho_0)/(1+z) \sim 10^{-3}$. In a comparison of their recent observations with the detailed calculations of the expected fluctuations from forming clusters of galaxies (Peebles and Yu, 1970), Carpenter *et al.* (1973) found that the upper limit to the observed fluctuations is less than the fluctuations that are expected by about a factor of two. However, if the intergalactic medium is hot and dense, then the optical depth produced by Thomson scattering may be large by $z \sim 10$ (Gunn and Peterson, 1965) and the fluctuations in the temperature of the cosmic background radiation would be considerably reduced.

Another effect which should contribute to structure in the distribution of matter, and to fluctuations in the temperature of the cosmic background radiation, is the finite size of the light horizon when matter and radiation decoupled. Only fluctuations on a scale smaller than that of the light horizon can be smoothed out, since communication is only possible with the portion of the Universe contained within the light horizon. Also, the smoothing process can only be effective while the radiation can be scattered by the matter. The proper distance to the light horizon is $r_H \sim c/H_0 \times (2/q_0)^{1/2}(1+z)^{-3/2}$, corresponding to ~ 150 kpc ($q_0 = 1$, $H_0 = 50$ km s^{-1} Mpc^{-1}) at $z \sim 1500$ when recombination took place. Material then on our light horizon is now at $r_H(1+z) \sim 200$ Mpc. If the smoothing process was effective over only $\leqslant 15\%$ of the distance to the light horizon at recombination, then we would expect to find structure on a scale of $\geqslant 30$ Mpc in the matter distribution. Structure appears to be present in the distribution of galaxies on a scale of ~ 30 Mpc.

However, the corresponding structure is not present at a similar amplitude in the 2.7 K cosmic background radiation. If the optical depth of the intergalactic medium is small, then the cosmic background radiation has not been scattered since the matter recombined, and background radiation from areas of the sky separated by

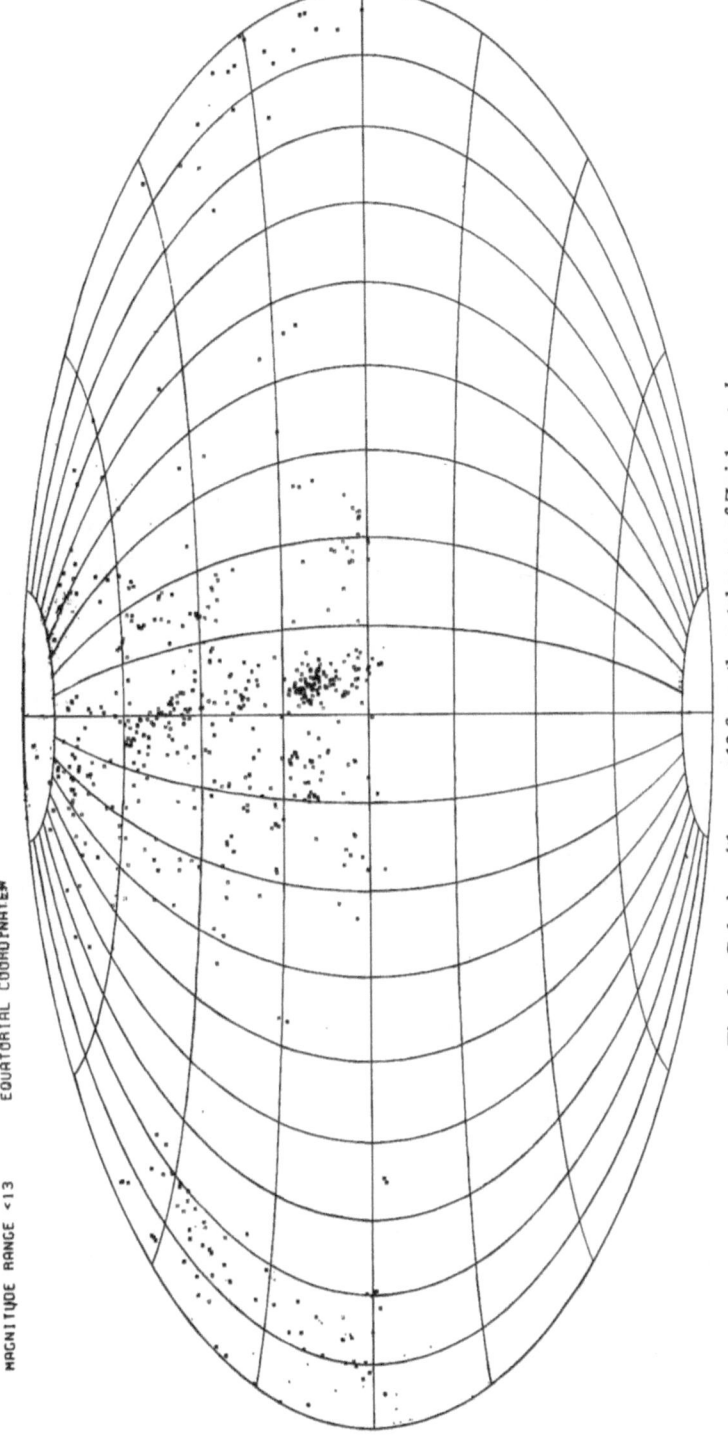

MAGNITUDE RANGE <13   EQUATORIAL COORDINATES

Fig. 3.   Galaxies with $m_{pg} < 13$ from the catalogues of Zwicky *et al.*

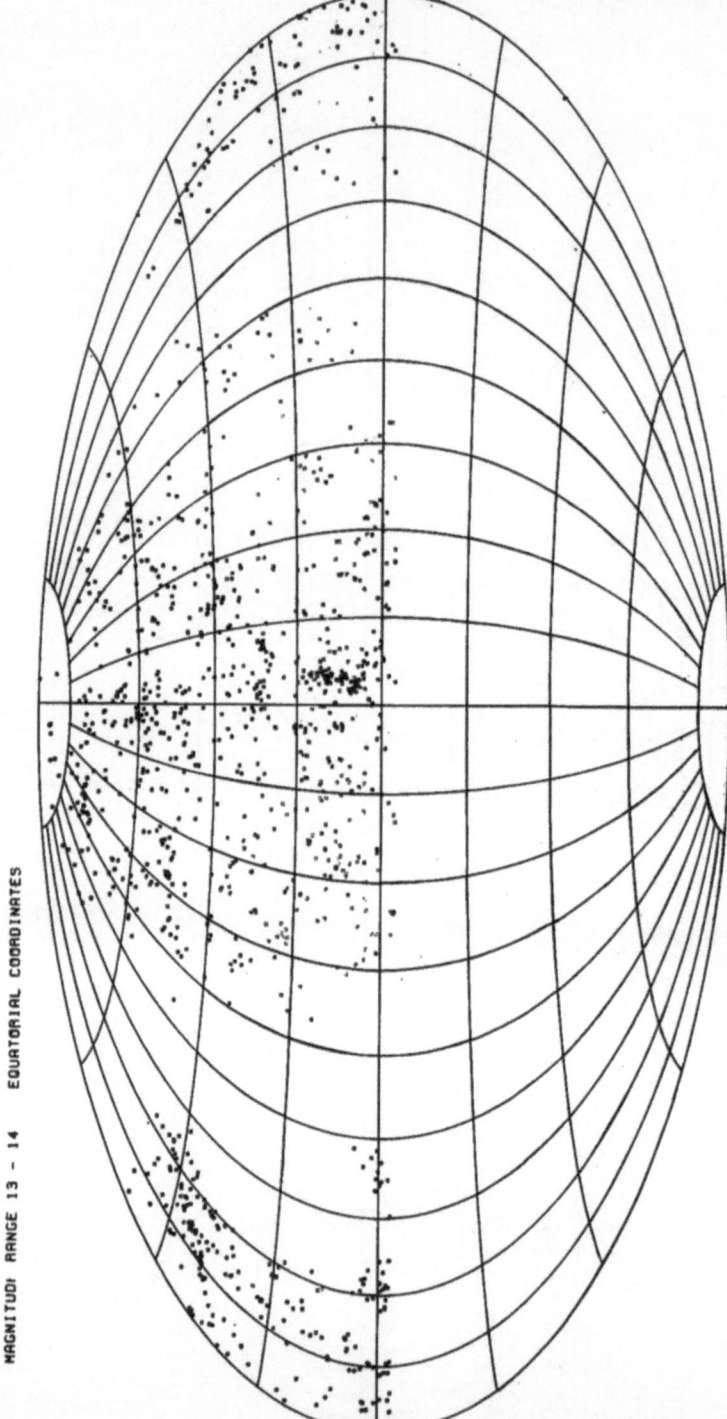

MAGNITUDE RANGE 13 - 14   EQUATORIAL COORDINATES

Fig. 4.   Galaxies with $13 \leqslant m_{pg} < 14$ from the catalogues of Zwicky *et al.*

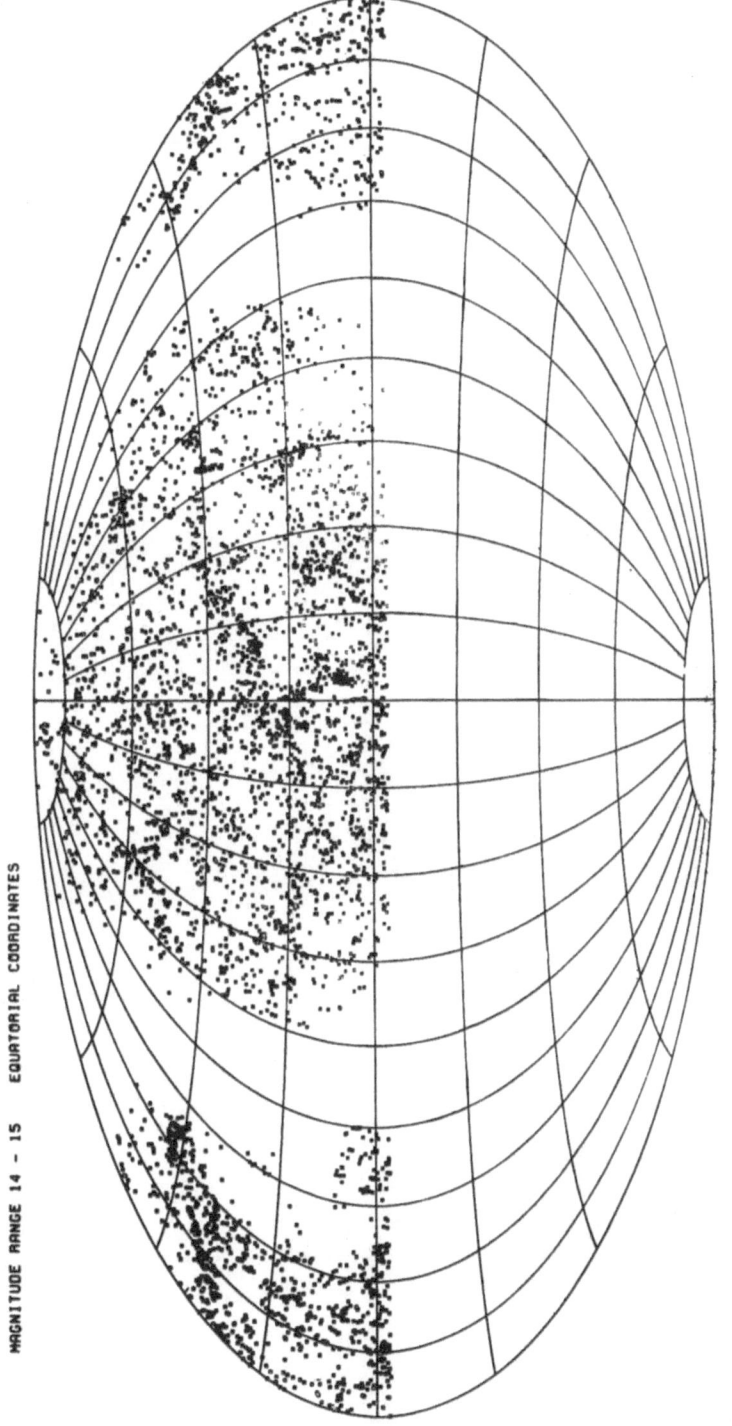

MAGNITUDE RANGE 14 – 15   EQUATORIAL COORDINATES

Fig. 5.   Galaxies with $14 \leqslant m_{pg} < 15$ from the catalogues of Zwicky *et al.*

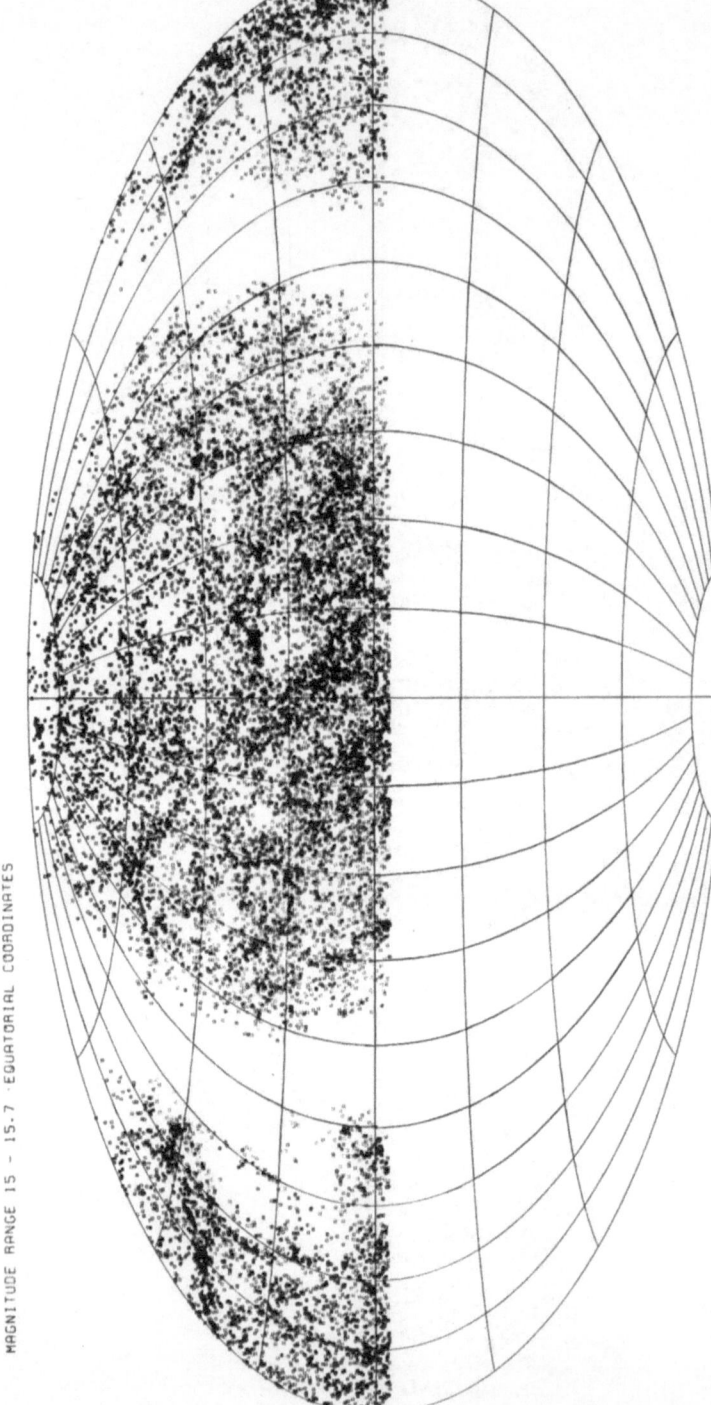

MAGNITUDE RANGE 15 - 15.7 · EQUATORIAL COORDINATES

Fig. 6. Galaxies with $15 \leqslant m_{pg} \leqslant 15.7$ from the catalogues of Zwicky *et al.*

more than $\theta \sim \sin^{-1}(2 q_0/z)^{1/2} \sim 2°$ comes from regions of the Universe separated by more than the light horizon at the time of recombination. Alternatively, if the intergalactic medium is optically thick by $z \sim 10$, the half-angle subtended by the light horizon when the radiation was last scattered is then $\theta \sim 25°$. Yet no temperature fluctuations greater than $\Delta T/T \sim 10^{-3}$ are observed, implying that the Universe is isotropic to better than 0.1%.

In the framework of general relativity, it is hard to avoid the conclusion that such an overall uniformity of the Universe was a special initial condition. Initial inhomogeneities tend to grow. Thus isotropy at a latter time implies isotropy at the beginning. Collins and Hawking (1973) found that the set of spatially homogeneous universes that approaches isotropy at infinite times is of measure zero in the space of all spatially homogeneous universes. As a way of justifying the special initial condition required to produce such a Universe, they argued that this may be the only type of Universe in which galaxies and intelligent life may form, and thus the Universe in which we live must be of this type.

I suggest that initial isotropy may come about through the action of quantum effects which force the early Universe into a structure of maximal symmetry. When the age of the Universe is less than $t \sim 10^{-44}$ s, the Compton wavelength of the mass contained within the light horizon is greater than the light horizon, and continuous fluid models based upon general relativity are inappropriate (Harrison, 1967). When the age of the Universe reaches $t \sim 10^{-23}$ s, the light horizon has expanded to the Compton wavelength of the π-meson and the massive hadron resonances that make up the early Universe may then begin to decay. Carlitz et al. (1973) found that large density fluctuations are produced during the hadron era, and that these are not completely damped out during the period of dissipation that occurs before recombination. In this picture, the present observable structures in the matter distribution, the superclusters and galaxies, are produced by the decay products of massive hadron resonances. The 2.7 K cosmic background radiation originates in the secondary gas of photons, leptons and baryons which builds up from the products of hadron decay.

Thus a forced initial isotropy would allow us to understand how regions of the Universe that were out of communication come to have the same temperature, and the large fluctuations produced by massive hadron resonances in the early Universe would provide the required perturbations for the formation of galaxies and clusters of galaxies.

It is a pleasure to acknowledge contributions from discussions with Dr R. Carlitz, Dr P. J. E. Peebles and Prof. F. Hoyle.

References

Abell, G. O.: 1972, in D. S. Evans (ed.), 'External Galaxies and Quasi Stellar Objects', *IAU Symp.* **44**, 341.

Carlitz, R., Frautschi, S., and Nahm, W.: 1973, *Astron. Astrophys.* **26**, 171.

Carpenter, R. L., Gulkis, S., and Sato, T.: 1973, *Astrophys. J. Letters* **182**, L61.

Collins, C. B. and Hawking, S. W.: 1973, *Astrophys. J.* **180**, 317.
de Vaucouleurs, G.: 1955, *Vistas Astron.* **2**, 1584.
Gunn, J. E. and Peterson, B. A.: 1965, *Astrophys. J.* **142**, 1633.
Harrison, E. R.: 1967, *Nature* **215**, 151.
Hauser, M. G. and Peebles, P. J. E.: 1973, *Astrophys. J.* **185**, 757.
Hubble, E.: 1929, *Proc. Nat. Acad. Sci.* **15**, 168.
Peebles, P. J. E. and Yu, J. T.: 1970, *Astrophys. J.* **162**, 815.
Peebles, P. J. E. and Hauser, M. G.: 1974, *Astrophys. J.*, Suppl. No. 252.
Penzias, A. A. and Wilson, R. W.: 1965, *Astrophys. J.* **142**, 419.
Shapley, H. and Ames, S.: 1932, *Harvard Ann.* **88**, No. 2.
Weinberg, S.: 1971, *Astrophys. J.* **168**, 175.
Zwicky, F., Herzog, E., Wild, P., Karpowicz, M., and Kowal, C. T.: 1960–68, *Catalogue of Galaxies and Clusters of Galaxies*, Pasadena, California Instit. Tech.

DISCUSSION

E. M. Burbidge: In the plot of 15–15.7 mag galaxies, there was a hole in the distribution which seemed to correspond to the high density region in the brighter galaxies in the Virgo cluster. Did it actually coincide?

Peterson: The region of low density lies 10° to the NE of the concentration of bright galaxies.

Tifft: There may be more incompleteness in the Zwicky catalogue at faint limits in *rich* regions, due to observer selection when counting.

Peebles: The Zwicky catalogue was only advertised to be complete to magnitude 15.5.

G. de Vaucouleurs: We have compared various recent catalogues of galaxies and are very puzzled that we see no trace of the supergalactic concentration of galaxies in the Zwicky catalogue although his absolute counts, the number of galaxies per square degree to his stated limit ($m = 15.5$), are in excess of the density extrapolated from counts to fainter limits, i.e. he finds a local excess, as would be expected if his counts are still affected by the Local Supercluster. But when we study the galaxy density vs. supergalactic latitude to compute the supergalactic concentration, it vanishes at the limit of the Zwicky catalogue. However, about ten years ago one of Dr Abell's students, Mr Carpenter, counted galaxies to a fainter limit ($m \simeq 16$) and still revealed a supergalactic concentration, (*Publ. Astron. Soc. Pacific.* **73**, 324, 1961) so there seems to be a discrepancy in the data.

Irving: Can these 'patches' be due to patchy galactic absorption?

Peterson: Peebles has correlated the Shane-Wirtanen catalogue ($m < 19$) with the Zwicky catalogue ($m < 15.5$) and finds that these low density regions are not due to galactic absorption since they change in size as the magnitude limit of the catalogue changes.

Abell: Zwicky himself (Herzog, E., Wild, P., and Zwicky, F.: 1957, *Publ. Astron. Soc. Pacific* **69**, 409) was very much aware that his galaxy counts indicate a shortage of faint galaxies in areas where there is a greater-than-average numerical density of brighter galaxies. He attributed this to intergalactic absorption by dust in the region of the brighter, nearer galaxies. Of course, this circumstance, as Dr Tifft suggests, may be a selection effect. At least, Zwicky was aware of the phenomenon and called attention to it.

DYNAMICS OF SUPERCLUSTERS AS THE MOST POWERFUL
TEST FOR THEORIES OF GALAXY FORMATION

L. M. OZERNOY

Theoretical Department, P. N. Lebedev Physical Institute, Academy of Sciences of
U.S.S.R., Moscow, U.S.S.R.

Abstract. Analysis of Brown's (1964, 1968) data on the orientation of galaxies inside superclusters confirms his conclusion that the directions of the major axes of the galaxies are significantly correlated. For a number of superclusters, the minor axes are found to be oriented along the direction of the flattening of the supercluster as a whole. By analogy with the Local Supercluster where the same phenomenon is associated presumably with the rotation established by de Vaucouleurs (1959), one may conclude that, in general, superclusters rotate. It is shown that rotation on scales as large as these cannot be explained by tidal or by any local effects, and must have a primeval nature analogous to the 3 K microwave background. Rotation on such scales contradicts the standard idea of galaxy formation from density fluctuations but favours the concept of primeval whirls which gave rise to pregalactic density inhomogeneities.

1. Introduction

When comparing the various theories of the formation of galaxies and their agglomerates with observations, the galaxies themselves are often used as a test, in particular the relationships between their main dynamical characteristics such as mass and radius, mass and angular momentum, etc. Since the galaxies have now reached a state such that the initial conditions are largely obliterated, these are not sensitive tests and it is difficult to determine the nature of the pregalactic structure in this way. For example, the conclusion about the necessity of pregalactic turbulence, on which von Weizsäcker (1951) has insisted, is based on comparatively subtle effects. It is therefore very desirable to find more clear and obvious traces of the initial conditions and seems reasonable that such traces manifest themselves principally on the scales of clustering of galaxy clusters, i.e. of second-order clusters (superclusters). These expectations are supported by an estimate based on observational data by de Vaucouleurs (1959) that the duration of the differential expansion of the Local Supercluster is of the order of 10^{10-11} yr. Consequently, at least some initial conditions which characterize the pregalactic medium may be expected to be 'frozen' into the dynamical properties of superclusters.

2. Existence of Superclusters

Although the existence of superclusters has been questioned for many years, an ever-growing number of papers indicate that superclusters are real. The most convincing proof has been given recently by Bogart and Wagoner (1973), who calculated the mean nearest-neighbour angular distance, $\langle \theta \rangle$, between the galaxy clusters belonging to the same Abell distance group, and compared it with the analogous value, μ, obtained from a random distribution of clusters. Their results show that,

John R. Shakeshaft (ed.), The Formation and Dynamics of Galaxies. 85–91. All Rights Reserved.
Copyright © 1974 by the IAU.

within the distance groups 5 and 6, $\langle \theta \rangle < \mu$ with very high significance, i.e. that clustering into superclusters really occurs. Specifically, $\langle \theta \rangle = 0.8\ \mu$, i.e. the cluster density in superclusters is twice as high as that in the surrounding regions.

It should be noted, however, that for clusters belonging to very different distance groups Bogart and Wagoner found $\langle \theta \rangle > \mu$, implying that clusters mutually separated along the line of sight seem to avoid each other on the sky. They could not explain this anticorrelation which is simply absurd from the physical point of view. At first glance, the lack of an explanation for this effect compromises the validity of their conclusions about the existence of superclusters but Bozhich and Ozernoy (1974) have shown that the inequality $\langle \theta \rangle > \mu$ for clusters of different distance groups, as well as the inequality $\langle \theta \rangle < \mu$ within the same distance groups, are corollaries of the same phenomenon, namely the clustering of galaxy clusters into superclusters.

The available data on isotropy of the 3 K microwave radiation provide growing evidence that the clustering of galaxy clusters into systems of higher order cannot be infinite as in Charlier's hierarchical models. At the same time the presence of second-order clusters does not contradict the upper limits on fluctuations of the microwave radiation, or of the X-ray background (Webster, 1972; Brecher, 1972).

3. Evidence in Favour of Rotation of Superclusters

3.1. ROTATION OF THE LOCAL SUPERCLUSTER

3.1.1. *Direct Evidence*

Although for a long time the high degree of flattening of the local Supercluster suggested that it was rotating, more detailed evidence of this rotation superimposed on the differential expansion was obtained only in 1958 by de Vaucouleurs (1959). This was based on the interpretation of the anisotropic radial velocity field of galaxies, and in this sense it is analogous to the proof of the rotation of our Galaxy given by Oort long ago. However, the assumed law of rotation and other parameters were chosen by de Vaucouleurs (1959) rather arbitrarily and were justly criticized by van Albada (1962).

According to de Vaucouleurs and Peters (1968), and de Vaucouleurs (1972), who gave a better founded discussion with revised parameters, the rotational velocity of the Local Group at the distance of 10 Mpc from the centre of the Local Supercluster is of the order of 400 ± 100 km s^{-1}, but the inaccuracy of this value is still too large to consider the rotation as indisputable.

3.1.2. *Indirect Evidence Based on the Orientation of Galaxies*

Reinhardt and Roberts (1972), using independent data on the position angles of spiral galaxies and on the decrease of the proportion of flat galaxies with super-galactic latitude, have shown recently that the orientation of their minor axes correlates with the direction of the poles corresponding to the flattening of the Local Supercluster. In other words, the angular momentum vectors of galaxies tend to coincide with the rotational axis of the Supercluster itself.*

3.2. ROTATION OF OTHER SUPERCLUSTERS

In contrast to the Local Supercluster, the radial velocities of galaxies belonging to other superclusters are not known accurately enough to reveal the rotation. Nevertheless, in addition to the significant flattening of many known superclusters, data of three other kinds also support the hypothesis of supercluster rotation: (i) the correlated orientation of galaxies in all superclusters investigated; (ii) the correlated orientation of radiosources; (iii) the rotation of some clusters. Let us consider these in more detail.

3.2.1. *Correlated Orientation of Galaxies*

Brown (1964, 1968) measured by eye the position angles of about 9000 galaxies on the plates of the *Palomar Sky Survey* and pointed out that the major axes of galaxies belonging to the supercluster in Pisces are predominantly oriented along the direction of extension of the supercluster itself. The same situation is observed in the supercluster in Ursa Major.

Reinhardt (1972), by re-analysing Brown's data, showed that for 11 representative groups of galaxies the hypothesis of non-random orientation is significant at below the 5% level, and for 9 of them at below the 1% level. The selection effects fail to explain these results.

Bozhich and Ozernoy (1974), using the same data, also came to the conclusion that the distribution of galaxy position angles is non-random, though with more modest significance level. Specifically, for four groups it appeared to be between 0.1% to 1%, in six groups from 1% to 5% and only in one group greater than 5%. These results are presented in Table I. In all the samples the conclusion about the

TABLE I

Significance levels (per cent) of the hypothesis that the position angles are distributed at random

Constellation	Apparent diameter	Flat galaxies	Spheroidal galaxies	All forms
Pisces		1.74	31.0	30.4
Virgo	$D > 65''$	0.21	2.0	3.5
Ursa Major	$D > 50''$	0.38	2.1	19.4
Ursa Major	$36'' < D < 50''$	0.52	3.1	8.8
Hydra	$D > 65''$	4.2	–	–
Hydra	$36'' < D < 65''$	0.12	1.2	0.1

* This statistical law permits, of course, large deviations. For example, the rotation axis of our Galaxy deviates from the direction of the supergalactic centre by 16°. Nearly the same situation takes place in the case of M31 whose rotation axis is aligned with that of our Galaxy within 11–18°. It is of interest that the directions of rotation of the Galaxy and of M31 are apparently the same, indicating that the angular momentum of the pregalactic substratum from which the Local Group of galaxies formed was significantly different from zero.

non-random distribution of position angles is much more significant for flattened galaxies than for spheroidal ones. When galaxies of all forms are taken together, it is then not possible to reject the hypothesis of a uniform distribution of position angles. The mixing of galaxies of different distance groups also lowers the significance of the non-uniformity found.

We considered also the distribution of position angles in the region which Brown names a remnant of the supercluster in Pisces. In contrast to the remaining 11 groups of galaxies, this region contains no excess concentration of galaxies. Although these galaxies are not divided into groups of different flattening, the distribution of their position angles is sharply non-uniform at the significance level 2×10^{-7} (!). Hence we may conclude that regions with a pronounced orientation of galaxies are not necessarily those of high galaxy density.

Although the non-uniformity in distribution of galaxy position angles in superclusters is significant, the predominant orientation of the axes is not very noticeable. It is possible that the orientation was much more significant at the very early stages of galaxy evolution, and was later smoothed by tidal forces.

3.2.2. *Correlated Orientation of Radiosources*

Thiel (1972) using 205 radiosources with $z < 0.1$ showed that they are oriented non-randomly at a significance level below 2%. Willson (1972) also found a clear correlation of major axes of radiosources on the scale of clusters and even of superclusters. Tight correlation between the orientation of clusters and their central supergiant cD-galaxies (Sastry, 1968) generating radio-components is evidence that the radiosource orientation is of physical nature. This orientation is probably connected with the correlated orientation of parent galaxies, i.e. finally with the rotation of clusters and superclusters, rather than with the metagalactic magnetic field. All the attempts to discover the latter on scales of superclusters are far from convincing.

3.2.3. *Rotation of Some Clusters*

If superclusters possess rotation indeed, at least some of their clusters ought to rotate (clearly, the inverse conclusion is not necessarily valid). Such rotations have been actually discovered (with differing reliabilities) from the analysis of radial velocities of galaxies, for the Coma cluster (Kalinkov, 1968a, b; Gorbachev, 1969), for the Hercules cluster (Kalinkov, 1971) and for some others.

For the Coma cluster the direction of the rotation axis is found to coincide with the minor axis of isodensity (Gorbachev, 1969). The system of clusters included in the Hercules supercluster is extended in a direction having a position angle $p = -10°$. It is interesting that the mean position angle of galaxy pairs in the clusters I–V of this system is nearly the same: $p = -3°7$ (Karachentsev, 1964). Hence, the direction of rotation, indicated by the orbital planes of galaxy pairs, correlates with the direction of proposed rotation of the whole Hercules supercluster.

The non-spherical form of the clusters which may be clearly seen from the data of the Zwicky catalogue is evidence for the wide spread of rotation of galaxy clusters.

Kostjuk (1971) found that there is only an insignificant increase in the mean sphericity when going from open to compact clusters.

A more detailed analysis of cluster rotation naturally requires more reliable and numerous data but, nevertheless, the above mentioned indirect evidence in favour of cluster rotation does not contradict the assumed rotation of superclusters.

4. Rotation of Superclusters as a Test for Theories of Galaxy Formation

The existence of rotation scales as large as galaxy clusters and, moreover, as super-clusters cannot be ignored by any theory of galaxy formation. Let us consider whether this rotation is compatible with the existing concepts.

4.1. CONCEPT OF ADIABATIC DENSITY PERTURBATIONS

As to galaxy clusters, their rotation, as was suggested by Peebles (1971), might be conditioned by the mutual tidal action of perturbations still at the stage of proto-clusters. However, when considered in detail, this idea encounters several serious difficulties:

(a) The clusters having minimal masses near to $M_d \approx 3 \times 10^{13} - 10^{14} M_\odot$ (those with smaller masses having dissipated in the course of gradual recombination, according to Peebles and Yu (1970), Chibisov (1972)) should possess the minimum rotation and their form should be close to spherical (Chibisov, 1972). However, the observed clusters of minimal mass (pairs, triplets, poor clusters etc.) have irregular forms and apparently quite large specific angular momenta.

(b) Such an angular momentum, as is shown by numerical estimates, could hardly be obtained in the tidal theory. These difficulties are especially obvious in the case of the Local Supercluster. The specific angular momentum of the orbital motion at the distance of the Local Group from the centre of the Supercluster ($R \sim 10$ Mpc) is equal to 10^{33} $(R/10$ Mpc$)$ cm^2 s^{-1}, more than 10^3 times that of the specific spin angular momentum of our Galaxy. Meanwhile, for the latter the tidal theory gives an estimate approximately one order less than that observed (Peebles, 1971). For more massive systems the difference between the predictions of tidal theory and the observations becomes still greater.

(c) Starting from the non-linear theory of adiabatic perturbations by Zel'dovich (1970), Doroshkevich (1972) showed that the angular velocity vectors of galaxies must be parallel to the plane of symmetry of the cluster. However, this conclusion is in sharp contradiction with the observed orientation of galaxies in superclusters, where the rotation axes of galaxies are mostly directed along the minor axis of a supercluster (see Section 3).

Consequently, the concept of galaxy formation which takes no account of whirls as an initial condition meets significant difficulties in explaining the origin of rotation at the later evolutionary stages on scales as large as clusters and especially super-clusters.

4.2. Concept of Whirl Perturbations

The correlation between directions of galaxy rotation on scales of about 100 Mpc means, in fact, that the rotation has a cosmological rather than local origin. Just the same statement serves as a basis for the whirl concept of the origin of galaxies and their clusters (see e.g., Ozernoy and Chibisov, 1970; Ozernoy, 1971; for a review, see Ozernoy, 1973). The quantitative parameters characterizing rotation of galaxy clusters and of the Local Supercluster may be explained by the whirl concept. Without going into details, let us note only the possibility of accounting for the flat form of superclusters by their rotation. In the literature (e.g. de Vaucouleurs and Peters, 1968; Reinhardt and Roberts, 1972) this is considered as hardly probable, due to the fact that the rotational period of the Local Supercluster exceeds greatly the age of the Metagalaxy. However, the ratio of the turn-over time, $T = 2\pi r/v$, to the cosmological time, t, is changed in the course of the cosmological expansion. Neglecting the difference between the rate of expansion of the Local Supercluster and that of the Meta-galaxy*, we have $r \propto (1+z)^{-1}$, $v \propto (1+z)$. Thus $T \propto (1+z)^{-2}$, while $t \propto (1+z)^{-3/2}$. The ratio T/t falls to unity approximately at the moment of recombination $(z \sim 10^3)$. It is clear that at such an early epoch neither tidal nor any local effects can produce rotation on the scales of interest. Consequently, the rotation of superclusters is of a primeval nature and their flat form is related, not to the present rotation, but to a far cosmogonical past.

5. Conclusion

As has been suggested above, for the choice between different theories of galaxy formation the most informative data are those connected with the largest agglomerates of galaxies, rather than with the galaxies themselves. Specifically, data concerning rotation of the Local Supercluster as well as other superclusters provide serious difficulties for the classical ideas on the origin of galaxies from primeval density fluctuations. These data are, on the contrary, strong evidence in favour of the whirl (turbulent) cosmogonical theory. The accumulation of further information concerning the kinematics of superclusters and, in particular, their rotation will not only influence our ideas on the cosmogony of galaxies, but will be of great importance for cosmology as a whole.

References

Bogart, R. S. and Wagoner, R. V.: 1973, *Astrophys. J.* **181**, 609.
Bozhich, S. P. and Ozernoy, L. M.: 1974, to be published.
Brecher, K.: 1972, *Astron. Astrophys.* **23**, 105.
Brown, F. G.: 1964, *Monthly Notices Roy. Astron. Soc.* **127**, 517.
Brown, F. G.: 1968, *Monthly Notices Roy. Astron. Soc.* **138**, 527.

* This simplification cannot be used in order to calculate, in the framework of the whirl theory, the present rotational velocity of the Local Supercluster. These calculations will be given elsewhere.

Chibishov, G. V.: 1972, *Astron. Zh.* **49**, 74.

de Vaucouleurs, G.: 1959, *Astron. Zh.* **36**, 977.

de Vaucouleurs, G.: 1972, in D. S. Evans (ed.), 'External Galaxies and Quasi-Stellar Objects', *IAU Symp.* **44**, 353.

de Vaucouleurs, G. and Peters, W. L.: 1968, *Nature* **220**, 868.

Doroshkevich, A. G.: 1972, *Astron. Zh.* **49**, 1221 (see also *Astrophys. Letters* **14**, 11, 1973).

Gorbachev, B. I.: 1969, *Astron. Tsirk. U.S.S.R.* No. 495, 7.

Kalinkov, M.: 1968a, *Dokl. Bolgar. Akad. Nauk* **21**, 621.

Kalinkov, M.: 1968b, *Astron. Tsirk. U.S.S.R.* No. 475, 4.

Kalinkov, M.: 1971, *Bolgar. Akad. Nauk. Bull. Sec. Astron.* **4**, 123.

Karachentzev, I. D.: 1964, *Izv. Akad. Nauk Arm. S.S.R.*, ser. fiz.-matem. **17**, 109.

Kostjuk, I. P.: 1971, *Problems of Cosmical Physics*, **6**, 205 (in Russian).

Ozernoy, L. M.: 1971, *Astron. Zh.* **48**, 1160.

Ozernoy, L. M.: 1973, in M. S. Longair (ed.), 'Confrontation of Cosmological Theories with Observational Data', *IAU Symp.* **63** (in press); Preprint No. 159 Lebedev Physical Institute.

Ozernoy, L. M. and Chibisov, G. V.: 1970, *Astron. Zh.* **47**, 769.

Peebles, P. J. E.: 1971, *Astron. Astrophys.* **11**, 377.

Peebles, P. J. E. and Yu, J. T.: 1970, *Astrophys. J.* **162**, 815.

Reinhardt, M.: 1972, *Monthly Notices Roy. Astron. Soc.* **156**, 151.

Reinhardt, M. and Roberts, M. S.: 1972, *Astrophys. Letters* **12**, 201.

Sastry, G. N.: 1968, *Publ. Astron. Soc. Pacific* **80**, 252

Thiel, M. A. F.: 1972, *Astrophys. Space Sci.* **17**, 39.

van Albada, G. B.: 1962, in G. C. McVittie (ed.), *Problems of Extragalactic Research*, Macmillan, New York, p. 411.

von Weizsäcker, C. F.: 1951, *Astrophys. J.* **114**, 165.

Webster, A. S.: 1972, *Nature* **238**, 20.

Willson, M. A. G.: 1972, *Monthly Notices Roy. Astron. Soc.* **155**, 275.

Zel'dovich, Ya. B.: 1970, *Astron. Astrophys.* **5**, 84; *Astrofizika* **6**, 319.

INTERGALACTIC MATTER AND RADIATION AND ITS BEARING ON GALAXY FORMATION AND EVOLUTION

G. R. BURBIDGE

Dept. of Physics, University of California, San Diego, P.O. Box 109, La Jolla, Calif. 92037, U.S.A.

Abstract. An up-dated review is given of the evidence for the presence of intergalactic matter and radiation in the Universe. It is concluded that the only important constituents which may make a sizable contribution to the total mass-energy are intergalactic gas and condensed objects with a very high mass-to-light ratio. If the QSOs are not at cosmological distances, cold atomic hydrogen may still be the most important constituent and may contribute much more mass than do the galaxies. The X-ray observations still do not unambiguously show that very hot gas is present, though it is very likely on general grounds that some hot gas is present in clusters of galaxies.

The question of whether or not large amounts of matter, enough to close the Universe, are present, remains unsettled. From the theoretical standpoint the answer depends almost completely on the approach taken to the problem of galaxy formation and to the cosmological model which is favoured.

1. Introduction

In this paper I shall first review the evidence concerning the different forms of mass-energy in the Universe, and then turn briefly to a discussion of the bearing of these results on our understanding of the formation and evolution of galaxies.

In an earlier paper prepared for *IAU Symp.* **44** (Burbidge, 1972) I gave an extensive discussion of the evidence for intergalactic matter and radiation as it appeared in 1970. More recently Field (1972) has also reviewed the evidence for the presence of intergalactic matter. The first part of this paper is therefore concerned with the up-dating of these earlier discussions.

It is convenient to discuss the evidence under several headings:

(1) Neutrinos,
(2) Relativistic particles,
(3) Gravitational radiation,
(4) Fields of electromagnetic radiation,
(5) Diffuse gas,
(6) Mass condensations.

Following the work of Sandage (1972) and of Abell and Eastmond (1968), we put $H_0 = 50$ km s^{-1} Mpc. This means that the critical density ($q_0 = \frac{1}{2}$) necessary to 'close' the Universe, or the steady-state value, is now $\varrho_c = 4.7 \times 10^{-30}$ g cm$^{-3} = 3 \times 10^{-6}$ particles cm^{-3}. On the other hand, the mass density which is present in the visible galaxies, ϱ_g, lies in the range $7\text{--}13 \times 10^{-32}$ g cm^{-3}, corresponding to a particle density of $4\text{--}8 \times 10^{-8}$ cm^{-3} (Noonan, 1972). The discrepancy between this value and the density required to close the Universe (a factor of 30–60) is what has given rise to much speculation concerning the presence of 'missing mass'. We shall discuss the various forms of mass-energy and ask how much they might contribute to the total energy balance of the Universe. At the same time, it should be remembered that if we do not live in a steady-state universe of the type described by Hoyle, or if we live in an open Friedmann universe with $0 < q_0 \ll 1/2$, there may not be such a large difference between the mass in galaxies and that in the Universe.

2. Neutrinos

As was pointed out in the earlier review (Burbidge, 1972) the experimental limits placed on the energy density of low-energy neutrinos from the cut-off in the β-decay spectrum are not severe. They are such that the neutrino energy density could still be many orders of magnitude above the critical density and not have been detected.

3. Relativistic Particles

We really have little idea as to what the energy density of relativistic particles amounts to. If cosmic rays are largely confined to galaxies, the mean energy density in the Universe is only about 10^{-16} erg cm$^{-3} \simeq 10^{-37}$ g cm^{-3}. If, on the other hand, the cosmic rays are universal with an energy density of 10^{-12} erg cm^{-3}, then the equivalent mass density $\simeq 10^{-33}$ erg cm^{-3}.

If the bulk of the cosmic rays has an extragalactic origin, it is likely that they are confined to clusters and superclusters, so that the universal energy density will be about 10^{-14} erg cm$^{-3} = 10^{-35}$ g cm^{-3} (Brecher and Burbidge, 1972a). Even the higher estimates only correspond to very small mass densities which do not have any appreciable effect on the evolution of the Universe.

The only point of some importance is that, if diffuse matter is present at a density $\gtrsim 10^{-30}$ g cm^{-3} in an evolving Universe, its interaction with a cosmic-ray flux with an energy density $\geqslant 10^{-12}$ erg cm^{-3} would give rise to a γ-ray flux greater than that detected. Thus we cannot have universal cosmic rays and a gas-filled Universe.

4. Gravitational Radiation

Over the last several years Weber's results suggesting that large amounts of mass $\geqslant 10^3 \ M_\odot$ yr^{-1} are being radiated in the form of gravitational waves from the galactic centre, have led to speculations that during their evolution galaxies might radiate a large amount of mass. However, it is proving difficult to confirm Weber's results, so that there is a serious question as to whether or not gravitational radiation has been detected.

Rees (1971) has argued that early in an evolving Universe there may have been a large amount of energy converted into long-wavelength gravitational waves, and that this may be present at critical density.

5. Fields of Electromagnetic Radiation

There are four distinguishable components of electromagnetic radiation in the Universe.

They are:

(a) starlight,

(b) nonthermal radio emission,

(c) diffuse X-ray and γ-ray background,

(d) microwave background radiation.

The best estimate of the energy density of starlight is still that given by Felten (cf. Burbidge, 1972) of about 1.5×10^{-14} erg cm^{-3} (1.5×10^{-35} g cm^{-3}). The energy densities of non-thermal radio emission and of the X-ray background are exceedingly small, but the X-ray background may provide evidence for the existence of hot gas. We shall discuss this later.

As far as the microwave background is concerned, there have been some recent developments. While since 1965 most people have thought it likely that the microwave background radiation was generated in a hot big-bang, the evidence that this is the case has not been unambiguous. This is due to the fact that while the measurements in the centimeter range fall very well on the Rayleigh-Jeans part of a black-body curve with a temperature of about 2.8 K, there has been considerable confusion as far as the measurements close to the peak of such a black-body curve near 1 mm are concerned. The indirect measurements using CN, CH and CH$^+$ all were compatible, or indicated a maximum near 1 mm, but the balloon and rocket measurements were in conflict with these, and in some cases with each other, indicating the existence of a much higher flux of radiation than that from a black-body radiation field (for references see Burbidge, 1971). It was these high measurements that led to the distinct possibility the radiation might not have arisen in a big-bang,* but that very large numbers of weak discrete sources were involved.

However, in the last six months, rocket results from Los Alamos (Williamson *et al.*, 1973) and from Cornell (Houck *et al.*, 1972) and balloon results from M.I.T. (Muehlner and Weiss, 1973), the last two groups being those which had earlier reported high fluxes, do not indicate the presence of any flux higher than that expected from a black-body radiation field at a temperature of about 2.8 K. If some of the earlier high measurements were correct at all, this flux must be variable and hence very local and not cosmological in origin.

6. Diffuse Gas

The most interest continues to centre on the possible existence of low density gas both in the outer parts of galaxies, in clusters of galaxies, and possibly even between clusters of galaxies. I last reviewed this situation some three years ago, and at that time stressed that there was no really compelling evidence for the existence of any appreciable amounts of gas anywhere outside galaxies. Field (1972) in his recent review has covered much of the same ground taking into account the most recent

---

* It should be stressed that the only really strong evidence for an evolving universe is the existence of a background flux which appears to have a black-body energy distribution.

X-ray data (up to 1971) and has given a slightly more optimistic discussion of the problem. I shall now up-date the discussion further.

6.1. DETECTION OF NEUTRAL ATOMIC HYDROGEN

The methods that have been used are to attempt to detect 21-cm absorption or emission in the spectra of radio galaxies or QSOs, or to detect $L\alpha$ absorption in the spectra of QSOs. As far as the 21-cm results are concerned, there is nothing new to report. The most reliable results are those of Penzias and Scott (1968) $n_H/T_E < 1.8 \times 10^{-8}$. Field (1972) concludes that $T_E \simeq 18\,\mathrm{K}$, so that $n_H < 3.2 \times 10^{-7}\,\mathrm{cm}^{-3}$, compared with a critical value of $2.8 \times 10^{-6}\,\mathrm{cm}^{-3}$. The limit based on the absence of a step longward of 1420 MHz, which would be due to intergalactic emission, is a weaker one and is about $4.7 \times 10^{-6}\,\mathrm{cm}^{-3}$. These results suggest that the neutral gas, smoothly distributed, can at most be about 10 times the mass in galaxies. A value close to this would imply a value of $q_0 \simeq 0.2$, which is certainly not ruled out when the real uncertainties associated with the attempts to determine q_0 by the direct method are evaluated. I stress this result because it is the most certain upper limit to the density of either neutral or ionized gas in the cosmos.

We turn next to the method based on the attempts to detect $L\alpha$ absorption in the spectra of QSOs with redshifts $\gtrsim 1.8$. I shall not repeat a discussion of the history of these attempts which have been described elsewhere (Burbidge and Burbidge, 1967; Burbidge and Burbidge, 1969; Bahcall, 1971). No evidence of an absorption trough in a QSO attributable to $L\alpha$ extending from the blue wing of $L\alpha$ emission to the atmospheric cut-off has ever been found, though a very large number of QSOs with large redshifts have by now been studied. As was stated in my previous review, either this means that the QSOs are not at cosmological distances, and this question is being discussed by Arp (p. 199) and Sargent (p. 195) at this conference, or the gas which is smoothly distributed is highly ionized. There is also the very remote possibility, discounted by almost everyone, that the space between the galaxies is essentially devoid of gaseous matter, $n(z \simeq 2) \leqslant 1.5 \times 10^{-11}\,\mathrm{cm}^{-3}$, so that all of the matter is condensed into discrete objects.

As is well known, I consider it very likely that the QSOs are comparatively close by, and if this is true we can get no information about the presence or absence of intergalactic gas from studying them. However, the majority has not yet accepted this view, and the absence of the $L\alpha$ trough marked the beginning of the many studies in which it was argued that the intergalactic gas must be very hot. We shall describe the evidence bearing on this possibility later.

What if the gas in intergalactic space is condensed into clouds? In this case, there is the possibility that the absorption spectra of typical clouds could be detected in the spectra of cosmologically distant QSOs.

It is well known that many QSOs with very large emission-line redshifts have rich absorption-line spectra. Since 1966 the question has been asked whether the absorption features arise in, or very close to the objects themselves, or whether they arise in the intervening medium. In addition to the optical observations, very recently Brown and Roberts (1973) have detected the first example of 21-cm absorption in the radio spectrum of a QSO. The object is 3C 286 which has an optical emission-line redshift $z \simeq 0.85$ while the absorption feature, if it has a rest wavelength of 21 cm, has a redshift of 0.69. In what follows I am borrowing heavily from a recent study (Burbidge and Burbidge, 1974) in which full references are given.

The general characteristics of the optical absorption spectra are:

(i) The majority of the absorption redshifts are very close to but less than the emission redshifts. However, some are very different.

(ii) The QSOs with absorption lines tend to have multiple absorption-line redshifts.

(iii) Many of the absorption lines are exceedingly sharp; in the case of PHL 957 the line widths are $\lesssim 30$ km s^{-1}, and the width of the 21-cm line in 3C 286 is less than 8.2 km s^{-1}.

(iv) In some objects, notably PHL 5200 and PHL 957, there are very broad absorption features. In them line widths of ~ 200 Å are found.

(v) There is considerable evidence that different redshift systems in some QSOs are connected by a line-locking mechanism suggesting that shells of gas are ejected from the QSOs by radiation pressure.

The two recently discovered QSOs with very large redshifts, OH 471 (Carswell and Strittmatter, 1973) and OQ 172 (Wampler *et al.*, 1973) ($z = 3.41$ and 3.53, respectively) have rather different spectra. OH 471 shows comparatively weak absorption, but a sharp cut-off in radiation shortward of the Lyman limit, while OQ 172 shows very many (> 130) sharp absorption lines, and considerable radiation beyond the Lyman limit. Since these objects have very similar redshifts, it would be hard to attribute the differences in the absorption spectra and the behaviour at the Lyman limit to the effects of an external intergalactic medium. It is more likely intrinsic. As far as the absorption spectra in general are concerned, we now summarize the arguments for the idea that they are due to intergalactic gas, and then for the idea that they arise in the QSOs themselves.

The arguments can be divided into two classes: General (statistical, etc.) and Detailed (individual objects; spectroscopic details).

6.1.1. *General Arguments*

In Favour of an Intervening Galaxy or Cloud

(a) If QSO redshifts are cosmological, one would expect that the line of sight to distant objects might intersect one or more intervening galaxies or intergalactic clouds if they exist.

(b) The absorption lines seen, for example, in PHL 938 are those predicted by Wagoner (1967) to appear if the absorption takes place in an intervening galaxy.

(c) The higher the redshift, the better the chance of seeing intervening absorption, and it is indeed the objects with $z \gtrsim 2$ which predominantly show absorption.

In Favour of QSO Production

(d) A histogram of distribution of ($z_{em} - z_{abs}$) shows a peak close to zero, meaning that most have $z_{em} \approx z_{abs}$. In a number of cases $z_{abs} > z_{em}$. These results strongly suggest that absorption lines are produced by gas in association with the QSO.

(e) If the emission redshifts are cosmological, we would expect, statistically, that all high-redshift QSOs would have one or several sets of absorption lines. In fact, it is found that (i) there are several objects having 5–8 sets of absorption lines; (ii) there are still some well-examined cases of objects with no absorption lines; (iii) there are objects such as 4C 5.34 and OQ 172 with a few sets of absorption lines, plus a large number of lines which appear to be only Lα, or Lα and Lβ. On the 'intervening cloud' theory, this would require the existence of dozens of clouds strung out between object and observer, spanning a large distance in some cases and none in others.

(f) The very broad absorption features in PHL 5200 (which is so far unique) are clearly associated with the object. Signs of incipient break-up of the broad absorption into multiple absorption suggests a sequence of events: ejection of optically thick gas at a range of velocities, 0–10^4 km s^{-1}; acceleration of some of the gas; onset of instability of gas leading to filament formation; stabilization of regions through which ions flow at particular velocities. These events can be understood in terms of a model with ejection followed by radiative acceleration and stabilization through gravity. An alternative model could be developed in which it is supposed that the gas shells lie at different levels in a gravitational potential well, with radiative balance.

Points (d), (e) and (f) appear to be quite definite in indicating that the absorptions are intrinsic to the QSOs. Several lines of observational investigation are suggested. They include:

Determining the smallest absorption redshift from the Lyman lines in 4C 5.34. The largest absorption redshift is very close to the emission redshift, $z = 2.88$. Can lines identifiable as Lα extend to the observational cut-off, i.e., to $\lambda = 3200$ Å? If so, they would extend to $z \approx 1.6$, and this would conclusively rule out the 'intervening cloud' hypothesis. How far do they extend in OQ 172, which has an even larger z_{em} of 3.5?

Deciding how many of the unidentified lines in the rich multiple absorption line objects such as PKS 0237-23 and PHL 957 are Lα, or Lα and Lβ.

Looking for secular changes in the spectrum of PHL 5200.

We now turn to more detailed arguments.

6.1.2. Detailed Arguments

In Favour of Intervening Galaxy or Cloud

(a) There is only one QSO for which the spectroscopic detail might be said to support the 'intervening' cloud hypothesis. It is PHL 938. However, even in this case, there are counter-arguments. We have already mentioned that the lines seen in PHL 938 were those predicted by Wagoner. The most cogent argument is that the Fe II lines at $z = 0.613$ come from several multiplets, and only those arising from the zero energy states are seen. None are present from the fine structure states at $z = 0.05$–0.11 eV. This means low-density gas, far enough removed from energy sources of UV or IR radiation not to have detectable population (cf. Bahcall and Wolf, 1968).

Against Intervening Galaxy or Cloud

(b) PHL 938 is a multiple-absorption object in having $z_{abs}=1.906$ and 0.613, even though the former shows only Lα.

(c) PHL 938 may have yet another system at $z=1.4$. Detailed comparison of its Fe II lines with satellite UV observations of stars shows that there are two unidentified lines in among the Fe II group around $\lambda=3800$ Å in PHL 938. These fit very well with C IV $\lambda\lambda$ 1548, 1551 at $z\simeq1.4$ (the ratio is correct to the 4th decimal place). Could this be a coincidence?

In Favour of QSO Production

(d) All sets of lines in different redshift systems in the multiple-z objects such as PKS 0237-23 are spectroscopically similar.

(e) Such differences as exist, e.g., in $1311+170$ (Strittmatter *et al.*, 1973), are explicable in terms of a different radiative flux on the gas.

(f) 3C 191, OQ 172, and probably $1331+170$ have excited fine-structure states of Si II populated.

(g) C IV $\lambda\lambda$ 1548, 1551 is a frequent doublet in absorption-line systems, but does not appear in interstellar gas in our Galaxy because the level of ionization required is too high.

(h) Coincidences in the wavelengths of absorption lines (absorption on the violet wing of the emission, one line at one z near-coinciding with a different line at different z) have *no explanation* in the 'intervening' hypothesis, but have a possible explanation (line-locking from radiative-gravity balance) in the QSO production hypothesis.

This has been a brief summary of the observational evidence. Finally, we mention briefly an argument of a semi-theoretical kind.

An argument which has been used against the idea that the absorptions are associated with the QSO is as follows: Studies of the excitation conditions in some cases have led to limits of several tens of kpc or more between the centers of emission in the QSOs and the absorbing clouds (McKee *et al.*, 1973). It is then argued that it is unreasonable to suppose that gas which is ejected can have maintained such a small velocity dispersion so far from the QSO. This argument itself is somewhat circular since it starts off by assuming a high luminosity derived from a cosmological distance whereas, if it is assumed that the QSOs are closer, their intrinsic luminosities are much lower and the radiation incident upon the gas is much less, and so the permissible limits to the distance between gas and source are correspondingly reduced.

A related argument has been made in the case of the 21-cm line in 3C 286, which is then attributed to the outer part of an intervening galaxy (Brown and Roberts, 1973). However, it can be seen from the calculations of Davidson (1972) among others that it is perfectly possible to have fairly dense clouds close to the QSOs and explain the features seen.

To conclude, it seems very unlikely that any of the absorption in the spectra of QSOs is attributable to the presence of an intervening medium.

6.2. DETECTION OF HIGHLY IONIZED GAS

The best chance of detecting ionized intergalactic gas would occur if it had a temperature high enough so that it emits thermal bremsstrahlung, with photon energies
high enough that they are not appreciably absorbed by gas in our Galaxy. This means
photons with energies $\geqslant 0.2$ keV, or gas temperature $\geqslant 2 \times 10^6$ K.

Over the last few years a diffuse X-ray flux has been detected over an energy range
from ~ 0.25 keV to a few MeV. Whether any of this flux is thermal bremsstrahlung from
hot intergalactic gas is still a matter of dispute. I shall discuss first the soft X-ray flux with
energy near 0.25 keV and then the possible origin of the harder X-rays. Field (1972),
Felten (1973), and most recently Silk (1973), have surveyed all but the most recent data.

6.2.1. *Possible Origins of the Soft X-Ray Flux*

It seems likely that there are at least two components of the soft X-ray flux, one
arising in the Galaxy and the other generated outside. Much attention has been
paid to various kinds of discrete source models which could account for the galactic
component, but this will not concern us here.

Felten (1973) and Hayakawa (1973) have concluded from a survey of the observations (some of which are in conflict), that the flux arising outside the Galaxy in the
direction of the poles ~ 500 photons $(\text{cm}^2 \text{ s sr keV})^{-1}$. It may be a real extragalactic
flux coming from great distances. The strongest objection to this hypothesis is the
fact that the Wisconsin group (McCammon *et al.*, 1971) have failed to observe
absorption of this flux by the Small Magellanic Cloud. Their result suggests either
that the flux arises between us and the Clouds, perhaps suggesting a hot halo model
but no extensive extragalactic component, or that sources in the SMC fill up the hole
caused by absorption of extragalactic flux coming from that direction. This latter
proposal does not appear likely, but on the other hand it is difficult to explain all of
the other observations without invoking the presence of some extragalactic component.
At the same time Kraushaar (1973) has pointed out that, if we accept this explanation,
it implies that normal galaxies could explain the whole of the observed background
if they radiate at the level required of the Small Magellanic Cloud. Consequently,
the extragalactic component would arise in normal galaxies and not in the extragalactic medium. Thus, even if it is present, it does not necessarily imply the existence
of hot intergalactic gas. Flux at this level of intensity would be radiation by a hot
intergalactic gas distributed uniformly, with $T \sim 4 \times 10^6$ K and a density approximately 0.3 of the critical density, i.e., $n_e = 10^{-6}$ cm^{-3}. Of course, if the sources arise
in a superposition of discrete components involving hot gas, perhaps gas in clusters
of galaxies (for later discussion), the mean density of the gas will be correspondingly
lower. (If the gas only fills 1% of the total volume, the density in this volume must
be ~ 3 times the critical density but the mean density will be only 3% of the critical
density.) Local models in which it is supposed that the flux arises in a hot halo, or in
hot gas in the Local Group $(R \sim 1 \text{ Mpc}, n \approx 3 \times 10^{-4} \text{ cm}^3, T \simeq 10^6 \text{ K})$ (Hunt and
Sciama, 1972), are also tenable.

With all of the present uncertainties, we see that the soft X-ray flux cannot be taken as a very reliable indicator of the presence of hot intergalactic gas.

6.2.2. *Possible Origins of the X-Ray Background with Photon Energies* $\geqslant 1$ keV

There is no question but that a highly isotropic X-ray flux with photon energy $1 \text{ keV} \leqslant E \leqslant 100$ MeV is present and has a truly extragalactic origin. The only mechanisms which are likely to be responsible for generating this flux are Compton scattering of relativistic electrons on the microwave background radiation, or thermal bremsstrahlung from a very hot gas and, at the high energy end, π° decay. There are several important questions that need to be answered before we can decide how important the X-ray background is from the point of view of studies of intergalactic matter. Not only must we identify the production mechanism, but we must also ask whether the background radiation is likely to be made up of discrete sources of known types.

As far as the mechanism generating the hard photons is concerned, the most important point is the shape of the spectrum. The proposals of Felten and Morrison (1966) and of Brecher and Morrison (1969) were that the Compton scattering of relativistic electrons on the microwave background was likely to be responsible. However, as was pointed out by Cowsik and Kobetich (1972), if there is a sharp enough bend in the spectrum as is perhaps the case near 40 keV, this may be difficult to explain if the relativistic electron spectrum does not have a bend as an intrinsic property. At the same time Brecher (1973) has correctly pointed out that, on the basis of what is presently known about synchrotron radio spectra and their parent electron spectra, the existence of bent electron spectra is not excluded.

On the observational side, it is clear that nothing can be concluded for certain until a very well determined spectrum of the background flux is measured. It still appears possible that there is very little change in the slope between ~ 10 keV and ~ 10 MeV (Pal, 1973; Pinkau, 1973) and, if this is correct, the Compton explanation is adequate and very plausible. However, despite the uncertainties in the data, many theoreticians and observers seem to have largely excluded the Compton explanation and have turned to the thermal bremsstrahlung model.

Field (1972) recently concluded that the present observational data are not too far in disagreement either with a hot gas model (uniform density) in a Friedmann universe (big-bang model) with an intergalactic gas temperature of 2×10^8 K (assuming $\gamma = \frac{5}{3}$, adiabatic expansion since $z \simeq 1$, and a closure density), or a Gold-Hoyle steady-state model ($T \simeq 6 \times 10^8$ K).

More recently, the results summarized by Schwartz and Gursky (1973) modify Field's conclusion and, in terms of a hot gas model in a Friedmann universe, they argue that the density is less than the closure density by a factor $\sim \sqrt{2}$, or that the temperature $\leqslant 3 \times 10^7$ K, or that the Hubble constant is only about 50 km s^{-1} Mpc^{-1}.

To summarize, with considerable uncertainty, the existence of the diffuse X-ray background can be adduced to be evidence in favour of the existence of very hot intergalactic gas, but in no sense do the observations prove that such a gas exists.

Even if the radiation is thermal bremsstrahlung, it is possible that the diffuse background is simply a sum of the radiation from discrete sources. We know that various types of extragalactic objects are discrete X-ray sources, though at present only a small sample of different types has been identified. The most prominent discrete sources are rich clusters of galaxies, but an attempt to add up the contributions from normal galaxies, radio galaxies, rich clusters, Seyfert nuclei, and QSOs led Schwartz and Gursky (1973) to the conclusion that they can only explain about 22% of the background from discrete sources, and the bulk of this is due to Seyfert galaxies based on a sample of one, NGC 4151.

6.2.3. *Other Observational Evidence for Gas in Clusters of Galaxies*

It has just been pointed out that a number of rich clusters of galaxies have been identified as extended X-ray sources (cf. Kellogg *et al.*, 1973). They may either be radiating X-rays which are due to Compton scattering, or they may be due to thermal bremsstrahlung, thus indicating the presence of gas in the clusters. Practically all of the clusters which have so far been identified also contain active radio galaxies so that, as was proposed by Brecher and Burbidge (1972b) [see also Burbidge (1973)], Bridle and Feldman (1972) and others, it is natural to suppose that X-rays will be generated through the Compton effect, since both relativistic electrons and the microwave background radiation are present. However, many workers have chosen to argue that the detection of extended X-ray sources indicates the presence of hot gas (cf. Silk, 1973). Attempts have been made to show that there is a correlation between the random motions of the galaxies in a cluster and the X-ray luminosity (Solinger and Tucker, 1972). While much has been made of the correlation $L_X \propto$ (velocity dispersion)^4, I do not find it very convincing since only a very small number of clusters (3 in the first instance) with good velocity dispersions are involved. These clusters are of different physical and dynamical types, while the velocity dispersions are not very different. This is why a correlation can only be found with a high power of Δv. Also theoretical work has been carried out based on the idea that the X-ray sources definitely show that hot gas is present (Yahil and Ostriker, 1973).

In fact, in my view, for the X-ray sources in clusters in general, no conclusion as to which mechanism is operating can yet be drawn. The best way of deciding this question is through studies of the X-ray spectra and measurements of the angular sizes and intensity profiles of the sources. In the simplest models we might expect that the Compton X-rays would have a spectrum of the form $I(E) \propto E^{-\alpha}$, while an exponential form with a single temperature would be expected if the X-rays are thermal bremsstrahlung. In more realistic models we expect considerable departures from these conditions. The problem encountered by some groups in handling the Gaunt factor (cf. Margon, 1973) has compounded the difficulties in interpretation. The majority of the clusters have been detected by the UHURU satellite and, in a recent paper, Kellogg *et al.* (1973) have shown that there is a wide range of X-ray luminosities and that either mechanism can be accommodated. The most closely studied cluster is the Coma cluster. In its central part a soft X-ray source $(E_X \leqslant 2 \text{ keV})$ has been discovered (Gorenstein *et al.*, 1973). The spectral data from these measurements together with the higher energy measurements suggest an exponential spectrum rather than an inverse power-law form. Gorenstein *et al.* therefore argue that the radiation most likely arises in a hot gas with $T \simeq 10^8$ K, $\varrho \simeq 10^{-27}$ g cm^{-3}. Even assuming a uniform density, the total mass of gas is far less than that required to bind the cluster.

The general conclusion from the data available on X-ray emission from clusters is that evidence for the presence of gas is ambiguous, but even if hot gas is present, it only contributes a small fraction of the total mass if the clusters are gravitationally stable.

Another approach to the problem of the existence of intergalactic or intracluster gas is through the study of extragalactic radio sources. Sources which have been identified with optical galaxies both inside and outside clusters frequently have two major components (which are structured) centred about the optical system. They have clearly been ejected from the parent galaxy and, since they have not freely expanded, they have either been contained by the pressure of an external medium or the components must contain enough mass to hold themselves together. A critical account of the various schemes which have been proposed has recently been given (De Young and Burbidge, 1973). Many authors who favour the ram-jet mechanism feel that the fact that it can explain some features of the sources is evidence for the existence of intergalactic gas, and with a judicious choice of parameters it has sometimes been claimed that the gas present is adequate to bind clusters or to equal the critical cosmological density. In fact, there is great uncertainty in these quantities and, if one goes to a model in which it is argued that the sources are generated from a number of coherent objects which are ejected from the galaxy, very little, if any, gas is required.

Further arguments along somewhat similar lines have been made following the studies of radio sources which appear to consist of active galaxies with radio trails (Miley *et al.*, 1972; Miley, 1973; Wellington *et al.*, 1973). Since these observations have only so far been interpreted by those who have used the observations on the *assumption* that an intergalactic medium is causing the effects observed, it is difficult to decide at this stage whether or not the results provide unambiguous evidence for the presence of gas.

Another line of evidence which also remains somewhat ambiguous is concerned with the high latitude 21-cm observations which have been interpreted by Oort (1971) as evidence that intergalactic gas is being accreted by the Galaxy. Other interpretations have been proposed, the most plausible of which is that this gas has been ejected from the Galaxy and is now falling back in.

Finally, we describe briefly evidence that gas is likely to be present following the evolution of galaxies.

If galaxies are formed by a process of condensation from a lower density medium, it is to be expected that some intergalactic gas is left behind. For those who are quite certain that galaxies formed by condensation, this provides the primary reason for believing in the existence of intergalactic gas. Oort (1970) has argued that only a small fraction ($\sim 1/16$) of the mass will condense into galaxies, leading to the view that 15/16, or approximately the closure density, is present in intergalactic gas. For the minority, including Ambartsumian, Hoyle and Narlikar, who believe that galaxies have evolved from very high density states, this cosmogonical argument for the presence of much intergalactic gas is hardly convincing.

Given that galaxies are indeed present, there are a number of processes which we know are operating which will tend to expel gas from them into the outside medium. These include:

(i) Large-scale explosions in the nuclear regions which are able to eject large masses at velocities considerably in excess of the escape velocity (cf. Burbidge, 1970).

(ii) Processes involving galactic winds (Mathews and Baker, 1971) which will drive gas from galaxies and eventually from clusters. Within clusters it is therefore reasonable to suppose that gas is present which has been ejected from galaxies and heated by the passage of other galaxies (Ruderman and Spiegel, 1971). If gas were present outside clusters, the converse process of accretion might occur (Gott and Gunn, 1972).

However, if the gas which is present in the intergalactic medium is only that which has been ejected from galaxies, it follows that the total mass involved is likely to be very much less than that presently condensed into galaxies, so that it will not approach the critical density.

7. Discrete Objects

It appears likely that there could still be much mass in the Universe in the form of dark discrete objects which have so far remained undetectable. This possibility has been discussed before by me and by others, and there is very little new that one can add. If some clusters of galaxies (like the Coma cluster) are bound, it is likely that this mass is in the form of evolved galaxies, dwarf systems, black holes or white holes.

Press and Gunn (1973) have recently discussed the possibility that, if the number of such objects is high enough (close to the critical density), it may be possible to detect them by the method of gravitational imaging.

8. Conclusions

There is still no very good direct evidence for the presence of significant amounts of mass-energy in forms other than that of discrete luminous objects, namely galaxies. Some intergalactic diffuse matter is likely to be present, but no strong evidence is available that its density approaches the critical value. My own view is that it is most likely that a significant amount of cold intergalactic hydrogen is present. As was stated earlier, if the quasi-stellar objects are comparatively close by, no limits are placed on the density of intergalactic gas from their absorption spectra, and the limit set by Penzias and Scott (1968) allows us to assume that quite a large mass – much greater than that in the galaxies – is present in the form of cold gas. It is also possible that we live in a universe in which $\varrho \ll \varrho_c$. Alternatively, as is well known, we cannot exclude the possibility that the bulk of the mass-energy is present in such exotic forms as neutrinos or black holes.

The X-ray observers should perhaps be discouraged from feeling that they ought to find evidence for large amounts of hot gas – it may simply not be there.

I am grateful to Margaret Burbidge for contributing to the discussion of absorption spectra of quasi-stellar objects. Extragalactic research at UCSD is supported in part by grants from the National Science Foundation and in part by NASA through grant NGL 05-005-004.

References

Abell, G. O. and Eastmond, S.: 1968, *Astrophys. J. Suppl.* **73**, 161.
Bahcall, J. N.: 1971, *Astron. J.* **76**, 283.
Bahcall, J. N. and Wolf, R. A.: 1968, *Astrophys. J.* **152**, 701.
Brecher, K.: 1973, *Astrophys. J.* **181**, 255.
Brecher, K. and Burbidge. G. R.: 1972a, *Astrophys. J.* **174**, 253.
Brecher, K. and Burbidge, G. R.: 1972b, *Nature* **237**, 440.
Brecher, K. and Morrison, P.: 1969, *Phys. Rev. Letters* **23**, 802.
Bridle, A. H. and Feldman, P.: 1972, *Nature (Phys. Sci.)* **235**, 168.
Brown, R. L. and Roberts, M. S.: 1973, *Astrophys. J. Letters* **184**, L7.
Burbidge, G. R.: 1970, *Ann. Rev. Astron. Astrophys.* **8**, 369.
Burbidge, G.: 1971, *Nature* **233**, 36.
Burbidge, G. R.: 1972, in D. S. Evans (ed.), 'External Galaxies and Quasi-Stellar Objects', *IAU Symp.* **44**, 492.
Burbidge, G. R.: 1973, in H. Bradt and R. Giacconi (eds.), 'X- and Gamma-Ray Astronomy', *IAU Symp.* **55**, 199.
Burbidge, G. and Burbidge, M.: 1967, *Quasi-Stellar Objects*, Freeman, San Francisco.
Burbidge, G. R. and Burbidge, E. M.: 1969, *Nature* **222**, 735.
Burbidge, G. R. and Burbidge, E. M.: 1974, in preparation.
Carswell, R. F. and Strittmatter, P. A.: 1973, *Nature* **242**, 394.
Cowsik, R. and Kobetich, E. J.: 1972, *Astrophys. J.* **177**, 585.
Davidson, K.: 1972, *Astrophys. J.* **171**, 213.
De Young, D. S. and Burbidge, G.: 1973, *Comments Astrophys. Space Phys.* **5**, 29.
Felten, J. E.: 1973, in H. Bradt and R. Giacconi (eds.), 'X- and Gamma-Ray Astronomy', *IAU Symp.* **55**, 258.
Felten, J. E. and Morrison, P.: 1966, *Astrophys. J.* **146**, 686.
Field, G.: 1972, *Ann. Rev. Astron. Astrophys.* **10**, 227.
Gorenstein, P., Bjorkholm, P., Harris, B., and Harnden, Jr., F. R.: 1973, *Astrophys. J. Letters* **183**, L57.
Gott, J. R. and Gunn, J. E.: 1972, *Astrophys. J.*, **176**, 1.
Hayakawa, S.: 1973, in H. Bradt and R. Giacconi (eds.), 'X- and Gamma-Ray Astronomy', *IAU Symp.* **55**, 235.
Houck, J. R., Soifer, B. T., Harwit, M., and Pipher, J. L.: 1972, *Astrophys. J. Letters* **178**, L29.
Hunt, R. and Sciama, D. W.: 1972, *Nature* **238**, 320.
Kellogg, E., Murray, S., Giacconi, R., Tananbaum, H., and Gursky, H.: 1973, *Astrophys. J. Letters* **185**, L13.
Kraushaar, W. L., Invited paper given at International Symposium and Workshop on Gamma-Ray Astrophysics, Greenbelt, June 1973.
Margon, B.: 1973, *Astrophys. J.* **184**, 323.
Mathews, W. G. and Baker, J. C.: 1971, *Astrophys. J.* **170**, 241.
McCammon, D., Bunner, A. N., Coleman, P. L., and Kraushaar, W. L.: 1971, *Astrophys. J. Letters* **168**, L33.
McKee, C. F., Tarter, C. B., and Weisheit, J. C.: 1973, *Astrophys. Letters* **13**, 13.
Miley, G. K.: 1973, *Astron. Astrophys.* **26**, 413.

Miley, G. K., Perola, G. C., van der Kruit, P. C., and van der Laan, H.: 1972, *Nature* **237**, 269
Muehlner, D. J. and Weiss, R.: 1973, *Phys. Rev. Letters* **30**, 757.
Noonan, T. W.: 1972, *Astrophys. J.* **171**, 209.
Oort, J. H.: 1970, *Astron. Astrophys.* **7**, 381.
Oort, J. H.: 1971, *Nature* **224**, 1158.
Pal, Y.: 1973, in H. Bradt and R. Giacconi (eds.), 'X- and Gamma-Ray Astronomy', *IAU Symp.* **55**, 279.
Penzias, A. A. and Scott, E. H.: 1968, *Astrophys. J. Letters* **153**, L7.
Pinkau, K.: Rapporteur paper given at International Cosmic Ray Conference, Denver, August 1973.
Press, W. H. and Gunn, J. E.: 1973, *Astrophys. J.* **185**, 397.
Rees, M. J.: 1971, *Monthly Notices Roy. Astron. Soc.* **154**, 187.
Ruderman, M. A. and Spiegel, E. A.: 1971, *Astrophys. J.* **165**, 1.
Sandage, A.: 1972, Proc. Mayall Symp., Tucson, Arizona 1971.
Schwartz, D. and Gursky, H.: Invited paper given at International Symposium and Workshop on Gamma-Ray Astrophysics, Greenbelt, June 1973.
Silk, J.: 1973, *Ann. Rev. Astron. Astrophys.* **11**, 269.
Solinger, A. and Tucker, W.: 1972, *Astrophys. J. Letters* **175**, L107.
Strittmatter, P. A., Carswell, R. F., Burbidge, E. M., Hazard, C., Baldwin, J. A., Robinson, L., and Wampler, E. J.: 1973, *Astrophys. J.* **183**, 767.
Wagoner, R. V.: 1967, *Astrophys. J.* **149**, 465.
Wampler, E. J., Robinson, L. B., Baldwin, J. A., and Burbidge, E. M.: 1973, *Nature* **243**, 336.
Wellington, K. J., Miley, G. K., and van der Laan, H.: 1973, *Nature* **244**, 502.
Williamson, K. D., Blair, A. G., Catlin, L. L., Hiebert, R. D., Loyd, E. G., and Romero, H. V.: 1973, *Nature (Phys. Sci)* **241**, 9.
Yahil, A. and Ostriker, J. P.: 1973, *Astrophys. J.* **185**, 787.

DISCUSSION

G. de Vaucouleurs: May I point out that this so-called average density of the Universe in the form of galaxies is a traditional fiction. One always gets the same number because the same galaxy counts to about the same faintest limit ($19 < m < 21$) are used over and over again so we are always dealing with the same volume of space. It does not prove that we are reaching a significant constant of universal value. Deeper galaxy counts ($m > 21$) are needed, with appropriate corrections, for a definite solution of this problem.

G. Burbidge: I have no comment on this point – perhaps Professor Oort would like to take it up?

Oort: In answer to the doubt expressed by de Vaucouleurs whether we can get any sensible estimate of the general density in the Universe furnished by galaxies, I would remark that in the first place one can get information on the luminosity function of galaxies from observations in nearby clusters. In the second place, the fact that we find approximate isotropy in the distribution of galaxies and clusters of galaxies if we go out to about 10^3 Mpc distance indicates that the volume surveyed is representative of the Universe as a whole (i.e. that in such a volume we have evened out the largest-size irregularities in the Universe).

G. de Vaucouleurs: Unfortunately the deepest galaxy counts to a well-calibrated limiting magnitude ($m = 19$, the Lick Survey) reach out to 200–300 Mpc only, which is not much larger than the scale of the largest size-irregularities so far detected (the Kiang-Saslaw analysis of the Abell clusters). Until much deeper counts are available the evidence that we have indeed reached a plateau of constant density in the $\log \varrho - \log R$ relation is subject to reasonable doubt.

Heidmann: I do not completely agree with de Vaucouleurs about the mean density of the Universe. Deep galaxy surveys lead to a density of a few $\times 10^{-31}$ g cm^{-3}. But this mean density is already reached for a volume within 10 Mpc radius centred on the Earth. This volume is already typical of the Universe at the gigaparsec scale. Of course, if this volume were centred on the Local Supercluster it would yield a higher density, as shown by de Vaucouleurs' density-radius relation.

G. de Vaucouleurs: 10 Mpc is much too small. Why don't you take just the first kiloparsec?

Baldwin: A search we have made at 21 cm (unpublished) suggests that the number density of intergalactic H I clouds is less than 3% of that of known galaxies for masses of $\sim 10^{10} M_\odot$, and for masses of $\sim 10^8 M_\odot$ is certainly no larger than that of known galaxies. For lower masses the absorption measurements of Penzias and Scott (*Astrophys. J. Letters* **153**, L7, 1968) provide the best limits.

Allen: It should be remembered that any emission measurement designed to detect intergalactic neutral hydrogen will fail if the spin temperature of the H I is in equilibrium with the 3 K background; H I at 3 K would be quite invisible in emission. Furthermore, if the gas is condensed into clouds, then it is possible that the available absorption measurements (on Virgo A, Cygnus A) have simply missed them.

G. de Vaucouleurs: If you take the ratio of hydrogen mass to total mass, or hydrogen mass to total luminosity for galaxies, there is a tendency for the fraction of the mass that is hydrogen to increase toward the later types, reaching an upper limit of about 10%. Why don't we see galaxies with 50%, 75% or even 100% of their mass in hydrogen?

G. Burbidge: I have no ideas about this one. I don't know how to make galaxies – that's my problem.

E. M. Burbidge: What about the compact galaxies which were called 'giant H II regions' by Sargent and Searle – were these not supposed to contain a large amount of neutral hydrogen?

Heidmann: There are extragalactic objects with a relative neutral hydrogen content higher than 10%, e.g. the Zwicky dwarf compact galaxies such as II Zw 40 in which the hydrogen mass is as large as the mass of stars.

G. de Vaucouleurs: How is the total mass obtained?

Heidmann: By a virial theorem estimate.

G. de Vaucouleurs: I don't believe that's reliable.

Toomre: Does Sargent believe such an estimate?

Sargent: Yes. I think the numbers are consistent with the mass fraction of hydrogen in some of these objects being almost 100%.

G. de Vaucouleurs: Only 20 yr ago, people were calculating the mass of the Large Magellanic Cloud by the same method, taking the 20 km s^{-1} width of the 21-cm hydrogen emission line as *the* velocity dispersion. This was complete nonsense because this is just the z velocity dispersion in a flat system, seen nearly face-on. As we know now, most of the kinetic energy is in the rotation; the virial mass is at best a lower limit of the true total mass.

Longair: Despite the wide range of models for the containment of radio source components involving ram pressure, a universal requirement is that $\varrho_e V^2 = U_{min}$, where ϱ_e is the ambient gas density, V the component velocity and U_{min} the minimum energy density in the component. It also applies to 'continuous flow' models such as the Rees model. Taking the largest values of U_{min} found in external radio components, and the maximum permissible values of V, i.e. the velocity of light, gives an absolute lower limit to ϱ_e of the order of the cosmological density. However, there is no guarantee that it refers to the intergalactic gas – rather, in many cases, it must refer to gas in clusters or even the outer regions of radio galaxies.

We have been investigating whether there exist real differences between sources reported to be in clusters and those in the general field and, whilst we have been hampered by the difficulty of distinguishing when a source is in a group or cluster, we find no evidence for any difference in the properties of double sources inside and outside rich clusters. I make a plea to the experts to help distinguish what types of clusters are associated with powerful radio sources since the data are at present extremely inhomogeneous.

G. de Vaucouleurs: One of our students at Texas, Mr F. Owen (now at NRAO), has been looking at radio sources in Abell clusters and his results, interpreted on the ram-pressure model, indicate that the density in the intergalactic medium decreases outwards from the centres of the clusters.

Longair: We obtain the opposite. When care is taken to avoid selection effects, there are no significant differences between sources in and out of clusters.

Kellermann: Is there any missing mass? Why should we worry about the critical density at all?

G. Burbidge: Why indeed?

Peebles: We can form, from Hubble's constant H and Newton's gravitation constant G, a number with units of density, H^2/G, and it seems a wonderful coincidence, and perhaps deeply significant, that the mean mass density of galaxies is roughly similar to this number. Of course one would like to see how close the coincidence is. Beyond this the 'critical cosmological mass density' is mainly theological.

Conway: The measurements of linear polarization in quasars and its fall-off at long wavelength, if interpreted as due to the Faraday effect, give values for the electron density within the components of the source. On the ram-pressure model one can give corresponding values *outside* the source, and these values come close to $5 \times 10^{-30} (1 + z)^3$ g cm^{-3}, i.e. close to the closure density. Of course, if the quasars are really in distant clusters of galaxies, this value is the cluster density.

Baldwin: Measurements of radio sources with steep spectra, associated with X-ray sources in clusters of galaxies, have shown them to be of relatively small angular sizes. This suggests that the origin of the X-rays is more likely to be a thermal mechanism than the inverse Compton process and puts a lower limit on the density of hot gas.

Gott: It is interesting that clusters like the Coma cluster contain such a small amount of hot intracluster gas. In reasonable cosmological models, material around the outskirts of the cluster should be bound to it and suffer infall into the cluster. Calculations show that, if $q_0 = \frac{1}{2}$ and if the missing density is supplied by intergalactic gas, then one would have expected more gas to have accumulated by now in the Coma Cluster than is actually observed. This argues against the presence of a closure density of intergalactic gas.

RECENT WESTERBORK OBSERVATIONS
OF HEAD-TAIL GALAXIES

G. K. MILEY, H. VAN DER LAAN, and K. J. WELLINGTON

Sterrewacht, Leiden, The Netherlands

Abstract. Some recent observations of head-tail galaxies are presented. The properties of these objects are reviewed and are shown to be consistent with the radio trail hypothesis.

1. Introduction

During the last few years, observations with the Cambridge One-Mile Telescope have defined a new class of relatively weak radio galaxies characterized by a peculiar elongated morphology (Ryle and Windram, 1968; Hill and Longair, 1971). The radio sources have a high brightness 'head' close to the optical galaxy and a narrow low brightness 'tail' which sometimes extends for many minutes of arc. All the head-tail galaxies at present known (Table I) appear to be associated with clusters of galaxies.

TABLE I

Head-tail radio galaxies

Object	Cluster	Identification
NGC 1265 (3C 83.1)	Perseus	Ryle and Windram (1968)
IC 310	Perseus	Ryle and Windram (1968)
WBK 0314+41	Perseus	Miley *et al.* (1972)
NGC 4869 (5C 4.81)	Coma	Willson (1971)
3C 129	Obscured	Hill and Longair (1971)
3C 129.1		Hill and Longair (1971), Miley (1973)
NGC 7385 (PKS 2247 + 11)	Zw 2247 + 11	Schilizzi (private communication) Ekers (private communication)

Here I shall first describe Westerbork observations of two of the most beautiful examples 3C 129 and NGC 1265. After comparing some characteristics which these objects have in common with 'normal' double radio galaxies, I shall conclude with some deductions about their nature and about the properties of the intergalactic medium. We shall see that all the data are consistent with the hypothesis that the galaxies are active radio sources and that the tails are merely trails left behind as these galaxies plough through a dense intra-cluster medium (Miley *et al.*, 1972).

2. 3C 129

Figure 1 shows a 1.4 GHz radio photograph of 3C 129. The radio tail extends for nearly half a megaparsec from the parent galaxy. Note the double nature of both the

John R. Shakeshaft (ed.), The Formation and Dynamics of Galaxies. 109–118. *All Rights Reserved.*

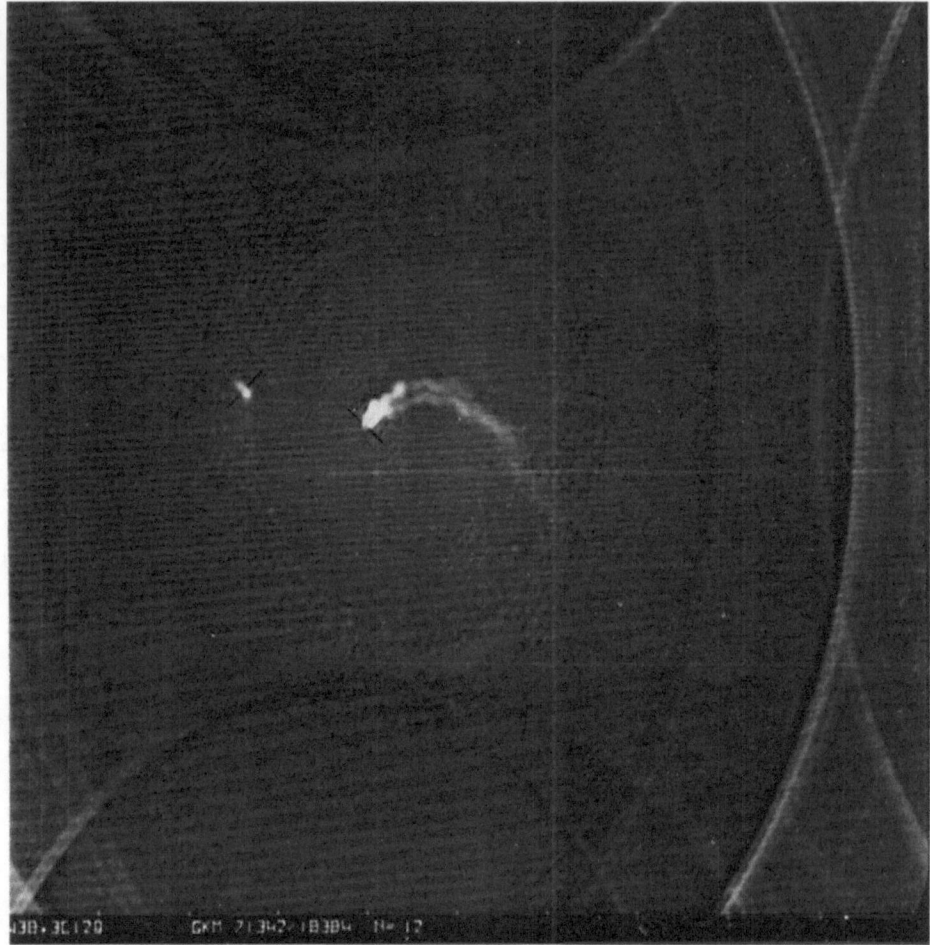

Fig. 1.   1.4 GHz radio photograph of 3C 129 (right) and 3C 129.1 (left) (from Miley *et al.*, 1972).
Bars indicate the position of the optical galaxy.

head and the tail. The polarization data are given in Figure 2 (Miley, 1973). The percentage polarization increases, and the radio spectrum gradually steepens, with distance along the tail (Figure 3). Compared with the head, the more relaxed tail has a higher polarization and steeper spectrum. It is interesting that for 'normal' double radio galaxies the more relaxed ones also appear to have relatively high polarizations and steep spectra (Miley and van der Laan, 1973).

3. NGC 1265

Figure 4 shows a 1.4 GHz radio photograph of the galaxy NGC 1265 in the Perseus Cluster and Figures 5 and 6 give the 1.4 GHz and 5.0 GHz maps superimposed on the

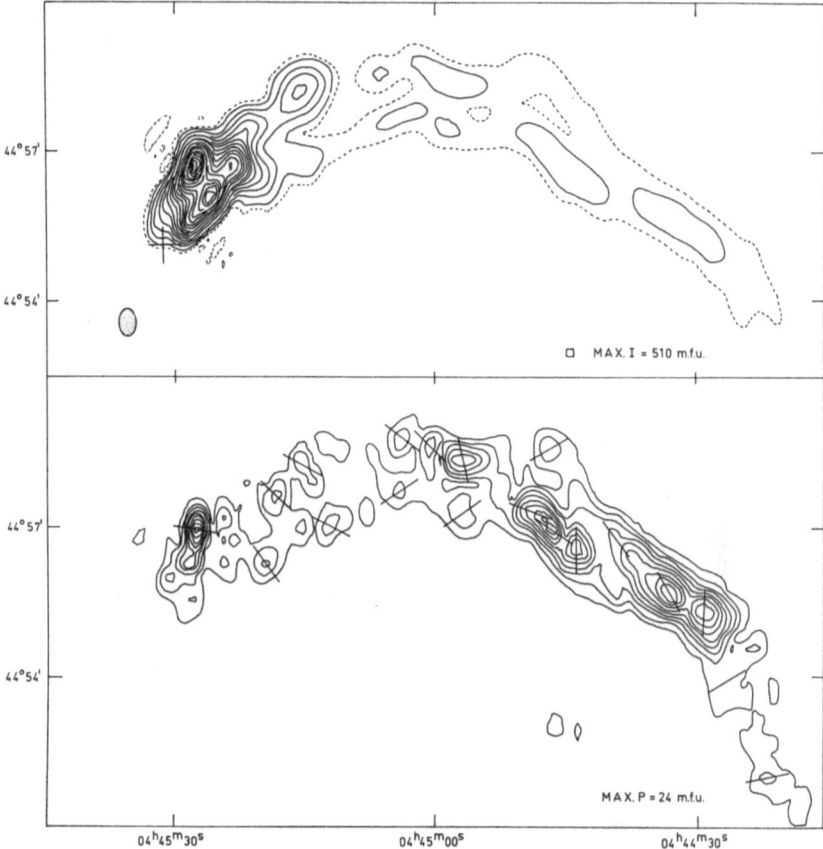

Fig. 2. The 1.4 GHz total intensity (top) and polarization (bottom) distributions of 3C 129 (from Miley, 1973).

Palomar prints. Here we also see the double nature of both head and tail. There is a compact opaque component near the centre of the optical galaxy. Note also the shape and discreteness of the head. Such a configuration is exactly what would be expected for a galaxy ejecting radio emitting blobs with a speed comparable to that of the galaxy (radial velocity ≈ 2300 km s^{-1}) every few million years (Wellington *et al.*, 1973).

4. Comparison with 'Normal' Radio Galaxies

The head-tail galaxies so far detected are intrinsically several hundred times weaker than the strongest 'normal' doubles. However, as we have seen, there are several characteristics common to both classes (Table II).

Because of these similarities it appears reasonable that head-tail galaxies should be explained within the context of theories of normal radio galaxies. The morphologies of 'normal' sources (Figure 7 of Miley and van der Laan, 1973) have been ex-

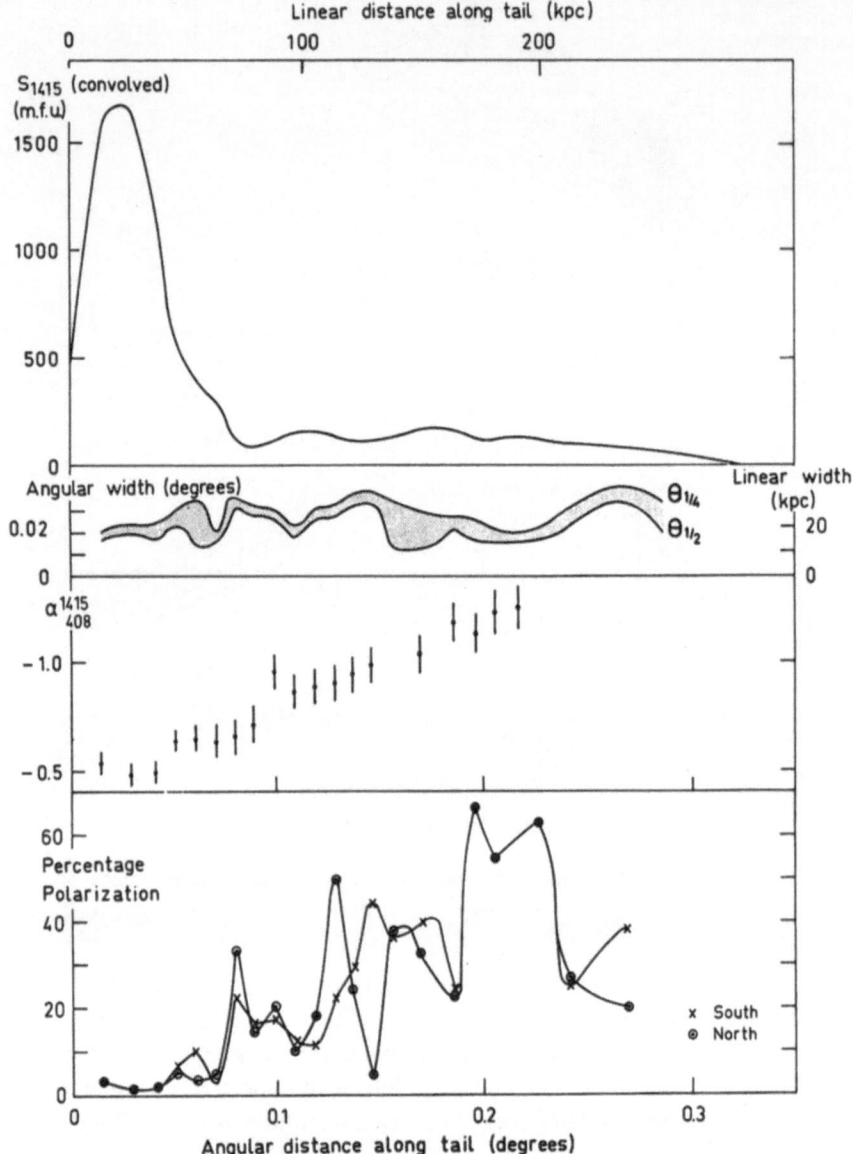

Fig. 3.   Variation of brightness with spectral index and percentage polarization derived from 1.4 GHz
measurements on 3C 129 (See Miley, 1973).

plained in terms of ram pressure confinement by an extragalactic medium (3C 61.1,
3C 274.1, 3C 284, 3C 390.3, 3C 184.1) and subsequent adiabatic expansion (3C 310,
3C 314.1). Jaffe and Perola (1973) have discussed such models for radio sources
ejected by moving galaxies. Their detailed radio trail models have been applied quite
successfully to 3C 129 and NGC 1265.

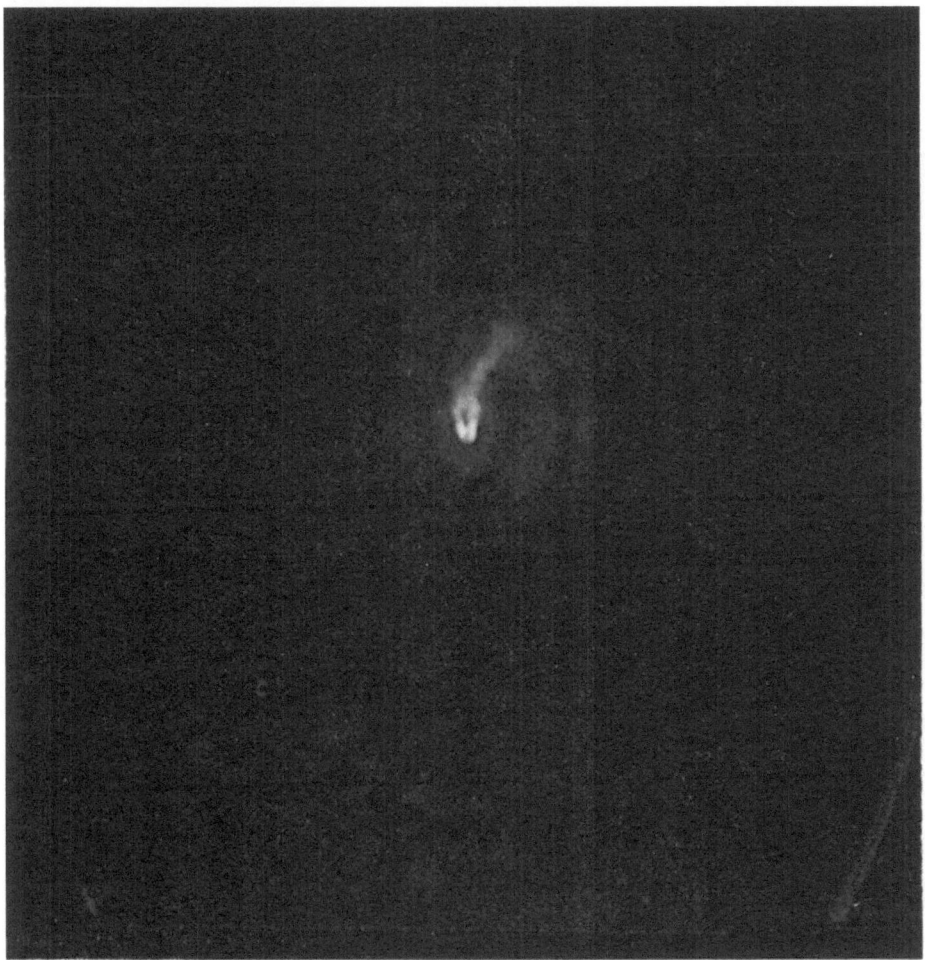

Fig. 4.   1.4 GHz radio photograph of NGC 1265 (from Miley *et al.*, 1972).

TABLE II

Similarities between 'normal' and 'head-tail' radio galaxies

Property	Normal	Head-tail
Morphology	Double	Double head + double tail
Polarization	Larger in more relaxed sources	Larger in more relaxed regions of source
Spectrum	Steeper for more relaxed sources	Steeper for more relaxed regions of source
Compact nuclear sources with flat spectra	Often present	Present in 3C 129, NGC 1265 and IC 310

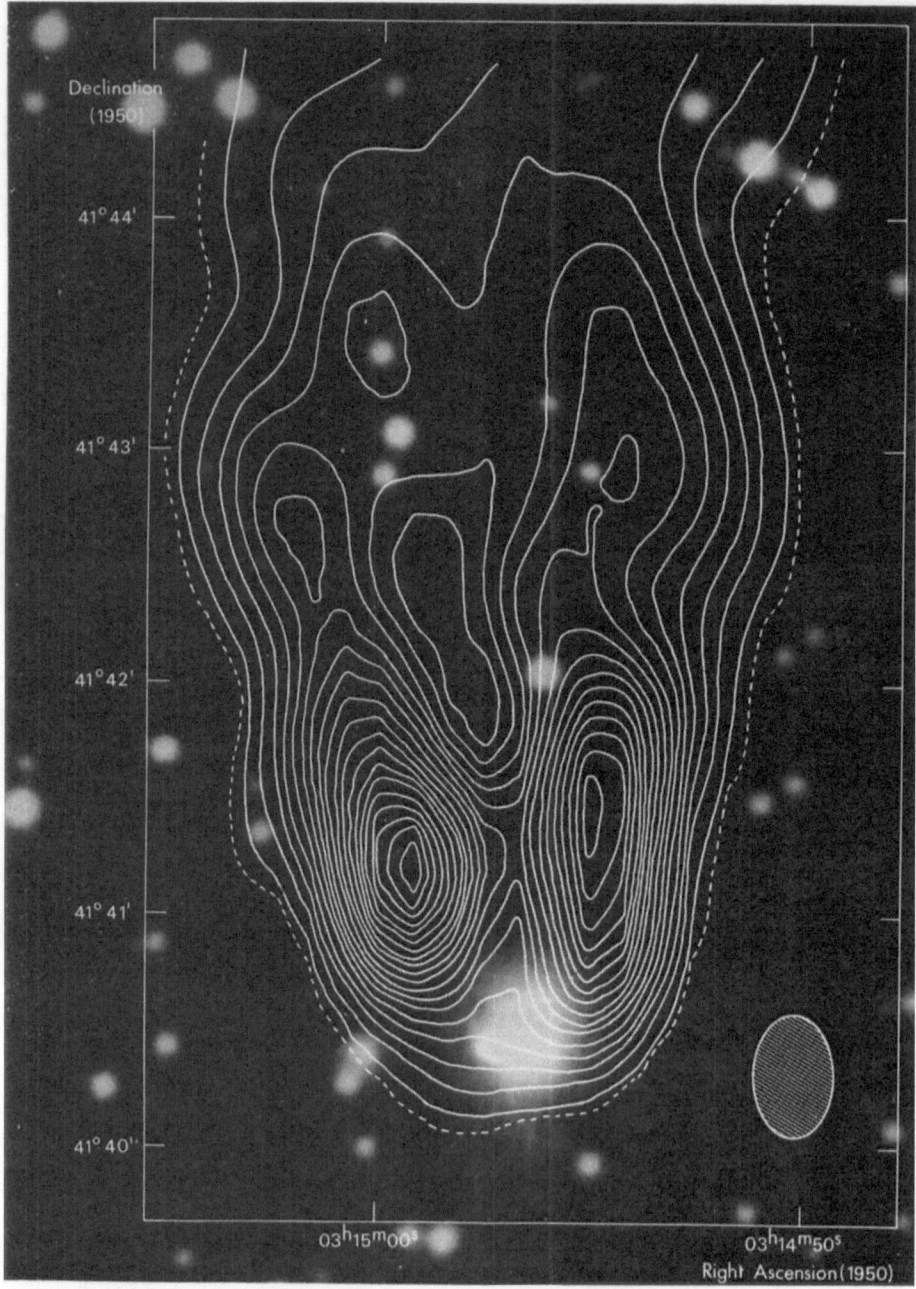

Fig. 5. 1.4 GHz contour map of the head of NGC 1265 superimposed on the blue Palomar Sky Survey print (from Wellington *et al.*, 1973).

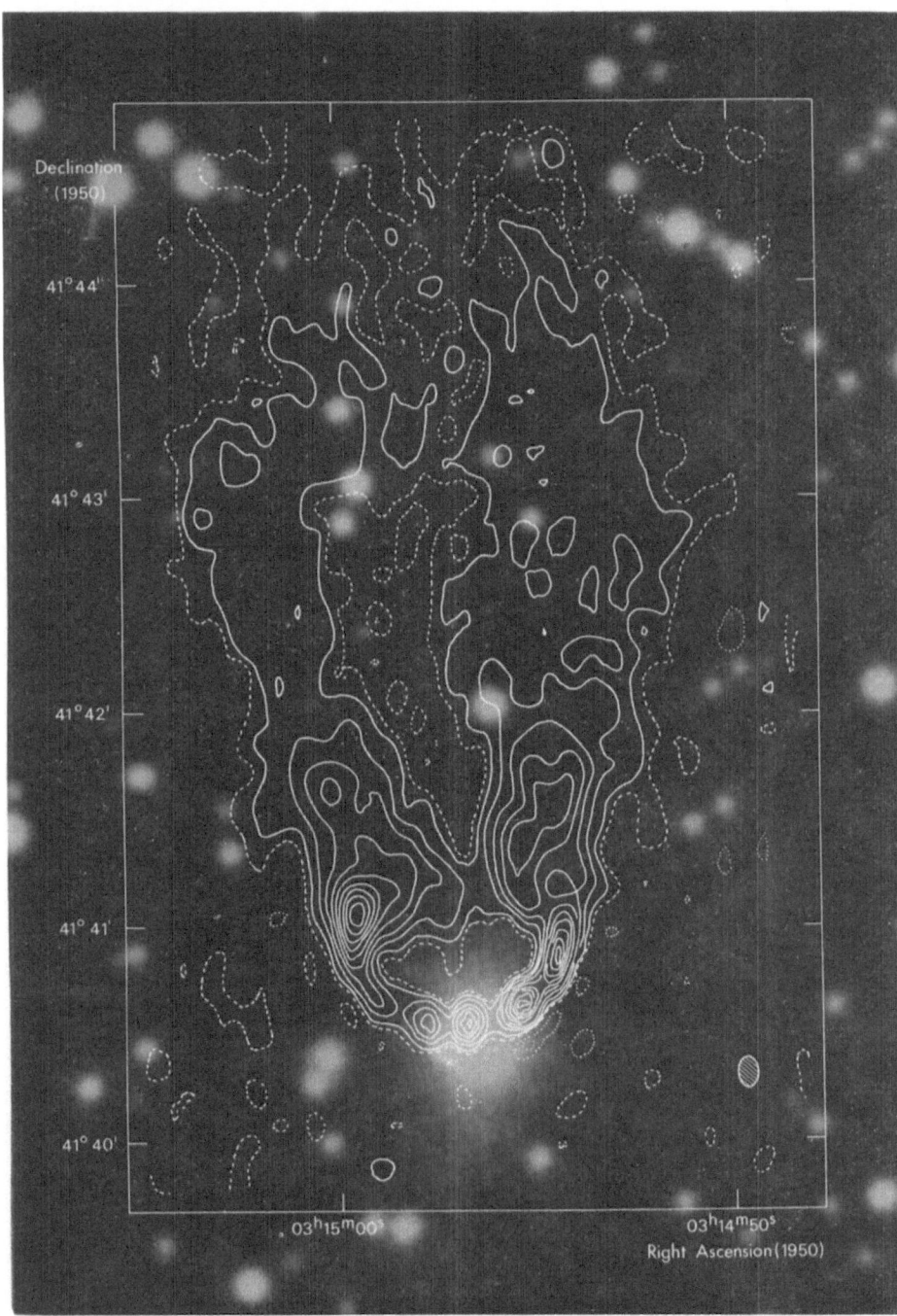

Fig. 6. 5 GHz contour map of the head of NGC 1265 superimposed on the blue Palomar Sky Survey print (from Wellington *et al.*, 1973).

G. K. MILEY ET AL.

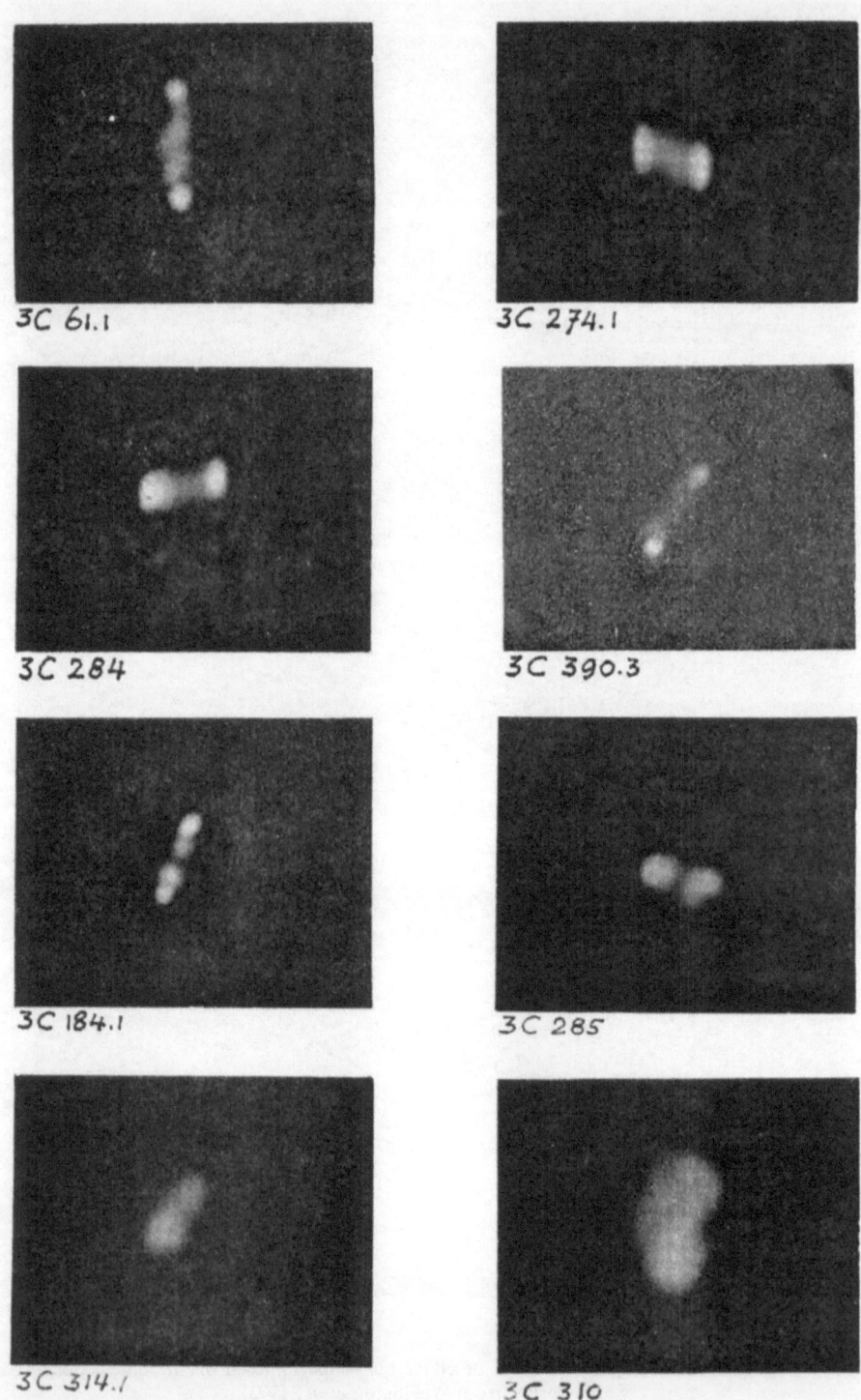

3C 61.1

3C 274.1

3C 284

3C 390.3

3C 184.1

3C 285

3C 314.1

3C 310

5. Conclusions

From the observations described above it is possible to draw a few tentative conclusions about the nature of head-tail galaxies:

(1) The 5 GHz map of NGC 1265 tells us that the nucleus of this galaxy explodes recurrently every few million years ejecting pairs of radio-emitting plasmoids. Energy releases $\gtrsim 10^{55}$ erg and ejection speeds of a few thousand kilometres per second are involved in each explosion. These ejection speeds are comparable with those now believed to occur in 'normal' radio galaxies (Mackay, 1973).

(2) The polarization and spectral observations suggest that the plasmoids evolve in a basically similar way to that of normal double radio sources, though possibly modified by the magnetosphere of the parent galaxy (Jaffe and Perola, 1973).

(3) The compact components which have been detected in NGC 1265 (Wellington et al., 1973), 3C 129 (Riley, 1973) and IC 310 indicate that between major eruptions the nuclei of the galaxies are actively 'simmering'.

The existence of radio trails in clusters provides convincing evidence for the presence of an intra-cluster medium. By assuming that the ram pressure exceeds the internal pressure within the source we can derive a lower limit for the density of the medium. The observations imply densities of at least 10^{-28} g cm^{-3} (Miley et al., 1972; Jaffe and Perola, 1973).

Detailed 5 GHz polarization maps of several head-tail sources are now being reduced. 5 GHz is a sufficiently high frequency that Faraday rotation should be small and the position angles close to those of the intrinsic electric vectors. Preliminary results on NGC 1265 show that the magnetic field in the tail is aligned *along* the direction of motion. This suggests that the magnetosphere drags behind the galaxy as suggested by Jaffe and Perola. Spectral comparisons should provide further insight into the physical processes involved. We have seen that such observations can yield valuable information on the evolution of radio sources and on the conditions within clusters.

References

Hill, J. M. and Longair, M. S.: 1971, *Monthly Notices Roy. Astron. Soc.* **154**, 125.
Jaffe, W. J. and Perola, G. C.: 1973, *Astron. Astrophys.* **26**, 423.
Mackay, C. D.: 1973, *Monthly Notices Roy. Astron. Soc.* **162**, 1.
Miley, G. K., Perola, G. C., van der Kruit, P. C., and van der Laan, H.: 1972, *Nature* **237**, 269.
Miley, G. K.: 1973, *Astron. Astrophys.* **26**, 413.
Miley, G. K. and van der Laan, H.: 1973, *Astron. Astrophys.* **28**, 359.
Riley, J.: 1973, *Monthly Notices Roy. Astron. Soc.* **161**, 167.

←
Fig. 7. 1.4 GHz radio photographs of eight simple double radio galaxies arranged roughly in order of the importance of the high brightness outer edges (from Miley and van der Laan, 1973). From top to bottom the pairs are 3C 61.1 and 3C 274.1, 3C 284 and 3C 390.3, 3C 184.1 and 3C 285, 3C 314.1 and 3C 310.
The brightest spot on each photograph is normalized to the maximum intensity on the map.

Ryle, M. and Windram, M. D.: 1968, *Monthly Notices Roy. Astron. Soc.* **138**, 1.
Wellington, K., Miley, G. K., and van der Laan, H.: 1973, *Nature* **244**, 502.
Willson, M. A. G.: 1971, *Monthly Notices Roy. Astron. Soc.* **151**, 1.

DISCUSSION

Longair: Some high-resolution (6″ arc) observations of sources with radio trails were published several months ago by Julia Riley (*Monthly Notices Roy. Astron. Soc.* **161**, 167, 1973). These indicate that 3C 129.1 is not a radio trail but a rather compact quadruple source centred on the galaxy. In 3C 129 there is a compact source associated with the galaxy, and complex structure in the high brightness regions at the beginning of the tail. It seems therefore that the properties of the trails are not as uniform as implied by Miley. For example, the radio spectral index of the trail in NGC 1265 does not steepen as it does in 3C 129. A catalogue of the differences is given in Riley's paper.

The existence of these compact sources indicates that the galaxies associated with the radio trails are active and the source of energy for the trails. One of the main reasons why we originally proposed that there might be an interaction in these systems is that every time we observe a radio trail, we observe an active radio galaxy of comparable radio luminosity in the same cluster. We do not find examples of isolated radio trails. This feature has still to be explained.

Miley: One is, of course, more likely to find an active galaxy in a cluster than in the general field. One of the problems is that one can only look in sufficient detail at comparatively nearby clusters and the statistics are poor.

Regarding 3C 129.1. I believe that this is indeed a head-tail source seen in projection. Riley's data merely imply that the head is complex, as is also the case for NGC 1265.

Longair: Since we are thinking mainly of radio sources of comparatively low luminosity, there are about 30 cluster sources in the 3C and 4C catalogues. In this connection can I ask Wal Sargent if there are redshifts for 3C 129 or 129.1 yet?

Sargent: No, I have not been able to determine the redshifts of 3C 129 and 3C 129.1 although I have observed both objects. Neither galaxy seems to have emission lines.

E. M. Burbidge: I assume that the density of the intra-cluster gas is too high for it to be cold gas; if it is hot gas, is the density consistent with X-ray measurements?

Miley: The value we derive is consistent with the X-ray measurements and implies temperatures of about 10^7 K.

Longair: The UHURU observations show that X-ray emission from the Perseus cluster is concentrated around NGC 1275 and does not extend as far as NGC 1265. There is also no definite evidence that the spectrum of the emission is that of thermal bremsstrahlung. It is therefore dangerous to use this X-ray emission as evidence for intergalactic gas in the vicinity of NGC 1265.

NEUTRAL HYDROGEN AT LARGE DISTANCES
FROM PARENT GALAXIES

R. D. DAVIES

University of Manchester, Nuffield Radio Astronomy Laboratories, Jodrell Bank, U.K.

Abstract. High sensitivity measurements in the vicinity of nearby galaxies have shown the existence of appreciable quantities of neutral hydrogen at large distances (2 to 5 Holmberg radii) from the centre. The galaxies studied were the M81/M82/NGC 3077 system, M31, M33, IC 342 and M51. In all cases the gas is at velocities consistent with being bound to the parent galaxy. Of these the M81 system contains the most circum-galactic neutral hydrogen. In several galaxies there is as much neutral hydrogen outside the Holmberg dimensions as inside.

1. Introduction

Neutral hydrogen surveys of nearby spiral galaxies show that the main neutral hydrogen concentrations lie in the regions of the well-developed spiral arms. Both pencil-beam and aperture synthesis observations place the bulk of the neutral gas within the galactic dimensions given by Holmberg (1958). There are, however, plausible reasons why neutral hydrogen may be more broadly distributed than the luminosity in a galaxy:

(a) Gas may lie in parts of the disk where the density is too low to allow star formation, a situation already known to exist in the Sd galaxy NGC 6822. This could be primordial gas with high angular momentum.

(b) Gas may be falling into a galaxy, either from intergalactic space or after ejection from the active nucleus of the galaxy itself (Oort, 1969).

(c) The gas may be intracluster gas which is gravitationally bound to one of the cluster galaxies.

(d) The gas may be torn from the inner spiral arms by the tidal interaction with a neighbouring galaxy. Tails and bridges which are formed in this way can lie far outside the main spiral structure (e.g. Toomre and Toomre, 1972).

The results described here bring together evidence obtained at Jodrell Bank for neutral hydrogen outside the optical confines of some nearby spiral galaxies.

2. The Observing Techniques

All the observations were made using a beam-switching technique with the Mark I 250 ft radio telescope, the system temperature being about 125 K. The signals were processed by a 256 channel autocorrelation receiver having an overall bandwidth of 5.0 MHz, which corresponds to a velocity range of 1060 km s^{-1}, and a velocity resolution of 7 km s^{-1}. Observations of IC 342 were taken with the Mark I radio telescope which was illuminated to give a beamwidth of $13' \times 17'$; the inner sidelobes were then about 5% of the main beam. All other observations were taken with the telescope after resurfacing, when the beamwidth was $12' \times 12'$ and the extended sidelobes less than 1%.

John R. Shakeshaft (ed.), The Formation and Dynamics of Galaxies. 119–128. All Rights Reserved.
Copyright © 1974 by the IAU.

3. The M81 Group

The M81 group includes the two I0 galaxies M82 and NGC 3077 within 1° of M81. Roberts (1972) has already shown that there is an extensive envelope of neutral hydrogen surrounding M81 which appears to form a bridge to M82. In the present survey, observations were taken every 6' in both RA and Dec over the whole region in which the aerial temperature was more than 0.2 K and Figure 1 shows the distribution of hydrogen spectra. Hydrogen is found in an area within a radius of 1°2 (70 kpc) of the centre of M81.

An obvious feature of the distribution is the extension of neutral hydrogen beyond

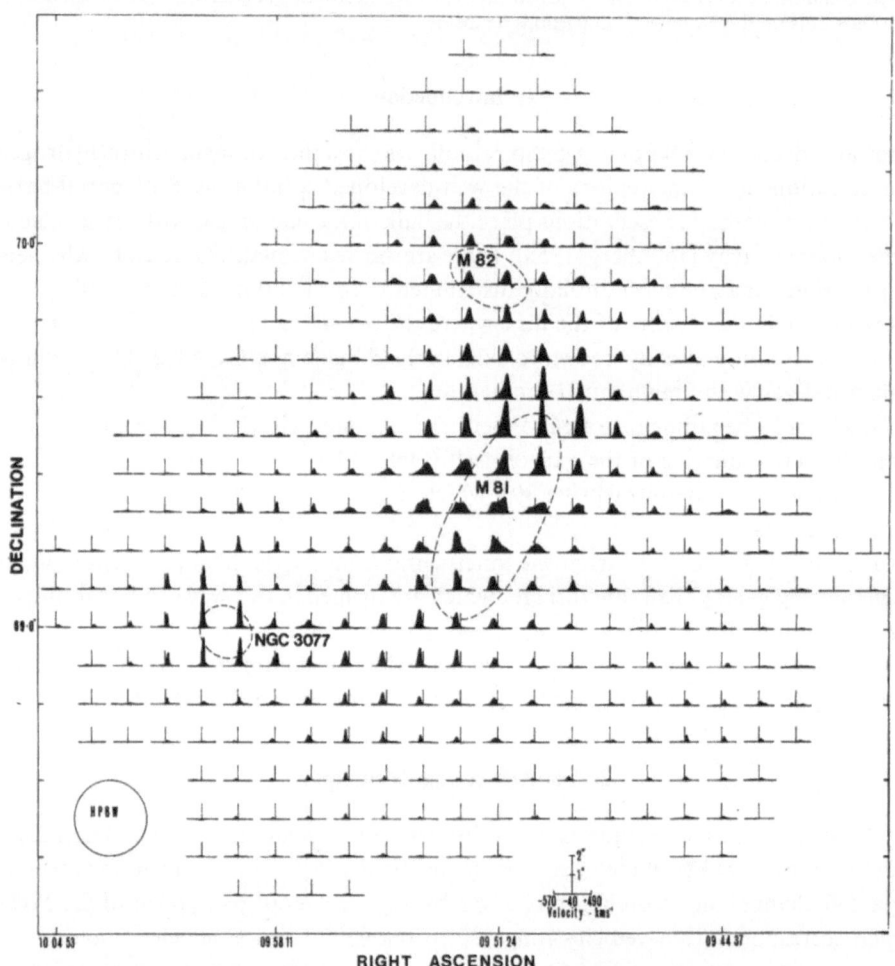

Fig. 1.   Neutral hydrogen spectra taken in the M81/M82/NGC 3077 group. The central velocity of each spectrum is −40 km s⁻¹ relative to the Sun; each spectrum extends ±530 km s⁻¹. The survey was continued in each direction until the signal fell to below 0.2 K in brightness temperature. Holmberg optical dimensions for each galaxy are shown by a broken line.

the Holmberg dimensions which are shown dashed in Figure 1. The hydrogen lies largely in extended envelopes around M81, M82 and NGC 3077 but a neutral hydrogen bridge connects M81 with its two companion galaxies. A further interesting feature is the presence of a neutral hydrogen companion on the SW side of M81 which can be traced over an area of $0°7 \times 0°5$ situated $0°7$ SW of the centre of M81. The gas in this companion is at a positive velocity relative to that in adjacent areas, which is rotating with M81.

It is believed that the distributed neutral hydrogen in the M81 group is indeed associated with the group and is not either an artefact of the observing system or foreground galactic emission. Emission in sidelobes of the beam would produce broad spectra at a level of a few hundredths of a degree, insignificant in comparison with the signals observed in Figure 1. The emission is very unlikely to be foreground emission since it follows the velocity distribution expected for M81, is centred on M81 and is not seen in comparison areas $2°$ away. The SW companion has a velocity of $+50$ km s^{-1} which is an unlikely velocity in this part of the sky ($l = 142°1$, $b = 40°9$).

The velocity field of the neutral hydrogen is significant. The bulk of the gas is at velocities which might be expected on the basis of a plausible rotation curve for M81; the gas near the NW major axis is at positive velocities and that near the SE major axis is at negative velocities relative to the centre. Thus, apart from the gas near NGC 3077 and the SW hydrogen companion, it appears to be co-rotating with M81 and is nearly in the same plane as M81.

A preliminary analysis of the data indicates that there is about twice as much neutral hydrogen outside the Holmberg dimensions of the three galaxies as is inside. All the bright neutral hydrogen emission of M81 mapped with the Westerbork synthesis telescope lies inside the Holmberg dimensions.

The present observations give a clear picture of the dynamics of NGC 3077. Its systemic velocity is $+15$ km s^{-1} and its major axis lies at p.a. $= 60°$. The neutral hydrogen diameter is $14'$ arc (as defined by the line of centroids of the iso-velocity maps of its emission) which is larger than the Holmberg dimensions of $8'8 \times 8'0$ arc.

The SW neutral hydrogen companion has a neutral hydrogen mass of $\sim 5 \times 10^7 \ M_\odot$ if it is situated at the distance of M81 (3.3 Mpc) and, with its observed velocity spread ($\sigma \approx 40$ km s^{-1}) and radius (4.3 kpc), does not appear to be gravitationally stable: a mass of $\sim 1 \times 10^9 \ M_\odot$ would be required to close such a system. It may be that the velocity dispersion is produced in localized knots which are small enough to be stable, and the whole system is rotating at some lower velocity.

The neutral hydrogen data are not alone in revealing matter outside the Holmberg diameters of the three objects. Two optical companions are seen in the close vicinity of M81, one $10'$ to the east, close to the minor axis, and another $30'$ to the south. Both lie within the neutral hydrogen envelope of M81. Arp (1965) has also shown the presence of faint optical emission beyond the Holmberg dimensions of M81.

4. M31

In the course of a study of the early-type companions of M31 a neutral hydrogen

object was discovered which may be a companion of M31. Its centre is at RA=
$=00^h35^m10^s$, Dec$=42°07'$ (1950) which places it $1°5$ from the centre and close to the
minor axis of M31. It has diameters at half-intensity of $\sim 0°6 \times 0°4$, with a major axis
approximately north-south. The cloud has a maximum aerial temperature of 0.5 K
and a mean velocity of -451 km s^{-1} relative to the Sun; for comparison M31 has a
systemic velocity of ~ -300 km s^{-1}. The full width at half-maximum intensity of the
spectrum is 22 km s^{-1} which corresponds to a velocity dispersion of 10 km s^{-1}.

The neutral hydrogen line integral through the centre of the cloud is 3×10^{19} cm^{-2},
implying a neutral hydrogen mass of 8×10^6 $M_\odot$ if the distance is that of M31. With
the value of velocity dispersion and diameter given above, this mass will stabilize the
cloud gravitationally.

All the evidence is consistent with the hypothesis that this cloud is a gravitationally
bound companion of M31. Its velocity and position would allow it to be in a bound
orbit. Alternative explanations do not appear attractive. It may be thought that it
could be a high velocity cloud (HVC), but its velocity of -451 km s^{-1} would be the
most extreme of any known. Another possibility is that it is part of the Magellanic
Stream (this volume p. 367), but again this seems unlikely since (a) the nearest part
of the stream lies $\sim 30°$ away and (b) the M31 cloud lies $20°$ off the projected axis of
the stream.

5. M33

A series of neutral hydrogen spectra were taken around the outside edge of the aperture synthesis map of M33 made by Wright et al. (1972). Neutral hydrogen had already
been shown to exist outside this area by de Jager and Davies (1971), Gordon (1971) and
Wright (1973). The present observations (Figure 2) cover an even wider area and are
at a higher sensitivity than those published previously. Neutral hydrogen with
$T_b \geqslant 0°2$ was found in all positions sampled within $1°$ of the centre of M33, but at
more than $1°3$ from the centre $T_b < 0.1$ K. The neutral hydrogen in the outer regions
is strongest at two positions (RA$=0^m0$, Dec$=-1°0$; RA$=-3^m0$, Dec$=+0°7$)
which lie within the tilted areas of M33 (de Jager and Davies, 1971).

The velocity of the gas in the vicinity of M33 is close to that which would be expected from an extrapolation of the isovelocity lines of the emission in the inner
regions. Consequently it is concluded that this material is rotating with M33 and is a
tenuous outer extension of the disk. The neutral hydrogen line integrals of the profiles
more than $0°3$ from the outer contour shown in Figure 2 range from 2 to 7×10^{19} cm^{-2}.
These correspond to a range of 0.02 to 0.06 cm^{-3} for the density if the gas is concentrated in a disk 200 pc thick.

6. IC 342

IC 342 is an Sc galaxy lying close to the galactic plane ($l=138°2$, $b=10°6$) where it is
heavily obscured by galactic dust. Neutral hydrogen was known to exist outside the
optical extent of $\sim 40'$ (corrected for 2.2 mag obscuration) as given by Rogstad et al.

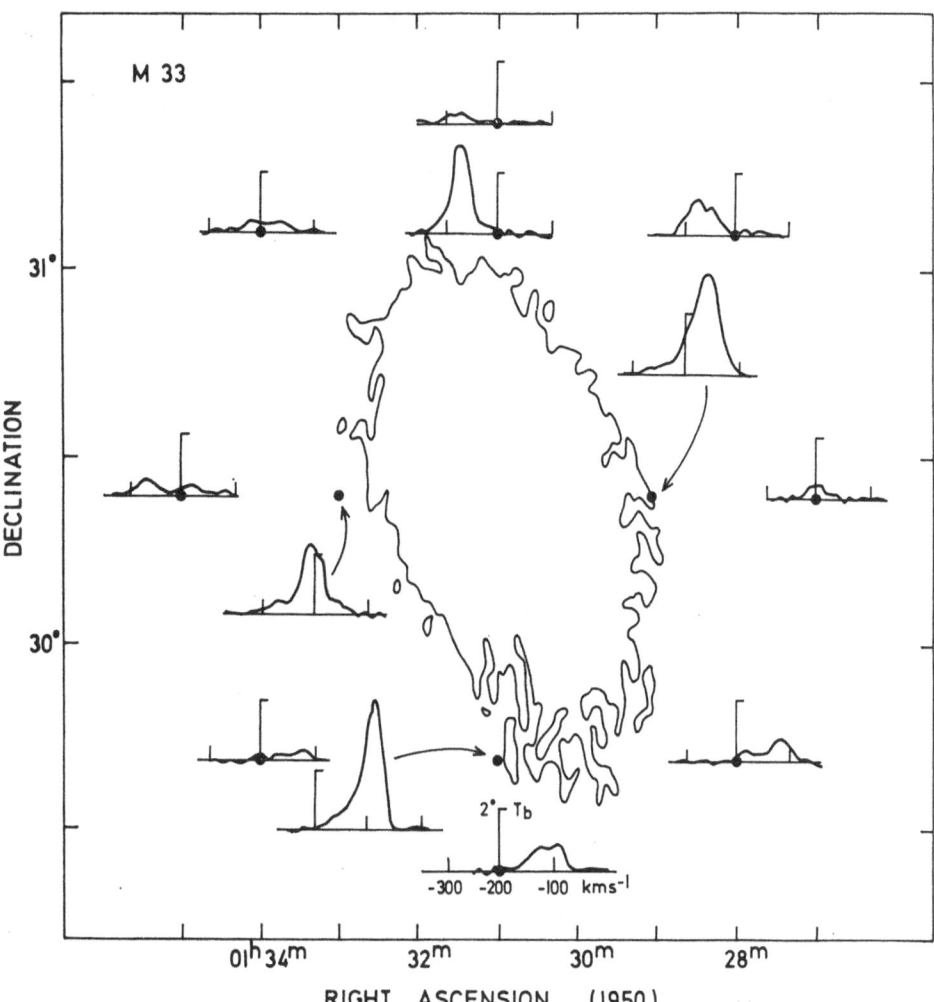

Fig. 2.   Neutral hydrogen spectra taken in the vicinity of M33 with an angular resolution of 12′ × 12′. The continuous contour is the outer contour (5×10^{20} cm^{-2}) of the aperture synthesis map of Wright *et al.* (1972). The vertical axis on each spectrum is drawn at a velocity of -200 km s^{-1} relative to the Sun, and the location of the beam centre is shown by a filled circle.

(1973). For example the aperture synthesis map published by Rogstad *et al.* shows neutral hydrogen over an area 50′ × 40′.

Observations using the 250 ft radiotelescope show extensive neutral hydrogen emission outside the confines of the aperture synthesis map, particularly on the north-western side. The situation is illustrated in Figure 3. Heavy lines show the loci of the maxima seen in the maps at adjacent velocities. An inner circle follows the region of strongest emission. The outer emission on the NW side lies ~25′ from the centre. A single velocity map illustrates the presence of both the inner emission and the gas

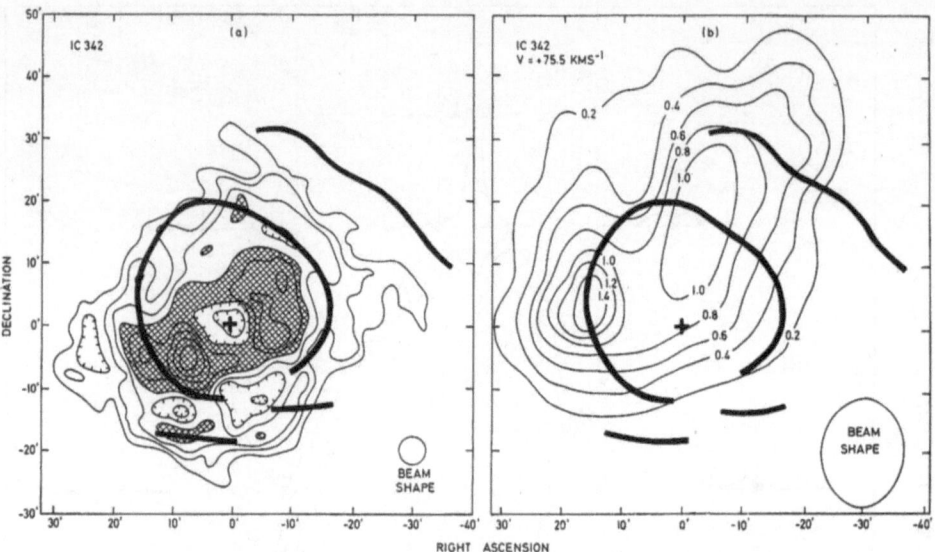

Fig. 3.   Neutral hydrogen in IC 342. (a) the light contours are the aperture synthesis map of Rogstad
et al. (1973) and the heavy lines are the loci of the centroids of isovelocity maps taken with a 13′ × 17′
arc beam. Neutral hydrogen will extend beyond these loci. (b) one constant velocity map of IC 342
superposed on the loci of centroids as described in (a). This illustrates the extent of the emission on
the NW side of IC 342.

on the NW side. Evidently this outer gas is in a distributed layer rather than in narrow
bright arms or else it would have been seen in the aperture synthesis maps. Its brightness temperature averaged over a 13′ × 17′ beam is about half that of the neutral
hydrogen in the inner area.

7. M51 (NGC 5194/5)

A survey of M51 has been made with the Mark IA radio telescope. Although the
telescope beamwidth (12′ × 12′) is comparable with the optical dimensions of the
galaxy (8.9 × 7.4) clear evidence was found for an extended neutral hydrogen component. This can best be seen on the SE side where the gas extends furthest from the
H I outline provided by the aperture synthesis map (Figure 4). The asymmetry of the
distribution is marked. An analysis of the observed spectra indicates that neutral
hydrogen extends 15′ from the centre of M51 at p.a. ∼ 120°. At this distance the mean
surface density is ∼ 2 × 10^{19} cm^{-2}. As in the case of IC 342, the gas in the outer
regions must be in a distributed form rather than in spiral arms or else it should have
been seen in the aperture synthesis map.

In the case of M51, gas is detected out to twice the Holmberg diameter
(8.9 × 7.4). The aperture synthesis dimensions given by Weliachew and Gottesman
(1973) are comparable with the Holmberg dimensions. Long exposure photographs
of M51 (Arp 1966) show optical luminosity out to ∼ 1.5 times the Holmberg dimensions. The neutral hydrogen in the outer envelope could well have been moved there

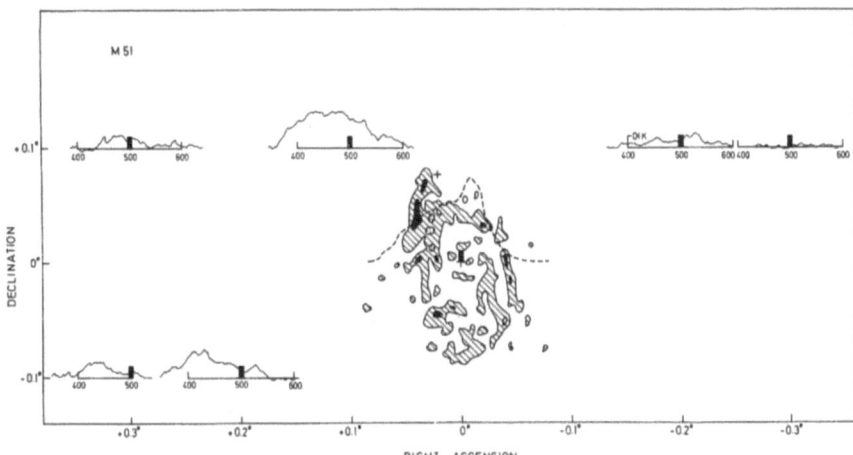

Fig. 4. Neutral hydrogen spectra taken with a resolution of $12' \times 12'$ in the vicinity of M51. The base of the filled rectangle represents the position of the beam centre for each spectrum and the height represents an aerial temperature of 0.1 K. The neutral hydrogen map taken with the Westerbork synthesis radio telescope is shown for comparison. Emission extends further to the east of M51 than to the west.

in the tidal interaction of NGC 5194 and 5195 as discussed by Toomre and Toomre (1972).

8. The Magellanic Clouds

Neutral hydrogen has been known for a long time to exist between the two Magellanic Clouds. No stars or dust are found in these regions (Hindman *et al.*, 1963). The recent observations by Mathewson, Cleary, and Murray (this volume, p. 367) have shown the existence of substantial amounts of neutral hydrogen extending in streamers from the Magellanic Clouds over large arcs in the sky. They can be traced from $l = 90°$, $b = -30°$, past the south galactic pole to the vicinity of the Magellanic Clouds then on to $l = 310°$, $b = 0°$. The geometry of these streamers and their velocity distribution argue strongly that they are associated with the Magellanic Clouds and have most likely been drawn from the Clouds during a recent encounter with the Galaxy.

It is clear that neutral hydrogen is widely spread throughout the Magellanic Cloud-Milky Way system. The streamers extend for at least 50 kpc and possibly even further from the Magellanic Clouds. Although the greater part of this gas is in the streamers, an appreciable fraction is also in isolated clouds. A further contribution is in the region lying between the clouds. Figure 5 shows the velocity distribution in a declination cut through the inter-Cloud region at $RA = 03^h 20^m$, to be published by Davies, Murray, Mathewson and Cleary. This emission is typical of the area between the two clouds with a surface density of $\sim 2 \times 10^{20}$ cm^{-2}.

9. Conclusions

Evidence has been presented for a widespread distribution of neutral hydrogen around

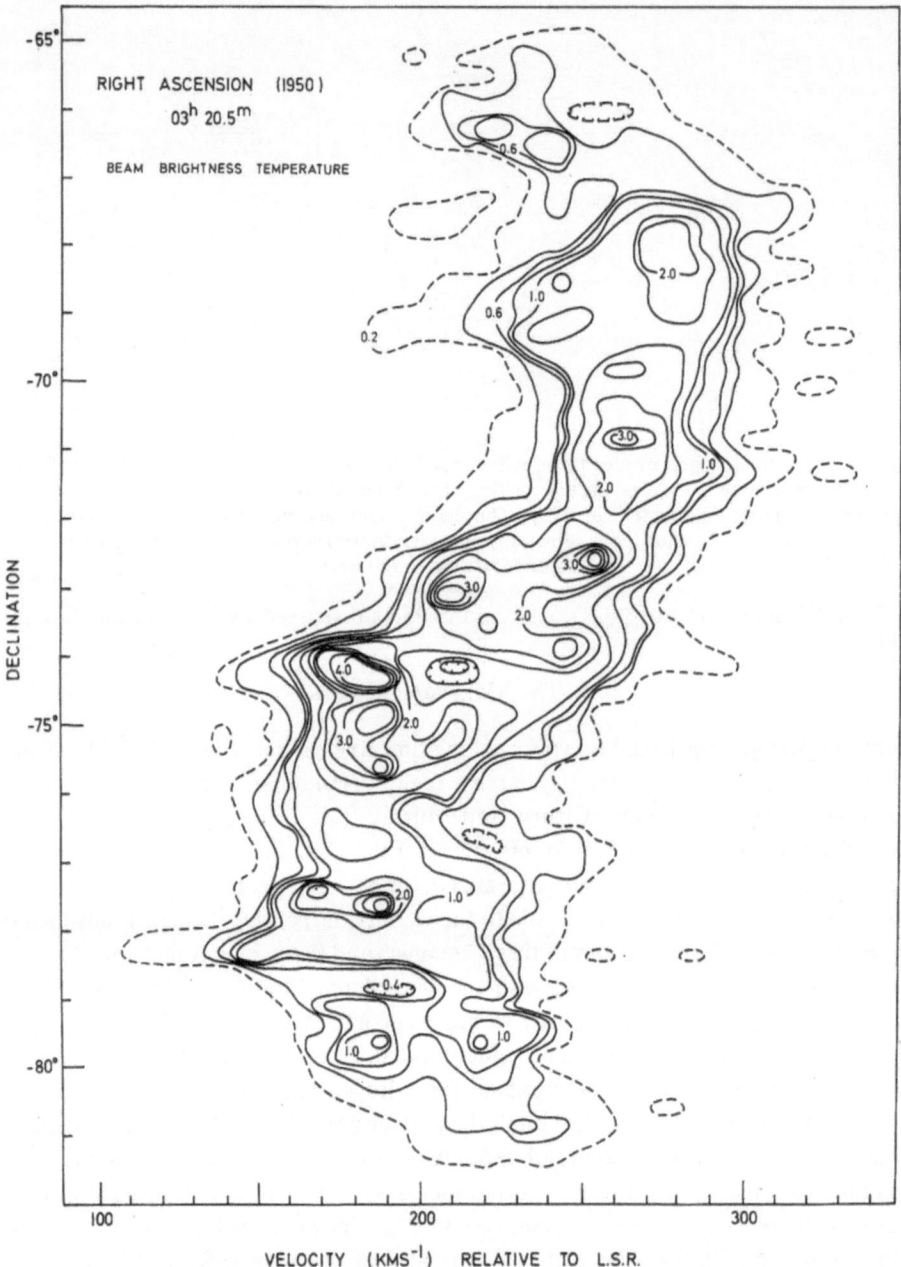

Fig. 5. A declination-velocity diagram taken at Right Ascension 03ʰ20ᵐ (1973.5); this is a line through the region between the two Magellanic Clouds. The diagram shows the extent of the bright neutral hydrogen emission. The systematic velocity trend with declination can also be seen.

some galaxies. In all cases where aperture synthesis maps have been made of these galaxies, the neutral hydrogen is found to extend further in the pencil-beam observations, which are more sensitive to extended features. This implies that the gas is distributed in broad features or in a more or less uniform layer outside the main spiral structure in these galaxies. The outer brightness levels in the pencil-beam surveys correspond to neutral hydrogen surface-integrals of $\sim 2 \times 10^{19}$ cm^{-2}.

In several cases the extended neutral hydrogen envelope may be causally related to the tidal interaction which is occurring between two galaxies at the present time or in the recent past. Galaxies in this category are M33 (see, for example, de Jager and Davies, 1971), the Milky Way-Magellanic Cloud system and the M51 (NGC 5194/95) system. The M81 system could conceivably be of this type but no strong evidence for a gravitational disturbance of the optical objects is found. It is not clear in detail how the gas could be perturbed into the outer parts of these galaxies, although as a general principle the material further from the centre of a subject galaxy is preferentially disturbed in a tidal interaction. Thus any material outside the main spiral structure will show more tidal distortion than the spiral arms themselves. Since the ratio of gas to star mass increases towards the outside edge of galaxies, the gas is likely to show the most severe tidal distortion.

The neutral hydrogen in the M81 system could have had quite a different history. It may never have been closely bound to a particular galaxy but may be orbiting in the general gravitational field of the cluster. The observations suggest that the gas is in a disk-like layer, since the velocity widths of the profiles are small and do not allow a large velocity component out of the plane of rotation of M81. Furthermore the mean velocity of the gas in each spectrum is comparable that with expected for galactic rotation.

Acknowledgements

I acknowledge the part played by M. W. Bright and R. J. Stephenson in obtaining and analyzing the data for IC 342 and the M81 group. M. N. Cleary, D. S. Mathewson and J. D. Murray collaborated with the author in observations of the Magellanic Cloud system with the Parkes 210 ft radiotelescope.

References

Arp, H.: 1965, *Science* **148**, 363.
Arp, H.: 1966, *Atlas of Peculiar Galaxies*, California Institute of Technology, Pasadena, U.S.A.
de Jager, G. and Davies, R. D.: 1971, *Monthly Notices Roy. Astron. Soc.* **153**, 9.
Gordon, K. J.: 1971, *Astrophys. J.* **169**, 235.
Hindman, J. V., Kerr, F. J., and McGee, R. X.: 1963, *Australian J. Phys.* **16**, 570.
Holmberg, E.: 1958, *Lund. Obs. Medd.*, Ser. II, No. 136.
Oort, J. H.: 1969, *Nature* **224**, 1158.
Roberts, M. S.: 1972, in D. S. Evans (ed.), 'External Galaxies and Quasi-Stellar Objects', *IAU Symp.* **44**, 12.
Rogstad, D. H., Shostak, G. S., and Rots, A. H.: 1973, *Astron. Astrophys.* **22**, 111.
Toomre, A. and Toomre, J.: 1972, *Astrophys. J.* **178**, 623.
Weliachew, L. and Gottesman, S. T.: 1973, *Astron. Astrophys.* **24**, 59.
Wright, M. C. H., Warner, P. J., and Baldwin, J. E.: 1972, *Monthly Notices Roy. Astron. Soc.* **155**, 337.

Wright, M. C. H.: 1973, *Astrophys. J.* **179**, 453.

DISCUSSION

Toomre: After this confirmation by Davies of Mort Robert's huge 21-cm disk around M81 and/or M82, those interested in numerology should recall that the *next* Messier galaxy, M83 = NGC 5236, was itself discovered about five years ago by B. M. Lewis (*Proc. Astron. Soc. Australia* **1**, 104, 1968) to possess a similar and possibly even larger exterior gas disk. Recent work by Rogstad (as yet unpublished) from Owens Valley not only confirms that disk but likewise provides much fascinating detail.

Lewis: How did you determine the systemic velocity of M82?

Davies: As the mean velocity of the 5 profiles centred on M82 and ± 6′ on each side.

Lewis: Are there any double-peaked profiles in the field around M82 or NGC 3077 which might distinguish the H I which belongs to each galaxy individually?

Davies: Some double peaked profiles were found in the region between M81 and NGC 3077 which help in separating the emission associated with these objects. No doubling was found between M81 and M82.

Allen: We have observed (paper in preparation) the edge-on system NGC 891 in the radio continuum and H I at 21 cm with the Westerbork Telescope in order to study the structure at large r and z. The angular resolution was 25″, and the optical image of the galaxy is about 12′ long. The radio continuum data show clearly a flattened radio halo extending up to 2′.5 (about 4 kpc) above the plane of the Galaxy along the rotation axis, and less extensive in the r direction than the optical image. The H I distribution is closely confined to the plane of the galaxy and extends slightly beyond the edge of the optical image in the r direction. If we were to view this galaxy more face-on, the size of the optical image would be smaller by a factor of about 1.5, and high-sensitivity H I observations would reveal H I far beyond the optical 'edge' of the Galaxy.

G. de Vaucouleurs: The effect of inclination on apparent diameters of galaxies was discussed at length by Heidmann, Heidmann and de Vaucouleurs (*Mem. Roy. Astron. Soc.* **75**, Parts 1–3, 1971). For the Reference Catalogue diameters: $\log D \simeq \log D$ (face-on) $+ 0.2 \log(D/d)$; for *isophotal* diameters at $\mu_B = 25.0$ mag (arc sec)$^{-2}$ the coefficient is more precisely 0.235.

Heidmann: Bottinelli (*Astron. Astrophys.* **10**, 437, 1971) has shown that, for galaxies of the same morphological type as M81, half of the hydrogen mass is contained within half the Holmberg radius. This result is in between what Allen just quoted for NGC 891 and Davies' value for M81.

THE EARLY HISTORY OF OUR GALAXY: DYNAMICS

K. C. FREEMAN

Mount Stromlo and Siding Spring Observatory, Research School of Physical Sciences, Australian National University, Canberra, Australia

Abstract. Dynamical aspects of the early history of our Galaxy are discussed, with emphasis on the two-component nature of disk galaxies. Topics include the Eggen, Lynden-Bell and Sandage picture of the Galaxy's collapse, dynamical problems of the disk, and the bulge component.

1. Introduction

The conventional picture of galaxy formation is that the protosystem, with dynamical time $\sim 10^8$ yr, condensed out of the intergalactic medium and became a stellar system. A key problem in galactic dynamics is to describe and understand how all this happened. There are two complementary approaches to the problem. One is the work on galaxy formation from the expanding Universe. The other is through galactic dynamics, probing backwards to the collapse epoch from what we know now about the internal kinematics and the chemical and mass distributions in galaxies. This knowledge is a composite of detailed data for the Milky Way, plus large-scale data for external systems, and we should not use either in isolation.

We start with the two component nature of the disk galaxies (Figures 1 and 2). We accept the two component picture because, as de Vaucouleurs pointed out (p. 1), the ratio of these two components in a particular galaxy can have any value from zero to infinity. The *disk* we believe is roughly in centrifugal equilibrium, and chemically fairly uniform: probably the only useful information left in the disk from the collapse phase is the angular momentum $\mathscr{H}$ and perhaps the mass-angular momentum distribution $\mathfrak{M}(h)$ (the mass with angular momentum per unit mass $< h$). The *bulge* or halo component has a wide range in chemical abundance, and a significant part of its support is from random stellar motions. The correlation observed of chemical and kinematical properties for halo stars in the Milky Way gives us hope that some residual information about the collapse phase remains.

We are searching for a formation picture: we do not have one yet, and we will be no closer at the end of this talk. The dynamical and chemical aspects of galaxy formation are intertwined: I will attempt to discuss mainly dynamical topics and Peimbert will talk about the chemical aspects.

2. Eggen, Lynden-Bell and Sandage (1962)

This classic paper is largely responsible for our present picture of galaxy formation. Eggen *et al.* (ELS, 1962) collected data for 221 well-observed nearby dwarf stars; they showed that e (orbital eccentricity), W (the velocity in the z-direction) and h (the angular momentum per unit mass) correlate with the ultraviolet excess $\delta(U-B)$, as shown in

John R. Shakeshaft (ed.), The Formation and Dynamics of Galaxies. 129–137. *All Rights Reserved.*
Copyright © 1974 *by the IAU.*

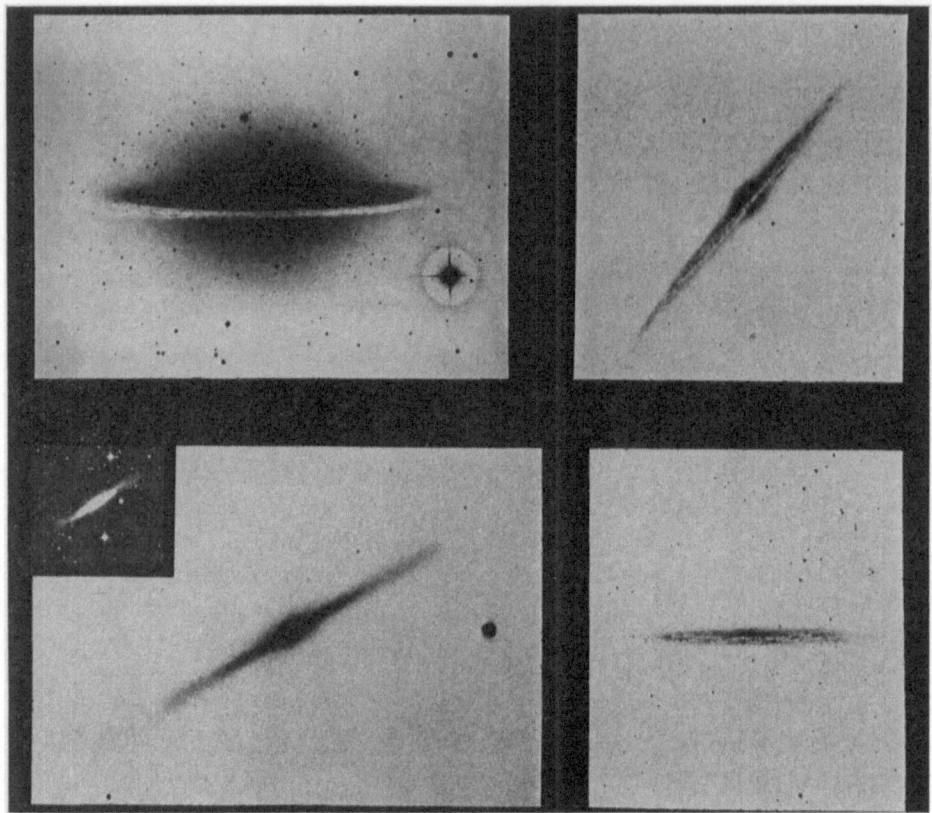

Fig. 1.   Four edge-on disk galaxies from the *Hubble Atlas*, to show their two-component nature.

Figure 3. Ultraviolet excess is a measure of metal abundance, and so presumably of age. In particular, for $\delta(U-B)<0.15$, $e<0.5$, $|W|<50$ km s^{-1} and $h>1500$ kpc km s^{-1} while stars with $\delta(U-B)>0.15$ have large e, low h and a wide range of $|W|$ values. They infer that the younger objects, with smaller $\delta(U-B)$, formed near the plane while the older metal-weak stars formed at any height: this leads to a formation picture in which the Galaxy collapsed to a disk, during or after the birth of the oldest stars. They also infer that this collapse was rapid; because e is an adiabatic invariant, a slow collapse from a rotating system would not produce the highly eccentric orbits seen in Figure 3. The scale of the collapse, in z, can be estimated from the $|W|-\delta(U-B)$ diagram. It is at least a factor 25, from $|z|_{max}\approx10$ kpc for the oldest stars to $|z|_{max}\approx400$ pc for the younger ones.

There are some uncertainties in the ELS picture. A rapid collapse is not the only way to produce the observed correlation of kinematical and chemical properties. Also, it is not clear how unique the $\delta(U-B)$-age relation is, even for a sample of nearby stars. For example, star formation would probably begin earlier in the dense inner parts of the protogalaxy, and there could have been several generations of star formation with

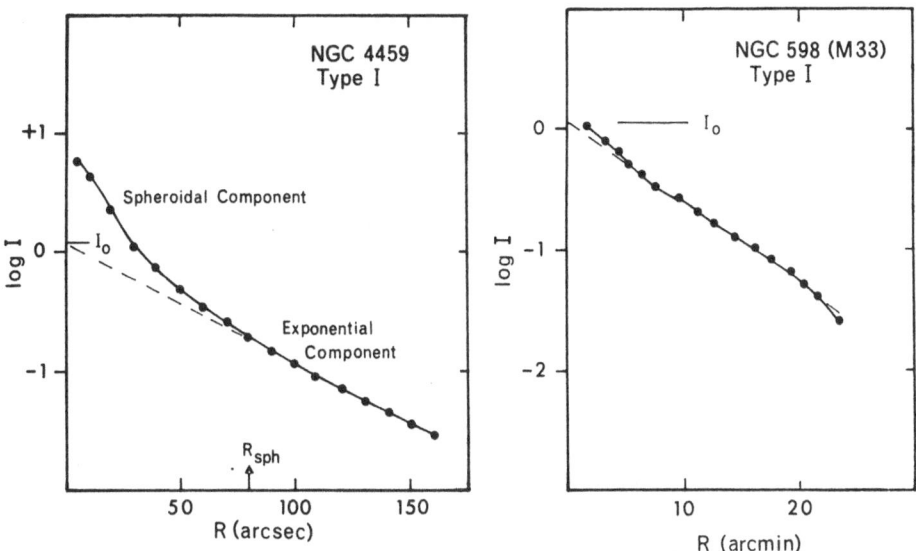

Fig. 2.   Two examples of luminosity profiles. I is the surface brightness and R the radius. Note how M33 has no significant spheroidal or bulge component.

consequent chemical enrichment before star formation even began in the outer parts of the system. This picture is reinforced by our knowledge that there is a chemical gradient in the halo (see Kinman, 1959a), and means that some *metal-rich* objects, formed in the inner parts of the system, could be *older* than some *metal-weak* stars seen now in the solar neighbourhood. This may be more plausible than having stars forming first in the most diffuse outer parts of the protosystem. We could even extend the picture to suggest that the low $\delta(U-B)$, low velocity objects of Figure 3 could have formed locally in the dense disk *before* the high $\delta(U-B)$ halo objects formed. In summary, I am saying that $\delta(U-B)$ could be a measure not of age only, but of age and place of formation.

Returning briefly to the possibility that some disk stars formed before the high-velocity halo stars, I want to point out that there is no obvious age problem in this concept. Sandage's (1970) estimate for globular cluster ages is $(10 \text{ to } 12) \times 10^9$ yr, while the old disk cluster NGC 188 has an estimated age of $(8 \text{ to } 10) \times 10^9$ yr: Eggen (1970) has found evolved disk stars which are certainly older than NGC 188. We are haggling about what happened in a time interval which is probably only a few $\times 10^8$ yr: nevertheless, details of the *order* of events within this short time interval are vital for understanding the galaxy formation process, so it is worth questioning whether the metal-weak stars really did form first.

Taking this argument one stage further, consider now galaxies like M33 and the LMC, in which the disk is dominant and the bulge component very weak indeed. In the ELS picture, the chemical enrichment of the disk stars occurs through element building in the first generation (halo) stars. However, in M33 and the LMC, the chem-

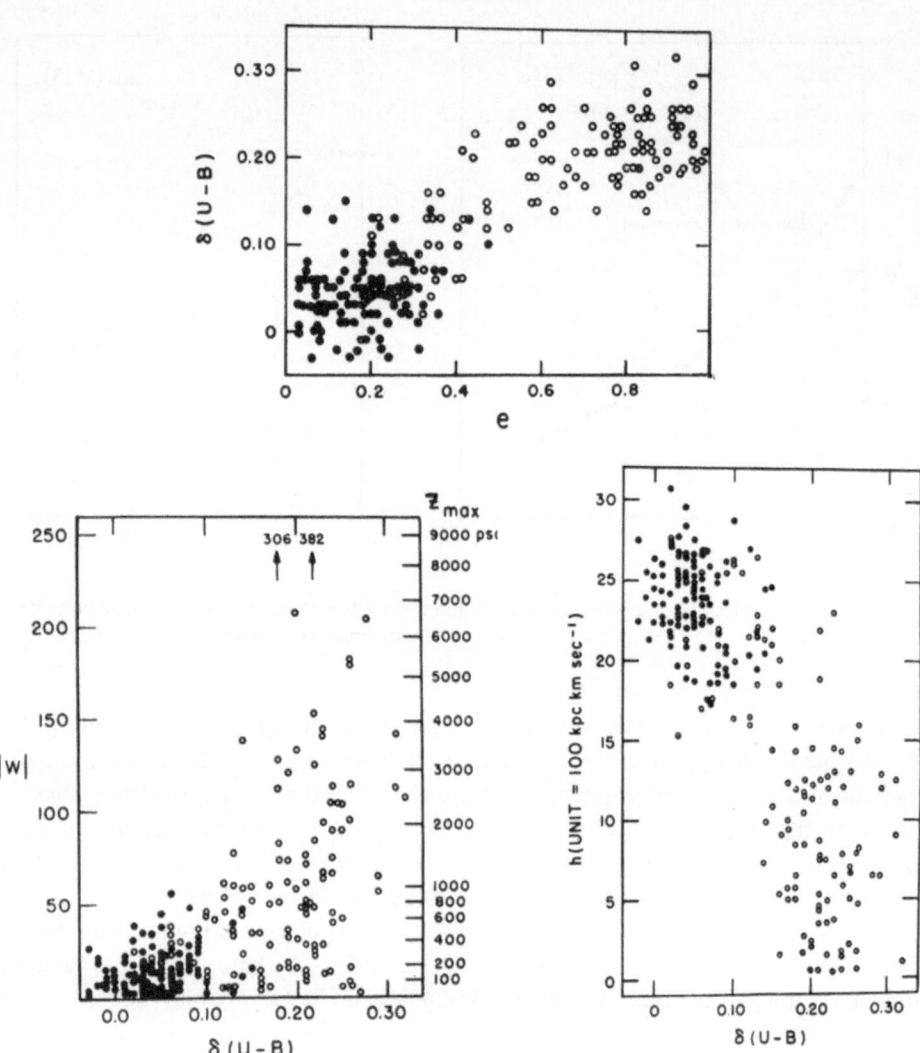

Fig. 3. Eccentricity, W and h against $\delta(U - B)$ for the ELS stars.

ical abundances in disk objects appear to be fairly close to the solar value: because their bulges are so weak, the disks in these galaxies probably processed their own heavier elements. Given that these disks can do it, maybe the disk of the Milky Way also did its own chemical enrichment independent of the halo population.

3. Dynamical Problems of the Disk

Now we consider the disk of the disk-like galaxies. Much of its interest for galaxy form-ation lies in its angular momentum content and distribution, which reflects that of the protocloud *if* the collapse was axisymmetric. Discussion of the disk usually begins from a stellar disk near centrifugal equilibrium. While there is plenty of evidence that the extreme Population I component in the disks of many galaxies is close to circular motion, such evidence for the massive *old disk* component is meagre, and depends on the small asymmetric drift of the old disk stars in the solar neighbourhood. Anyhow, for a start we consider the disk to be cold and flat.

Surface photometry of disk galaxies shows that the disk has *almost always* the exponential surface-brightness distribution seen in Figure 2. From the uniformity of the colour in the disk, it seems likely that the surface *density* distribution $\mu(R)$ is also exponential, i.e. $\mu(R) = \mu_0 e^{-\alpha R}$ in the disk. The origin of this characteristic distribution is not yet really understood. It has two interesting features for formation theory: (i) The density scale μ_0 is observed to be approximately constant from galaxy to galaxy (Freeman, 1970): since the disk mass $\mathfrak{M} = 2\pi\mu_0/\alpha^2$ and its angular momentum $\mathscr{H} = 1.109 \, (G\mathfrak{M}^3/\alpha)^{1/2}$, it follows that $\mathscr{H} \propto \mathfrak{M}^{7/4}$: since $\mathscr{H}$ and $\mathfrak{M}$ are probably conserved through the collapse, this 7/4 law would hold also for the protodisks. Note that the exponent is close to the value 5/3 which follows from elementary theory. (ii) For the exponential disk alone, the mass-angular momentum distribution $\mathfrak{M}(h)$ is almost identical to that for a rigidly rotating sphere of uniform density (or to the rigidly rotating disk D with surface density $\mu(R) = \mu(0) \, (1 - R^2)^{1/2}$ that results from projecting this sphere on to the plane of rotation). Figure 4 shows $\mathfrak{M}(h)$ for the two disks and also their surface density distributions: the two disks have the same $\mathfrak{M}$ and almost the same $\mathscr{H}$ and $\mathfrak{M}(h)$. This shows how $\mathfrak{M}(h)$ alone does not uniquely define the disk, as Mestel (1963) has pointed out: the particular $\mathfrak{M}(h)$ alone does not establish the exponential distribution. Two comments on $\mathfrak{M}(h)$: (a) If $\mathfrak{M}(h)$ is primordial and

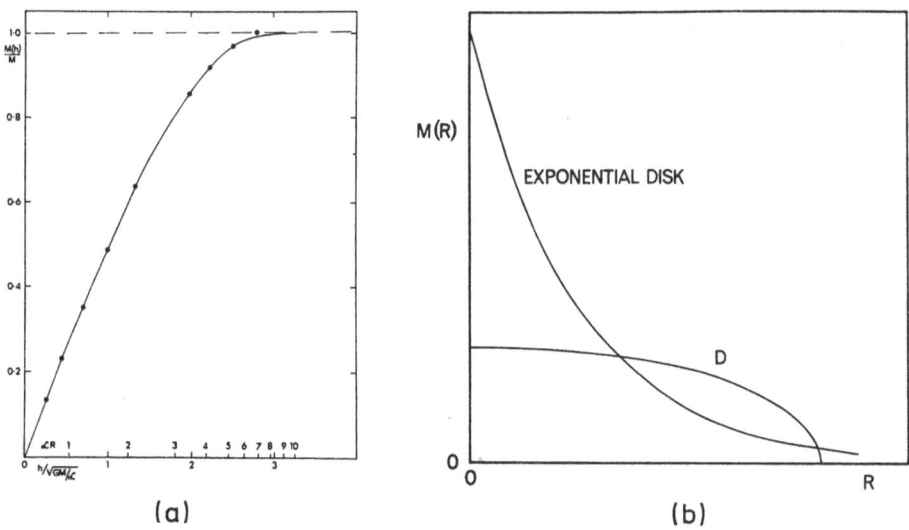

Fig. 4. (a) The full curve is $\mathfrak{M}(h)$ for the exponential disk. The points show $\mathfrak{M}(h)$ for the disk D described in the text. (b) Surface density distribution for an exponential disk and a disk D with the same $\mathfrak{M}$ and almost the same $\mathscr{H}$ and $\mathfrak{M}(h)$.

the collapse was axisymmetric, then all protospirals had a fairly similar $\mathfrak{M}(h)$. On the other hand, if $\mathfrak{M}(h)$ was established by internal torques during the collapse phase, then there existed a very efficient process that drove $\mathfrak{M}(h)$ to the sphere-like distribution during this phase. (b) There do exist a few spiral galaxies in nearby rigid rotation. These would have the characteristic $\mathfrak{M}(h)$ if their mass distributions were like the disk D. We are studying some of these systems now at Mount Stromlo.

All this discussion is partly irrelevant if the disks are not really close to centrifugal equilibrium. The evidence suggesting that the disk could be hot comes from theory. Hohl's (1971) N-body study began with 10^5 particles distributed with the same $\mu(R)$ as the disk D but with enough velocity dispersion σ to suppress local axisymmetric instabilities according to Toomre's rule: $\sigma = \sigma_{\min} = 3.36\ G\mu/\varkappa$, where $\varkappa$ is the local epicyclic frequency. He found that the disk was unstable to a large-scale bar-like mode and, by the time the system had settled down, the velocity dispersion had increased to (2 to 6) $\sigma_{\min}$; Ostriker and Peebles (1973) considered this further, asking: *how can we account for the apparent stability of the Galaxy?* They collected various examples of rotating disk-like systems, including the MacLaurin spheroids, Kalnajs' (1972) exact self-consistent disk, and their own three-dimensional 150 to 500-body integrations. They pointed out that each of these examples was stable only for the ratio $T_{\mathrm{mean}}/|\Omega| \lesssim$ $\lesssim 0.14$: here Ω is the potential energy and T_{mean} is the kinetic energy associated with ordered motion, so from the virial theorem the ratio must lie between 0 and 0.5. This means that these systems are stable only if a large part of their kinetic energy comes from random motions. There are then two ways to stabilize the galactic disk: (i) the disk is really hotter than we believe, with random stellar velocities at least as large as the rotational velocity; (ii) the halo mass *interior* to the disk is comparable to the disk mass (this increases $|\Omega|$).

Is either of these ways readily compatible with observation? We see flat systems, like the one in Figure 1, with no significant visual bulge component. Their flatness argues against a high velocity dispersion in the disk, unless this dispersion is very anisotropic with the ratio of component in the plane to component normal to the plane typically 200 km s^{-1} to 30 km s^{-1}. The smallness of the bulge component argues against the presence of a massive stabilizing bulge, unless the M/L ratio for the bulge is very large, well in excess of 100. There is evidence that M/L for the bulge is not so large: (i) the Cambridge results for M31, reported by Baldwin at this meeting (p. 139), suggest that its bulge has $M/L \approx 10$. (ii) Illingworth's unpublished results for globular clusters, which are probably fairly typical examples of the halo population, give M/L values of about 2, even for the most concentrated clusters which would be least affected by the galactic tidal field.

In summary, while theory demands $T_{\mathrm{mean}}/|\Omega| \lesssim 0.14$ for the stability of disk systems, we see systems in which the bulge is apparently weak and the velocity dispersion apparently small (compared to the rotational velocity): their value of $T_{\mathrm{mean}}/|\Omega|$ would then be significantly larger than 0.14. This is obviously an urgent problem which will hold back our understanding of the formation and dynamics of the disk component until it is sorted out.

4. The Bulge Component

If the two-component picture for disk galaxies is valid, we can regard the bulge component as an elliptical galaxy modified by the presence of the disk: the primary evidence for this is the $R^{1/4}$-dependence of the surface-brightness distribution in the bulges of some nearby systems. Since the galactic halo contributes so much to our present fragmentary knowledge about galaxy formation, it seems vital to get some understanding of the present dynamical state of spheroidal systems. Carrick at Mount Stromlo is now completing a systematic program of surface photometry and dynamical model construction to see if the bulges can be understood as modified elliptical systems.

Some further encouragement for interpreting bulges as modified ellipticals comes from comparing their stellar content. We know that (i) the spectral energy distributions for the bulges of the brighter spirals and the giant ellipticals are similar; (ii) the M/L value of 10 for the bulge of M31 mentioned earlier is close to the value for most normal ellipticals; (iii) the line strengths and colours change radially for many bulges and ellipticals in a way comparable with the abundance gradient in the galactic halo.

It would be very useful to know the $\mathcal{H}/\mathfrak{M}$ ratio for classes of halo objects: note that a large asymmetric drift for a class of objects does not necessarily mean a low value of $\mathcal{H}/\mathfrak{M}$, because the halo population is so extended spatially. Estimates for RR Lyrae stars and globular clusters are probably the most reliable. Kinman (1959b) found an asymmetric drift of 167 ± 30 km s^{-1} for the globular cluster system, and deduced that its $\mathcal{H}/\mathfrak{M}$ was close to the mean value for the whole Galaxy. On the other hand, the most metal-weak RR Lyrae stars have a drift of 220 ± 23 km s^{-1} (see Oort, 1965), and it seems unlikely that their $\mathcal{H}/\mathfrak{M}$ could be so high. The $\mathcal{H}/\mathfrak{M}$ ratio could be relevant to the reason why these objects formed so early in the life of the Galaxy.

Finally, I have a brief comment on the nature of the collapse of the bulge, and how it halted. There appear to be two main pictures. Firstly the Lynden-Bell (1967) picture, in which star formation occurs while the system is far from equilibrium; the bulge then relaxes to its present distribution through mean field relaxation. Secondly, the Larson picture (p. 191), in which the protosystem gets hot enough before star formation to be in equilibrium. Both pictures have their difficulties; again, until we understand better what triggers off star formation, it will be difficult to decide which is right.

5. Conclusion

There are two immediate problems. The first, from Section 3, is the reason for the apparent stability of the disk. The second is the old problem of the process that triggers off star formation in the protogalaxy. There may be a new clue. We now know that the bright blue clusters in the LMC, such as NGC 1866, are like globular clusters in every way but in their stellar content: they are a mere 10^7 to 10^8 yr old. There are no young clusters like these in the Galaxy. If we could understand why these globular clusters can form now in the LMC but not in the Galaxy, we would be closer to a solution of the star formation problem.

References

Eggen O. J.: 1970, *Vistas in Astronomy* **12**, 367.
Eggen, O. J., Lynden-Bell, D., and Sandage, A.: 1962, *Astrophys. J.* **136**, 748.
Freeman, K. C.: 1970, *Astrophys. J.* **160**, 811.
Hohl, F.: 1971, *Astrophys. J.* **168**, 343.
Kalnajs, A.: 1972, *Astrophys. J.* **175**, 63.
Kinman, T. D.: 1959a, *Monthly Notices Roy. Astron. Soc.* **119**, 538.
Kinman, T. D.: 1959b, *Monthly Notices Roy. Astron. Soc.* **116**, 559.
Lynden-Bell, D.: 1967, *Monthly Notices Roy. Astron. Soc.* **136**, 101.
Mestel, L.: 1963, *Monthly Notices Roy. Astron. Soc.* **126**, 553.
Oort, J. H.: 1965, in A. Blaauw and M. Schmidt (eds.), *Galactic Structure*, vol. V of *Stars and Stellar Systems*, University of Chicago Press, p. 455.
Ostriker, J. P. and Peebles, P. J. E.: 1973, *Astrophys, J.* **186**, 467.
Sandage, A.: 1970, *Astrophys. J.* **162**, 841.

DISCUSSION

Larson: It seems to me that the Eggen-Lynden-Bell-Sandage (ELS) argument for a collapse time of 2×10^8 yr is incorrect, since it is based on the assumption that stars are formed in essentially circular orbits. If the stars are formed in a cloud collapsing on a free-fall time-scale, they would have large radial velocities and be in highly eccentric orbits: if so, this would destroy the basis of the ELS argument. The time 2×10^8 yr is inconsistent with the initial radius of 50 kpc suggested by ELS, which implies a free-fall time of about 5×10^8 yr.

My second comment relates to the question of whether the collapse was axisymmetric. I don't know of any reason why the collapse should be axisymmetric, but even if it were, I have some preliminary calculations of axisymmetric protogalaxy collapse which show that the result is the formation of a thin, uniform disk which would be unstable to non-axisymmetric modes. Thus strict axisymmetry would be destroyed in any case, as would be the strict conservation of angular momentum for each mass element.

Oort: I do not know of any observational evidence opposing the idea that a large fraction of the mass of a galaxy is contained in a halo. From the data on stars in the vicinity of the Sun we see that the only *known* stars that contribute significantly to the mass of the halo are intrinsically very faint subdwarfs. As pointed out already in my article (p. 455) in Vol. V of *Stars and Stellar Systems*, there is nothing to oppose the idea that subdwarfs below the limit to which they have so far been found could make up a halo that would contain most of the mass of the Galaxy.

Freeman: The work of Rodgers *et al.* (in preparation) suggests that the mass contribution of objects from the L.F.T. (Luyten's five-tenths) catalogue down to $M = +11$ is very small. These objects have halo-type chemistry and kinematics. It doesn't *prove* anything but is indicative.

Oort: Who knows what the missing mass is made up of? It may be very faint sub-dwarfs.

Freeman: How faint would you want to go?

Oort: Perhaps to $M = +15$.

Freeman: Baldwin's work on the rotation curve of M31 (p. 139) is also indicative.

E. M. Burbidge: From the work of Murray and Sanduleak (*Monthly Notices Roy. Astron. Soc.* **157**, 273, 1972) and others at the Royal Greenwich Observatory, it looks as though the missing mass in the solar neighbourhood may be accounted for by M dwarfs in the disk population.

van der Kruit: If the exponential disks have the same μ_0 for all galaxies, it is only necessary to specify α to find the rotation curve parameters. Is this in contradiction with Brosche's result that the maximum rotation velocity and the radius at which this occurs are almost completely independent?

Freeman: No. The rotation curve is determined by the field of the exponential plus bulge components, and Brosche's result follows because the mass ratio of these two components varies from galaxy to galaxy.

Mestel: Two comments: (i) Miller, Prendergast, and Quirk (in W. Becker and G. Contopoulos (eds), *The Spiral Structure of our Galaxy, IAU Symp.* **38**, 365, 1970) found that in order to keep their disk models cool they needed to have powerful dissipation, e.g. by shock formation in gas clouds, which thermalize and radiate away much of the gravitational energy released during instability. Could this be the answer to the difficulty of Ostriker and Peebles? (ii) In discussing star formation during collapse of the proto-galaxy, one should distinguish between the initial collapse and the final settling down into a flattened structure. A local density increase can amplify by its self-gravitation (cf. Christopher Hunter's work (*Astrophys. J.* **135**, 594, 1962; **139**, 570, 1964) on fragmentation in a collapsing sphere) while still *retaining* its energy of z-motion, presumably forming ultimately a globular cluster, with its characteristic large-amplitude z-motion. The disk population by contrast would form from gas which has lost the bulk of its z-energy before break-up into stars.

Wright: In 1969 Mrs Heidmann (*Astrophys. Letters* **3**, 153) showed that $H \propto M^{5/3}$, while your result indicates that $H \propto M^{7/4}$. Would you like to comment on this?

Secondly, if one puts both these results in a mass-radius form then they are respectively $M \propto R^3$ and $M \propto R^2$. Additionally, for elliptical galaxies Fish (*Astrophys. J.* **139**, 284, 1964) has shown that $M \propto R^2$ (essentially). It seems, therefore, rather dismaying that centrifugal equilibrium and the rise of electron opacity (as considered by Fish) lead to such similar results even though the physical processes are so different.

Freeman: $(7/4-5/3)/(513)$ is only 0.05, and I doubt whether my data or Mrs Heidmann's could differentiate between these two exponents. You realise also that there is a near-indeterminateness in the *observational* estimations of this exponent which makes these estimations yet more uncertain than they appear.

Miller: There are some technical points that must be treated carefully in interpreting stability results from n-body calculations. In particular, numerical instabilities may mask physical effects. Our own n-body calculations gave similar results to Hohl's; I discussed these points in some detail three years ago at Cambridge (*IAU Colloq.* **10**), so won't go into detail here. However, recent numerical experiments have shown that we can sort out some of these effects. A strictly axisymmetric n-body integration showed a stability threshold and confirmed Toomre's numerical values, leading to the conclusion that we can build stable n-body integrations and that the failure to show a stability threshold in our (and Hohl's) calculations is most likely due to non-axisymmetric modes.

Some care is required in applying these stability criteria. The paper of Ostriker and Peebles is based on maintenance of stability in an axisymmetric system. But no-one believes that our Galaxy is axisymmetrical. It is possible that a non-axisymmetric system is more stable than an axisymmetrical system and one could argue that spiral features arise from some such process. A crucial question is what are the consequences of an instability? Does a system gently shift to a more stable configuration, or does something catastrophic happen? So far, we have no answers to these questions.

Gott: Traditionally, the velocity envelope of the highest velocity stars has been taken to indicate the escape velocity for the Galaxy, and then the ratio of escape velocity to circular velocity is found to be less than the Keplerian value of $\sqrt{2}$. From this it has been concluded that the mass in the Galaxy is primarily in a disk configuration. Since the presence of a massive spheroidal halo would presumably push this ratio toward the Keplerian value, it may be possible to place upper limits on the mass of a spheroidal halo component. It would be worthwhile in this context to re-examine the usual assumption that the velocity envelope of the highest velocity stars does in fact indicate the escape velocity.

THE DISTRIBUTION OF MASS-TO-LIGHT RATIO
WITH RADIUS IN M31 AND M33

J. E. BALDWIN

Cavendish Laboratory, Cambridge, England

Abstract. New H I observations indicate that the mass-to-light ratios in the disks of M31 and M33 are independent of radius.

In recent years several studies have suggested that in many late-type spiral galaxies the mass-to-light ratio, M/L, increases with radius in their outer parts. If confirmed, this result would have profound significance for any discussion of the formation and evolution of galaxies. It is therefore important to question whether the result is correct. Freeman (1970) has found that the observational data are consistent with uniform values of M/L for those galaxies where the luminosity distribution can be represented well by an exponential disk model. Since comparatively large changes in the mass distribution in the outer parts of galaxies lead only to small changes in the rotation curve, both precise measures of the apparent rotation curve and knowledge that the departures from circular motion are small are required if this result is to be established securely.

H I observations of M31 and M33 have been made with the Cambridge Half-Mile Telescope with an angular resolution of 2′ and sufficient sensitivity to determine the rotation curves to an accuracy of a few km s^{-1}. In M33 (Warner *et al.*, 1973) it was found that the apparent value of M/L is 4.9, almost independent of radius out to 30′ (6 kpc) beyond which there are important deviations from circular motion. In M31 (Emerson and Baldwin, 1973) the observations show differences from earlier work both by the absence of a low minimum in the rotation curve near 10′ radius and by a steady fall in circular velocity with radius beyond 60′. The observations are consistent with a simple model in which the apparent M/L ratio of the spheroidal population seen in the light distribution has a value of 25 and the exponential disk 12.5. The deviations between the model rotation curve and the observed mean rotation curve are smaller than the differences between the two rotation curves from the NE and SW major axis data. When corrected for absorption, the M/L values for the disks of M33 and M31 are 3 and 5 respectively. The results for these two galaxies, which have the best determined rotation curves, suggests that the constancy of the M/L ratio may be a common property of the disks of spiral galaxies.

Unfortunately, the good fit of the observations to the models cannot be used as evidence excluding the existence of massive halos with large M/L ratios, mentioned earlier in this meeting (p. 134) as a means of stabilizing galactic disks. Such halos with a suitable density law with radius could, for example, have rotation curves which mimic very closely those of exponential disks. One test which might now be applied is to check whether the mass density in the disk models at $z=0$ deduced from the rota-

John R. Shakeshaft (ed.), The Formation and Dynamics of Galaxies. 139–140. All Rights Reserved.
Copyright © 1974 by the IAU.

tion curve is consistent with the measured layer thickness and velocity dispersion of the H I.

References

Emerson, D. T. and Baldwin, J. E.: 1973, *Monthly Notices Roy. Astron. Soc.* **165**, 9P.
Freeman, K. C.: 1970, *Astrophys. J.* **160**, 811.
Warner, P. J., Wright, M. C. H., and Baldwin, J. E.: 1973, *Monthly Notices Roy. Astron. Soc.* **163**, 163.

DISCUSSION

Freeman: I take the point about the many ways of constructing mass distributions to produce observed rotation curves. However, if one makes the very reasonable assumption that M/L is uniform within the bulge component, then your relatively low M/L must follow.

G. de Vaucouleurs: There is some justification for your assumption that M/L is constant in the disk of M33, which has a constant colour index; but I doubt very much that M/L is constant in the spheroidal component of M31, because the colour index and metallicity are known to decrease from $B - V \simeq +1.0$ in the core, to $B - V \simeq +0.6$ in the halo. Models with variable M/L have been calculated for M31 by J. Einasto (*Tartu Astron. Obs. T.*, No. 40, 1972).

Pishmish: The dip in the rotation curve of M31 which you indicated as a strange feature may not be such a strange feature after all since, according to the hydrodynamical equations of a steady-state stellar system, the gravitational force is not only balanced by the centrifugal force (velocity of rotation) but also by the dispersion of the velocities of the objects of which the rotational velocity is being observed. Hence it would be advisable to determine the dispersion of velocities at these points of depression. Only after this is done and negative results obtained might one search for another explanation for the dip.

Baldwin: That possibility may arise in other galaxies but for M31 we now have good evidence that the dip is not a feature of the true rotation curve.

van den Bergh: How sure are we that the innermost emission regions observed by Rubin and Ford are in fact H II regions and not supernova remnants (which might have quite a large velocity dispersion)?

Baldwin: I have no detailed knowledge of their observations. It may be a possibility.

THE EARLY HISTORY OF OUR GALAXY:
CHEMICAL EVOLUTION

MANUEL PEIMBERT

Instituto de Astronomía, Universidad Nacional Autónoma de México, Mexico

Abstract. A general review is given of chemical abundance determinations; particular emphasis is given to abundances of galactic and extragalactic metal-poor objects since presumably they represent the abundances of the primeval material from which our Galaxy was formed. The following results are stressed: (a) most of the helium present in the galaxies of the local group as well as in other galaxies was produced before these objects were formed, (b) the heavy elements were produced mainly as the result of stellar evolution, (c) there is a chemical abundance gradient in our Galaxy and, by analogy with other galaxies, it is expected to be steeper near the nucleus, (d) the carbon and oxygen content of our Galaxy increased at a rate different from the metals, reaching their present abundance earlier than the other heavy elements, and (e) the increase of the iron abundance in the disk of our Galaxy with time has been small while that of carbon is negligible; furthermore, as a group the super-metal-rich stars correspond to the old disk population. Several models of galactic chemical evolution are reviewed.

1. Introduction

It is not possible to cover all the research papers relevant to the chemical history of the Galaxy in a review of this nature, therefore I have been forced to be selective in the topics covered. I will not discuss the light elements, with the exception of hydrogen and helium. There are two recent review papers on the light elements by Reeves *et al.* (1973) and by Tinsley (1973). A very important element as regards its abundance is deuterium which, according to Hoyle and Fowler (1973) and Colgate (1973), can be produced not only in the big-bang but also by spallation in envelopes of super-massive objects or in supernovae. However, the energy required to produce the galactic deuterium by this method is very large and consequently it is more likely that this element has been produced in the big-bang. At the other end of the atomic mass range, the information that can be obtained from the radioactive elements is very limited and it will only be mentioned that from the ^{232}Th/^{238}U and the ^{235}U/^{238}U ratios found on Earth the formation of radioactive elements appears to have been more pronounced in the early stages of our Galaxy than later on (Schramm, 1972).

In Section 2 we review galactic abundance determinations relevant to the chemical evolution of our Galaxy, but discuss first some results obtained from other galaxies, which might bear on this problem. The latter offer two advantages over the local observations: firstly, we can observe galaxies in which star formation has not been very prominent so that their chemical abundances might be more representative of the original material from which galaxies were formed, and secondly we can observe at different distances from the centre and thus study possible chemical abundance gradients.

In Section 3 we will discuss recent theories on production sites and in Section 4 review models of the evolution of the Galaxy proposed to explain the observations.

John R. Shakeshaft (ed.), The Formation and Dynamics of Galaxies. 141–156. All Rights Reserved.

2. Observations

2.1. INTERSTELLAR MATTER IN EXTERNAL GALAXIES

2.1.1. H II *Regions and Nuclei of Galaxies*

In order to understand the evolution of the Galaxy one of the key elements to study is helium. An overall examination of the He/H ratio is thus essential. The most reliable determinations are those obtained from normal H II regions observed optically. Peimbert and Spinrad (1970a) derived very accurate N(He)/N(H) ratios in five extra-galactic objects which yielded an average value of ~0.10. To obtain a value as close as possible to the primeval one it is clear that objects for which the contamination due to stellar evolution has been small should be studied. Therefore Peimbert and Spinrad (1970b) obtained the He/H ratio in NGC 6822, an irregular galaxy of the local group, slightly metal deficient. More relevant results were obtained by Searle and Sargent (1972), who determined the He/H abundance ratio in IZw 18 and IIZw 40, two blue compact galaxies which are metal deficient, particularly the former, and by Dufour (1973) on the Magellanic clouds, especially interesting for the study of the chemical evolution of our Galaxy, since the three objects probably originated from material with the same primeval chemical composition. A summary of the abundances derived from metal poor H II regions is presented in Table I. It is clear from this table that the He/H abundance ratio is nearly the same for all objects and similar to that found in H II regions of the solar neighbourhood.

Another problem of interest that can be tackled is the presence of chemical abundance gradients across the disks of spiral galaxies. Peimbert (1968) argued that the nitrogen abundance in the nuclei of spiral galaxies was higher than that of H II regions of the solar neighbourhood. Searle (1971) established the existence of N/H and O/H abundance gradients across the disks of spiral galaxies; the former being steeper than

TABLE I

Abundances of extragalactic metal-poor gaseous nebulae[a]

	[He/H]	[O/H]	[N/O]	[Ne/O]	Reference
NGC 6822 V	−0.02	−0.23	−0.55		[1]
LMC	−0.03	−0.30	−0.45	0.00	[2]
NGC 5471	−0.01	−0.42		+0.17	[3]
II Zw 40	−0.03	−0.50		0.00	[3]
NGC 185-1	+0.15	−0.83			[4]
SMC	−0.02	−1.01	−0.39	−0.08	[2]
I Zw 18	+0.03	−1.14		−0.14	[3]

[a] The abundance ratios are logarithmic, relative to those of the Orion Nebula by Peimbert and Costero (1969).
[1] Peimbert and Spinrad (1970b).
[2] Dufour 1973.
[3] Searle and Sargent (1972)
[4] Jenner *et al.* (1973).

the latter. Benvenuti *et al.* (1973) established the presence of a N/S abundance gradient extending throughout the disks of spiral galaxies but more pronounced near the nuclei (see Table II). Independent evidence, based on stellar absorption features, in favour of the existence of abundance gradients close to and in the nuclei of galaxies has also been presented by McClure (1969), Spinrad *et al.* (1971), and Spinrad *et al.* (1972).

TABLE II

Chemical abundances[a] at different distances from galactic nuclei

Object	H	N	S	N/S
M51 (Nucleus $r \leqslant 1.25''$)				1.53
M81 (Nucleus $r \leqslant 1.25''$)				1.42
M51 (Nucleus $r \leqslant 3.5''$)				1.07
M81 (Nucleus $r \leqslant 3.5''$)				0.98
M33 (inner H II Regions)				0.48
M8	12.00	7.50	7.14	0.36
Orion	12.00	7.63	7.50	0.13
M101 (outermost H II Regions)				-0.35

[a] Given in $\log N$ (Benvenuti *et al.*, 1973).

2.1.2. *Planetary Nebulae*

Jenner *et al.* (1973) have obtained the relative hydrogen, helium and oxygen abundances in NGC 185-1, a planetary nebula in NGC 185, for which the O/H ratio is lower by about an order of magnitude, and the He/H ratio is similar to those of H II regions and planetary nebulae in our Galaxy.

Sanduleak *et al.* (1972) found that the 6584/Hα ratios in planetary nebulae (and probably also the H II regions) of the SMC are very small and they conclude that nitrogen must be deficient in the planetary nebulae. This is in agreement with the abundance determinations of Dufour (1973) and implies that the planetary nebulae in the SMC may be similar in their chemical composition to that in M15.

2.1.3. *Quasars*

It is well known that some quasars exhibit very faint helium emission lines that might imply an intrinsically low helium abundance (Osterbrock and Parker, 1966; Wampler, 1967, 1968; Bahcall and Kozlovsky, 1969a, b; Peimbert and Spinrad, 1970a). However, the derived He/H abundance ratios are strongly model dependent, so these abundance determinations are not reliable at present (Burbidge *et al.*, 1966; Williams, 1971; MacAlpine, 1972; Jura, 1973).

2.2. INTERSTELLAR MATTER IN OUR GALAXY

2.2.1. H II *Regions*

Some of the most reliable determinations of the He/H abundance ratios from optical emission lines have been made by Peimbert and Costero (1969). The objects observed are mainly H II regions of the solar neighbourhood, where interstellar reddening is small.

He/H abundance ratios have also been determined from radio observations for many H II regions at different distances from the centre of our Galaxy (see for example Churchwell and Mezger, 1970). The agreement between the radio and optical determinations for the objects in common is very good. It should be mentioned that there are three H II regions within 150 pc of the centre of our Galaxy where $N(He^+ + He^{++})/N(H^+)$ is less than 0.03 and thus considerably smaller than the $N(He)/N(H)$ value determined in all the H II regions observed optically. Unfortunately, the radiation field is not known and it is not possible to estimate the amount of neutral helium inside these H II regions (Churchwell and Mezger, 1973; Huchtmeier and Batchelor, 1973). From the arguments presented in this review the helium abundances in these regions are expected to be normal. There are at least three possible explanations for the presence of large amounts of neutral helium inside these H II regions: (a) a luminosity function near the nucleus of our Galaxy different from that of the solar neighbourhood, in the sense that it does not contain stars of spectral type earlier than O9; (b) the existence of dust particles which absorb more efficiently those photons able to ionize helium than those able to ionize hydrogen; and (c) cloud collisions with relative velocities in the 30 to 60 km s^{-1} range such that hydrogen is ionized but not helium. At present there are arguments in favour of each of these hypotheses and it is not possible to decide among them.

The nitrogen, oxygen and sulphur abundances relative to hydrogen have been determined in H II regions of the solar neighbourhood and are very similar to the solar abundances (Peimbert and Costero, 1969; Goldberg et al., 1960; Lambert, 1968; Lambert and Warner, 1968).

2.2.2. *Planetary Nebulae*

From a photoelectric study of planetary nebulae of the solar neighbourhood, Peimbert and Torres-Peimbert (1971) found that the He/H and the O/H abundance ratios were very similar to those of H II regions of the solar neighbourhood, while the N/O abundance ratio was found to be a factor of four higher. These two results are consistent with the ideas, firstly that the relative H, He and O abundances are representative of the matter from which planetary nebulae were formed and secondly that the envelopes of planetary nebulae enrich the N/H abundance ratio of the interstellar medium.

Two planetary nebulae of Population II have been studied in detail: 49+88°1 (Miller, 1969) and K648, a member of the globular cluster M15 (O'Dell et al., 1964; Peimbert 1972). Both planetary nebulae show a He/H ratio similar to that of H II regions, while they are underabundant in oxygen and neon by a factor of about 10. The case of K648 is particularly interesting because the Fe deficiency derived from the globular cluster starlight is of about two orders of magnitude, and thus it appears that O and Ne are not correlated with the iron abundance. These results are presented in Table III.

TABLE III

Abundances of galactic objects[a]

	[He/H]	[O/H]	[N/O]	[Ne/O]	[S/O]	[A/O]	References
K 648 (M15)	0.00	−1.12	< −0.22	+0.17			[1]
49 + 88°1	+0.11	−0.78	+0.77	−0.27			[2]
Planetary Nebulae of the Solar Neighbourhood	+0.04	−0.03	+0.64	0.0			[3]
Novae	+0.18	+0.84	+1.23		−0.26		[4]
RS Ophiuchi	+0.63	+1.02	+0.75		−0.52		[5]
Cas A		> +1.64	< −1.44		+0.11	+0.05	[6]

[a] Relative to those of Orion Nebula by Peimbert and Costero (1969).
[1] Peimbert (1972).
[2] Miller (1969).
[3] Peimbert and Torres-Peimbert (1971).
[4] Pottasch (1959).
[5] Pottasch (1967).
[6] Peimbert (1971).

2.2.3. Novae

Since the distribution of galactic novae corresponds to that of Population II objects, it is important to find out whether or not their observed abundances are those of the original material from which they were formed. Pottasch (1959, 1967) has determined chemical abundances from emission lines originating in shells of novae. He found for six objects that $[O/H] \sim +1.0$ and $[N/H] \sim +2.0$ and, by comparing RS Oph with the solar corona, that $[Fe/H] \sim +0.6$. Pottasch pointed out that, since iron is not produced by the nova system itself, this was an argument against the production of the other overabundant elements by the nova during the course of its evolution; however, a considerable fraction of hydrogen in this object has been converted into helium with the consequence that hydrogen has been depleted relative to the other elements and the iron abundance might be normal, with the only enriched elements being He, N and O and possibly C.

Mustel (1971) and Arkhipova (1973) find from a study of absorption lines that in DQ Herc and HR Del most elements show normal abundances but CNO are overabundant, by about two orders of magnitude in the former case and about an order and a half in the latter.

From the previous discussion it follows that at present it is not possible to derive the primordial abundances of these objects and that at least the observed H, He, C, N and O abundances have been affected by stellar evolution.

2.2.4. Supernova Remnants

The hypothesis that heavy-element enrichment of the interstellar medium is largely due to exploding supernovae can be tested by studying the chemical composition of supernova remnants. Observations of the Crab Nebula (Woltjer, 1958) show that, under the assumption of case *B* for the hydrogen lines, helium is overabundant by a factor of at least 2. However, the hydrogen lines might be closer to case A than to case B and consequently the helium overabundance might not be as large. The remnant of Cas A is more interesting because the velocity of expansion is considerably higher; it has been found that, in the system of fast moving knots, oxygen, sulphur and argon are overabundant by at least a factor of 30 with respect to hydrogen and nitrogen (Peimbert and van den Bergh, 1971; Peimbert, 1971).

2.3. STARS AND STELLAR SYSTEMS IN OUR GALAXY

2.3.1. *Metal Poor Stars*

Although Bw and sdB stars show very weak helium lines corresponding to low photospheric helium to hydrogen abundance ratios, various arguments suggest that these results are not reliable indicators of their internal composition and we therefore cannot use these objects to obtain information about their original He/H abundance ratios (Baschek *et al.*, 1972).

Some early results on the abundance of the elements from carbon to iron in metal-poor stars indicated that within a factor of two the relative abundances of the heavy elements were similar to those of the Sun even if their Fe/H ratio differed by as much as three orders of magnitude. However, recent and more accurate results imply that the situation is more complicated.

Wallerstein (1962) in a study of 31 G dwarfs found that manganese is deficient with respect to the other iron peak elements: chromium, nickel and iron. He also found that the stars having large velocities with respect to the local standard of rest are metal-poor but apparently α-rich with respect of iron, i.e. they show excesses of Mg, Si, Ca and Ti. Similar results were also presented by Spite (1968).

In a recent discussion on the chemical evolution of the Galaxy, Wallerstein (1971) reviewed the evidence in favour of the [α/Fe] enrichment in metal-poor stars and showed that sodium and aluminum did not increase in abundance until [Fe/H] reached -2.0 but subsequently increased rapidly to a total similar to that of iron. Since explosive carbon burning yields nuclei from neon through aluminum, Wallerstein suggests that this sudden increase in the sodium and aluminum abundances is due to a relatively higher efficiency of explosive carbon burning possibly caused by a higher initial abundance of CNO elements.

Hearnshaw (1972a, 1972b) analyzed 19 disk G subgiants, older or nearly as old as NGC 188. His results on Mg, Ca and Ti, as well as those on Mn, confirm the results by Wallerstein (1962). Furthermore the iron-deficient (disk) stars are generally no more than marginally carbon deficient, a result which should be compared with those of Cohen and Strom (1968) and Pagel and Powell (1966) who obtained [C/Fe]$=0.0$ and -0.1 for the two extremely metal-poor subdwarfs HD 140283 and HD 25329. The relationships found by Hearnshaw are presented in Table IV.

From a study of seventy G and K stars, mostly giants, Conti *et al.* (1967) concluded that the oxygen content of our galaxy increased at a rate different from that of the metals, reaching its present abundance sooner than the other heavy elements. The

TABLE IV

Results from 19 old subgiants[a]

Element	$d[M/Fe]/d[Fe/H]$
C	−0.7
Ca, Ti	−0.3
Fe, Cr, Ni	0.0
Mn	+0.8

[a] Hearnshaw (1972).

extremely metal-poor giant star HD 122563 (Wallerstein *et al.*, 1963; Pagel, 1965; Wolffram, 1972) has been recently studied to obtain the CN (Sneden, 1973) and O (Lambert *et al.*, 1974) abundances with the results $[C/Fe] = -0.4$; $[N/Fe] = +1.2$ and $[O/Fe] = +0.6$. These values indicate that probably some carbon was transformed into nitrogen through nuclear reactions in the stellar interior and that subsequently the material was brought up to the surface; the oxygen overabundance, however, could very well be primeval and supports the results of Conti *et al.* (1967) that the oxygen content increased earlier than the iron content of our galaxy.

Powell (1972) obtained the iron to hydrogen ratios of 93 late F dwarfs within 25 pc of the Sun from their ultraviolet excesses. The ages of these stars have been estimated from their positions in the H-R diagram. This study and that of Hearnshaw (1972a, 1972b) indicate that there has been a modest enrichment of Fe in the disk since the formation of the Galaxy, the average increase being by a factor of 4 and the scatter at any one age covering a range of a factor of 2 to 3. This result is in contradiction with the work of Bond (1970) who, from the study of 69 metal-poor stars of the solar neighbourhood, finds that for stars with $[Fe/H] \gtrsim -1$ there is little or no correlation between galactic orbital eccentricity and chemical composition. Moreover, Hearnshaw does not find any significant increase in the carbon abundance which is a more representative element of the total Z value.

In an excellent review paper on population effects in the Galaxy (Pagel, 1972; see also Pagel, 1970), it is suggested that for halo stars in our Galaxy a correlation of the type $[N/Fe] = [Fe/H]$ exists. This correlation is based on Gmb 1830, ν Ind, μ Cas and HD 6833. Tomkin (1972) also finds for Gmb 1830 that $[Fe/H] \sim -1.3$ while $[N/H] \sim -2.0$.

2.3.2. *Super-Metal-Rich Stars*

Torres-Peimbert and Spinrad (1971), from the chemical abundance determinations by Spinrad and Taylor (1969) and the theoretical evolutionary models by Torres-Peimbert (1971a), found that the mean age of nine SMR K giants is 6×10^9 yr and is larger than those of the old galactic clusters M67 and NGC 188 (Torres-Peimbert, 1971b). Torres-Peimbert and Spinrad were the first to point out that SMR stars as a group correspond to the old disk population. Hearnshaw (1972a, b) also finds that SMR stars are relatively old and that they do not represent the tail-end of the abundance distribution in young disk stars. A similar result is obtained by Williams (1972) from ages and kinematical properties of SMR stars. He finds that they are associated with the metal-poor stars having $[Fe/H] \sim -0.5$ which suggests that the interstellar medium was poorly mixed during and immediately after the collapse of the Galaxy.

Janes and McClure (1972), from CN observations in 799 K giant stars, present evidence for a radial gradient in CN strength. This feature is correlated with $[Fe/H]$ and implies an increase of $[Fe/H] \sim 0.23$ between 15 and 5 kpc, from which they conclude that the heavy element abundance depends more on the position at birth of the star in the galactic disk than on its age. A similar result is obtained by Grenon (1972) who finds that SMR stars are old, have elliptical orbits and were probably born at a location closer to the galactic centre than their present position.

2.3.3. Globular Clusters

Van den Bergh (1965, 1967) and Sandage and Wildey (1967) suggested that there are at least two parameters affecting the characteristics of the horizontal branch in the HR diagram of globular clusters. The parameters most often considered have been the He/H abundance ratio and the Z value represented by the Fe abundance. The use of age as an independent parameter seems to be in contradiction with the idea that the globular clusters were formed during the rapid collapse phase of the galaxy and with the estimated duration of 2×10^8 yr for such a collapse (Eggen *et al.*, 1962). However, the recent findings that $N(He)/N(H) \sim 0.10$ in the interstellar matter of several galaxies with widely different metal abundances have led several authors to search for other parameters to replace Y.

Rood (1972) has studied this problem and suggests that anomalous C-M diagrams in our Galaxy and in other galaxies could be explained by variations in age of the order of 1×10^9 yr; according to him, a further parameter could be the mass loss between the red giant phase and the horizontal branch. Hartwick and McClure (1972), from CN observations of red giant stars in globular clusters, have suggested that, in addition to Fe/H, a second independent parameter could be the nitrogen abundance.

In Table V we show some CNO abundances derived from different objects, and it is clear from this table as well as from the discussion in this section that for many objects there is no good correlation between the Fe and the CNO abundances Furthermore Simoda and Iben (1968, 1970) have demonstrated the importance of CNO abundances (particularly that of oxygen) in the evolution of population II stars; since most of the Z value is due to CNO elements and not to the Fe group, it follows that for studies of galactic and stellar evolution it is very important to know the CNO abundances. Consequently it is indeed quite possible that a second parameter to explain the HR diagrams of globular clusters is the CNO abundance.

3. Theoretical Production of Chemical Elements

3.1. BIG BANG

All the well-studied H II regions in our Galaxy and in other galaxies point to

TABLE V

CNO abundances[a]

	[Fe/H]	[C/H]	[N/H]	[O/H]	References
HD 122563	−2.7	−3.1	−1.5	−2.1	[1, 2, 3]
HD 140283	−2.2	−2.2			[4]
K 648 (M15)	−2.1		< −1.34	−1.12	[5]
HD 25329	−1.3	−1.4	−0.8		[6]
ν Indi	−1.2		−1.9		[7]
Gmb 1830	−1.1		−2.0		[7, 8]

[a] Relative to the solar ones.
[1] Wolffram (1972).
[2] Sneden (1973).
[3] Lambert et al. (1974).
[4] Cohen and Strom (1968).
[5] Peimbert (1972).
[6] Pagel and Powell (1966).
[7] Pagel (1972).
[8] Tomkin (1972).

$N(\text{He})/N(\text{H}) = 0.10 \pm 0.02$ and there is at present no clearcut observation that implies a considerably lower or higher primeval helium abundance for any object. Therefore the helium abundance in galaxies and in particular in ours should be the result of a very general process. Calculations of nucleosynthetic processes (Wagoner et al., 1967) suggest that a simple version of the big bang provides the best explanation. These computations also show that, for the range of baryonic density comparable with observations, the big bang can be responsible only for the abundances of D, ^{3}He, ^{4}He, and possibly ^{7}Li in our Galaxy and not for the cosmic abundances of the rest of the elements.

3.2. SUPERMASSIVE OBJECTS

To explain the energy requirement of strong radio sources Hoyle and Fowler (1963a, b) proposed the existence of supermassive objects (SMO) with masses in the range 10^5 to 10^8 $M_\odot$ which release energy by violent explosions from galactic nuclei. It was suggested, moreover, that these objects could also produce the helium and metal abundances observed in extreme Population II stars.

Wagoner et al. (1967) made several models of SMO whose gravitational collapse is reversed at some specified temperature They found that in the more realistic cases where protons were initially present the computations predicted similar abundances for C^{12} and C^{13} but these particular models are ruled out by the value $C^{12}/C^{13} > 3$ determined for eight extremely metal-poor stars by Cohen and Grasdalen (1968).

Bisnovatyi-Kogan (1968) studied exploding supermassive stars in the 10^5 $M_\odot$ range and, in the only interesting cases, most of the hydrogen was converted into helium. Moreover, Wagoner (1969) computed the element production in SMO for expansion rates of the order of 10^4 to 10^7 times the gravitational expansion. Regardless of the physical plausibility of these expansion rates, there is no observational evidence for massive gaseous clouds with a very high He/H ratio, hence apparently ruling out such models. There are available several recent computations of SMO, but in most cases there are no detailed evolutionary models and Population I abundances have been assumed.

Silk and Siluk (1972) have suggested that the primordial gas out of which the Galaxy condensed may have been significantly enriched in heavy elements by material ejected from quasi-stellar objects, the assumption being that these heavy elements would be formed in SMO. The observational objections mentioned above would seem to rule out this theory also. In a modified theory, in which the heavy elements in quasars are produced by normal stars, we would still be faced with the problem of the time scale for element formation and the practically unknown chemical abundances and rates of mass ejection from quasars.

To summarize, it appears that SMO are not responsible for the helium and metal abundances observed in Population II stars.

3.3. NORMAL STARS

Burbidge *et al.* (1957) proposed that the synthesis of the elements takes place in stars. According to them, the galaxy condensed from a cloud of hydrogen and the heavier elements were produced in several generations of stars that were formed, evolved and later spread part of the processed matter into the interstellar medium.

It has been predicted from stellar evolution models (Iben, 1964, 1967) that nitrogen is enriched through the CN cycle and convected to the surface when the stars leave the main sequence and travel to the giant branch; furthermore Paczynski (1970) has suggested that stars less massive than about $4M_\odot$ lose only the material above the hydrogen-burning shell. The nitrogen enrichment effect has been confirmed observationally in the envelopes of planetary nebulae. These objects seem to provide a large fraction of the nitrogen enrichment of the interstellar medium and may be partly responsible for the nitrogen gradient observed across the disks of spiral galaxies (Torres-Peimbert and Peimbert, 1971). Both planetary nebulae and mass loss from red giants will only affect appreciably the carbon and nitrogen abundances of the interstellar medium.

Detailed calculations of explosive nucleosynthesis (Arnett, 1969; Truran and Arnett, 1970; Arnett *et al.*, 1971) have indicated that the material expelled in the detonation events which give rise to supernovae will typically be composed of elements from carbon to iron. Furthermore Arnett (1971) obtains an excellent agreement between the theoretical predictions and the solar abundances. These results strongly suggest that the chemical enrichment of our Galaxy is mainly produced by stars.

The apparent different pace of enrichment of CO and the elements of the iron group can be explained if the efficiency of helium burning followed by mass loss, relative to explosive silicon burning, was higher in the early generations of stars than in the later ones. This could be due to less massive supernovae and/or to a lack of seed elements for the onset of silicon burning in the early stages of galactic evolution.

4. Models of Galactic Chemical Evolution

From the frequency with which main sequence stars of different metal abundance occur in the disk, van den Bergh (1962) pointed out that the rate of metal creation in the Galaxy has decreased much faster than the rate of star formation. A result similar to van den Bergh's was obtained independently by Schmidt (1963) who noted that, under the assumption of a constant initial mass function (IMF), it was not possible to explain the lack of metal-poor stars in the solar neighbourhood; he therefore postulated that the IMF initially contained more massive stars than it does now. This scarcity of metal-poor stars in the disk has been the key issue in all the ideas and models both in favour of and against the chemical evolution of the Galaxy.

A more elaborate model based on the three basic assumptions made by Schmidt (1963, namely: no mass infall in the disk from the halo or from the intergalactic medium, an IMF containing more massive stars than at present, and a homogeneous collapse) was developed by Truran and Cameron (1971) who suggested that in the early halo nearly all the stars produced were more massive than 5 $M_\odot$. They were able to explain the lack of metal-poor stars in the disk of the Galaxy, and in their model the metal abundance increases monotonically with time.

Talbot and Arnett (1971), on the assumption of a constant IMF and no mass infall, have produced models in which the metal abundance of the interstellar medium initially increases with time but can decrease later or level off to an asymptotic value.

On the basis of the proposed interpretation by Oort (1970) of the 21-cm observations of high-velocity clouds in our Galaxy, Larson (1972a, b) has considered mass infall from the intergalactic medium to galaxies and its effect on galactic evolution. He suggests that during the later stages in the formation of galaxies there is approximate equilibrium between the rate of infall of gas and the rate at which gas is transformed into stars, so that the amount of interstellar matter in the Galaxy remains approximately constant with time. On this assumption he finds that the chemical composition of the disk is almost independent of time and of any initial element synthesis in the Galaxy, a result which may account for the observed fact that to a first approximation the heavy element abundance, Z, in the solar neighbourhood seems to be almost constant as a function of time. A higher proportion of massive stars would then be needed towards the nuclei of galaxies to explain the observed abundance gradients.

Quirk and Tinsley (1973) have assumed that stars are formed very efficiently until the gas content reaches equilibrium at its present value (10^9 yr), and thereafter the birthrate just equals the rate at which gas enters the system from stellar mass-loss or from infall of intergalactic matter. They propose a time variation of the IMF, with a higher proportion of massive stars in the past than at present. The models predict an initial increase of metallicity of the interstellar gas with time, followed by a later decline as the infalling primeval material dilutes the initial enrichment.

The hypothesis that the IMF contained more massive stars when the Galaxy was formed has been recently under attack. Van den Bergh (1972) pointed out that the luminosity function of disk stars suggests that the stellar birthrate function initially contained relatively *more* stars of low mass than it does now. Moreover, from the results of Woolley *et al.* (1971) for the luminosity function of stars within 25 pc of the Sun, Unsöld (1972) finds that the luminosity function of the unevolved stars is the same for the disk and the halo, from which he concludes that it is highly improbable that the halo IMF had originally a very large peak for stars with 5 to 10 times larger masses. Unsöld (1969, 1972) has used this argument, together with the observed uniformity of the relative abundances of the heavier elements, to support his theory that there has not been appreciable chemical evolution in our Galaxy due to individual stars and that the heavier elements originated in a gigantic explosion at a very early stage of evolution of the galactic nucleus.

From the near uniformity of the chemical composition in the interstellar gas of Sb and Sc galaxies, Searle (1972) also argues that it is unlikely that the IMF was very different from the present one in the early stages of evolution of our Galaxy. He has suggested that the IMF is the same everywhere and at all times, and explains the observed low fraction of metal-deficient stars in the disk by proposing that, contrary to Schmidt's (1963) hypothesis, *the collapse of the galaxy was inhomogeneous*, with enhanced star formation in the regions of high metal content. From ratios of stellar mass to gas mass of 1 to 100 in Sb and Sc galaxies, he predicts the present day metal-to-hydrogen ratio in the interstellar medium to be within the narrow limits $0.4\,Z_\odot \leqslant Z \leqslant$ $\leqslant 4Z_\odot$ in accordance with the observations. The predicted Z values are very insensitive to the details of the collapse and on whether primordial material is still falling into the Galaxy or whether it ceased to fall long ago.

Talbot and Arnett (1973) too have obtained better agreement with observations from a model of constant IMF, an inhomogeneous collapse and enhanced star formation in the regions of high metal content.

Ostriker and Thuan (1973) have made detailed computations assuming a substantial amount of matter falling into the disk from a massive galactic halo. From a constant IMF (Salpeter's) they are able to explain the lack of metal-poor stars in the disk as well as the existence of some old metal-rich stars.

Finally, it should be mentioned that to produce a unique solution, or to narrow down the number of possible solutions, to the evolution of our Galaxy it will be necessary not only to study the chemical abundances in detail but also the photometric and the dynamical properties of the models (see for example Tinsley, 1972; Larson, 1973; Biermann and Tinsley, 1974; Ostriker and Thuan, 1973).

References

Arkhipova, V. P.: 1973 (in preparation).
Arnett, W. D.: 1969, *Astrophys. J.* **157**, 1369.
Arnett, W. D.: 1971, *Astrophys. J.* **166**, 153.
Arnett, W. D., Truran, J. W., and Woosley, S. E.: 1971, *Astrophys. J.* **165**, 87.
Bahcall, J. N. and Kozlovsky, B.: 1969a, *Astrophys. J.* **155**, 1077.
Bahcall, J. N. and Kozlovsky, B.: 1969b, *Astrophys. J.* **158**, 529.

Baschek, B., Sargent, W. L. W., and Searle, L.: 1972, *Astrophys. J.* **173**, 611.
Benvenuti, P., D'Odorico, S., and Peimbert, M.: 1973: *Astron. Astrophys.* **28**, 447.
Biermann, P. and Tinsley, B. M.: 1974, *Astron. Astrophys.* **30**, 1.
Bisnovatyi-Kogan, G. S.: 1968, *Astron. Zh.* **45**, 74.
Bond, H. E.: 1970, *Astrophys. J. Suppl.* **22**, 117.
Burbidge, E. M., Burbidge, G. R., Fowler, W. A., and Hoyle, F.: 1957, *Rev. Mod. Phys.* **29**, 547.
Burbidge, G. R., Burbidge, E. M., Hoyle, F., and Lynds, C. R.: 1966, *Nature* **210**, 774.
Churchwell, E. and Mezger, P. G.: 1970, *Astrophys. Letters* **5**, 227.
Churchwell, E. and Mezger, P. G.: 1973, *Nature* **242**, 319.
Cohen, J. G. and Grasdalen, G. L.: 1968, *Astrophys. J. Letters* **151**, L41.
Cohen, J. G. and Strom, S. E.: 1968, *Astrophys. J.* **151**, 623.
Colgate, S. A.: 1973, *Astrophys. J. Letters* **181**, L53.
Conti, P. S., Greenstein, J. L., Spinrad, H., Wallerstein, G., and Vardya, M. S.: 1967, *Astrophys. J.* **148**, 105.
Dufour, R. I.: 1973, *Bull. Am. Astron. Soc.* **5**, 324.
Eggen, O. J., Lynden-Bell, D., and Sandage, A. R.: 1962, *Astrophys. J.* **136**, 748.
Goldberg, L., Müller, E. A., and Aller, L. H.: 1960, *Astrophys. J. Suppl.* **5**, 1.
Grenon, M.: 1972, in G. Cayrel de Strobel and A .M. Delplace (eds.), L'âge des étoiles', *IAU Colloq.* 17, Paris-Meudon Observatory, Ch. LV.
Hartwick, F. D. A. and McClure, R. D.: 1972, *Astrophys. J. Letters* **176**, L57.
Hearnshaw, J.: 1972a, *Mem. Roy. Astron. Soc.* **77**, 55.
Hearnshaw, J.: 1972b, in G. Cayrel de Strobel and A. M. Delplace (eds.), 'L'âge des étoiles', *IAU Colloq.* 17, Paris-Meudon Observatory, Ch. XLI.
Hoyle, F. and Fowler, W. A.: 1963a, *Monthly Notices Roy. Astron. Soc.* **125**, 169.
Hoyle, F. and Fowler, W. A.: 1963b, *Nature* **197**, 533.
Hoyle, F. and Fowler, W. A.: 1973, *Nature* **241**, 384.
Huchtmeier, W. K. and Batchelor, R. A.: 1973, *Nature* **243**, 155.
Iben, I., Jr.: 1964, *Astrophys. J.* **140**, 1631.
Iben, I., Jr.: 1967, *Astrophys. J.* **147**, 624.
Janes, K. A. and McClure, R. D.: 1972, in G. Cayrel de Strobel and M. Delplace (eds.), 'L'âge des étoiles', *IAU Colloq.* 17, Paris-Meudon Observatory, Ch. XXVIII.
Jenner, D. C., Ford, H. C., and Epps, H. W.: 1973, *Bull. Am. Astron. Soc.* **5**, 13.
Jura, M.: 1973, *Astrophys. J.* **181**, 627.
Lambert, D. L.: 1968, *Monthly Notices Roy. Astron. Soc.* **138**, 143.
Lambert, D. L. and Warner, B.: 1968, *Monthly Notices Roy. Astron. Soc.* **138**, 181.
Lambert, D. L., Sneden, C., and Ries, L. M.: 1974, *Astrophys. J.* **188**, 97.
Larson, R. B.: 1972a, *Nature* **236**, 21.
Larson, R. B.: 1972b, *Nature Phys. Sci.* **236**, 7.
Larson, R. B.: 1973, *Bull. Am. Astron. Soc.* **5**, 320.
MacAlpine, G. M.: 1972, *Astrophys. J.* **175**, 11.
McClure, R. D.: 1969, *Astron. J.* **74**, 50.
Miller, J. S.: 1969, *Astrophys. J.* **157**, 1215.
Mustel, E. R.: 1971, *Sci. Inf. Astron. Council Natl. Acad. Sci. U.S.S.R.* **19**, 32.
O'Dell, C. R., Peimbert, M., and Kinman, T. D.: 1964, *Astrophys. J.* **140**, 119.
Oort, J. H.: 1970, *Astron. Astrophys.* **7**, 381.
Osterbrock, D. E. and Parker, R. A. R.: 1966, *Astrophys. J.* **143**, 268.
Ostriker, J. P. and Thuan, T.: 1973, private communication.
Paczynski, B.: 1970, *Acta Astron.* **20**, 47.
Pagel, B. E. J.: 1965, *Roy. Observ. Bull.* No. 104.
Pagel, B. E. J.: 1970, *Quart. J. Roy. Astron. Soc.* **11**, 172.
Pagel, B. E. J.: 1972, in G. Cayrel de Strobel and A. M. Delplace (eds.), 'L'âge des étoiles', *IAU Colloq.* 17, Paris-Meudon Observatory, Ch. XLVII.
Pagel, B. E. J. and Powell, A. L. T.: 1966, *Roy. Observ. Bull.* No. 124.
Peimbert, M.: 1968, *Astrophys. J.* **154**, 33.
Peimbert, M.: 1971, *Astrophys. J.* **170**, 261.
Peimbert, M.: 1972, *Mém. Soc. Roy. Sci. Liège*, 6e série, **5**, 307.
Peimbert, M. and van den Bergh, S.: 1971, *Astrophys. J.* **167**, 223.

Peimbert, M. and Costero, R.: 1969, *Bol. Obs. Tonantzintla Tacubaya* **5**, 3.
Peimbert, M. and Spinrad, H.: 1970a, *Astrophys. J.* **159**, 809.
Peimbert, M. and Spinrad, H.: 1970b, *Astron. Astrophys.* **7**, 311.
Peimbert, M. and Torres-Peimbert, S.: 1971, *Astrophys. J.* **168**, 413.
Pottasch, S. R.: 1959, *Ann. Astrophys.* **22**, 412.
Pottasch, S. R.: 1967, *Bull. Astron. Inst. Neth.* **19**, 227.
Powell, A. L. T.: 1972, *Monthly Notices Roy. Astron. Soc.* **155**, 483.
Quirk, W. J. and Tinsley, B. M.: 1973, *Astrophys. J.* **179**, 69.
Reeves, H., Audouze, J., Fowler, W. A., and Schramm, D. N.: 1973, *Astrophys. J.* **179**, 909.
Rood, R. T.: 1972, in G. Cayrel de Strobel and A. M. Delplace (eds.), 'L'âge des étoiles', *IAU Colloq.* 17, Paris-Meudon Observatory, Ch. XX.
Sandage, A. and Wildey, R.: 1967, *Astrophys. J.* **150**, 469.
Sanduleak, N., MacConnell, D. J., and Hoover, P. S.: 1972, *Nature* **237**, 28.
Schmidt, M.: 1963, *Astrophys. J.* **137**, 758.
Schramm, D. N.: 1972, 'Nucleo-Cosmochronology', preprint of a paper presented at the Cosmochemistry Symposium held in Cambridge, Mass. Aug. 14–16, 1972.
Searle, L.: 1971, *Astrophys. J.* **168**, 327.
Searle, L.: 1972, in G. Cayrel de Strobel and A. M. Delplace (eds.) 'L'âge des étoiles', *IAU Colloq.* 17, Paris-Meudon Observatory, Ch. LII.
Searle, L. and Sargent, W. L. W.: 1972, *Astrophys. J.* **173**, 25.
Silk, J. and Siluk, R. S.: 1972, *Astrophys. J.* **175**, 1.
Simoda, M. and Iben, I. Jr.: 1968, *Astrophys. J.* **152**, 509.
Simoda, M. and Iben, I. Jr.: 1970, *Astrophys. J. Suppl.* **22**, 81.
Sneden, C.: 1973, *Astrophys. J.* **184**, 839.
Spinrad, H., Gunn, J. E., Taylor, B. J., McClure, R. D., and Young, J. W.: 1971, *Astrophys. J.* **164**, 11.
Spinrad, H., Smith, H. E., and Taylor, D. J.: 1972, *Astrophys. J.* **175**, 649.
Spinrad, H. and Taylor, B. J.: 1969, *Astrophys. J.* **157**, 1279.
Spite, M.: 1968, *Ann. Astrophys.* **31**, 269.
Talbot, R. J. Jr. and Arnett, W. D.: 1971, *Astrophys. J.* **170**, 409.
Talbot, R. J. Jr. and Arnett, W. D.: 1973, *Astrophys. J.* **186**, 69.
Tinsley, B. M.: 1972, *Astron. Astrophys.* **20**, 383.
Tinsley, B. M.: 1973, in D. N. Schramm and W. D. Arnett (eds.), *Explosive Nucleosynthesis*, University of Texas Press, Austin, Texas.
Tomkin, J.: 1972, in G. Cayrel de Strobel and A. M. Delplace (eds.), 'L'âge des étoiles', *IAU Colloq* 17, Paris-Meudon Observatory, Ch. L.
Torres-Peimbert, S.: 1971a, *Bol. Obs. Tonantzintla Tacubaya* **6**, 3.
Torres-Peimbert, S.: 1971b, *Bol. Obs. Tonantzintla Tacubaya* **6**, 113.
Torres-Peimbert, S. and Peimbert, M.: 1971, *Bol. Obs. Tonantzintla Tacubaya* **6**, 101.
Torres-Peimbert, S. and Spinrad, H.: 1971, *Bol. Obs. Tonantzintla Tacubaya* **6**, 15.
Truran, J. W. and Arnett, W. D.: 1970, *Astrophys. J.* **160**, 181.
Truran, J. W. and Cameron, A. G. W.: 1971, *Astrophys. Space Sci.* **14**, 179.
Unsöld, A.: 1969, *Science* **163**, 1015.
Unsöld, A.: 1972, in J. Xanthakis, B. Barbanis, L. Mavridis and J. Hadjidemetriou (eds.), *First European Astronomical Meeting under the Auspices of the IAU*, Springer-Verlag (in press).
van den Bergh, S.: 1962, *Astron. J.* **67**, 486.
van den Bergh, S.: 1965, *J. Roy. Astron. Soc. Can.* **59**, 151.
van den Bergh, S.: 1967, *Publ. Astron. Soc. Pacific* **79**, 460.
van den Bergh, S.: 1972, in D. S. Evans (ed.), 'External Galaxies and Quasi-Stellar Objects', *IAU Symp.* **44**, 1.
Wagoner, R. V.: 1969, *Astrophys. J. Suppl.* **18**, 247.
Wagoner, R. V., Fowler, W. A., and Hoyle, F.: 1967, *Astrophys. J.* **148**, 3.
Wallerstein, G.: 1962, *Astrophys. J. Suppl.* **6**, 407.
Wallerstein, G.: 1971, *Astron. Soc. Pacific Leafl.* 500.
Wallerstein, G., Greenstein, J. S., Parker, R., Helfer, H. L., and Aller, L. H.: 1963, *Astrophys. J.* **137**, 280.
Wampler, E. J.: 1967, *Publ. Astron. Soc. Pacific* **79**, 210.

Wampler, E. J.: 1968, *Astrophys. J.* **153**, 19.

Williams, P. M.: 1972, in G. Cayrel de Strobel and A. M. Delplace (eds.), 'L'âge des étoiles', *IAU Colloq.* **17**, Paris-Meudon Observatory, Ch. LIII.

Williams, R. E.: 1971, *Astrophys. J. Letters* **167**, L27.

Wolffram, W.: 1972, *Astron. Astrophys.* **17**, 17.

Woltjer, L.: 1958, *Bull. Astron. Inst. Neth.* **14**, 39.

Woolley, R., Pocock, S. B., Epps, E., and Flinn, R.: 1971, *Roy. Observ. Bull.*, No. 166

DISCUSSION

E. M. Burbidge: Is there any controversy still about the super-metal-rich stars?

Peimbert: Yes, but the main points that I want to make are (i) that these objects are fairly old and (ii) that many astronomers agree on the existence of at least some super-metal-rich stars.

Rodgers: Well determined super-metal-rich stars, e.g. 31 Aql, δ Pov, all have old disk orbits, implying that their birthplaces were nowhere near the galactic nucleus.

Danziger: I believe there is at least one planetary nebula in the solar neighbourhood with a helium abundance significantly greater than the average for planetary nebulae. Frogel, Persson and I have recently analyzed the spectrum of the peculiar southern planetary NGC 6302, and we obtain a value of $[N(He^+) + N(He^{++})]/N(H^+) \approx 0.19$–$0.20$.

Przybylski: Last year I obtained high dispersion spectra of three stars in the Magellanic Clouds. In two of them, HD 6884 (SMC) and HC 269781 (LMC), the helium abundances are normal.

Similar results were obtained previously for two other extragalactic stars, HD 32034 (LMC) and HD 7583 (SMC).

van Woerden: You gave abundances for Cas A. Do they refer to the fast condensations or to the stationary ones?

Peimbert: The abundances refer to the fast moving knots. We believe that the material in the fast condensations was ejected during the supernova explosion, while the stationary condensations were expelled before the explosion and correspond to the outer layers of the presupernova object. The chemical abundances of these two systems are very different; the stationary condensations seem to be nitrogen rich while the fast condensations are definitely nitrogen deficient.

Oort: Why did you prefer Cas A to the Crab Nebula in the discussion of abundances in supernova shells? As the expansion of the Crab nebula has not been appreciably decelerated it could not have been seriously contaminated with interstellar gas.

Peimbert: To obtain abundances from forbidden lines, a knowledge of the electron temperature is needed. For Cas A we have some information on the temperature from observations of the [O III] 4363/5007 line intensity ratio. There are no published determinations of the electron temperature for the Crab Nebula so it is not possible to estimate the abundance of the heavy elements. In the case of the He/H abundance ratio the temperature dependence of both recombination lines almost cancels out, consequently the helium overabundance reported by Woltjer (1958) for the Crab Nebula seems to be real.

Freeman: Concerning the second parameter for the morphology of the horizontal branch in globular clusters: Does anyone know of any horizontal branch models for which [N/H] or [CNO/H] were varied independently of [Fe/H]?

Hartwick: Vandenberg and I have computed some initial horizontal branch models and find that a CNO overabundance of approximately two could explain the NGC 7006 red horizontal branch.

Rodgers: Because of the dominant influence of CO on free O or C, one must be wary of using CNO as a lump sum parameter to explain observed CN features.

E. M. Burbidge: It makes good sense that there should be no correlation between N, C, and O abundances, either in individual objects or in regions in galaxies, because they are produced by different nuclear processes. N is the prime indicator of hydrogen-burning in the CNO cycle in stars $\gtrsim 2\,M_\odot$ (its production and final ratio to C is dependent on the temperature in the hydrogen-burning region). In regions in galaxies, as G. Burbidge pointed out many years ago, since helium production is swamped by the high overall abundance of helium it is again N that is the prime indicator of H-burning in aggregates of high luminosity stars.

A. V. Peterson: In connection with Margaret Burbidge's remark, the O-type subdwarfs HZ 44, $+25°$ 4655, HD 127493, show very strong He and N lines while the carbon and oxygen lines are very weak. Model atmosphere and abundance calculations give the He fractions by *number* for these stars as 0.38, 0.91, 0.50 respectively. The CNO *mass* fractions (e.g. $\mu_c N_c / \Sigma \mu_i N_i$) show carbon down by a factor of about 30, oxygen down by a factor of about 10, and nitrogen up by about a factor of 10 relative to solar abundances (Peterson, A. V.: 1970, Ph.D. Thesis, Cal. Inst. Tech.). Thus these abundances imply that hydrogen-burning has proceeded via the CNO cycle in these stars.

Kerr: Mass loss from evolved stars is important for the chemical evolution of a galaxy. G. R. Knapp and I have recently carried out high-sensitivity 21-cm observations of a giant elliptical, NGC 4472, and of a number of globular clusters, and we have still been unable to find neutral hydrogen in any of these systems. For NGC 4472 we can set a limit of $8 \times 10^7 \, M_{\odot}$, and for the globulars M22 and M4 our limits are about one solar mass. These limits are well below the values that might be expected from conventional mass-loss rates. Other explanations need to be considered, but it seems quite possible that the mass-loss rate for evolved stars is lower than that normally accepted.

Arp: Could gas have been blown out of NGC 4472?

Kerr: Not obviously.

Rickard: Dave Phillips at Cerro Tololo, and I at ESO, have used image-tube spectrographs to look for Hα and other hydrogen recombination lines in several 'open' globular clusters. So far we haven't detected any emission above the sky background and there seems to be little interstellar gas in the ionized form.

Kerr: In discussing Churchwell and Mezger's report of a very low abundance of ionized helium in some H II regions near the galactic centre, Dr Peimbert suggested as one of several possible explanations that the helium might be shielded by dust from the ionizing radiation. P. D. Jackson of Maryland points out in a forthcoming paper that there is at least one H II region near the Sun which has a low helium abundance, and this happens to be a very dusty nebula. This analogy supports the view that the helium in the regions near the centre is mainly neutral, due to shielding by dust.

Robinson: Can one say what fraction of the interstellar gas has been through nuclear processing in stars?

Peimbert: An estimate is possible using a model of galactic chemical evolution and studying the changes in parameters like D/H, ^{3}He/H or ^{4}He/H. At present the results are strongly model dependent.

DIFFERENCES BETWEEN GALAXIES

SIDNEY VAN DEN BERGH

*David Dunlap Observatory, University of Toronto, Ontario, Canada, and
Institute of Astronomy, University of Cambridge*

Abstract. It is pointed out that the metallicity of galactic and extragalactic globular clusters correlates with the density of the regions in which they were formed. Within individual galaxies the regions of highest density have the highest metallicity.

Some of the differences in family traits between the globulars in the Andromeda Nebula, the Galaxy and the Magellanic Clouds might be understood if the M31 cluster system is $\sim 1 \times 10^9$ yr *older* and the SMC cluster system $\sim 1 \times 10^9$ yr *younger* than galactic globular clusters.

1. Evidence for Age Differences Between Galaxies

It is usually assumed that the differences between galaxies along the Hubble sequence are due primarily to differences in initial conditions which led to variations in their rate of evolution, rather than to some being younger than others. There is, however, some evidence which suggests that age differences might also be important. Perhaps the strongest such evidence comes from the study of globular clusters associated with members of the Local Group. Intercomparison of the morphology of the colour-magnitude diagrams of globular clusters in the Galaxy and in the Magellanic Clouds shows a number of significant differences. The most striking of these is that Small Cloud clusters of intermediate metal abundance ($\Delta V \sim 2.5$) have a much stronger population on the red side of the RR Lyrae variable gap than do clusters of similar metal abundance in the Galaxy (see Table I). According to computations by Rood (1973) such a difference could be due to either of the following two causes: (1) the hydrogen abundance in the Small Cloud might be a few percent greater (i.e. the helium abundance lower) than it is in galactic globular clusters, or (2) the globular clusters in the Small Magellanic Cloud might be $\sim 1 \times 10^9$ yr *younger* than are their galactic counterparts (see Figure 1). Presently available observational data (cf. Peimbert p. 141) do not favour the view that the helium abundance in emission nebulae in the Magellanic Clouds differs from that in the Galaxy, although the small change that is required to account for the differences between the globulars in the Clouds and in the Galaxy probably cannot yet be ruled out with certainty.

Spectral scans of a few strong-lined globulars in M31 by Spinrad and Schweizer (1972) appear to show that these clusters exhibit quite strong hydrogen lines. This suggests that these globulars have strongly developed blue horizontal branches despite the fact that their metallicity is high. According to Rood's computations such an effect could be understood if the globulars in M31 are $\sim 1 \times 10^9$ yr *older* than are their galactic counterparts. This might possibly indicate that high mass (and high density) galaxies such as M31 collapsed somewhat earlier than did galaxies of lower density and lower mass.

Observations of galactic globular clusters show that many of the metal-poor glob-

TABLE I

CM diagrams of cloud clusters

	M_V(H.B.)	M_V(R.G.)[b]	ΔV[c]	HB[d]	$(B-V)_{0max}$[e]	Reference
SMC						
NGC 121[a]	19.5	16.8	2.7	R	1.6	Tifft (1963)
NGC 339	?	16.0:	?	?	2.2:	Gascoigne (1966)
NGC 361	?	16.5	?	?	2.2	Arp (1958)
NGC 419	19.3:[f]	16.8	2.5:	R	4.1	Walker (1972b)
L 1	?	17.0	?	?	2.2	Gascoigne (1966)
K 3	19.1	16.6	2.5	R	2.3	Walker (1970)
LMC						
NGC 1466[a]	19.1	16.3	2.6	B	1.5	Gascoigne (1966)
						Hodge (1960a)
NGC 1783	19.2	16.8	2.4	B	1.8	Gascoigne (1962)
NGC 1841	19.6	16.6	3.0	B	1.7	Gascoigne (1966)
NGC 1846	?	16.0	?	?	2.5	Hodge (1960a)
NGC 1978[a]	?	16.4	?	?	2.0	Hodge (1960b)
NGC 2231	?	16.5:	?	?	1.6:	Gascoigne (1966)
NGC 2257[a]	19.2	16.1	3.1	B	1.8	Gascoigne (1966)
						Walker (1972a)

[a] Cluster contains RR Lyrae variables.
[b] Magnitude of giant branch at $(B-V)_0 = 1.4$
[c] Difference between M_V(H.B.) and M_V(R.G.)
[d] R – most of stars on red side of RR Lyrae gap, B – most stars on blue side of gap.
[e] Colour of reddest cluster giant star. In some clusters the reddest star might be a field object.
[f] The colons indicate uncertain values.

ulars at distances $\gtrsim 50$ kpc from the galactic nucleus have red horizontal branches. For $R = 50$ kpc the *free-fall* time-scale* is $\sim 1 \times 10^9$ yr. According to Rood's calculations this is sufficient to account for the observed differences between the horizontal branch morphology of distant globular clusters and that of globulars nearer to the galactic centre.

An evolutionary picture in which the metal-poor globular clusters in the outer halo of the Galaxy are younger than are the clusters nearer to the galactic nucleus is consistent with the model of galactic evolution that has been proposed by Larson (1969).

Two dwarf spheroidal galaxies are known to have red horizontal branches of the NGC 7006 type. According to the calculations by Rood (1973) this indicates that these dwarf spheroidals are slightly younger than the oldest galactic Population II objects. A similar suggestion has previously been made by Castellani *et al.* (1973) who based their conclusion on the properties of the RR Lyrae variables in dwarf spheroidal systems.

---

* It should be emphasized that the free-fall time-scale represents a lower limit to the actual collapse time-scale. This is so because the collapse of the Galaxy might have been delayed by turbulent pressure or by radiation pressure from a quasar-like galactic nucleus.

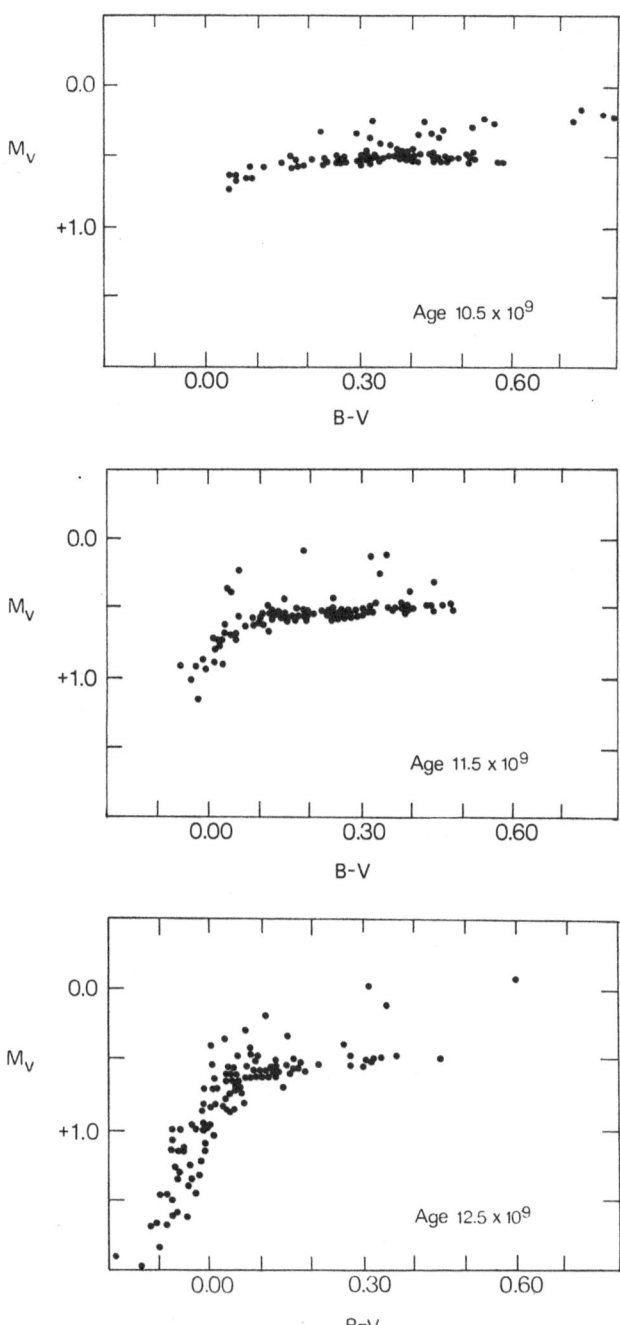

Fig. 1.   Dependence of the morphology of the globular cluster horizontal branch on age (in years) for a cluster with $X = 0.75$ and $Z = 0.001$ according to Rood (1973). Note that the luminosity level of the horizontal branch is insensitive to age.

2. Differences Between the Globular Cluster Systems Associated
with Different Galaxies

The globular cluster systems associated with different galaxies exhibit a number of striking differences. Perhaps the most fundamental of these is that the mean metallicity of globular cluster families differs from galaxy to galaxy. The data in Table II seem to show a correlation between the mean metallicity of a cluster family and the mass (or mean density) of the galaxy with which it is associated. All of the globulars associated with the Fornax dwarf spheroidal system appear to be metal-poor. The certain globular clusters (i.e. those that contain RR Lyrae stars) in the Magellanic Clouds, such as NGC 121, 1466, 1978 and 2257, all appear to be moderately metal-poor. The globular clusters associated with the Galaxy exhibit the entire range from extreme metal deficiency to metal abundances close to that of the Sun. The globular clusters near the Andromeda Nebula exhibit the same range in line strength as do those associated with the Galaxy. The mean metallicity of the M31 globulars is, however, considerably greater than is that of the clusters associated with the Galaxy. Finally, recent observations by Ables *et al.* (1973) appear to show that the globular clusters associated with the giant elliptical galaxy M87 are even redder (and hence presumably metal richer) than are those in M31.

Within the Galaxy, metal-poor globulars appear to have formed predominantly in low density halo regions, whereas metal-rich globulars are mainly found in the high density inner disk and central bulge of the Galaxy. Taken at face value these observations suggest that the metal abundance of globular clusters was mainly determined by the density of the regions in which they were formed.

Intercomparison of the globular clusters in the Galaxy, in the Magellanic Clouds and in M31 provides strong evidence for the conclusion that clusters in different galaxies exhibit differing 'family traits'. For example the well-known correlation between the morphology of the horizontal branch and metallicity of galactic globular

TABLE II

Relation between mass, line-strength and colour
of globular clusters

Object	$\mathfrak{M}/\mathfrak{M}_\odot$	$\langle L \rangle$	$\langle B - V \rangle_0$
Fornax[a]	1.3×10^7	0	0.67[b]
Galaxy	1.3×10^{11}	4	–
Andromeda	3.1×10^{11}	8	0.73[c]
M87	$\sim 1 \times 10^{13}$	–	0.8[d]

[a] $(\mathfrak{M}/\mathfrak{M}_\odot)/(L_V/L) = 1$ assumed.
[b] No galactic reddening correction applied.
[c] Values refer only to halo clusters (van den Bergh, 1969).
[d] Colours from Ables *et al.* (1973).

clusters does not seem to be valid for M31 or for the Magellanic Clouds. Furthermore, many of the globular clusters in the Magellanic Clouds appear to contain exceedingly red stars. Presumably these are mild carbon stars situated at the tip of the asymptotic giant branch. Probably these stars have managed to mix the carbon formed in their interiors to their surfaces. No stars of this type have been discovered in any galactic globular clusters. Stars with $B - V > 2.0$ have also been observed in the Sculptor dwarf spheroidal system (Hodge, 1965). Additional support for the notion that the Population II component in the Magellanic Clouds may exhibit some similarity to the Population II observed in dwarf spheroidals is provided by two types of observations of W Virginis stars: (1) no long-period ($P > 10$ days) W Virginis stars have so far been discovered in any dwarf spheroidal system. In this respect the dwarf spheroidals resemble the Magellanic Clouds in which W Virginis stars appear to be exceedingly rare. (2) Van Agt (1967) has shown that the short-period W Virginis stars (BL Herculis stars) in dwarf spheroidal systems obey a different period-luminosity relation than do those that occur in galactic globular clusters. Photoelectrically-calibrated photographic observations of a 1.43 day SMC field variable near NGC 121 (Tifft 1963) show that this object must lie on the dwarf spheroidal period-luminosity relation for BL Herculis stars rather than on that for galactic globular clusters.

3. Relation Between Metallicity and Mean Density in Galaxies

Within individual galaxies there is clear-cut evidence for metal abundance gradients. The highest metallicity occurs in the dense cores of galaxies (McClure, 1969). Progressively lower heavy-element abundances are observed as regions of lower density are approached. The correlation between mean metallicity and absolute magnitude (and hence presumably mean density) which is observed in elliptical galaxies (Faber, 1973) also seems to hold for spirals and irregulars. A number of lines of evidence (see Section 4) appear to indicate that young stars in the Magellanic Clouds are mildly metal-deficient compared to those which are currently being formed in the solar vicinity of the Galaxy. Among systems containing a young population component, the correlation between density (or absolute magnitude) and metallicity manifests itself in its most extreme form in the blue dwarf galaxies that have recently been studied by Searle and Sargent (1972).

The relation between metallicity and absolute luminosity is, within the accuracy of the data, the same for elliptical galaxies located inside and outside of clusters (McClure and van den Bergh, 1968). This suggests that either (1) ellipticals in the field have been ejected from clusters or (2) the evolutionary history of a galaxy is determined exclusively by its initial conditions and not by its environment.

It would be very important to gain some theoretical understanding of the reason for the correlation between the metallicity of a stellar system and its density. Such a correlation might be understood if the luminosity function of star formation (Salpeter function) in dense regions favoured high-mass stars that evolved to produce heavy elements.

4. Differences among Young Stellar Populations

Differences in metallicity (but not in helium abundance) are strongly suggested by a number of recent studies of young stellar populations in galaxies. Probably the most direct evidence for differences in chemical composition between galaxies comes from observations of H II regions. Such observations show that the emission regions in the dwarf galaxy NGC 6822 (Peimbert and Spinrad, 1970) are deficient in nitrogen and oxygen by factors of 6 and 1.7 respectively compared to H II regions near the Sun. In its most extreme form this phenomenon manifests itself in the very low luminosity galaxies studied by Searle and Sargent (1972) in which some heavy elements are ~ 10 times less abundant than they are in the Orion Nebula.

Direct observations of the metal abundances in the most luminous stars in the Magellanic Clouds have so far produced rather confusing results. Perhaps the best that can be said is that currently available observational data are not inconsistent with the view that young stars in the Magellanic Clouds are mildly metal-deficient compared to their galactic counterparts. In a series of papers on open clusters in the Magellanic Clouds, published in the late 1950's, Arp was able to show that young and intermediate age clusters in the Clouds exhibit a number of striking differences from their galactic counterparts. Presumably such differences imply that the evolutionary tracks of highly evolved young stars are quite sensitive to chemical composition.

Additional evidence for composition differences between young stellar populations in galaxies of different type may be obtained from observations of the magnitudes of the brightest red and blue stars of Population I. Table III lists the magnitudes of the brightest red and blue stars in all late-type galaxies for which photoelectrically-calibrated magnitudes and accurate distances are available. The data in the Table are

TABLE III
Magnitudes of brightest stars in galaxies

Name	M_V	Brightest blue star	Brightest red star	ΔV	Reference
Galaxy[a]	-20.4:[b]	5.68	7.69	2.01	Johnson and Morgan (1955)
NGC 2403	-19.3	18.25[c]	19.98[d]	1.73	Tammann and Sandage (1968)
LMC	-18.5	9.11	10.5:[e]	1.39	van den Bergh (1968)
SMC	-16.8	10.13	10.8:[e]	0.67	van den Bergh (1968)
NGC 6822	-15.7	16.61	16.65	0.04	Kayser (1967)
IC 1613	-14.8	17.15	16.5:[d][e]	-0.65	Sandage (1971)

[a] Data refer to h and χ Persei only.

[b] M_V is derived from $M_B = -19.7 \pm 0.3$ (van den Bergh, 1972) and $B - V = 0.7$ from $V(\text{rot})_{\max} \simeq 250$ km s^{-1} (van den Bergh, 1971).

[c] $B - V = 0.0$ assumed.

[d] Refers to maximum light.

[e] $B - V = 2.0$ assumed.

plotted in Figure 2. This figure shows that there is a tight correlation between ΔV, the *difference* in visual magnitude of the brightest blue and red stars, and the absolute magnitude of the galaxy in which these stars occur. Inspection of the data in Table III shows that the absolute magnitudes of the brightest blue stars in galaxies are very sensitive to the absolute magnitude (and hence presumably to the total Population I content) of the galaxy in which they occur. As Tammann and Sandage (1968) have already pointed out, the magnitudes of the brightest red stars do *not* appear to depend on the luminosity of their parent galaxy. Observations such as those by Searle and Sargent (1972) suggest that low ΔV values go with low metal abundance. This view is confirmed by observations of M33. Searle (1971) has shown that the heavy element abundance in this galaxy decreases with increasing distance from its nucleus. This suggests that ΔV should be smaller in the outer regions of M33 then it is near the nucleus of this galaxy. This is confirmed by Walker (1964) and by Madore (1971) who find that the ratio of the number of bright blue stars to the number of bright red stars decreases with increasing distance from the centre of the Triangulum Nebula. Similarly Hartwick (1970) has been able to show that the ratio of blue to red supergiants in the direction of the galactic centre is greater than it is in the anti-centre direction.

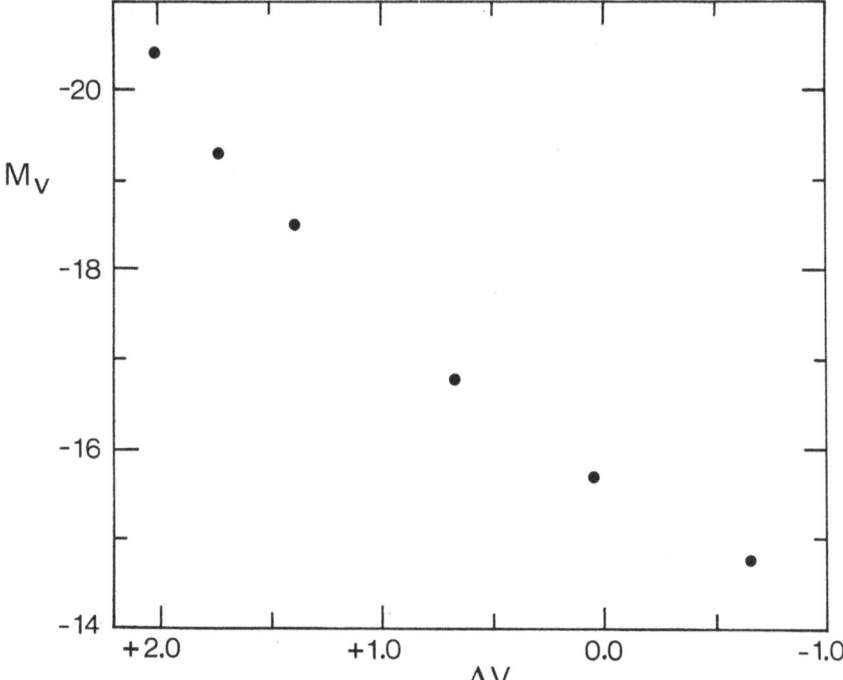

Fig. 2. The figure shows that ΔV, the difference in visual magnitude between the brightest blue and red stars, is strongly dependent on the absolute magnitude M_V of the galaxy in which they occur. Since the upper end of the luminosity function of galaxies is very sparsely populated, the very small dispersion in the figure is almost certainly fortuitous.

In summary it appears that there is now strong observational evidence in support of the conclusions that: (1) chemical composition differences exist between different galaxies and (2) radial composition gradients exist within individual galaxies. The trend of the data is such that *galaxies with a high mean density exhibit a higher metal abundance than do galaxies with a low mean density. Similarly, within individual galaxies the regions with the highest heavy element abundance also appear to be the regions of highest density.*

5. Variable Stars and Chemical Abundance

It has long been known (van den Bergh, 1958) that the mean period of Cepheids interior to the Sun is longer than is that of Cepheids in the anti-centre direction. This observation indicates that Cepheids were formed from interstellar material that had a radial composition gradient. The sense of this gradient suggests that the interstellar gas in the outer part of the galactic disk (where Cepheid periods are similar to those in the Large Magellanic Cloud) has a lower metal abundance than does the gas in the inner region of the Galaxy. Such a trend is strongly supported by the observations of H II regions in a number of nearby external galaxies. These observations clearly show a higher heavy element abundance in the inner parts of galaxies than they do in their outer regions (Peimbert, 1968; Searle, 1971).

Additional evidence in support of the view that the evolutionary tracks of young high-evolved stars are quite sensitive to differences in chemical composition is provided by the observation that the period-frequency distribution of Cepheid variables in the Magellanic Clouds differs radically from that in the Galaxy. The data in Table IV show that the Cepheids in the Small Magellanic Cloud have a much shorter median period than do those in the Galaxy. The Cepheids in the Large Cloud are seen to have a median period that is intermediate between that of Cepheids in the Galaxy and in the Small Cloud.

TABLE IV

Periods of Cepheids (Gaposchkin, 1971)

	Galaxy	LMC	SMC
Median $\log P$ (day)	0.78	0.64	0.42

It is not yet certain whether the period-maximum amplitude relation for Cepheids in the Magellanic Clouds is similar to that for galactic Cepheids. Originally Arp and Kraft (1961) appeared to have shown that the Cepheids in the Magellanic Clouds had a period-maximum amplitude relation different from that which is observed near the Sun. If correct, this conclusion would have far-reaching consequences because it might indicate that the period-luminosity relation of Cepheids is not the same in all galaxies. More recently Shaltenbrand and Tammann (1970) have found that there is no evidence for a systematic difference between the period versus maximum-pulsation-amplitude relations in the Galaxy and in the Magellanic Clouds. This question has recently been re-examined by Madore (unpublished). His results are plotted in Figures 3 and 4. These figures show that it is not yet possible to state with certainty whether the period versus pulsation-amplitude relations in the Magellanic Clouds and in the Galaxy are in fact the same.

The Cepheid period-luminosity-amplitude relation of Sandage and Tammann (1971) yields distance moduli $(m-M)_B = 18.91$ and $(m-M)_B = 19.35$ for the LMC and SMC respectively. According to Graham (1973), the mean of the median magnitudes of the RR Lyrae variables in these systems are $\langle \dot{B} \rangle = 19.22$ in the LMC and $\langle \dot{B} \rangle = 19.55$ in the SMC. From these data, $\langle M_{\dot{B}} \rangle = +0.31$ in the LMC and $\langle M_{\dot{B}} \rangle = +0.20$ in the SMC (See Table V). These values are significantly *brighter* (Sandage, 1972) than the

TABLE V

Absolute magnitudes of RR Lyrae variables

	LMC	SMC
$(m-M)_B$ (Sandage and Tammann, 1971)	18.91	19.35
$\langle B \rangle$ (Graham, 1973)	19.22	19.55
$\langle M_{\dot{B}} \rangle$ (RR)	+0.31	+0.20
$\langle M_{\dot{B}} \rangle$ (RR) (van Herk, 1965)	+0.87 ± 0.22	

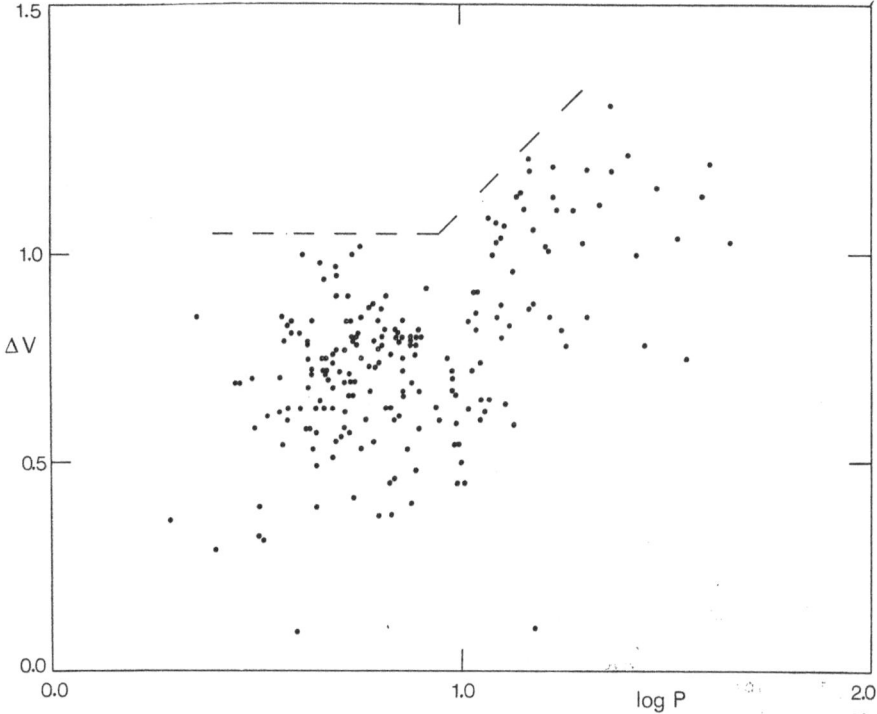

Fig. 3.   Period-amplitude relation for galactic Cepheids. The dashed line shows the adopted upper envelope for galactic Cepheids.

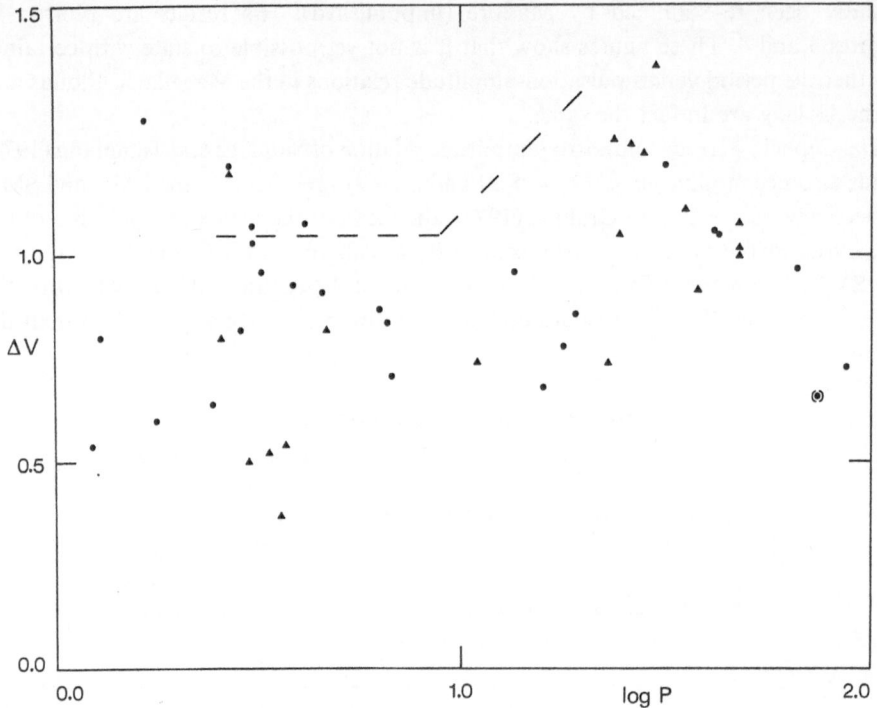

Fig. 4.   Period-amplitude relation for the SMC (●) and LMC (▲). Also shown is the upper envelope to the *galactic* period-amplitude relation.

value $(M_B^i) = +0.87 \pm 0.22$ m.e. which van Herk (1965) obtained for galactic RR Lyrae variables. Taken at face value, these results indicate that either:

(1) the RR Lyrae variables in the Magellanic Clouds are brighter by ~ 0.5 mag than their galactic counterparts or

(2) the Cepheid period-luminosity-amplitude relation in the Clouds is ~ 0.5 mag fainter than it is in the Galaxy.

I am grateful to Drs Newell and Rood for making unpublished material available to me.

References

Ables, H. D., Newell, E. B., and O' Neil, E. J.: 1973, *Paper Presented at the Los Angeles Meetings of the Astron. Soc. Pacific.*
Arp, H. C.: 1958, *Astron. J.* **63**, 487.
Arp, H. C. and Kraft, R. P.: 1961, *Astrophys. J.* **133**, 420.
Castellani, V., Giannone, P., and Renzini, A.: 1973, in J. D. Fernie (ed.), 'Variable Stars in Globular Clusters and in Related Systems', *IAU Colloq.* **21**, 217.
Faber, S. M.: 1973, *Astrophys. J.* **179**, 731.
Gaposchkin, C. P.: 1971, in A. B. Muller (ed.), *The Magellanic Clouds*, D. Reidel Publ. Co., Dordrecht-Holland, p. 34.
Gascoigne, S. C. B.: 1962, *Monthly Notices Roy. Astron. Soc.* **124**, 201.

Gascoigne, S. C. B.: 1966, *Monthly Notices Roy. Astron. Soc.* **134**, 59.
Graham, J. A.: 1973, in J. D. Fernie (ed.), 'Variable Stars in Globular Clusters and in Related Systems', *IAU Colloq.* **21**, 120.
Hartwick, F. D. A.: 1970, *Astrophys. Letters* **7**, 151.
Hodge, P. W.: 1960a, *Astrophys. J.* **132**, 341.
Hodge, P. W.: 1960b, *Astrophys. J.* **132**, 346.
Hodge, P. W.: 1965, *Astrophys. J.* **142**, 1390.
Johnson, H. L. and Morgan, W. W.: 1955, *Astrophys. J.* **122**, 429.
Kayser, S. E.: 1967, *Astron. J.* **72**, 134.
Larson, R. B.: 1969, *Monthly Notices Roy. Astron. Soc.* **145**, 405.
McClure, R. D.: 1969, *Astron. J.* **74**, 50.
McClure, R. D. and van den Bergh, S.: 1968, *Astron. J.* **73**, 1008.
Madore, B. F.: 1971, University of Toronto (M.Sc. Thesis).
Peimbert, M.: 1968, *Astrophys. J.* **154**, 33.
Peimbert, M. and Spinrad, H.: 1970, *Astron. Astrophys.* **7**, 311.
Rood, R. T.: 1973, *Astrophys. J.* **184**, 815.
Sandage, A. R.: 1971, *Astrophys. J.* **166**, 13.
Sandage, A. R.: 1972, *Quart. J. Roy. Astron. Soc.* **13**, 202.
Sandage, A. R. and Tammann, G. A.: 1971, *Astrophys. J.* **167**, 293.
Schaltenbrand, R. and Tammann, G. A.: 1970, *Astron. Astrophys.* **7**, 289.
Searle, L.: 1971, *Astrophys. J.* **168**, 327.
Searle, L. and Sargent, W. L. W.: 1972, *Astrophys. J.* **173**, 25.
Spinrad, H. and Schweizer, F.: 1972, *Astrophys. J.* **171**, 403.
Tammann, G. A. and Sandage, A. R.: 1968, *Astrophys. J.* **151**, 825.
Tifft, W. G.: 1963, *Monthly Notices Roy. Astron. Soc.* **125**, 199.
van Agt, S. L. T. J.: 1967, *Bull. Astron. Inst. Neth.* **19**, 275.
van den Bergh, S.: 1958, *Astron. J.* **63**, 492.
van den Bergh, S.: 1968, *J. Roy. Astron. Soc. Can.* **62**, 145.
van den Bergh, S.: 1969, *Astrophys. J. Suppl.* **19**, 145.
van den Bergh, S.: 1971, *Publ. Astron. Soc. Pacific* **83**, 663.
van den Bergh, S.: 1972, *Astron. Astrophys.* **20**, 469.
van Herk, G.: 1965, *Bull. Astron. Inst. Neth.* **18**, 71.
Walker, M. F.: 1964, *Astron. J.* **69**, 744.
Walker, M. F.: 1970, *Astrophys. J.* **161**, 835.
Walker, M. F.: 1972a, *Monthly Notices Roy. Astron. Soc.* **156**, 459.
Walker, M. F.: 1972b, *Monthly Notices Roy. Astron. Soc.* **159**, 379.

DISCUSSION

Graham: The present data do not indicate any difference between the mean absolute magnitude of the Magellanic Cloud RR Lyrae stars and that of the RR Lyraes in the Galaxy. I find that a time averaged mean visual magnitude of $+0^m5$ seems to fit well in all three cases (*IAU Colloq.* **21**, 1973).

van den Bergh: The data in Table V give the median magnitudes that you obtained for the RR Lyrae variables in the globular clusters in the Magellanic Clouds. As you point out, these values are probably too bright by 0.1 or 0.2 mag due to the effects of background enhancement. This small overestimate of the luminosity of the RR Lyrae variables in the Clouds is almost exactly compensated for by the fact that photoelectric observations of galactic RR Lyrae variables (Fitch, W. S., Wisniewski, W. Z., and Johnson, H. L.: 1966, *Lunar and Planetary Lab. Communications* No. 71) show that van Herk (1965) used a magnitude system which was about 0.2 mag too bright for the galactic RR Lyrae stars. There remains therefore a discrepancy of between two and three standard deviations between the absolute magnitudes of the galactic RR Lyrae stars and the absolute magnitudes of the cluster type variables in each of the Clouds that is obtained *by using the period-luminosity-amplitude relation* of Sandage and Tammann (*Astrophys. J.* **167**, 293, 1971). This discrepancy is virtually removed if the period-luminosity-colour relation (Sandage, A. R. and Tammann, G. A.: 1969, *Astrophys. J.* **151**, 531), rather than the period-luminosity-amplitude relation, is used to estimate the distance moduli of the Clouds. According to Sandage and Tammann (1969) the P-L-C relation yields distance moduli

$(m - M)_B$ of 18.65 and 19.05 for the LMC and SMC respectively. These values yield $\langle M_{\dot{B}}\rangle_{RR} = 0.57$ for the LMC and $\langle M_{\dot{B}}\rangle_{RR} = 0.50$ for the SMC, values which differ only marginally from van Herk's luminosity for galactic RR Lyrae variables. The most straightforward interpretation of this result is that the period-luminosity-colour relation for cepheids yields a more nearly correct distance to the Clouds than does the period-luminosity-amplitude relation.

Gascoigne: It has been suggested by Christy (1971, in A. Muller (ed.), *The Magellanic Clouds*, D. Reidel Publ. Co., p. 136) and Iben (*Ann. Rev. Astron. Astrophys.* **5**, 606, 1967) that the SMC cepheids are metal-deficient relative to those in the Galaxy. Changing the metal content changes the mass-luminosity and $T_{\text{eff}} - (B - V)$ relations, and it can be shown that changing Z from 0.02 to 0.005 makes cepheids at a given period and colour fainter by 0.3–0.4 mag. In the present context this brings the SMC nearer by about 0.35 mag in the modulus.

The observed PLC relations for cepheids in the Galaxy or Magellanic Clouds are satisfied with a scatter better than 0.1 mag, observational errors included. This implies a very high degree of chemical homogeneity in these cepheids.

Sargent: There are two pieces of evidence regarding the possible existence of young galaxies which you did not discuss. The first concerns Ring Galaxies, of which about one dozen are known. These are galactic sized objects with diameters of around 10 kpc. They have no optical nucleus; the rings appear to be composed of giant H II regions. Kinematic work (unpublished) by Lynds, by Speigel and Thys and by myself shows that the rings are expanding (or contracting, depending on which side is closest) on a timescale of about 10^8 yr. Recent studies of two rings by Searle and myself (unpublished) also indicate a time scale of a few time 10^8 yr for the dominant stellar populations of these objects.

The second piece of evidence concerns compact clusters of galaxies – Seyfert's Sextet is a typical example – in which the galaxies are almost touching one another. In such systems the timescale for the galaxies to collide and, presumably, to tear one another apart, is much less than the Hubble time. Aspects of this problem have been discussed by Peebles in his book *Physical Cosmology*.

van den Bergh: Cannot Seyfert's Sextet be interpreted as a cluster, initially larger, which has ejected some of its members and thereby lost kinetic energy, so shrinking to its present size?

Sargent: Numerical experiments suggest that the probability of this being the case is very low.

Tifft: Do not Seyfert's Sextet galaxies show spectral characteristics of *old stars* –, i.e. strong HK absorption lines?

Sargent: Yes, some of the galaxies show emission lines in addition to H and K absorption from old stars, but generally the spectra give no sign that the galaxies are young.

Heidmann: We made 21-cm line observations of twelve Haro galaxies (*Astron. Astrophys.* **29**, 217) and found a very strong correlation between mean optical surface brightness and neutral hydrogen mass-to-luminosity ratio: when the brightness varies from 19 to 21 mag (arc sec)$^{-2}$, this ratio increases from 0.03 to 1.0 in solar units. We think this could be accounted for by flashes of stars occurring in Haro galaxies, such as those advocated by Sargent and Searle for Zwicky compact galaxies.

Lewis: What effect does your work on the absolute magnitudes of bright blue stars and the period-luminosity relation of cepheids have upon their role as distance indicators to nearby galaxies such as NGC 2403?

van den Bergh: Since there is now some suspicion that the cepheid period-luminosity relation may depend on chemical composition, it would be most prudent to compare the galactic period-luminosity relation only with that in other luminous galaxies of types Sb or Sc.

Present evidence seems to suggest that bright blue variables may form a continuum between the brightest stable main sequence stars and supernovae of type II. If this conclusion is correct then the blue Hubble-Sandage variables are not suitable as precision distance indicators.

DWARF CONTENT OF OLD GALAXY POPULATIONS

A. E. WHITFORD

Lick Observatory, Board of Studies in Astronomy and Astrophysics
University of California, Santa Cruz, Calif., U.S.A.

Abstract.* Scanner observations of five galaxies in the region near 9910 Å showed that the strength of the dwarf-sensitive Wing-Ford band is below the level of detection. This result is at variance with the predictions of strongly dwarf-enriched population models and in accord with the evidence for giant-dominated radiation at 2.3μ reported by Baldwin *et al.* (1973) on the basis of the strong CO band found in three galaxies. In the absence of any evidence for a power-law mass function, the consequences of a rounded function are examined.

A cool M-star component in the stellar population of elliptical galaxies and the central regions of Sb spirals is required by the strong infrared radiation found in wide-range spectrophotometric studies (e.g. Stebbins and Whitford, 1948; Johnson, 1966). Spinrad and Taylor's (1971; hereafter ST) population models for the nuclear regions of M31 and M81, based on extensive scanner measurements of the continuum and spectral features, identified this cool component with a dwarf-enriched lower main sequence. Observational evidence came mainly from the dwarf-sensitive Na feature at $\lambda\lambda$8183–8195. The more luminous E galaxies and the central regions of the more luminous Sb's form a homogeneous group with similar colours and strong spectral features (Faber, 1972).

The unidentified molecular feature at about 9910 Å, first noted by Wing and Ford (1969) in the spectra of late dwarfs, comes at a wavelength where the cool M-star component contributes a larger fraction of the total light relative to the giant K0 III component dominant in the violet (Morgan and Mayall, 1957). ST's model calculations show that contribution of the M stars is 22% at 8190 Å, 33% at 9910 Å, and 47% at 2.3μ. The preliminary scanner investigation of the strength of the Wing-Ford band (Whitford, 1972) has been extended to include 35 K and M stars, reaching dwarfs as late as dM8. The luminosity discrimination is found to increase with advancing type and to provide strong differentiation for types M5 and later.

No detectable indication of a dwarf population could be found from scanner measurements of the strength of the Wing-Ford band in 5 galaxies: the nuclear regions of the Sb spirals M31 and M81, and the ellipticals M32, NGC 4472, and NGC 4486. This result is in accord with the observations of Baldwin *et al.* (1973), who found that the giant-sensitive CO band at 2.3μ appeared at full strength in the nuclear regions of M31 and M81.

ST's scanner detection of the Na feature at a level well above the range of observational uncertainty stands in opposition to the results at longer wavelengths. Additional observations by a technique that eliminates the need to calibrate out competing sideband effects are desirable. A re-examination of the failure to find the Na doublet on

---

* A more detailed report is in preparation, to be submitted to the *Astrophysical Journal*.

photographic spectra of galaxies at 200 Å mm^{-1} (Whitford, 1966) showed that the line strength to be expected from the ST dwarf-enriched model for M31 would have been below the detection limit. Equivalent widths of the Na lines were determined on stellar spectra taken at 32 Å mm^{-1} for this analysis.

Table I shows a comparison of the average observed index Δ (9916) for galaxies and the indices predicted by population models. The index, which measures the absorption in magnitudes in a 32 Å scanner band, had observational mean error of ± 0.008 for M31, ± 0.010 for M81, and ± 0.024 for NGC 4472 and NGC 4486.

TABLE I

Model predictions

Galaxy	Model	M-star Component	$\mathfrak{M}/L_V$	$\Delta(9916)$
M31	ST	$\mathfrak{M}^{-3.6}$ to $\mathfrak{M} = 0.1$	43.5	0.054
M81	ST	M stars partly giants	27	0.038
NGC 3379	ST	rounded at $\mathfrak{M} = 0.15$	33	0.033
M31	mod. ST	$\mathfrak{M}^{-3.0}$ to $\mathfrak{M} = 0.1$	13	0.027
M31	mod. ST	$\mathfrak{M}^{-2.0}$ to $\mathfrak{M} = 0.1$	2.4	0.018
M31	mod. ST	M stars all giants	0.8	0.017
Mean observed index, 5 galaxies				0.012

Notes: M stars in models all dwarfs except as noted. ST: Spinrad-Taylor model as published. Mod. ST: Modified ST model.

The contradiction between the observations and the ST model for M31 must be considered in the light of the current downward revision of $\mathfrak{M}/L$ values of galaxies. Adoption of the new value for the velocity dispersions in the nuclear region of M31 found by Morton and Thuan (1973) would lower the ratio from $\mathfrak{M}/L_v = 43.5$ to $\mathfrak{M}/L_v = 13$. New observations of velocity dispersion in ellipticals (Morton and Chevalier, 1973) have not included the giant E's here considered. Existing dispersion data and results on E galaxies from the Page (1961) double-galaxy method suggest that for the giant E's $15 < \mathfrak{M}/L_v < 30$, assuming $H = 50$ km s^{-1} Mpc^{-1}.

The table shows that a modified M31 model based on a power-law mass function that combines a satisfactory value of $\mathfrak{M}/L$ and a predicted Δ (9916) index compatible with observations is difficult, but perhaps marginally possible. Only a modest reduction in the exponent in the mass function $N(\mathfrak{M}) \propto \mathfrak{M}^{-x}$ from the high value $x = 3.6$ in the ST model is possible if all mass is assumed to be in the form of stars.

If a high proportion of late dwarfs is no longer required to satisfy the observed spectral features, however, there would appear to be grounds for questioning the power-law mass function. There is no obvious physical process that would suddenly stop fragmentation at some particular low-mass cut-off. $\mathfrak{M}/\mathfrak{M}_\odot = 0.1$ was adopted in the ST model. Kumar (1972) has pointed out that the known stars near the Sun of mass $\mathfrak{M}/\mathfrak{M}_\odot < 0.08$, even though destined after a short life to become black dwarfs, could be significant in the mass budget of the solar neighbourhood. The crowding of the mass into the stars nearest the cut-off in a 'sawtooth' power-law mass function appears artificial. In the ST model for M31, the M8V stars at $\mathfrak{M}/\mathfrak{M}_\odot = 0.1$ represent 67% of the mass.

Larson (1973) has proposed a probabilistic fragmentation theory that results in a gaussian mass-distribution function $\theta(\mathfrak{M}) = \mathfrak{M}N(\mathfrak{M})$. In the solar neighbourhood the stars contributing the greatest mass per unit interval of $\log\mathfrak{M}$ are those near $\mathfrak{M}/\mathfrak{M}_\odot = 1.4$, and the observed rounding off at low masses (e.g. Hartmann, 1970) finds a natural explanation. The reduced proportion of M5–M8 dwarfs, relative to a power-law mass function, would yield a $\mathfrak{M}/L$ ratio too low for the type of galaxies here considered.

Larson has pointed out that, under the different conditions of star formation near the nucleus of a massive galaxy, the Jeans mass would be reduced and the peak of the mass distribution would be pushed toward a lower mass than that of the peak in the solar neighbourhood, thus increasing the fraction of the mass in lower main-sequence stars. The high mean density in the nuclear region would work in this direction. The shock waves mentioned by Larson as influencing the local density during initial collapse would presumably not be a major factor during the formation of the second-generation metal-rich stars that dominate the populations of massive galaxies. The absence of coolants in the metal-poor medium from which the first-generation stars formed during the initial collapse would favour a high proportion of short-lived supermassive stars, according to Truran and Cameron (1971). They suggest that these stars would leave a residue of black holes that would account for any missing mass.

Tinsley (1973) has interpreted the result from the CO band observations as pointing toward rapid luminosity evolution in old galaxy populations. If confirmed by similar observations of giant-E galaxies of the type used as test objects in the Hubble redshift plot, a correction Δq_0 to the apparent observed value of q_0 would result in a true value of q_0 near zero. Tinsley's analytical relation between x and the evolution rate applies particularly to the mass range $1 < \mathfrak{M}/\mathfrak{M}_\odot < 1.2$, i.e. to the stars at the main-sequence turnoff during the look-back time. The Wing-Ford band test is most sensitive to dwarfs of mass $0.1 < \mathfrak{M}/\mathfrak{M}_\odot < 0.2$, and the CO band appears to be most sensitive with respect to dwarfs of mass $\mathfrak{M}/\mathfrak{M}_\odot < 0.5$. In view of the quite significant changes in slope in Larson's (1973) physically plausible mass distribution model over a range $\Delta \log\mathfrak{M} \simeq 0.5$ to 0.7, there would appear to be some uncertainty in estimating the local slope of the mass function at the mass where the evolution rate is determined from observed upper limits to the numbers of stars of much lower mass, the M dwarfs.

References

Baldwin, J. R., Danziger, I. J., Frogel, J. A., and Persson, S. E.: 1973, *Astrophys. Letters* **14**, 1.

Faber, S. M.: 1972, *Astron. Astrophys.* **20**, 361.

Hartmann, W. K.: 1970, *Mem. Soc. Roy. Sci. Liège* **19**, 49.

Johnson, H. L.: 1966, *Astrophys. J.* **143**, 187.

Kumar, S. S.: 1972, *Astrophys. Space Sci.* **17**, 219.

Larson, R. B.: 1973, *Monthly Notices Roy. Astron. Soc.* **161**, 133.

Morgan, W. W. and Mayall, N. U.: 1957, *Publ. Astron. Soc. Pacific* **69**, 291.

Morton, D. C. and Chevalier, R. A.: 1973, *Astrophys. J.* **179**, 55.
Morton, D. C. and Thuan, T. X.: 1973, *Astrophys. J.* **180**, 705.
Page, T.: 1961, *Proc. 4th Berkeley Symposium Math. Statistics* **3**, 277.
Spinrad, H. and Taylor, B. J.: 1971, *Astrophys. J. Suppl.* **22**, 445.
Stebbins, J. and Whitford, A. E.: 1948, *Astrophys. J.* **108**, 413.
Tinsley, B. M.: 1973, *Astrophys. J. Letters* **184**, L41.
Truran, J. W. and Cameron, A. G. W.: 1971, *Astrophys. Space Sci.* **14**, 179.
Whitford, A. E.: 1966, 'Spectral Classification and Multicolour Photometry' *IAU Symp.* **24**, 19.
Whitford, A. E.: 1972, *Bull. Am. Astron. Soc.* **4**, 230.
Wing, R. F. and Ford, K.: 1969, *Publ. Astron. Soc. Pacific* **81**, 527.

DISCUSSION

Van den Bergh: To how large an area do your observations of NGC 4472 and NGC 4486 refer?
   Whitford: The scanner entrance aperture had the angular dimensions 12″ × 60″.

Larson: The results of some detailed galaxy evolution models computed by Beatrice Tinsley and myself (in preparation) show that the steep power-law mass spectrum proposed by Spinrad and Taylor is very difficult to reconcile with the various observed photometric properties of elliptical galaxies (*UBV* colours, etc.). The shallower Salpeter mass spectrum, and the gaussian mass spectrum which I suggested, both result in predicted colours which are in satisfactory agreement with observations. I think that our results are consistent with those of Dr Whitford.

   Whitford: Yes, though I think there may still be trouble with the mass-luminosity ratio. It is of course possible to squeeze in a much larger mass for a given total light by assuming that the maximum of the Larson gaussian distribution function comes at an appropriately small fraction of the solar mass. One has to analyze whether a maximum consistent with the observed value of M/L will not at the same time predict so much light from low-mass dwarfs as to be inconsistent with the negative results from the dwarf-sensitive indicators. Black holes may turn out to be unavoidable.

E. M. Burbidge: Which are the crucial Spinrad-Taylor measurements giving the high M/L ratio – which ones ought to be rechecked?

   Whitford: The greatest weight was given to the strength of the dwarf-sensitive Na 8190 feature. Sandra Faber has shown that the distribution of mass among the M dwarfs is not well determined by the Spinrad-Taylor observations of M31, and that a lower M/L value is not excluded. Their 32 Å scanner band included a significant amount of absorption by water vapour, subject to night-to-night variations. Giant-sensitive TiO absorption in the sidebands added a second complication that had to be allowed for. A check by a technique that isolated and measured only the strength of the Na doublet lines would be desirable.

   G. de Vaucouleurs: Have you considered the possibility that strong infrared radiation like that observed from Seyfert galaxies could fill in the absorption in the Wing-Ford band expected from dwarf stars?

   Whitford: I am not sure about the angular extent of the infrared sources in the Seyferts. The abnormal optical effects are confined to an area around the nucleus smaller than that accepted by the scanner. Johnson's infrared observations of ellipticals showed an energy curve falling off at 3.4μ in a way that was compatible with a reasonable M-star component, with no unexplained excess. In any case, a rather peculiar infrared source is required if it is to fill in the Wing-Ford band and leave the CO band at 2.3μ at full strength, as observed by Baldwin *et al.*

THE FORMATION OF DISKS BY INELASTIC
COLLISIONS OF GRAVITATING PARTICLES

Applications to the Formation of Galaxies

ANDRÉ BRAHIC

Observatoire de Paris-Meudon, Université de Paris VII, France

Abstract. The numerical study of a gravitating system of colliding particles has many potential applications, for instance the formation of flat galaxies, the formation of the solar system and the evolution of Saturn's rings. Preliminary results are presented for the galactic case. The system tends towards a final equilibrium state and it seems that such a collision mechanism can flatten a protogalaxy.

Since the time of Descartes, Laplace and Poincaré, turbulence has played a central role in cosmological speculation and in theories of the formation of galaxies. It appears not unreasonable to suppose that the early Universe was statistically uniform on a macroscopic scale – as suggested by the microwave background – but disordered on the small scale. Therefore we shall assume that a protogalaxy consists of a small number of clouds or turbulence eddies, whose positions and velocities are initially randomly distributed. Each cloud describes an orbit in the galactic field and the system loses energy through inelastic collisions between clouds.

Poincaré (1911) showed qualitatively that in such a system the following things happen:

(1) a central condensation will be formed and, in order to conserve angular momentum, the outer part of the system will expand;

(2) the system as a whole will flatten into a plane perpendicular to the initial angular momentum vector;

(3) the orbits of the bodies will become increasingly more circular.

Brosche (1970) made a quantitative and analytical model of this process with some approximations and found that the Hubble sequence could be thought of as an angular momentum sequence at constant mass. He also noted that a detailed N body calculation could be expected to improve this kind of model.

With the help of M. Hénon, I am at present developing a general program for this kind of problem. It is a numerical simulation of a gravitating system of particles with inelastic collisions. Such a program has many potential applications other than the formation of galaxies, for instance the dynamics of Saturn's rings, the formation of the solar system and also the evolution of the nuclei of galaxies. We may note other work in this context: by Spitzer and Saslaw (1966), Spitzer and Stone (1967) and Sanders (1970) who have studied the nuclei of galaxies; McCrea (1960), Alfvén and Arrhenius (1970) and Urey (1972) who have studied the formation of the solar system and planets; and Jeffreys (1947), Bobrov (1970) and Cook *et al.* (1972) who have studied Saturn's rings.

Calculations of this sort have so far only been carried out by Ulam (1968) and by

John R. Shakeshaft (ed.), The Formation and Dynamics of Galaxies. 173–179. *All Rights Reserved.*
Copyrigh © 1974 by the IAU.

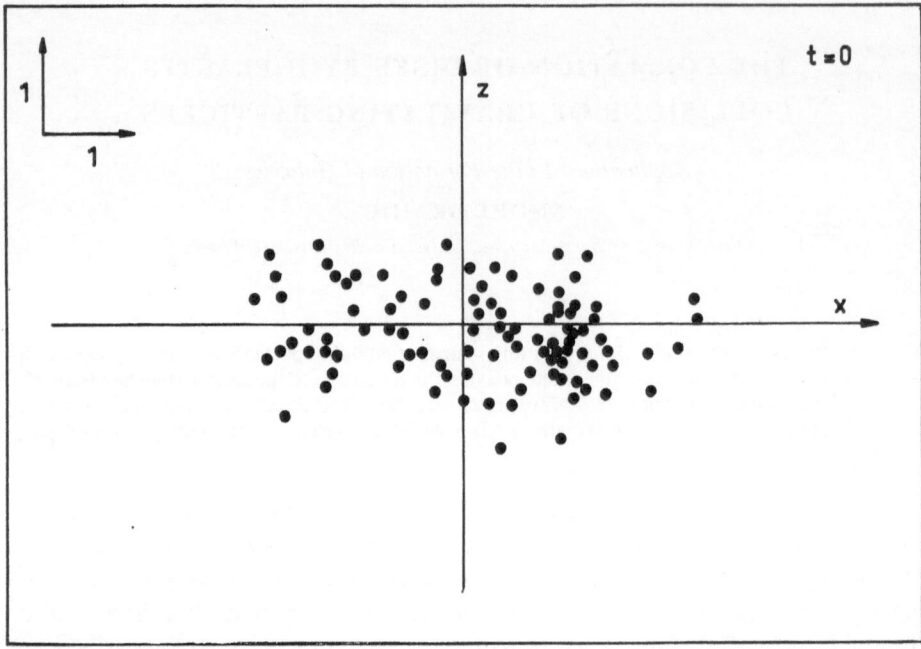

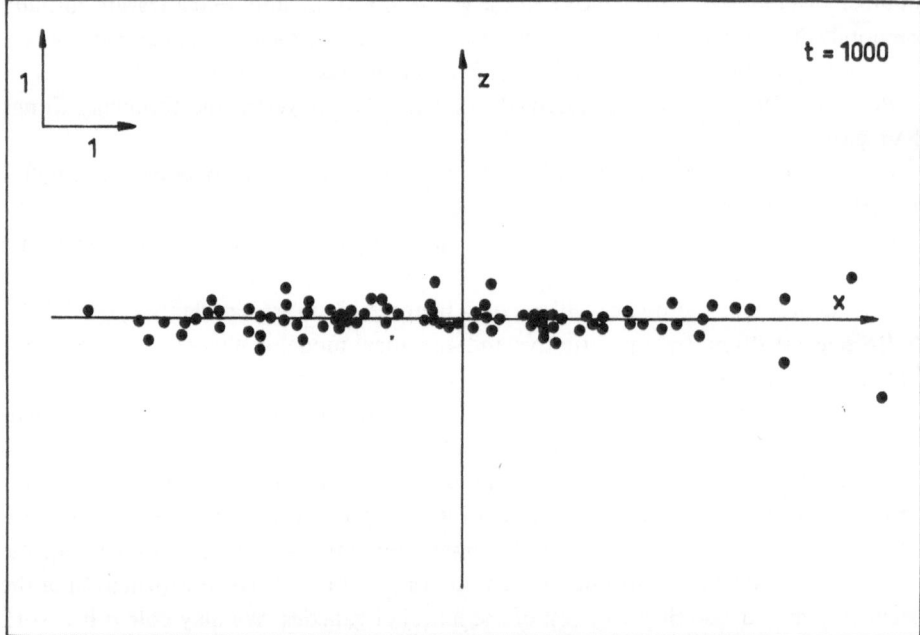

Fig. 1.   Projections onto a plane parallel to the initial angular momentum vector at the initial time and at the end of the computation. The number N of clouds is equal to 100. Each cloud has the same radius $R = 0.07$ and the same mass. The coefficient of elasticity k is equal to 0.3. The maximum value of the initial inclinations of the orbits is equal to 0.5 radian. The unit of time corresponds approximately to the mean time necessary for a cloud to move through one radian on its orbit. After 2500 collisions, one out of 100 clouds has escaped and sixteen out of 100 clouds have fallen on the central body of radius 0.1.

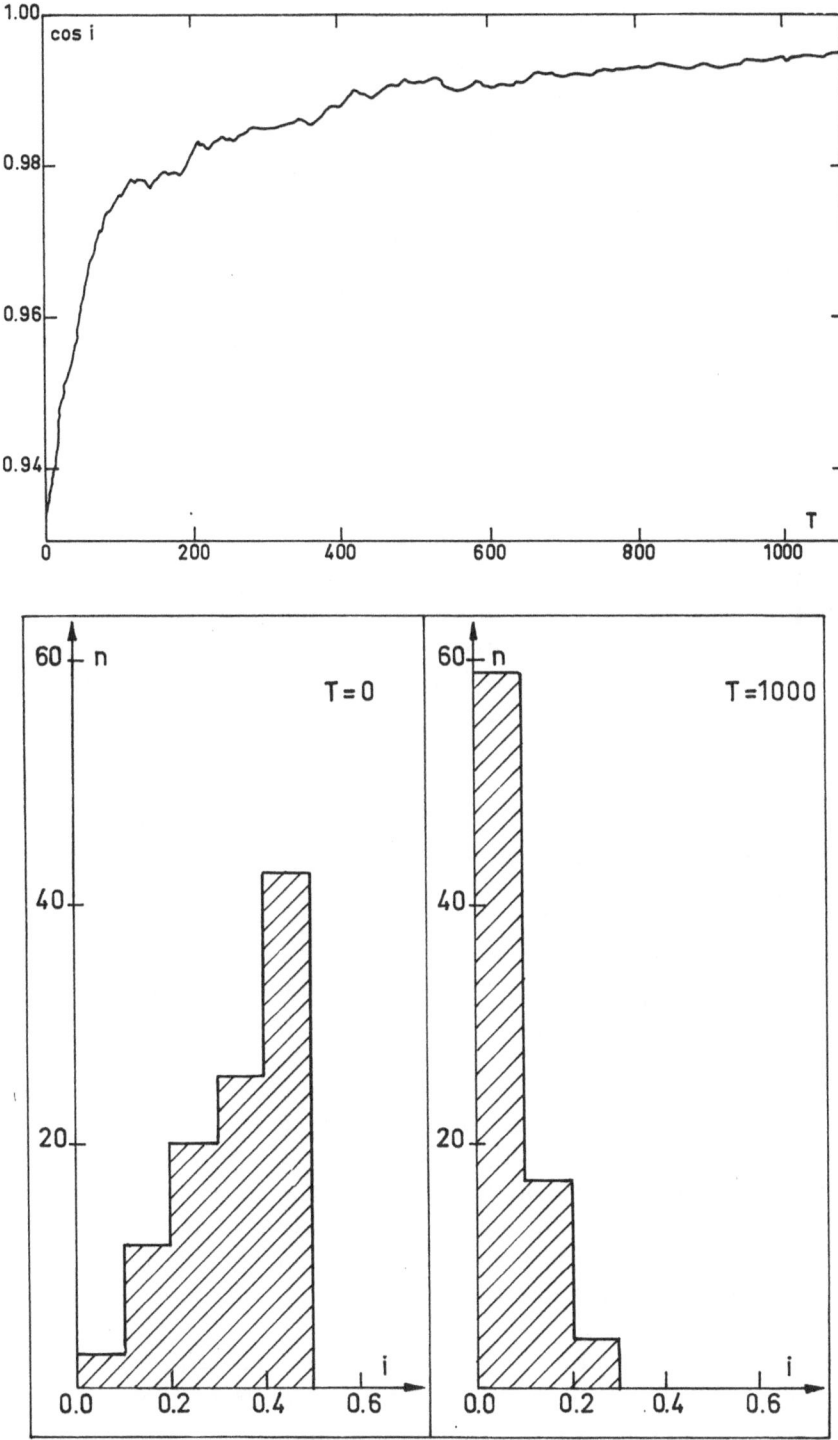

Fig. 2.   Top: The variation as a function of time of the mean cosine of the inclination *i*. Bottom: Histograms of *i* at the initial time and at the end of the computation.

Trulsen (1972). Ulam was interested in the nuclei of galaxies; he used a small number of bodies and, moreover, assumed that they coalesced into a single body after collision. Trulsen studied the dynamics of jetstreams and the formation of planets. Corresponding numerical experiments have been made in molecular dynamics by Alder and Wainwright (1959, 1960), Rahman (1964) and Verlet (1967, 1968) and have stimulated significant progress. A very similar approach has also been applied by us to the case of Saturn's rings (Brahic 1973).

We thus consider a system of clouds moving in a given gravitational field and interacting through inelastic collisions. We neglect the mutual attraction of the clouds and we study numerically the evolution of the system. I shall not here enter into the technical details of the rather intricate calculations, which will be given in a paper in preparation. Let me say only that the principal difficulty is to know whether two clouds will in fact collide. This problem is comparatively simple in molecular dynamics because particles move in straight lines between collisions, but in our case each cloud describes a galactic orbit.

The time scale t_c of evolution of the system is of the order of the mean time between collisions for one cloud. Therefore it is inversely proportional to the number N of clouds and to their geometrical cross-section which is itself proportional to the square of the mean radius R of a cloud. In fact, if we change either N or R, we simply make the evolution go slower or faster. Consequently, a proper choice for N and R permits us to follow realistically the evolution of our system using a minimum of computer time, albeit using a reduced number of larger bodies. However, the procedure becomes unsatisfactory is R is too large, since the gravitational force would then not be the same for two colliding particles, thereby introducing distortion which would increase towards the centre of the system. Too small a number of clouds would also be unsatisfactory because statistical fluctuations would be too large.

I am currently studying a sequence of numerical models and investigating their astrophysical implications. Results have been obtained for the simplest case in which cloud orbits are keplerian around a central mass point. Positions and velocities at any given time are obtained from Kepler's equation. In a collision the grazing component of velocity is conserved and the perpendicular component is multiplied by a coefficient k which lies between 0 corresponding to the completely inelastic case and 1 corresponding to the elastic case. The initial conditions were set up by selecting at random the six elements of the keplerian orbit in such a way that trajectories were all ellipses lying between two spheres centred on the central mass point and with inclinations lying between 0 and some maximal value. We have assumed that clouds on hyperbolic trajectories escape at once and, for technical reasons, that clouds near to the centre are captured by the centre of mass.

The four figures show an example of the evolution of the system. We can see in Figure 2 that the system has flattened considerably after 2500 collisions, but it is not yet completely flat. Figure 2 shows the change of inclination with time. Figure 3 shows a spread of radius: the system extends. We can see in Figure 4 that, at first, the eccentricities increase because a thermal equilibrium is established between radial and

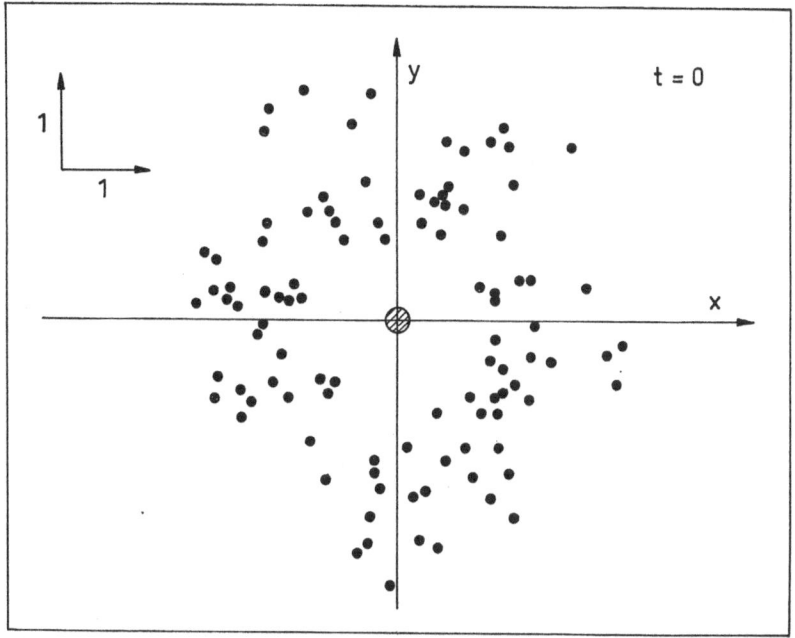

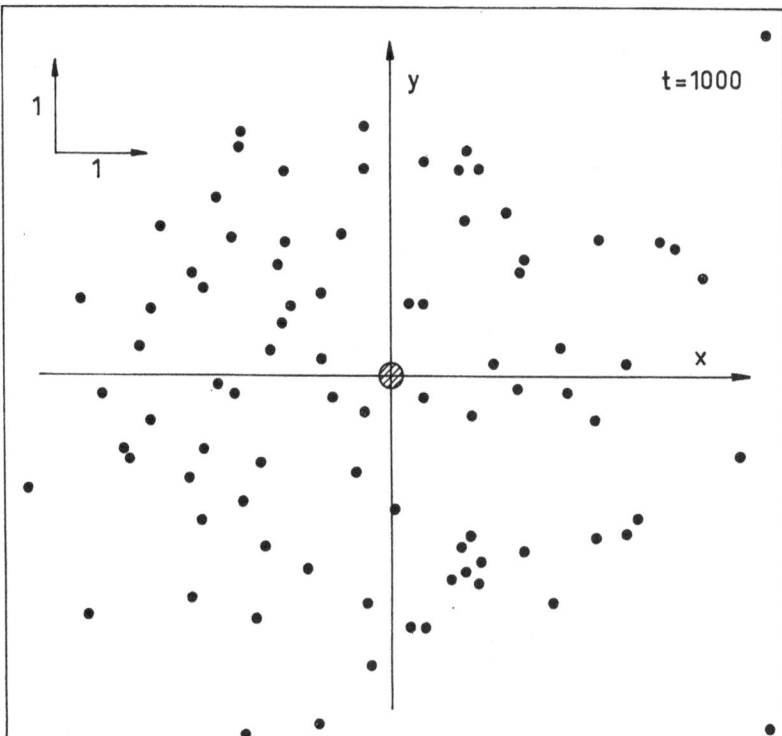

Fig. 3.   Projections onto a plane perpendicular to the initial angular momentum vector. Initial trajectories are all ellipses lying between two spheres of radius 1 and 3 respectively, centred on the central mass point.

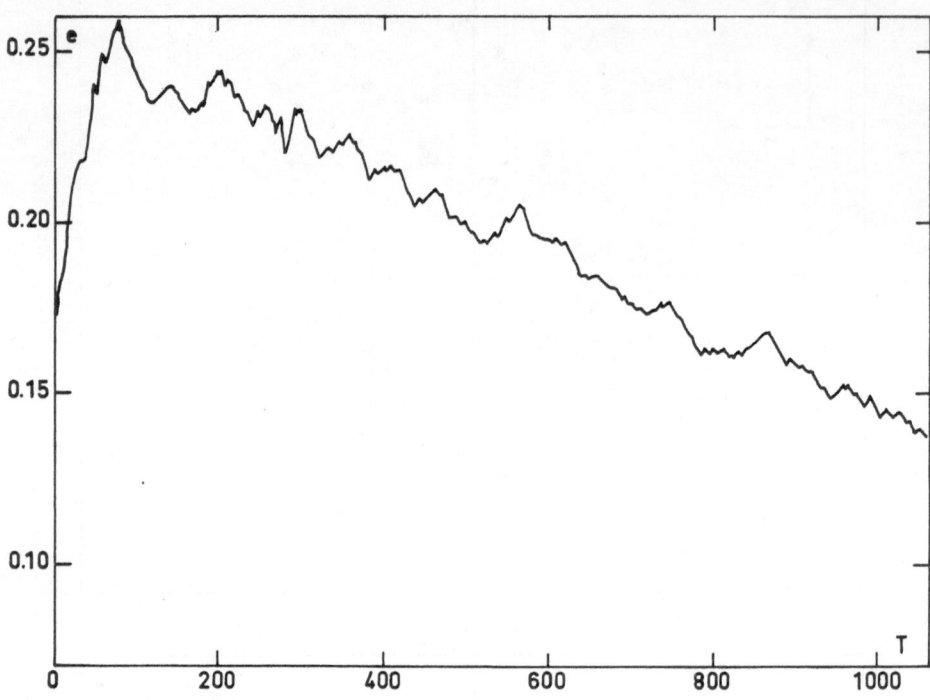

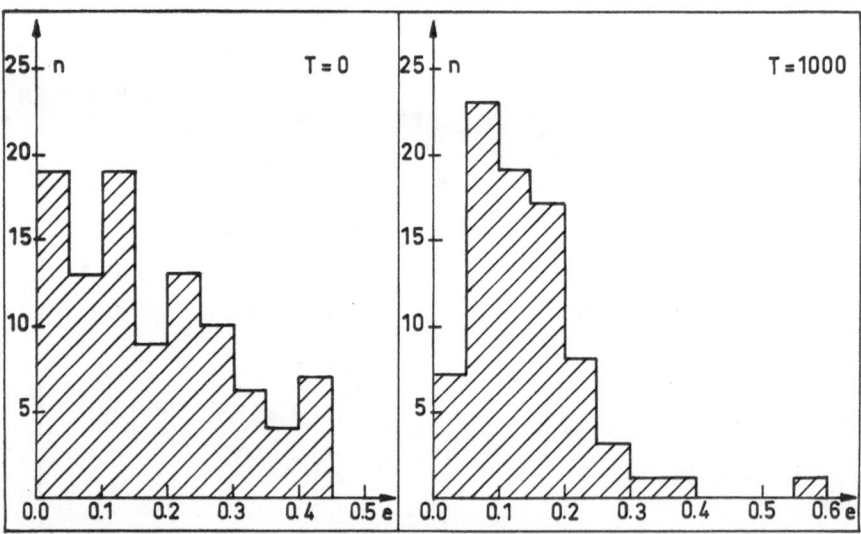

Fig. 4.   The variation as a function of time of the mean eccentricity of the orbits and
the corresponding histograms.

vertical velocities. After that the orbits tend to become more and more circular.

Of course, this first model is probably too crude for a realistic description of a galaxy. But the results already suggest that:

(1) the collision rate decreases with time and the system tends towards a final equilibrium state;

(2) such a mechanism can flatten a protogalaxy. The flattening is already important when each cloud has suffered about ten collisions.

Subsequent models will include fragmentation of clouds during each collision, and also the use of a more realistic field.

References

Alder, B. J. and Wainwright, T. E.: 1959, *J. Chem. Phys.* **31**, 459.
Alder, B. J. and Wainwright, T. E.: 1960, *J. Chem. Phys.* **33**, 1439.
Alfvén, H. and Arrhenius, G.: 1970, *Astrophys. Space Sci.* **8**, 338; **9**, 3.
Bobrov, M. S.: 1970, in *Surfaces and Interiors of Planets and Satellites*, Academic Press, London and New York, p. 377.
Brahic, A.: 1974, in Y. Kozai (ed.), 'The Stability of the Solar System and of Small Stellar Systems', *IAU Symp.* **62**, 83.
Brosche, P.: 1970, *Astron. Astrophys.* **6**, 240.
Cook, A. F., Franklin, F. A., and Palluconi, F. D.: 1973, *Icarus* **18**, 317.
Jeffreys, H.: 1947, *Monthly Notices Roy. Astron. Soc.* **107**, 260.
McCrea, W. H.: 1960, *Proc. Roy. Soc. London* A256, 245.
Poincaré, H.: 1911, *Leçons sur les hypothèses cosmoniques*, Hermann, Paris, Chapter 5, p. 86
Rahman, A.: 1964, *Phys. Rev.* **136A**, 405.
Sanders, R. H.: 1970, *Astrophys. J.* **159**, 1115.
Spitzer, L. and Saslaw, W. C.: 1966, *Astrophys. J.* **143**, 400.
Spitzer, L. and Stone, M. E.: 1967, *Astrophys. J.* **147**, 519.
Trulsen, J.: 1972, *Astrophys. Space Sci.* **17**, 241; **17**, 330; **18**, 3.
Ulam, S. M.: 1968, *Bull. Astron.* **3**, 265.
Urey, H. C.: 1972, *Astrophys. Space Sci.* **16**, 311.
Verlet, L.: 1967, *Phys. Rev.* **159**, 98.
Verlet, L.: 1968, *Phys. Rev.* **165**, 201.

DISCUSSION

Abell: Can you scale your calculations to an actual galaxy? That is, with conditions you might expect for a protogalaxy, how long would those ten or so collisions (per condensation) take to flatten a system of galactic mass and dimensions?

Brahic: The only thing which can be said is that the flattening is important when every cloud has experienced about ten collisions. The mean number of collisions experienced by one cloud depends on the initial number and dimension of the clouds. It seems reasonable to assume that in a realistic case each cloud will experience one to four collisions per revolution. In that case, the flattening will occur in about 10^9 yr.

Savedoff: How does the flattening time depend upon the parameter k of your model?

Brahic: For k below 0.4 or 0.5 one gets the same effect, but for k near 0.7 there is no flattening.

DYNAMICS OF ROTATING STELLAR SYSTEMS:
COLLAPSE AND VIOLENT RELAXATION*

J. RICHARD GOTT III

California Institute of Technology, Pasadena, Calif., U.S.A.

Abstract. Axisymmetric systems, having differing amounts of initial angular momentum and containing 2000 stars (mass points) each, are allowed to collapse under their own gravitational attraction. Collapse and violent relaxation are found to lead to the formation of equilibrium structures after only a few free-fall times. The systems contain a sufficient number of mass points so as to be effectively collisionless and the equilibrium structures are consistent with accurate satisfaction of the stationary collisionless Boltzmann equation. Rotating equilibrium models produced in this way resemble elliptical galaxies, with elliptical isophotes. With cosmologically reasonable amounts of initial angular momentum, E0 to E4.5 galaxies can be produced. The equilibrium models show rotation curves in general agreement with that expected from violent relaxation theory. In particular the models show a central region with solid-body rotation. Further out, the systems become differentially rotating. When infall effects are included one can produce models with extended envelopes in good agreement with observed elliptical galaxies. It is suggested that the crucial factor in determining whether an elliptical or a spiral galaxy is formed is whether or not star formation is complete by the time the proto-galaxy reaches its point of maximum collapse.

I shall discuss several dynamical models I have made for the formation of elliptical galaxies, models which cover the collapse and violent relaxation phase leading to the formation of the equilibrium galaxy. The approximation of an axisymmetric gravitational field is utilized to simplify the N-body calculation and thus allow treatment of many more stars than would otherwise be possible. In particular, sufficient stars may be treated to achieve a system which is effectively collisionless.

The procedure can be outlined as follows. Our rotating systems are axisymmetric and contain 2000 mass points (henceforth called stars) of equal mass. To derive the gravitational potential Φ due to these stars, we divide space into a network of toroidal cells centred on the rotation axis of the system and having approximately square cross-sections. The mass within each cell is assumed to be concentrated on the central ring of the cell and the values of the potential are then calculated on a series of rings forming the corners of the toroidal cells. The potential within each cell is found by linear interpolation from the values at the cell's four corner rings. At the end of each time step, this approximate gravitational potential is used to find the force on each star, the stellar trajectories are then advanced and the process is repeated. The network of toroidal cells is set up by constructing a spherical coordinate system (r, θ, ϕ) centred on the galaxy and whose axis $\theta = 0$ is the rotation axis of the galaxy. At the origin we establish a small reflecting sphere of radius R_s. Let $\alpha = \pi/18$ or an angle of $10°$. Then the cell C_{ij} contains all points with coordinates in the ranges:

$$(i - 1)\,\alpha < \ln\,(r/R_s) < i\alpha, \quad (j - 1)\,\alpha < \theta < j\alpha, \quad 0 \leqslant \phi < 2\pi. \tag{1}$$

* Supported in part by the National Science Foundation [GP-36687X, GP-28027].

John R. Shakeshaft (ed.), The Formation and Dynamics of Galaxies. 181–190. All Rights Reserved.
Copyright © 1974 by the IAU.

Each cell has an angular diameter of $10°$ in θ and an outside radius approximately 19% larger than the inside radius. This gives each cell an approximately square cross section. In our calculations we use 24 layers of 18 cells each: $C_{ij}(i=1, 24; j=1, 18)$. These cells cover a spherical volume of radius $\exp(24\alpha) R_s \simeq 64 R_s$. Stars passing beyond this outer radius are assumed to escape; stars hitting the inner reflecting sphere are reflected elastically from its surface. Calculation of the potentials and advance of the trajectories of the 2000 stars requires approximately 79000 operations per elementary time step compared with $3N^2 = 12 \times 10^6$ operations per time step required for an N-body calculation with $N = 2000$.

Our models are started at the cosmological epoch of maximum expansion of the protogalaxy. We then follow the collapse and violent relaxation for four models, I, II, III, IV with increasing amounts of initial angular momentum. Each of the models begins as a system of 2000 stars distributed randomly so as to produce a uniformly dense sphere of stars of total mass M_{gal} and radius R_0. R_0 thus represents the maximum radius of expansion of the protogalaxy. If such a uniform sphere of stars were set in uniform (solid-body) rotation with angular rate $\Omega_0 = (GM_{gal}/R_0^3)^{1/2}$, then instantaneously it would be holding itself up against gravity in its equatorial plane. The stars in each galaxy are given systematic velocities V_ϕ so that the initial state is one of uniform rotation with angular rate Ω_i, where for Model I, $\Omega_i = 0$ (no rotation); for Model II, $\Omega_i = (\frac{1}{4})^{1/2} \Omega_0$; for Model III, $\Omega_i = (\frac{1}{2})^{1/2} \Omega_0$; and for Model IV, $\Omega_i = (\frac{3}{4})^{1/2} \Omega_0$. Superimposed upon this systematic rotation, the stars are given small random velocities with a Maxwellian distribution so that $\langle V_r^2 \rangle^{1/2} = \langle V_\theta^2 \rangle^{1/2} = \langle (V_\phi - \langle V_\phi \rangle)^2 \rangle^{1/2}$. Initially $\langle V_r \rangle = \langle V_\theta \rangle = 0$. In Models I, II, and III the random velocities were chosen so that the initial virial coefficient for the random kinetic energy, $|2T_{random}/W|$, was equal to one half, where $W = -(3GM_{gal}^2/5R_0)$. In Model IV the energy in the random velocities was reduced from that above by a factor of 2.

Let us now also define some useful quantities. Let $\varrho_0 = (3M_{gal}/4\pi R_0^3)$ be the standard density and $V_0 = (3GM_{gal}/5R_0)^{1/2}$ be the standard virial velocity. Let $E_b = -(T+W)$ be the binding energy for each model. Define for each model the energy radius $R_E = (3GM_{gal}^2/10E_b)$ which is the characteristic radius of a system in virial equilibrium with binding energy E_b. Define a characteristic oscillation time $T_c = 2\pi (R_E^3/GM_{gal})^{1/2}$. T_c is twice the free-fall time of a galaxy of mass M_{gal}, binding energy E_b, and initial radius $r_{max} = 2R_E$ (see Spitzer, 1968). The age of the Universe when $r = r_{max}$ is thus $(\frac{1}{2}) T_c$; the age of the Universe at the end of the collapse is T_c. The times in each model are measured in terms of T_c for that model.

Models I–IV possess reasonable amounts of cosmological rotation such as might be obtained through tidal interactions with neighbouring protogalaxies (cf. Peebles, 1971). These rotating spheres of stars are then allowed to collapse under their own gravitational force. During the collapse phase the potential oscillates violently, producing the Lynden-Bell violent relaxation, and one finds for these rotating systems, just as for spherical systems, that they reach relaxed equilibrium configurations after only a couple of free-fall times.

Let us examine our expectations for the equilibrium models. Consider a non-rotating sphere of stars with no initial kinetic energy and a radius of R_0. Then $-E_b = T + W$ with $W = -(3GM_{gal}^2/5R_0) = -E_b$ and $T = 0$. Let it collapse and undergo violent relaxation. When virial equilibrium is reached, $2T + W = 0$ and $-E_b = T + W$ so $W = -2E_b$, or twice its original value, and the equilibrium system will be spherical with a characteristic radius of $\sim (\frac{1}{2}) R_0$. Now consider a rotating galaxy of radius R_0 and $\Omega = \Omega_0$ so that it is initially holding itself up against gravity in the equatorial plane. This is surely an upper limit on the amount of rotation we would expect cosmologically. Let this galaxy collapse. It continues holding itself up in the equatorial plane, leaving $r \sim R_0$ in that direction. Rotation does not inhibit collapse perpendicular to the plane, so to first approximation the collapse proceeds perpendicularly as in the

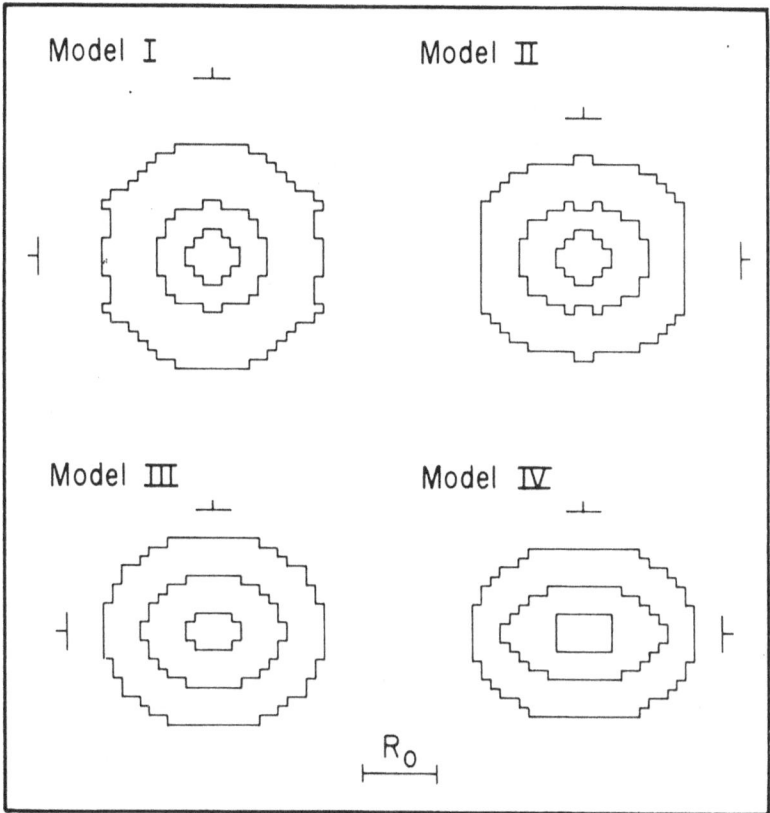

Fig. 1. Isophotes for the equilibrium galaxies. The diagram represents the appearance of these galaxies to a distant observer whose line of sight lies in their equatorial plane. The rotation axis of each galaxy is vertical and the equatorial plane is horizontal. The innermost isophote represents a projected surface density of $I_0 = 0.8\, M_{gal}/R_0^2$. The next two isophotes represent intensities of 0.1 I_0 and 0.01 I_0, while the tick marks indicate the extent of the 0.001 I_0 isophote. The isophotes are computed from a grid in one quadrant only, thereby giving the diagrams a four-fold symmetry. The scale is given at the bottom: all four galaxies were initially spheres of radius R_0 before undergoing collapse and violent relaxation. Model I had no initial rotation and is therefore spherical (E0). Models II–IV had progressively larger amounts of initial rotation, and have produced progressively more elliptical equilibrium galaxies (E1.5, E2.5, E4.5).

spherical case to give $r \sim (\frac{1}{2}) R_0$. These approximations indicate an equilibrium galaxy with a 2 to 1 axial ratio (i.e., an E5 galaxy). Thus, with the maximum reasonable amount of initial rotation, we get an equilibrium galaxy with a flattening of at most E5. Only in dissipative systems with gas present can we produce flat systems such as spiral galaxies.

In Figure 1 are shown the isophotes for the equilibrium galaxies. They represent (as do later data) time averages over the period $(\frac{3}{2}) T_c < t < (\frac{11}{2}) T_c$. The equilibrium models are indeed elliptical galaxies, with isophotes that are elliptical in shape, and Models I–IV correspond approximately to E0, E1.5, E2.5, E4.5 galaxies, respectively. Our general expectations are confirmed: Model I, with no rotation, is quite spherical and the models with greater initial angular momentum are more flattened but not flatter than E5. Peebles' (1971) favoured value for angular momentum gained by gravitational interactions would give nearly spherical galaxies (Model II), but his upper bound would allow production of E3 and E4 galaxies. Earlier in this conference Dr de Vaucouleurs summarized (p. 1) the observational data on the intrinsic flattening in elliptical galaxies. The results show elliptical galaxies to have an intrinsic

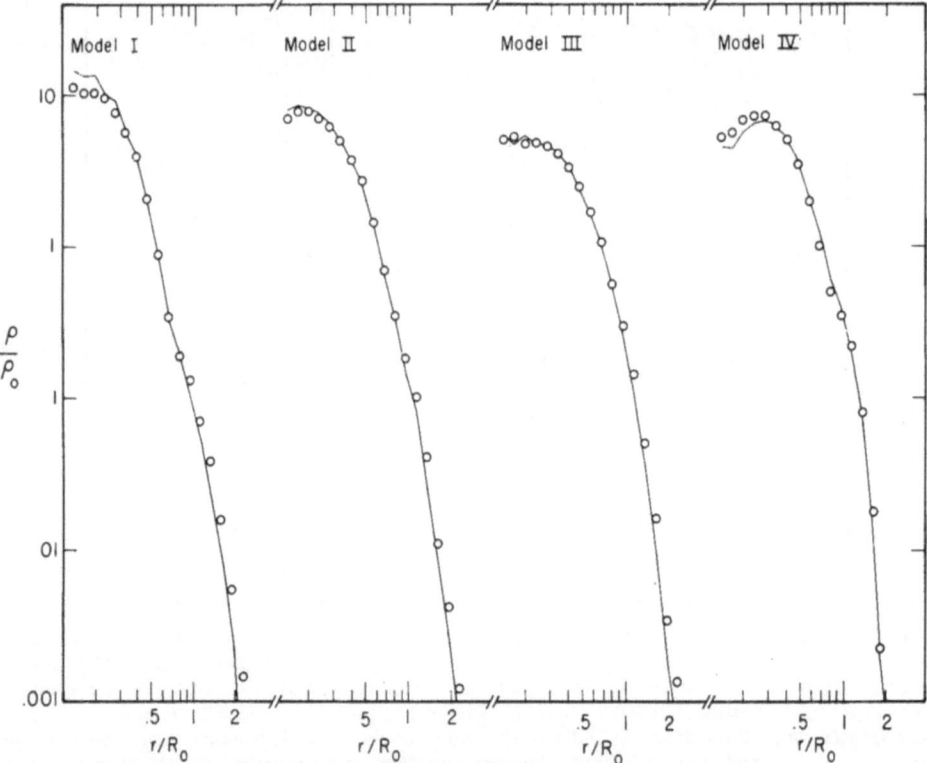

Fig. 2.   Stellar density in the galactic plane as a function of radius for the equilibrium models. The open circles represent the actual runs of density with radius while the solid line represents that expected from satisfaction of the stationary collisionless Boltzmann equation. $\varrho_0 = 3M_{\text{gal}}/4\pi R_0^3$ is a standard pre-collapse density which is the same for all four models.

flattening between E0 and E5 with only a few E6's and no E7 galaxies, in qualitative agreement with what we predict for dissipationless collapse models.

In Figure 2 the stellar density in the galactic plane as a function of distance from the centre is plotted for each model. The data points are densities from successive cells lying on the galactic plane. The solid line shows the density expected from satisfaction of the stationary, collisionless Boltzmann equation:

$$-\frac{\partial \Phi}{\partial r} = -\frac{\langle V_\phi \rangle^2}{r} + \langle V_r^2 \rangle \left[\frac{\partial \ln \varrho}{\partial r} + \frac{\partial \ln \langle V_r^2 \rangle}{\partial r} + \frac{1}{r} \left(1 - \frac{\langle V_\theta^2 \rangle}{\langle V_r^2 \rangle} \right) + \right.$$
$$\left. \frac{1}{r} \left(1 - \frac{\langle (V_\phi - \langle V_\phi \rangle)^2 \rangle}{\langle V_r^2 \rangle} \right) \right]. \tag{2}$$

To compute the solid lines in Figure 2 we use our approximate potential and evaluate all velocities and velocity dispersions (within cells) as a function of radius with derivatives replaced by differences. We solve for $\ln \varrho$ as a function of r, and normalize. Very good agreement is found between the actual runs of density and those predicted from satisfaction of the collisionless Boltzmann equation. This demonstrates that the equilibrium models are stationary and essentially collisionless.

Rotation curves for the equilibrium models are presented in Figure 3. Significantly, all three rotating models show a central region of solid-body rotation. Lynden-Bell (1967) has shown that the statistically most probable distribution function for a rotating system is one of solid-body rotation with isothermal Maxwellian residual velocities. Thus our models show that the violent relaxation is efficient enough to produce this 'most probable' solid-body rotation at least in a central region. Also plotted in Figure 3 are rotation curves calculated by Lynden-Bell (1967), taking into account partial relaxation effects. Our models show very good general agreement with Lynden-Bell's theoretical prediction. Also presented for comparison are rotation curves from a self-consistent model by Prendergast and Tomer (1970) which assumes a distribution function with Maxwellian residual velocities and a tidal energy cut-off. The model's behaviour is not in agreement with that predicted by Prendergast and Tomer (1970) whose sharp tidal cut-off artificially drags their rotation curves down to zero at a certain radius.

Figure 4 shows the velocity dispersions for Models I, II, and IV. In Model I which is non-rotating we see a core region where the velocity distribution is isotropic and isothermal. This is to be expected due to the violent relaxation. Outside the core we find a halo where the velocity dispersion is largest in the radial direction. In the rotating models, isotropy of the random velocities in the centre is not achieved nearly as well as in the spherical system, particularly in the equatorial plane where the velocity dispersion perpendicular to the plane is the largest component. This is not unreasonable physically for a system which has collapsed primarily in the direction perpendicular to the plane. The departures from isotropy in the centre become progressively larger as we go toward more rapidly rotating systems. A more detailed presentation of these model results appears in Gott (1973).

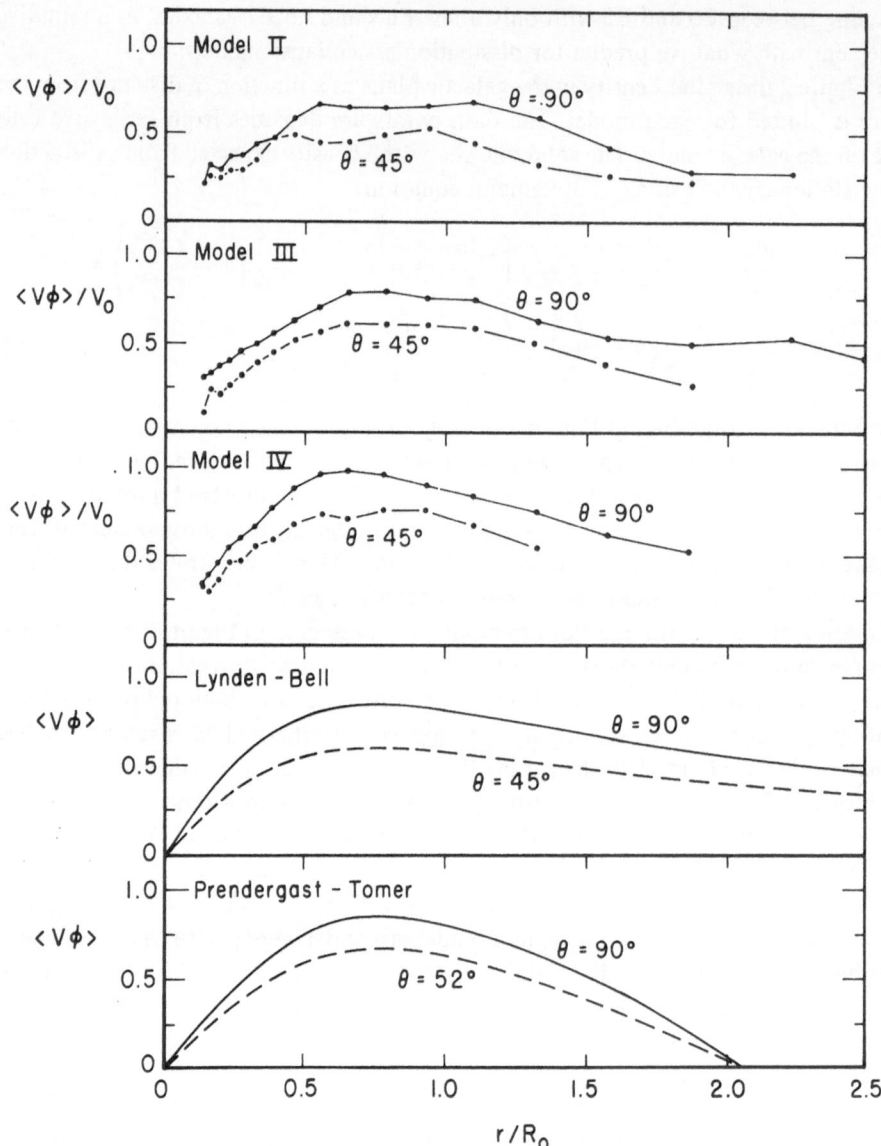

Fig. 3.   Rotation curves. $\theta = 90°$ is in the equatorial plane, $\theta = 45°$ is along a line halfway between the equatorial plane and the axis of rotation. Results from the three rotating models are presented. They show approximate solid-body rotation in the centre and differential rotation further out. For comparison are two normalized theoretical curves: Lynden-Bell (violent relaxation and estimates of partial relaxation effects) and Prendergast and Tomer (self-consistent equilibrium models with Maxwellian random velocities and a tidal energy cut-off).

The models we have been considering are quite simple ones. They are produced by the collapse of an isolated sphere of stars. A truly realistic model should include not only the protogalaxy itself, but should consider the protogalaxy in a complete cosmological setting. In addition to the protogalaxy itself, there will be outlying, still expanding material which is cosmologically bound to it and which will suffer infall into the protogalaxy after the protogalaxy has completed its collapse (cf. Gott and Gunn, 1971). This infall material will give the galaxy a less tightly bound, extended envelope. I have recently completed a series of models which take into account these infall effects and for which the angular momentum of the protogalaxy is derived directly from a tidal interaction with a neighbouring protogalaxy. The results of these calculations are, I think, quite promising. They show that with infall one can produce ex-

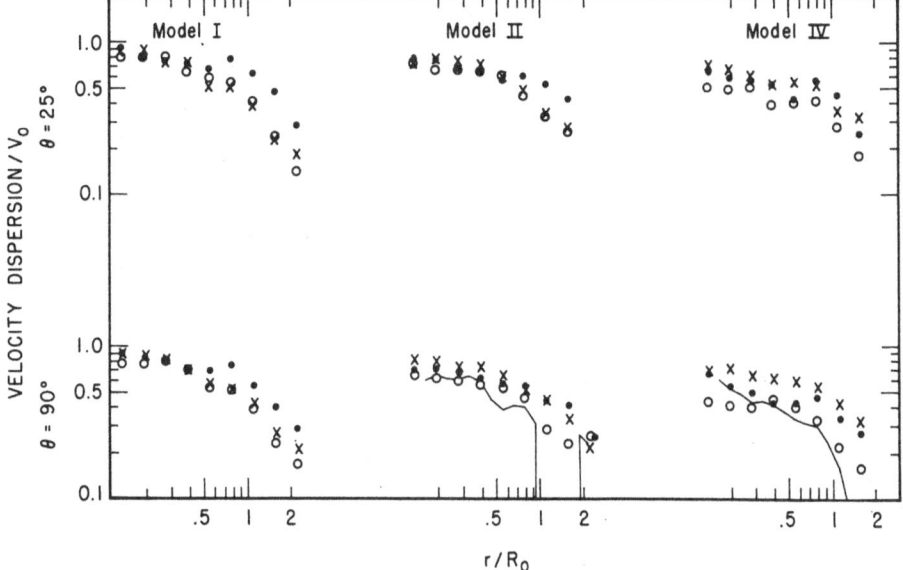

Fig. 4. Velocity dispersion as a function of radius from the centre. $\theta = 90°$ is in the galactic plane, $\theta = 25°$ is along a line making an angle of $25°$ with the rotation axis. A spherical coordinate system (r, θ, ϕ) is used.

● $= \langle V_r^2 \rangle^{1/2}$,

× $= \langle V_\theta^2 \rangle^{1/2}$,

○ $= \langle (V_\phi - \langle V_\phi \rangle)^2 \rangle^{1/2}$.

$V_0 = (3GM_{gal}/5R_0)^{1/2}$ is a standard velocity which is the same for all systems. Since Model I is spherical, its curves for $\theta = 90°$ and $\theta = 25°$ are similar. It has a very isotropic velocity distribution near the centre; far from the centre the halo stars dominate with their preferentially radial dispersions. A similar but smaller halo effect can be seen in Model II ($\theta = 25°$). The more rapid the rotation, the less isotropic the velocity distributions become. Better isotropy is achieved near the rotation axis than in the galactic plane. For the two rotating models, the solid curve represents the expected value of $\langle (V_\phi - \langle V_\phi \rangle)^2 \rangle^{1/2}$ (given $\langle V_r^2 \rangle^{1/2}$ and the rotation curve) under the assumption of a Schwarzschild velocity distribution everywhere and satisfaction of the stationary collisionless Boltzmann equation. In Model II, $\langle V_\phi \rangle$ falls off slightly faster than r^{-1} for a short stretch so the solid curve drops to zero there. Results for Model III are intermediate between Models II and IV and are not presented.

tended envelopes like those observed in actual elliptical galaxies. Also it seems possible to produce model galaxies with isophotal eccentricities in their envelopes similar to those observed in real galaxies.

It should be emphasized that all of these models have no gaseous dissipation in them. That is, they represent models in which star formation is essentially completed by the time the galaxy reaches its point of maximum collapse. Such early star formation might take place in the primeval globular clusters, as suggested by Peebles and Dicke (1968).

There are several reasons to think that elliptical galaxies might be formed in just this manner. First, as I have mentioned, the observed eccentricities of elliptical galaxies are in good qualitative agreement with those predicted from a dissipationless collapse. Second, in elliptical galaxies there is an observed tendency for the isophotes to become rounder as one approaches the centre of the galaxy. (Note, for example, behaviour of this sort in Models II and IV.) If gaseous dissipation were important, we would expect the inner isophotes of the galaxy always to be flatter than the outer ones, which is not the case. Finally, the Lynden-Bell violent relaxation operating in a free-fall collapse without dissipation offers an attractive way to produce the relaxed, ellipsoidal configurations of stars we observe in elliptical galaxies.

I have made other models of collapsing protogalaxies which include both gas and stars. These give the expected results. A rotating sphere composed of stars and gas clouds (having inelastic collisions) was allowed to collapse, with the gas clouds slowly being turned into stars as the collapse proceeded. During the collapse phase, the stars and gas moved together as a unit and the sphere proceeded through a series of flattened ellipsoids until it became a flat pancake as the material reached the plane.

Lynden-Bell (1967) has shown that a uniformly rotating, uniform density fluid sphere collapses homologously through a series of uniformly rotating MacLaurin ellipsoid configurations. The gas clouds do not dissipate any significant amount of energy during the homologous collapse, because each cloud sees clouds in its neighbourhood moving with similar velocity to itself. When the material reaches the plane, this situation changes: the stars pass through the plane on their free trajectories, undergo violent relaxation, and form a relaxed spheroidal component similar to that seen in the elliptical galaxy models we have discussed. The gas, however, dissipates its energy quickly in the vertical direction to form a thin disk. The critical factor is the amount of gas left at the time of maximum collapse. Whatever gas is left at this time will dissipate its energy and form a disk. If some of this gas subsequently turns into stars, the new stars will also be confined to the disk.

Thus, we have a reasonable scenario for the formation of elliptical and spiral galaxies. In elliptical galaxies the star formation is essentially complete by the time of maximum collapse and a spheroidal galaxy is produced. If star formation is not completed by the time of maximum collapse and substantial amounts of gas are left over, then gaseous dissipation is important and a flat, disk-like spiral galaxy is produced. In this conference (e.g. p. 130) it has been emphasized that spiral galaxies show both disk and spheroidal components. We can find a continuous sequence from ellipticals with no disk, to spirals with a spheroidal nuclear bulge, to spirals with no observable spheroidal component. The value of $M_{disk}/M_{spheroidal}$ varies smoothly along the sequence. With the present model it is easy to see how such a sequence would originate: by varying the initial star formation rate we could have any desired fraction of left-over gas at the time of maximum collapse and therefore produce any given ratio of $M_{disk}/M_{spheroidal}$.

In summary then, we have presented dynamical models for the formation of elliptical galaxies. The equilibrium elliptical galaxy is produced by a simple free-fall collapse and violent relaxation process. The present picture offers a natural mechanism for forming elliptical and spiral galaxies. The early star formation rate is seen as the key factor in determining which type of galaxy is produced.

References

Gott, J. R., III: 1973, *Astrophys. J.* **186**, 481.
Gott, J. R., III and Gunn, J. E.: 1971, *Astrophys. J. Letters* **169**, L13.
Lynden-Bell, D.: 1967, *Monthly Notices Roy. Astron. Soc.* **136**, 101.
Peebles, P. J. E.: 1971, *Astron. Astrophys.* **11**, 377.
Peebles, P. J. E. and Dicke, R. H.: 1968, *Astrophys. J.* **154**, 891.
Prendergast, K. H. and Tomer, E.: 1970, *Astron. J.* **75**, 674.
Spitzer, L., Jr.: 1968, *Diffuse Matter in Space*, Wiley and Sons, New York.

DISCUSSION

Contopoulos: Did you take into account encounters between neighbouring stars?

Gott: Calculations show that placing the stars in these toroidal cells has the effect of making the stars gravitate as if they were rings, with a cut-off in their gravitational attraction if they approach closer than half the cell-width. In other words, once two stars are within the same toroidal cell they exert no more gravitational attraction on each other. The toroidal cells each have a nearly square cross-section with a cross-sectional width and breadth approximately 19% as large as their radius. The volume of space into which the galaxy is placed is divided into 432 such toroidal cells. Encounters between stars (i.e. encounters between rings) are thus handled in a natural fashion. The systems contain a sufficient number of stars so that the effects of two-body encounters are insignificant and the systems can be regarded as effectively collisionless.

Miller: Similar calculations have been run by plasma physicists for studies of plasma confinement, using about 120 000 pseudo-particles in the form of rings in cylindrical coordinates. These calculations have been done by Morse *et al.* at Los Alamos (in B. Alder *et al.*, *Methods in Computational Physics*, Academic Press, N.Y., **9**, 213), by Killeen *et al.* at Livermore (*J. Comput. Phys.* **11**, 360, 1973), and by the group at Culham. For ellipticals, such ring models should be satisfactory.

Mark: Can you please repeat again your evidence that the evolution is collisionless? You do not have very many rings.

Gott: First, theoretical estimates were made of the effective two-body relaxation time for this particular numerical scheme with $N = 2000$. These results indicated that $T_R \sim 120 \, T_c$, where T_c is the characteristic orbital time of the systems. We study the models only out to a time $t = 5.5 \, T_c$, so the systems are effectively collisionless over this time-scale. After the systems reached their equilibrium configurations, the individual stellar energies were plotted to watch the random walk in individual stellar energy due to two-body encounters. Values of $\langle (\Delta E)^2/E^2 \rangle$ for stars over the observed period yielded values of $T_R \sim 60 \, T_c$, consistent with the theoretical estimates and indicating an effectively collisionless system. Additional checks included the fact that no deepening of the central potential well (or change in any of the equilibrium model parameters) was observed between $t = (3/2) \, T_c$ and $t = (11/2) \, T_c$. Finally, there is the check that the equilibrium models accurately satisfy the stationary collisionless Boltzmann equation as indicated in Figure 2.

Carrick: The rotating models of Prendergast and Tomer (*Astron. J.* **75**, 674, 1970) have velocity dispersions in each of three dimensions equal near the centre, since this region is nearly isothermal. In your models, for stronger rotation, the three are significantly different, even close to the centre. Is this a statistical effect?

Gott: No. The predominance of the velocity dispersion perpendicular to the galactic plane for locations in the plane is shown consistently for a range of radii in each of the three rotating models. The effect seems physically reasonable for systems that have suffered their main collapse perpendicular to the plane. It becomes progressively more pronounced as one goes toward the more rapidly rotating models for which the violent relaxation is simply not effective enough to produce isotropic velocity dispersions.

Tifft: Some galaxies in the Coma cluster (see Rood, H. J. and Baum, W. A.: 1962, *Astron. J.* **73**, 442) show a variation of ellipticity with radius having no obvious systematic pattern; some elliptical galaxies are also known to show rotated axes of isophotes. Could you comment on this?

Gott: In the outer envelopes of elliptical galaxies, violent relaxation is not so effective and it takes longer for the outer envelopes to reach equilibrium. Presumably, differences in initial conditions may manifest themselves as irregularities in individual cases.

Photometry of a number of elliptical galaxies by van Houten indicates that the majority of ellipticals have envelopes in which the isophotal eccentricity remains nearly constant with increasing radius. Some ellipticals, e.g. NGC 205, show rotated axes of isophotes in their outer parts which are probably due to tidal distortion (by M31 in the case of NGC 205).

G. de Vaucouleurs: Two comments: (i) the Coma cluster objects are not good tests because many are not E, but L galaxies, often barred, and the high space density causes much tidal interaction. (ii) Among the best observed large E galaxies that are comparatively isolated, the ellipticity of the isophotes is never constant; it increases from the nucleus to a maximum which agrees roughly with the Hubble *n* index (En), then decreases slowly in the corona (see Hubble, *Astrophys. J.* **71**, 231, 1930; Oort, *Astrophys. J.* **91**, 273, 1940; de Vaucouleurs, *Ann. Astrophys.* **11**, 287, 1948; *Hdb. der Phys.* **53**, 322, 1959; Liller, *Astrophys. J.* **132**, 306, 1960; van Houten, *Bull. Astron. Inst. Neth.* **16**, 1, 1961). Homothetic ellipsoids are *not* satisfactory models of E galaxies. Are these effects observed in your models?

Gott: It's hard to tell for the outermost isophotes because these regions take a long while to relax. It's not difficult to get different ellipticities. If one takes a spherical non-rotating cloud which collapses, then one expects the final galaxy to have a radius, in virial equilibrium, about half that of the initial cloud. Now by postulating differing amounts of angular momentum, the maximum being such that there is no collapse in the plane of rotation, one can get ellipticities between E0 and E5, but not significantly flatter.

G. de Vaucouleurs: I think observers would say that they've found a few E6 galaxies, such as NGC 670, 1209, 4386, 4564, 4660, 4697, 4865, 5028, 6877 and 6909.

MODELS FOR THE FORMATION AND EVOLUTION
OF SPHERICAL GALAXIES

RICHARD B. LARSON

Yale University Observatory, New Haven, Conn. 06520, U.S.A.

Abstract. Detailed dynamical model calculations based on a conventional collapse picture of galaxy formation, and conventional assumptions concerning star formation and stellar evolution, are found to be able to reproduce satisfactorily the basic structural and photometric properties of elliptical galaxies. The quasar phenomenon may be identifiable with the formation of the nucleus of a giant elliptical galaxy.

Because the basic structural properties of elliptical galaxies must have been largely determined at the time of formation, they can be properly accounted for only by fairly detailed dynamical models describing the galaxy formation process. Also, various cosmological studies and the interpretation of objects like the quasars require a knowledge of how galaxies form and evolve with time. For these reasons, an attempt has been made to construct models of the formation and evolution of spherical galaxies, with the hope that the results will apply also to elliptical galaxies which are not too highly flattened. The basic hydrodynamical method used in these calculations is that developed earlier in *Monthly Notices Roy. Astron. Soc.* **145**, 405, 1969, with the incorporation of the effects of stellar mass loss and heavy element production and of some new assumptions for the star formation and gaseous dissipation rates. The photometric properties of these models have been computed as a function of time and radial coordinate by B. M. Tinsley. This work will be described in detail elsewhere, e.g. *Monthly Notices Roy. Astron. Soc.* **166**, 585, 1974, and therefore only a brief summary of the salient features of the models is presented here.

In a typical case, the calculations begin with a spherical protogalaxy of mass $10^{11} M_\odot$ and radius 50 kpc; in contrast to the calculations described by Gott in this volume (p .181), we assume here that the protogalaxy is initially entirely gaseous. The gas is assumed to possess large-scale internal turbulent or streaming motions whose dissipation through collisions allows the system to become highly centrally condensed; the dissipation time scale is assumed to be approximately equal to the local dynamical time scale for the system. The gas is continually transformed into stars during the collapse, and it is assumed that the characteristic time scale for star formation is also approximately equal to the dynamical time scale, the justification being that star formation may be triggered by the compression or shock fronts produced when large-scale gas streams collide. The evolution of the galaxy is followed for 10^{10} yr, after which time the residual gas content is reduced to only a few times $10^6 M_\odot$. With reasonable choices for the parameters of the model, the resulting projected stellar density distribution gives a close fit to the observed surface brightness distribution in typical elliptical galaxies like NGC 3379. Assuming a conventional Salpeter stellar mass spectrum, the models also reproduce satisfactorily other ob-

John R. Shakeshaft (ed.), The Formation and Dynamics of Galaxies. 191–194. *All Rights Reserved.*
Copyright © 1974 by the IAU.

served properties of typical elliptical galaxies, such as the nuclear velocity dispersion (~ 200 km s^{-1}), the mass-to-light ratio (~ 20), the approximate metal abundance ($Z \sim 0.02$) and the $B-V$ colour (~ 0.97).

An important general prediction of these models is a steep metal abundance gradient in the nuclear region, qualitatively similar to the abundance gradients observed by McClure, Spinrad, and others. The abundance gradient is produced by the continuing inflow of residual gas through an already formed background of stars, which causes the heavy elements lost from the stars to be swept inward and concentrated at the centre of the galaxy. A further important result is that the metal abundance distribution in the nuclear region approaches an approximate steady state, so that most of the stars at any point are formed with nearly the same metal abundance, and relatively metal-poor stars are quite rare. Thus the observed scarcity of metal-poor stars in the solar vicinity, which is difficult to account for with conventional 'one zone' models of galactic evolution, can be explained naturally by a dynamical picture in which the galactic disk is built up by the continuing inflow of gas into a region in which an approximate steady-state metal abundance has been established. These results, together with the results mentioned above for the structure of the models, provide strong support for the conventional collapse picture of galaxy formation and show that more exotic models, such as galaxy formation by ejection from a nuclear source, are not required to account for any of the observed properties of typical elliptical galaxies.

The total star formation rate in the present models reaches a maximum at a time about 1.0–2.5×10^9 yr after the big bang, corresponding to a redshift of the order of $2-5$, with considerable uncertainty. For a model of mass $10^{11}\ M_\odot$ the maximum star formation rate is of the order of $10^2\ M_\odot$ yr^{-1} and the corresponding maximum stellar luminosity is of the order of $3 \times 10^{11}\ L_\odot$; these numbers scale with the mass and so could be 1 to 2 orders of magnitude higher for massive elliptical galaxies. Thus, forming galaxies are predicted to have redshifts and luminosities comparable with those of quasars. This result, together with the fact that during the later stages of the galaxy formation process the residual gas and star formation activity become more and more strongly concentrated in the nucleus of a galaxy, suggests that the quasar phenomenon may be identifiable with a stage of the galaxy formation process when residual gas is concentrating at the centre to form a highly condensed nucleus. The present models do not directly account for the non-thermal nature or the variability of quasar radiation, but they provide a natural justification for those quasar models which postulate that large numbers of pulsars or similar condensed objects are formed in a dense ambient gas in the nucleus of a galaxy. This is expected to occur in a forming galactic nucleus, since a high rate of star formation in the nucleus will result in a high rate of supernova and pulsar production.

In models where the boundary is allowed to expand with time, the inflow of gas into the nuclear region may continue at an appreciable rate for up to 10^{10} yr or longer, owing to the long time required for the outermost layers of the protocloud to collapse. Thus it is possible that, in some cases, star formation and related activity may continue or recur in the nucleus of a galaxy long after the formation of the bulk of the galaxy is completed. This may help to explain why nuclear activity is observed in a number of presumably old galaxies at relatively small redshifts.

DISCUSSION

Rickard: From your diagram of the growth of metals as a function of time, one can see that the metal abundance *decreases* in the later stages in the *outer* region of the protogalaxy. What is the reason for this?

Larson: This effect is found also in some of the non-dynamical models of other authors and is due simply to the fact that, after the initial phase of rapid star formation and metal enrichment, most of the gas in the outer region is supplied by mass loss from relatively old low-mass stars which do not contribute additional metals; thus the metal abundance of the ejected gas is essentially the average metal abundance with which the halo stars were formed, and this is significantly smaller than the peak value.

G. de Vaucouleurs: As an observational test of your models, would you expect the fraction of E galaxies with emission lines in their nuclei to increase with distance (look-back time)? Or the ratio [N II]/Hα to be expected to vary appreciably in the last 10^9 yr?

Larson: The models predict that the amount of gas in a galaxy decreases as a function of time, and one would thus expect to see, at least on the average, more gas and stronger emission lines in galaxies at larger redshifts. One might also expect some variation with time in the emission line ratios; in particular, a relative strengthening of nitrogen lines due to the continuing production of nitrogen. However, my models do not yet provide predictions for individual elements.

Cox: Have you included the energy input to the residual gas by supernovae? I think Mathews and Baker (*Astrophys. J.* **170**, 241, 1971) found that, with such low gas densities, the gas cannot radiate the energy input and an outflow of galactic wind occurs.

Larson: These models do not include the supernova energy input, and that is one of the main deficiencies which I hope to correct. It may be that some or all of the residual gas is blown out during the later stages of evolution. It may also be that intense quasar activity in a forming galactic nucleus could have interesting effects in blowing gas out of the system and possibly setting a limit to further gas inflow and star formation; these are problems that remain to be investigated.

Disney: Would you please comment on the sensitivity of the results to the assumed initial conditions? Are not computations dominated by free-fall strongly dependent on the initial conditions, in particular the initial density distribution?

Larson: In hydrodynamic collapse calculations for interstellar gas clouds and protostars, it has generally been found that the results are not strongly dependent on the details of the initial conditions, e.g. the initial density distribution, and I think that the same conclusion holds here. From experiments with the models, it seems that the structure of the resulting galaxy is determined more by the physical processes occurring during the collapse, particularly the gaseous dissipation and star formation processes, than by the details of the initial or boundary conditions. However, these latter do affect the time scale of the collapse and the rate of continuing inflow of residual protogalactic gas throughout the evolution of the galaxy.

Wright: Your result of QSO luminosities and redshifts after the collapse of a protogalaxy follows, I believe, simply because the free-fall time for reasonable protogalactic densities corresponds to a value of $z \approx 2$. At an A.A.S. meeting two years ago (*Bull. Am. Astron. Soc.* **4**, 267, 1972) I noted this result but concluded it was simply an amusing coincidence since I didn't believe that the extremely spherical collapse needed to get a very small emitting region was realistic.

Larson: It is, of course, hardly likely that real collapsing protogalaxies would be strictly spherical; probably the situation would be chaotic and irregular, and to the extent that my models are applicable at all they would provide only a smoothed, spherically symmetrized representation of the complicated and chaotic real situation. Nevertheless, the observed presence of highly condensed nuclei in elliptical galaxies is an indication that, even though the collapse may not be closely spherical, protogalaxies *are* able to develop highly condensed cores; thus I think that the spherical collapse models are not completely without reality. It is true that the region of active nuclear star formation in the models is one to two orders of magnitude larger than the 10 pc size which one infers for the region of intense activity in quasars, but again it should be kept in mind that the models describe only a smoothed average of the condensation and star formation processes which in reality are probably highly inhomogeneous and confined to atypically small, dense regions; it is plausible that quasar activity would be most prominent in such regions.

THE REDSHIFTS OF EXTRAGALACTIC OBJECTS

WALLACE L. W. SARGENT

Hale Observatories, California Institute of Technology, Carnegie Institution of Washington, U.S.A.

Abstract. We review the evidence for a general shift-distance law for galaxies. This is shown directly by a redshift-distance diagram recently prepared by Sandage and Tammann for Sc I galaxies. The distances are measured from H II region diameters and from the diameters of the Sc I galaxies themselves for the more distant systems. We also show the redshift-apparent magnitude diagrams for several objects which are identifiable over a large range in distance and which prove to have a small range in absolute magnitude: these are supernovae of type I (Kowal and Sargent, 1973, unpublished), brightest cluster galaxies (Sandage, 1972a), and radio galaxies (excluding N-type galaxies) (Sandage, 1972b). Plots of redshift versus angular size for brightest cluster galaxies (Sandage, 1972c) also have the correct slope for this to be a distance effect. In summary, the tightness of the correlations for the various diagrams listed implies that there is a very close correlation between distance and redshift for galaxies out to at least $z = \Delta\lambda/\lambda \sim 0.2$ and that there is no evidence for redshift anomalies among the objects in these diagrams.

The interpretation of the cosmological redshift as a Doppler shift due to the expansion of the Universe is established only by indirect arguments (Peebles, 1971). A repetition of the test proposed by Hubble and Tolman (1935), which involves a correlation of surface brightness and redshift, would be important in this connection.

We discuss possible examples of redshift anomalies in galaxies. These include (a) compact groups of galaxies – Stephan's Quintet, VV172, Zwicky's triple system and Seyfert's Sextet – which each have one highly discrepant member; (b) Tifft's (1972a, 1972b, see also p. 243) work on band structure in the redshift-magnitude diagrams for clusters of galaxies; (c) the systematic difference in redshift between spirals and ellipticals in the Virgo cluster (Holmberg, 1961; Tammann, 1972); (d) Arp's work on galaxies with apparent connections to high redshift companions – NGC 7603, NGC 772; (e) Arp's work on the redshifts of dwarf companions to large spiral galaxies (Arp, 1971; Lewis, 1971). We find that there is no compelling reason to believe that redshift anomalies are responsible for any of the phenomena exhibited by these systems. There have been no proper statistical studies which would eliminate the possibility of chance projections of background or foreground galaxies being responsible for some anomalies. In particular Lynds' (1972) work shows convincingly that the discrepant object, NGC 7320, in Stephan's Quintet is due to such a chance superposition. In the case of Tifft's work the fact that the band structure observed for the Virgo cluster is nearly orthogonal to that claimed for the Coma cluster makes his interpretations highly questionable.

We next consider the redshifts of N-type galaxies. The conventional view is that the N-galaxies are 'mini-Quasars' although it is also possible that the N-galaxies and

John R. Shakeshaft (ed.), The Formation and Dynamics of Galaxies. 195–198. All Rights Reserved.
Copyright © 1974 by the IAU.

quasars form a sequence of increasing non-cosmological contribution to the redshift. Sandage (1973) showed that it is possible to decompose the light of N-galaxies into a non-thermal component and a component with colours similar to those of a normal galaxy. Moreover, in a plot of the redshift of the N-galaxy versus the magnitude of the 'galaxy' component, the points fall on the line defined by normal radio ellipticals. Thus there is no evidence for a non-cosmological contribution to the redshifts of N-galaxies out to $z \sim 0.2$. The same result is found for some individual N-type systems (3C 371, III Zw 2, B 264, RN 8) which are in clusters or which have normal companion galaxies with the same redshift.

Finally we discuss the present evidence which bears on the quasar distances. The evidence points strongly, but not conclusively, to cosmological distances for these objects. The work by Gunn and his associates on quasars in clusters with the same redshift (5 cases out of 7 so far investigated) and Kristian's (1973) evidence for galaxies around quasars are particularly compelling. At the same time the features which have contradicted the cosmological viewpoint have tended to disappear with the passage of time. Among these we may cite (a) the distribution of redshifts over the sky is no longer thought to be highly anomalous (Wills, 1972); (b) according to Burbidge (1973) there is no longer any reason to believe in anomalous peaks or in periodicities in the distribution of quasar redshifts; the remaining anomalies are associated with peculiar emission-line objects, such as radio galaxies, that are not quasars; (c) the anomalous proximity of four 3CR quasars to bright galaxies in the *Reference Catalogue* has not been found in other samples of quasars and galaxies; (d) the chain of quasars near NGC 520 pointed out by Arp (1970) looks much less impressive when all quasars in that region of the sky are considered.

Just before the Symposium, Wampler *et al.* (1973) discovered that the radio source 4C 11.50 is a double quasar 4.8″ apart. The components are 17^m and 19^m and have redshifts $z = 0.435$ and $z = 1.901$, respectively. At first it appeared that there was an absorption line corresponding to Mg II $\lambda 2800$ in absorption in the high redshift object at a wavelength corresponding to the low redshift. Even more surprising there seemed to be weak emission features corresponding to the high redshift in the low redshift object. Later observations did not reveal these emission features. However, the close proximity of the two objects has a low probability on the cosmological hypothesis where it would have to be accidental. A large number of such pairs would be hard to account for in conventional terms.

References

Arp, H. C.: 1970, *Astron. J.* **75**, 1.
Arp, H. C.: 1971, *Nature Phys. Sci.* **231**, 103.
Burbidge, G. R.: 1973, *Mitt. Astron. Gesellsch.*, Nr. 34, p. 19.
Holmberg, E.: 1961, *Astron. J.* **66**, 620.
Hubble, E. and Tolman, R. C.: 1935, *Astrophys. J.* **82**, 302.
Kristian, J.: 1973, *Astrophys. J. Letters* **179**, L61.
Lewis, B. M.: 1971, *Nature Phys. Sci.* **231**, 13.

Lynds, C. R.: 1972, in D. S. Evans (ed.), 'External Galaxies and Quasi-Stellar Objects', *IAU Symp.* **44**, 376.

Peebles, P. J. E.: 1971, *Comments Astrophys. Space Sci.* **3**, 173.

Sandage, A. R.: 1972a, *Astrophys. J.* **178**, 1.

Sandage, A. R.: 1972b, *Astrophys. J.* **178**, 25.

Sandage, A. R.: 1972c, *Astrophys. J.* **173**, 485.

Sandage, A. R.: 1973, *Astrophys. J.* **180**, 687.

Tammann, G.: 1972, *Astron. Astrophys.* **21**, 355.

Tifft, W. G.: 1972a, in D. S. Evans (ed.), 'External Galaxies and Quasi-Stellar Objects', *IAU Symp.* **44**, 367.

Tifft, W. G.: 1972b, *Astrophys. J.* **175**, 613.

Wampler, E. J., Robinson, E. L., Baldwin, J. A., Hazard, C., and Jauncey, D. L.: 1973, *Nature* **246**, 203.

Wills, D.: 1972, *Nature Phys. Sci.* **238**, 70.

DISCUSSION

G. Burbidge: We (Burbïdge, G. R. and O'Dell, S. L.: 1973, *Astrophys. J.* **183**, 759) have recently re-done the work by Bahcall and Hills (*Astrophys. J.* **179**, 699, 1973) on the redshift-magnitude relationship for the brightest QSOs and we get a slope of about 4 rather than their value of 5. The slope turns out to depend critically on the presence of 3C 273. If it is left out, the slope is only two, which shows – that the slope is not well-determined at all.

Sargent: – that 3C 273 ought to be put in!

G. Burbidge: Although the recent work on the peaks at $z = 0.06$ and 1.95 in the number vs redshift distribution suggests that they are not absolutely established on a statistical basis, I still believe that there's something significant about them.

Bolton: Two comments:

(i) regarding Kristian's evidence on underlying galaxies. I looked at these objects on the Sky Survey prints and in all but one case I believe the objects should have been called galaxies in the first place.

(ii) On the close QSOs, Peterson and I have found five cases in the southern sky of pairs of QSOs within about 30″ of each other. One of each pair is identified with a radio source.

Sargent: In reply to those two points:

(i) If you are referring to the objects near the line representing the Hubble relationship, then what you say is to be expected (on the cosmological interpretation of redshifts). The important fact, which must be explained by those who hold the non-cosmological interpretation of redshifts, is why those objects *away* from the line do *not* show evidence of a 'fuzz' around them.

(ii) Wampler estimates that there are about 10^6 QSOs in the sky with magnitudes brighter than 19. The probability is therefore about 10^{-4} of finding one of these lying within 5″ of a particular QSO and, correspondingly, one expects over the whole sky about 100 pairs within 5″ of each other. If the QSOs selected are bright ones then, of course, the expected number of pairs goes down.

Heidmann: Arp's statistics on the redshift differences between main and companion galaxies have recently been redone by Bottinelli and Gouguenheim (*Astron. Astrophys.* **26**, 85, 1973) using all of the 50 groups in de Vaucouleurs' Chapter 17 (*Stars and Stellar Systems*, vol. 9). Arp's result is plainly confirmed, the difference being 90 km s^{-1} with a 30 km s^{-1} mean error; so the zero value is excluded at the 3σ level.

Sargent: My disagreement is not about the ability of astronomers to subtract one number from another, but whether the numbers are right. The systematic effects must be looked at very carefully.

Heidmann: These have been investigated by Bottinelli and Gouguenheim and I refer you to their paper.

Abell: Calculation of the probability of finding a particular configuration of objects in the sky is very tricky. When a particular pattern is *already observed*, one cannot say that it is as unlikely to have happened by chance as the probability of it having occurred at random if that pattern were described in *advance* of its discovery. Suppose, for example, one numbered 1000 pieces of paper as 1 to 1000, and put them into a box; then suppose one paper is drawn at random and is found to be, say, number 633. The chance that no. 633 would be drawn, if named in advance, is 0.001, but that is the same probability as any other number, and one *had* to be drawn.

In the case of Wampler's discovery of a 19^m quasar 5″ from a 17^m quasar, Sargent has reminded us that one would expect 100 examples of two quasars within 5″ of each other. One such pair happens to have been discovered to date.

G. Burbidge: There is also Stockton's pair (*Nature Phys. Sci.* **238**, 37, 1972).

Arp: The statement by Sargent about the line of QSRs from NGC 520 should be corrected. I plotted only faint ($V < 17^m$) QSRs; he plotted also PHL – radio-quiet objects of all apparent magnitudes, which are not a complete sample and from which no conclusions can be drawn.

If you restrict yourself to a homogeneously selected sample but over a larger area than I originally considered, then the two QSRs on the other side of the chain of four QSRs appear significant. I never discussed them because they were further out and one or two others fell off the chain. But of the two QSRs in a line opposite the original chain, one has a redshift almost exactly equal to a quasar in the original chain. Sargent's diagram makes it look as though there is a cluster of QSOs around NGC 520 – I never mentioned this situation because of the incompleteness of this sample – but if Sargent believes the sample is complete he should explain the apparent clustering around NGC 520.

Sargent: Do you wish me to reply to that, or can we go for a beer?

EVIDENCE FOR NON-VELOCITY REDSHIFTS –
NEW EVIDENCE AND REVIEW

HALTON ARP

*Hale Observatories, Carnegie Institution of Washington, California Institute of Technology,
Pasadena, Calif., U.S.A.*

Abstract. Evidence for non-velocity redshifts in quasars and galaxies is reviewed. It is shown that all current statistical tests favour the association of at least some quasars with relatively nearby galaxies. It is shown that a statistically significant number of quasars falling close to faint galaxies have intermediate redshift (between $z = 0.4$ and 1.8). This confirms the previous result that those are the intrinsically most luminous quasars and keeps open the possibility that less luminous quasars can appear projected at large distances around very nearby galaxies.

It is shown that in four individual cases where quasars fall, projected, closest to bright galaxies, the galaxies show evidence of physical interaction. New evidence for perturbations of the inner isophotes of NGC 4319 by Markarian 205 and by a radio source on the opposite side is presented.

Evidence for systematically higher redshifts for compact and peculiar companion galaxies is reviewed. The intrinsic redshift-morphology relation for clusters of galaxies is commented upon and the relation to Tifft's work is noted. As a summarizing diagram, the individual associations of high redshift quasars and companions are used to show an empirical continuity of observed characteristics between compactness (taken to be a measure of youth) and excess redshift.

Finally some possible theoretical explanations for intrinsic redshifts are mentioned.

1. Introduction

One of the first and major results of radio astronomy was to show that many radio sources tended to pair across galaxies from which, it was presumed, they had been ejected. (The original paper is Jennison and Das Gupta (1953), but for summaries see Maltby *et al.* (1963) and Matthews *et al.* (1964)). When some individual radio sources identified with optical objects were claimed to be associated with galaxies of brighter apparent magnitude, however, considerable scepticism was encountered. The difficulty was that the optically-identified radio sources were characteristically quasars or galaxies of much higher redshift than the galaxy with which they were supposed to be associated (for summary see Arp, 1971a).

Because the quasars are the highest redshift objects now known, they pose in the purest form the problem of whether the conventional redshift-distance relation is applicable to all celestial objects. Therefore the following review will examine first the current evidence for and against quasars being at cosmological distances.

2. Associations of Quasars with Galaxies

There are two kinds of evidence associating quasars with nearby galaxies: (a) statistical and (b) individual connections to, or perturbations of, adjoining galaxies.

2.1. STATISTICAL EVIDENCE

The first evidence suggested that some quasars were associated with violently dis-

John R. Shakeshaft (ed.), The Formation and Dynamics of Galaxies. 199–224. All Rights Reserved.

rupted or exploding galaxies (Arp 1966, 1967). Subsequently analysis of a complete and homogeneous sample of 3 CR and Parkes quasars (QSRs) indicated that these radio bright QSRs (down to 5.0 and 1.5 f.u. at 408 MHz) were associated with bright Shapley-Ames galaxies (Arp, 1970).

Most recently, an extremely thorough statistical analysis of the relation of QSRs and bright galaxies was made by Burbidge *et al.* (1971) and Burbidge *et al.* (1972). My own evaluation of the present situation would be that the density of QSRs has been investigated preliminarily out to the order of 20° around bright galaxies. Bright here means objects in the *Shapley-Ames Catalogue* (1932) which is complete to about $m_{pg} = 12.5$ mag and the *Reference Catalogue of Bright Galaxies* (de Vaucouleurs and de Vaucouleurs, 1964), which is complete to about $m_{pg} = 13.0$ mag. Arp (1970) showed that the faint $(m_v \geqslant 17.0$ mag) QSRs, from the complete 3CR and *Parkes Catalogue* identifications, fell on average $5°8$ away from galaxies brighter than $m_{pg} = 10.5$ mag, whereas the same number of QSRs randomly distributed on the sky fell an average of $10°7$ from these bright galaxies. Since both the bright galaxies and the faint QSRs were distributed in roughly similar and distinctive patterns over the north galactic hemisphere, it was assumed in that study that this was the reason for the smaller average distance between the two sets of objects. But Burbidge *et al.* (1971) showed that essentially just four QSRs which fell very close ($<7'$) to bright galaxies represented a very strong excess over that predicted on the basis of a random distribution of QSRs with respect to those galaxies. It is not clear, therefore, without additional investigation, whether the 3CR and Parkes QSRs are significantly associated out to distances of, say, 6° or more, or whether the association is primarily caused by a relatively few QSRs associated very closely, say within 7', where Burbidge *et al.* (1971) found such an outstandingly high density of QSRs.

In later papers, Bahcall *et al.* (1972), Hazard and Sanitt (1972), and Burbidge *et al.* (1972) tested associations of QSRs with fainter galaxies. No statistically significant correlations were found. The latter paper pointed out, however, that several selection effects, such as faintness and closeness of QSRs to faint galaxies, might bias against discovery and fail to reveal an association if it were present.

Most recently Browne and McEwan (1973) have used improved optical identifications for Parkes radio sources greater than 0.35 f.u. Two new QSRs found within $1°7$ and $1°2$ of Zwicky Catalogue galaxies (Zwicky *et al.*, 1961, 1965) lower the probability of finding accidentally the observed number of QSRs close to these galaxies to something less than 5%. Although this is of marginal significance, the new quasars are found less than $2°5$ from galaxies, which gives weight to the view that they are like the examples of 3C quasars physically associated within 7' of bright galaxies but just removed to greater distances.

I would like to discuss further this question of why a greater percentage of QSRs are not found close to fainter galaxies. In the first place, even on conventional grounds, the figure of 1.79 galaxies/sq. deg. as bright as 15.7 mag (Browne and McEwan, 1973) must overestimate to some extent the number of *distant* galaxies. In the light of Section 3 to follow – where a number of faint, high-redshift systems are indicated to be relatively nearby – the areal density of distant galaxies in fact might be reduced appreciably. Secondly I note that a large percentage of QSRs may be extremely close to us and be projected at relatively large distances from very bright galaxies. In the original Arp (1970) paper, evidence was found for the QSRs in the apparent magnitude range $m_v = 14$ to 16 mag to be associated with very nearby, Local Group galaxies. The QSRs in the $m_v = 17$ to 19 mag range were primarily associated with galaxies in the $m_{pg} = 9$ to 11 mag range. Therefore it is clear that if one investigates galaxies in the $m_{pg} = 13$ to 15 mag range, as in these later studies, one would only expect QSRs generally in the $m_v = 21$ to 23 mag range – beyond the usual discovery capability for QSRs.

Looked at from the standpoint of redshifts, that early paper showed the low redshift QSRs ($z = 0.2$ to 0.4) to have generally low luminosities, then higher luminosities in the $z = 0.4$ to 1.5 range, and drastically lower luminosities again in the $z = 1.8$ to >2 range. It is interesting to note in this respect that the five QSRs in the paper by Burbidge et al. (1972) are associated with what would be termed fainter galaxies in the present context. Everyone of them falls between $z = 0.5$ to 1.4 in redshift. If the distribution of QSR redshifts is taken from Barbieri et al. (1967), the probability of getting a redshift in this range by accident is 0.47, and the chance of getting all five in this range is only 0.02. In Table I of the present paper we have added,

TABLE I

Close association between quasars and galaxies

Object Pair	m	z	$r(')$
3C 455	19.7	0.543	0.4
NGC 7413	15.2	0.033	
3C 232	15.8	0.534	1.9
NGC 3067	12.7	0.005	
3C 268.4	18.4	1.400	2.9
NGC 4138	12.1	0.004	
3C 275.1	19.0	0.557	3.5
NGC 4651	11.3	0.003	
3C 309.1	16.8	0.904	6.2
NGC 5832	13.3	0.002	
2020-370	–	–	0.3
Spiral galaxy	–	1.1	
PHL 1226	–	0.404	0.9
IC 1746	14.5	–	
3C 270.1	18.6	1.519	5.1
pec ring galaxy	(17)	–	
0159-11	16.4	0.68	39
IC 1767	(15)	–	
Mark 132	15	1.75	45
NGC 3079	11.9	0.041	
3C 254	18.0	0.734	126
Mayall's pec. object	(15)	0.035	

below these five associations, seven more associations of QSRs with individual galaxies, which the author thinks are among most probable associations. Now the redshifts are still all restricted between $z=0.4$ and 1.8, or just in the brightest luminosity range for QSRs as was discussed earlier. The chance of the entire list of twelve being drawn accidentally between the above redshift limits from a list containing the normal distribution of redshifts is less than 0.007.

In summary then, the question of why every QSR is not associated closely with some galaxy seems to be answered by the result that many QSRs are relatively close to us and are projected at considerable distances on the sky from the galaxies from which they originate. This is particularly true of the fainter QSRs where we find an increasing percentage of redshifts $z>1.8$ which represent QSRs of drastically lower luminosity and therefore considerably closer by.

As for the important, and by now much investigated, question of whether the quasars are associated with clusters of galaxies, a brief summary of the current situation might be the following: A number of searches have been made around quasars of low redshift ($z<0.36$) for clusters of galaxies at the same redshift. Three possible examples of bona fide quasars near groups or clusters of the same redshift have been found, but Burbidge and O'Dell (1973) claim that, in view of the uncertainty in number density of less populous clusters, these cases are not statistically significant. On the other hand, two independent investigations have turned up evidence that quasars with redshifts $z>0.3$ are slightly correlated with clusters of redshift $z<0.2$ (Bogart and Wagoner 1973; Bahcall et al., 1973). This correlation was not considered statistically significant, but it would be of interest to see whether the significance would be improved by taking only nearer clusters or more luminous quasars, as discussed above.

2.2. Association of individual quasars with galaxies

In order to search for evidence of a perturbation or connection, it is necessary to look carefully at those cases where the quasar falls closest to an adjoining galaxy. The four cases currently known of closest association are:

(1) 2020–370, where the quasar falls 21″ from a small spiral galaxy (Peterson and Bolton, 1972); (2) 3C 455, where the quasar falls 23″ from NGC 7413 (Arp et al., 1972); (3) Markarian 205, where the quasar falls 42″ from the large spiral NGC 4319 (Weedman, 1970); and (4) PHL 1226, which falls 55″ from the spiral galaxy IC 1746 (Burbidge et al., 1971). Of these cases, the first is unfavourable for study because the galaxy is small and, due to its extreme southern declination, good plates are not available.

In the second case, an isodensity tracing of the galaxy, NGC 7413, and the quasar 3C 455 is shown in Figure 1. Even though the galaxy looks like a dynamically relaxed, regular E galaxy, there is a conspicuous (and very unusual) perturbation of the inner isophotes of the galaxy in a northeasterly direction, the same approximate direction as the quasar. As we will discuss in the succeeding cases, the amount of rotation expected in the time since ejection from the inner regions of the galaxy would easily account for this displacement in direction. As far as the direct photographs show, the object is a normal, Population II galaxy in which it is not usual to find elongated luminous forms (although, as the jet from the nucleus of M87 demonstrates, some E galaxies at some times appear to be capable of ejecting luminous matter, which is in a physical state not well understood at present.)

In the third case, the quasar Mark 205 (at $z=0.07$) is close to a spiral galaxy NGC 4319 ($z=0.006$). Arp (1971b) claimed there was a luminous filament connecting the two objects both in the continuum and in Hα. Lynds and Millikan (1972) found a feature in continuum light but doubted its interpretation as a connection. Both Lynds and Millikan, and Adams and Weymann (1972), did not find a connection in Hα, while Ford and Epps (1972) not only did not find a feature at Hα but also concluded that none existed. Arp (1973b) showed that the Lynds-Millikan continuum feature coincided exactly in position and shape with the claimed emission connection. Figures 2 and 3 show two more Hα photographs I have just obtained, each of which reveals the luminous connection.

Another new result of considerable interest concerns a radio source found just north of NGC 4319 by van der Kruit (1971). He originally identified the radio source

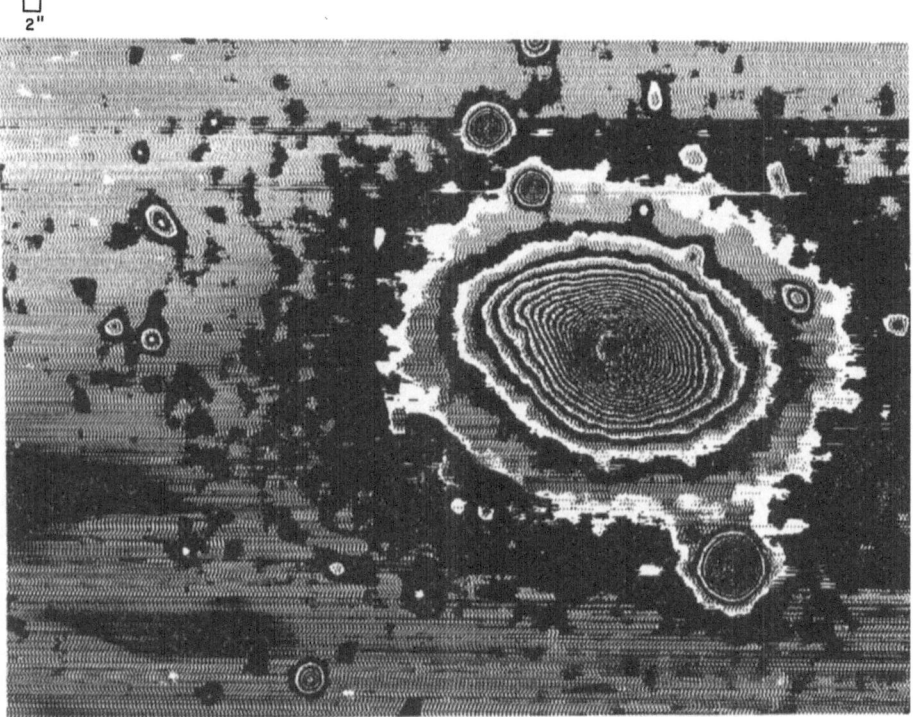

Fig. 1.   Isodensity tracing of NGC 7413 (E galaxy) and the QSR, 3C 455 (star image immediately north-east of galaxy). Perturbation of isophotes is seen in general direction of QSR but slightly offset in direction. North is at the top and east at the left, as also in Figures 2–8.

with an optically-bright star, but measures of photographic plates by Sulentic (un-published) establish the radio source actually to be 1.2 s east and 15″ south of that star. Figure 4 shows, by a small cross, the location of the radio source relative to NGC 4319. The new position of the radio source places it within a few arc seconds of a moderately small galaxy, which then is possibly close enough to be a candidate for an optical identification. The existence of this radio source, however, takes on added significance when the following new results on the inner isophotes of the spiral galaxy are considered.

Initial isophotal analysis of the interior regions of NGC 4319 with GALAXY at Edinburgh (Arp and Pratt, unpublished) showed that a perturbation of the interior regions pointed in the general direction of Markarian 205. Figure 5 gives the result of further analysis with a Joyce-Loebel isodensity tracer at the Jet Propulsion Laboratory in Pasadena on a 20-minute exposure, 103a-J plate obtained with the 200-in. reflector (Arp and Sulentic, unpublished). It is seen that the major axis of the projected inner regions of NGC 4319 deviates considerably from that of the outer regions. It is seen further that from the inner regions, roughly along the line of

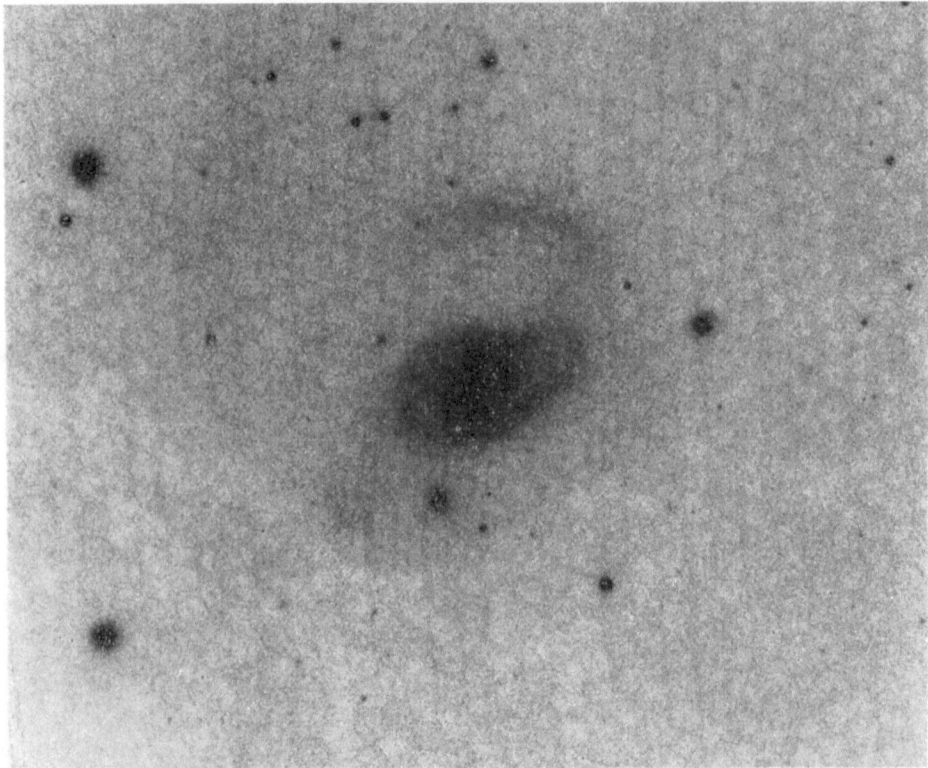

Fig. 2.   Photograph of NGC 4319 (S galaxy) and Markarian 205 (compact object 42″ south). Hα interference filter with new 90mm image tube on 200-in. Palomar reflector. Faint luminous feature confirmed going north from Markarian 205 to galaxy in same place as previous connections photographed.

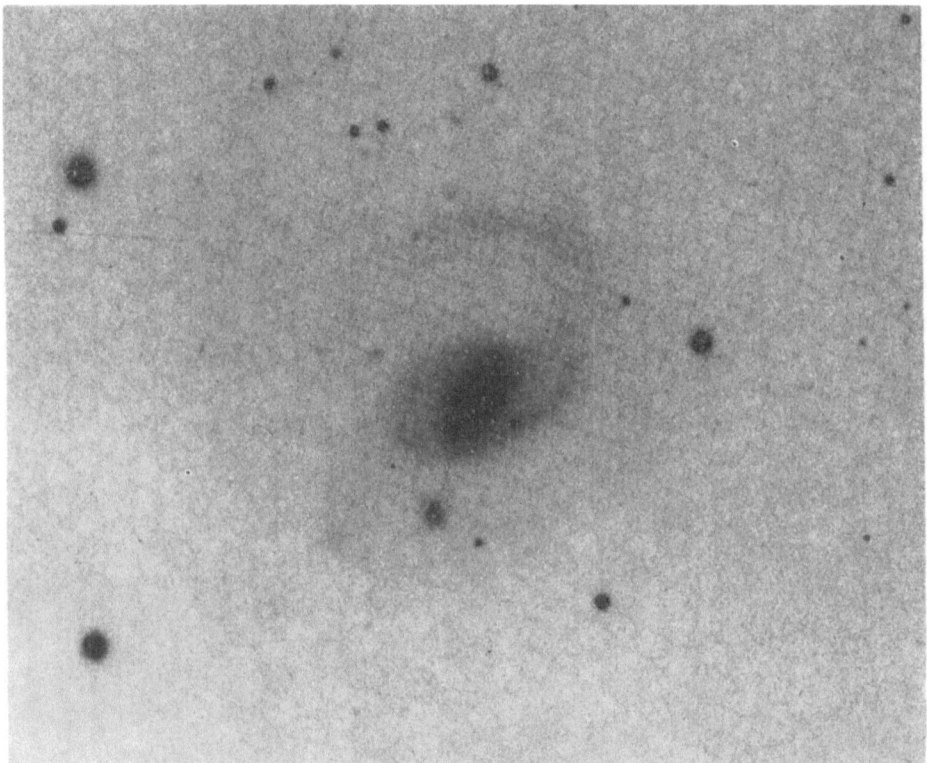

Fig. 3.   A different photograph with same technique as in Figure 2 to show how a second, independent plate registers the feature.

elongation, there are perturbations throughout a number of contour lines extending both north and south-east. In Figure 5 we have drawn dashed lines from the centre of the galaxy southward to Markarian 205 and north-eastward to the radio source on the other side of the galaxy. Then we have rotated these lines clockwise (in the direction of the rotation of the galaxy as projected on the plane of the sky) by 30°. Figure 5 enables the viewer to judge how accurately these lines now pass out from the centre of NGC 4319 directly along the disturbances in the isocontour lines in both directions.

   The interpretation of these observations fulfils earlier conclusions that material objects are ejected outward from galactic nuclei in roughly opposite directions (see Arp, 1972a for summary). In the present case the model is one in which the quasar and the radio source were ejected 10^7 yr ago with a velocity of about 1000 km s^{-1}. Passage outward perturbed the inner regions of the galaxy in a direction which has now swung around about 30° in the direction of rotation. We hypothesize that the ejection was somewhat out of the plane of the spiral so that the outer regions of NGC 4319 are not much disturbed. (It should be noted that the connection to Markarian 205 discussed initially does not necessarily have any physical relation to the perturbation of the inner regions. The connection could, for example, be a small amount of material drawn out behind Mark 205 and seen projected down onto the plane of NGC 4319.)

The fourth case mentioned in the introduction to this section was that of PHL 1226 and IC 1746. The initial discovery of this quasar, which lies less than 1' from a moderately bright spiral galaxy, was made on *Palomar Sky Survey prints* (Burbidge *et al.*, 1971). On those prints the quasar seems attached to the galaxy by a luminous bridge. But later photographs with the 200-in. telescope in good seeing showed that a small galaxy image lay between the quasar and the large galaxy. Although a very suspicious configuration, no continuous connection was apparent between the quasar and galaxy. Now, however, new and extensive isodensity traces of two different photographic plates (Figures 6 and 7) reveal a number of significant features (Arp and Sulentic, unpublished):

(a) Again, the major axis of the inner isophotes of the spiral galaxy is considerably rotated from the axis of the outer isophotes.

(b) The inner spiral arm of the main galaxy, and the fainter continuation of that arm, seem to be extended toward the peculiar galaxy and the quasar.

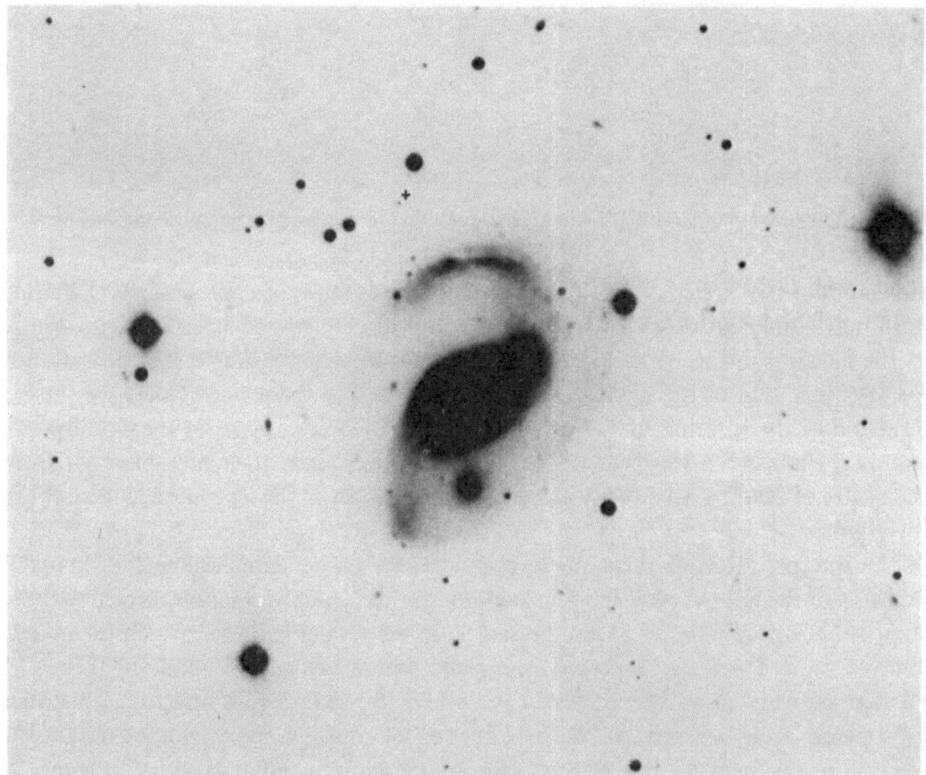

Fig. 4.   Deep photograph of NGC 4319 and Mark 205 in continuum wavelengths (IIIa-J plate with no filter). Photograph shows connection and also position of radio source north-east of NGC 4319 (marked by a cross).

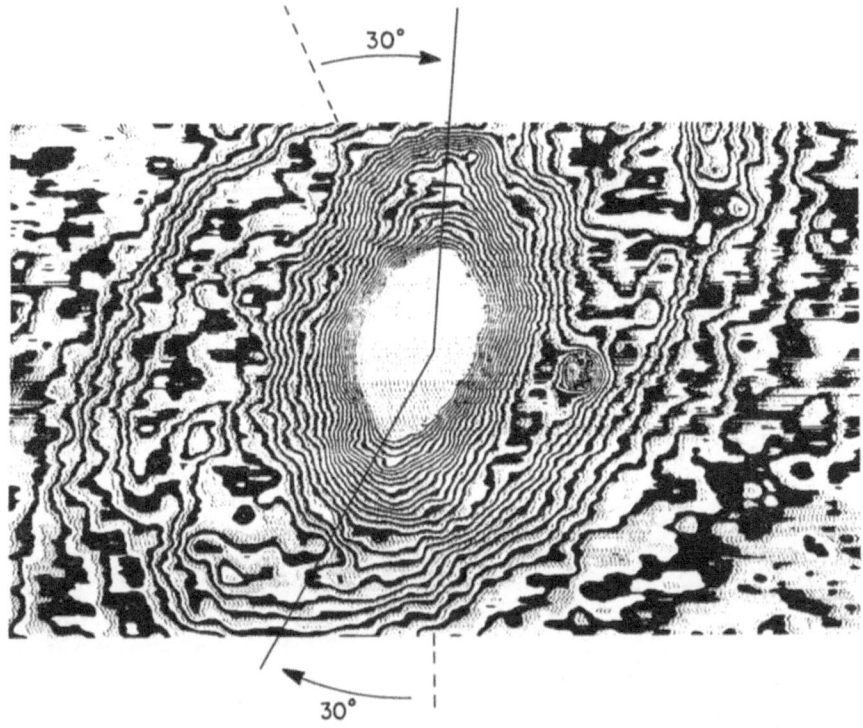

Fig. 5. Isodensity trace of short exposure of NGC 4319 showing perturbation of inner isophotes. Dashed line to north-east is drawn to radio source, dashed line south drawn to Mark 205. After rotation by 30° in direction of rotation of galaxy, the dashed lines are then drawn in solid. Note extension of innermost isophotes along solid lines.

(c) The outer isophotes of the spiral galaxy appear to envelop the peculiar galaxy and the quasar.

(d) A small, peculiar 'line'-like galaxy just north-west of the quasar has wings coming off on either side both toward and away from the quasar, suggesting some form of interaction.

(e) Finally, but most important, on both plates though particularly on the short-exposure plate taken in good seeing conditions, a connection of luminous material can be seen between the quasar and the small peculiar galaxy lying in the direction of the spiral.

Figure 6 shows the isophotal traces of the deepest plate, which illustrates conclusions (b), (c), (d), and (e). Figure 7 presents the lighter exposure, which illustrates conclusions (a), (b), and (e).

The redshifts of the small peculiar galaxies which seem to be interacting with the quasar are not known. But there would be no precedent for their having redshifts as high as $z = 0.404$, which would be necessary in order to match that of the quasar. The importance of the result, therefore, is that once more we have observational evidence for the physical association of quasars with objects of much smaller redshift.

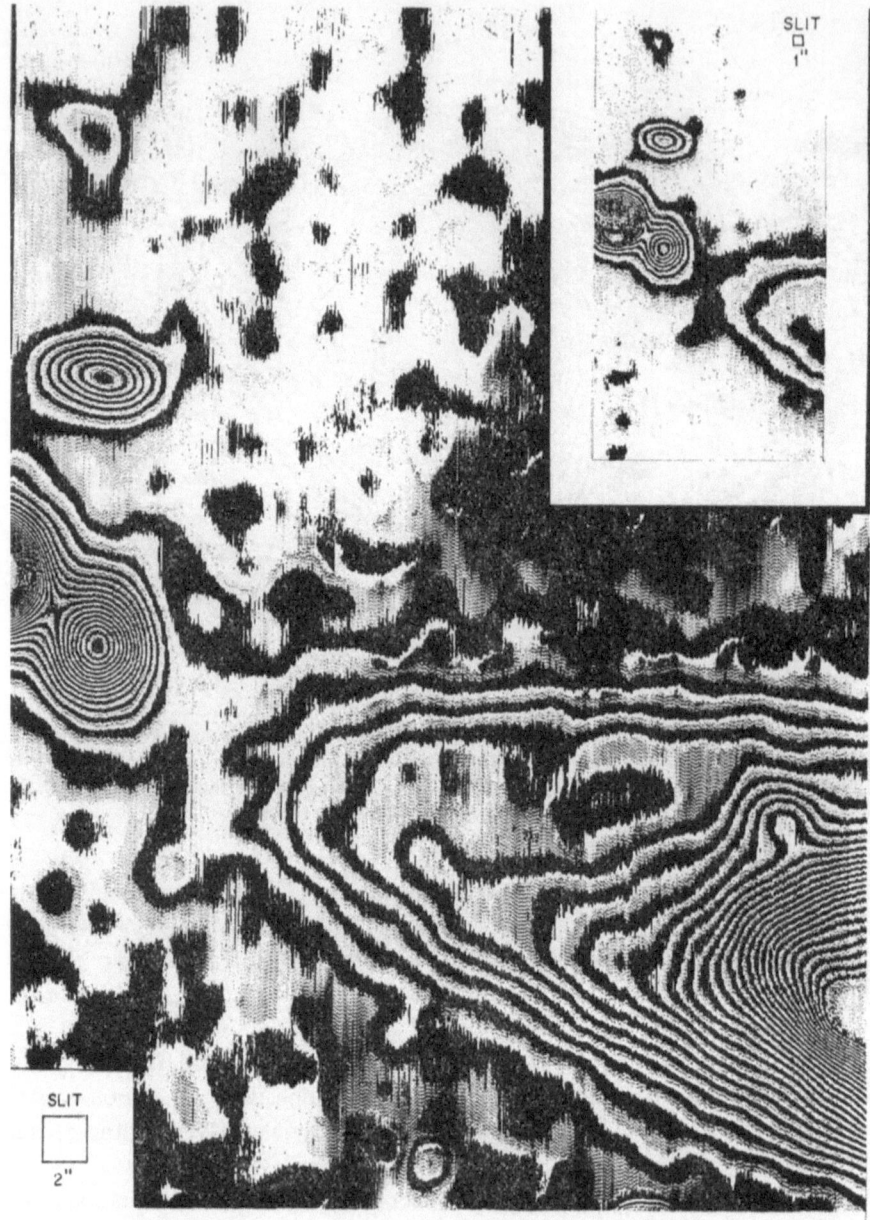

Fig. 6.   Isodensity trace of photograph of region between the edge of the spiral galaxy (IC 1746) and the quasar PHL 1226 and peculiar galaxy. Note appearances of connection between latter two objects, envelope connecting the two to IC 1746 and perturbation of small galaxy north-west of the quasar.

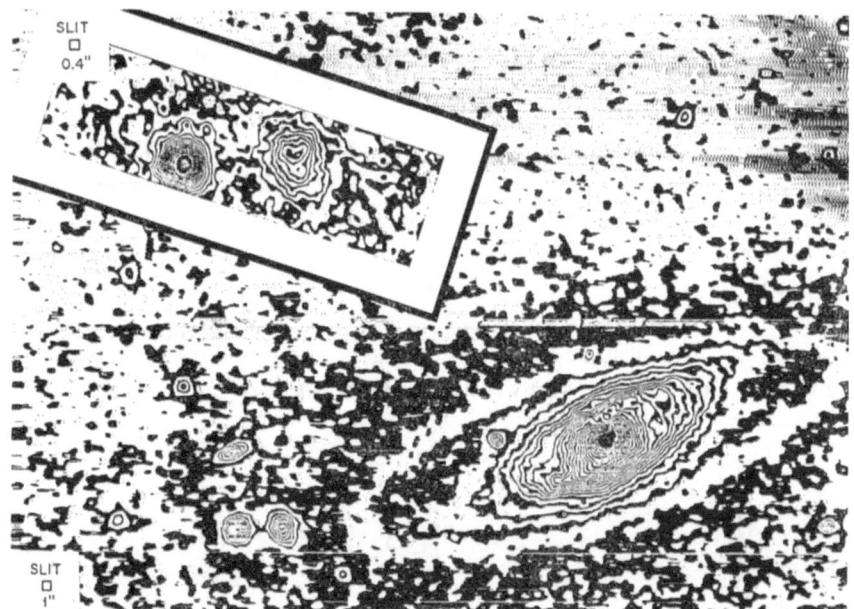

Fig. 7.   Isodensity trace of shorter exposure shows bridge between quasar and peculiar galaxy more clearly.

In summary, therefore, we can state that of the three closest cases which could be investigated in detail, all reveal evidence for interaction of the quasar with the galaxy.

3. Association of High Redshift Galaxies with Low

Further compelling evidence for the existence of anomalous redshifts is the association of small, high-redshift galaxies with nearby, low-redshift systems. There are many examples of high-redshift companions (see reviews by Arp, 1971a and 1973b) but only a few individual associations will be discussed here, and the rest summarized.

3.1. MULTIPLE INTERACTING SYSTEMS

The most famous case of a multiple interacting system is Stephan's Quintet. A series of papers (summarized in Arp, 1973a) provides evidence that all members of the Quintet, including NGC 7319 at $z = 6700$ km s^{-1} and other members to $z = 5700$ km s^{-1}, are associated with the large, nearby Sb spiral NGC 7331, which has a redshift of $z = 800$ km s^{-1}. The observational evidence is in the form of visible interaction between high and low redshift members (NGC 7319 and NGC 7320), identically-sized H II regions in high and low redshift members (NGC 7318 and NGC 7320), luminous filaments in the region between NGC 7331 and Stephan's Quintet, and excessive numbers of radio sources in the region between NGC 7331 and Stephan's Quintet. In connection with this latter point, and in consonance with the introductory remarks about conventional associations of radio sources with galaxies, Figure 8 illustrates a particularly strong line of radio sources, probably unresolved, running southward from NGC 7331, including Stephan's Quintet and going beyond to the south (Kaftan-Kassim and Sulentic, to be published). It is also clear from the diagram of this region that there is a group of high-redshift galaxies right around NGC 7331, then a group of high-redshift galaxies around the low-redshift NGC 7320 in the Quintet, and finally a comparable group of high-redshift galaxies north of NGC 7331 on the opposite side from Stephan's Quintet. There is even a fairly strong radio source in the northern group that forms a radio pair across NGC 7331 with the radio source in Stephan's Quintet.

In an attempt to determine the distance independently of redshift, two separate measurements have recently been made of H I in the high-redshift members of the

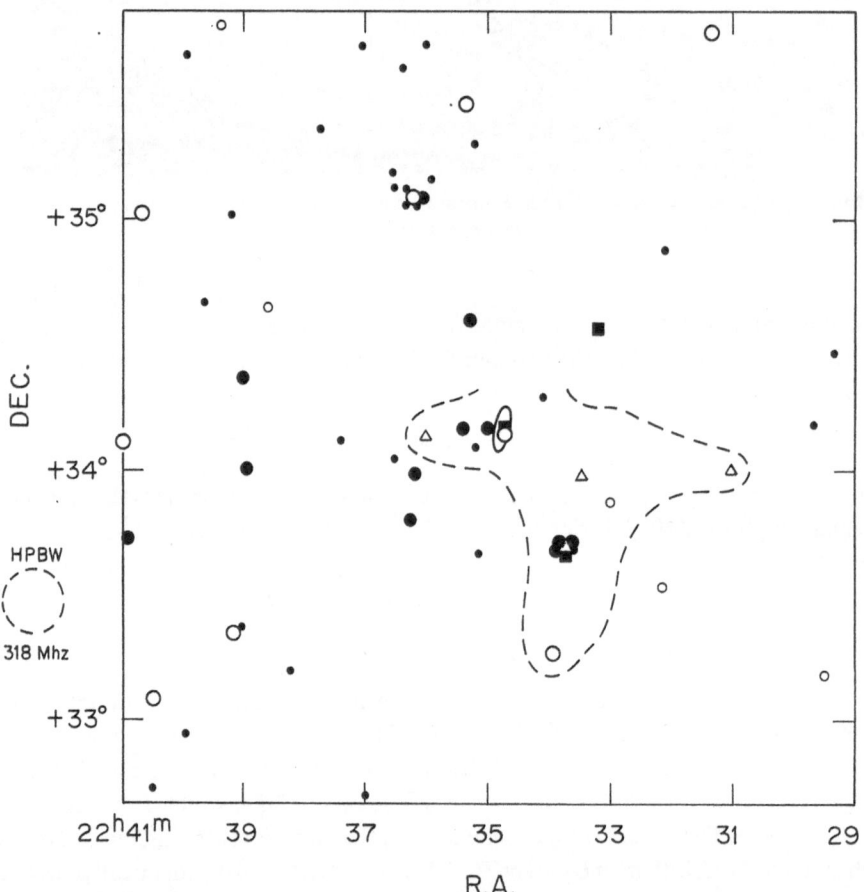

Fig. 8. NGC 7331 region: Open circles represent Bologna 408 MHz sources (small $S < 0.5$ f.u.; large $S > 0.5$ f.u.). Solid boxes, circles, and dots represent Zwicky catalogue galaxies brighter than 14, 15, and fainter than 15 magnitudes, repectively. Triangles and dashed line are the preliminary results of observations of Kaftan Kassim and Sulentic at Arecibo at 318 MHz.

Quintet. (H I in the low-redshift member, NGC 7320, was measured by Allen in 1970.) Everyone agrees that NGC 7320 is at the distance of the low-redshift NGC 7331. But Balkowski *et al.* (1973; see also Balkowski *et al.*, this volume, p. 237) claim that the HI measures with the Nançay radio telescope indicate that the high-redshift NGC 7319 cannot be at its conventional redshift distance but is probably at the closer distance of NGC 7320. They get $D=22^{+15}_{-9}$ Mpc. From a quite similar HI signal observed with the NRAO 300-ft radio telescope, Shostak (1974) gets NGC 7319 at a larger distance, $D=47^{+55}_{-26}$ Mpc, possibly at the conventional redshift distance. (The distance of NGC 7319 with the new Hubble constant of $H=50$ km s^{-1} Mpc^{-1} would be about $D=120$ Mpc.) Although the NRAO and Nançay determinations of the distance to NGC 7319 agree within their own quoted probable errors, there is considerable disagreement as to whether the object is more likely to be at the 10 or 11 Mpc distance of NGC 7331 and NGC 7320 or at the redshift distance of 120 Mpc.

The major difference between the two studies of NGC 7319 seems to be the body of data against which the NGC 7319 measures were calibrated. It is not clear which body of data is to be preferred, but it should be mentioned that Shostak calibrated against Roberts' (1969) H I measures of nearby galaxies, whereas the French project used data by Balkowski (1972). Roberts took distances for nearby galaxies from de Vaucouleurs' group membership assignments or values based on the assumption of a Hubble constant of 100 km s^{-1} Mpc^{-1}, whereas none of Balkowski's (1972) distances were based on the Hubble law. The distance for NGC 7319 derived in the French study should, therefore, be independent of the assumption of the validity of the Hubble law.

This point is particularly significant when one considers that the distance derived for NGC 7319 depends on the morphological classification assigned, i.e., Sb, Sbc, Sc, etc. Not only is there uncertainty in this assignment, but there is uncertainty about the standard Sc galaxies against which it would be calibrated if it were so classified. Recent work has raised some question as to whether Sc galaxies have generally a component of intrinsic redshift (Jaakkola, 1971). Also Sc I (luminosity class I) galaxies show very peculiar redshift distribution properties (Rubin *et al.*, 1973). We shall proceed to the discussion of these effects, but they are mentioned here in order to introduce the question of whether there might be systematic errors in the parameters of the Sc or Sc I classes.

To finish the discussion of multiple interacting systems, it can be stated that there are other groups like Stephan's Quintet, which may be used to test the proposition that such groups are *characteristically* associated with large, low-redshift galaxies. A list of the best examples has been assembled by Arp (1973a). Depending on whether we choose bright galaxies to a limit $m_H=12.2$ mag or galaxies out to 75' from the disturbed groups, or include the first six or seven most interacting groups, it is found that, omitting Stephan's Quintet, the remaining systems have a probability of only 10^{-4} to 10^{-6} of being accidental. In other words, if these high-redshift systems are not characteristically associated with large, low-redshift galaxies, then we should be able to find about one thousand multiple interacting systems not near large galaxies. It is my belief that anyone familiar with the frequency of these peculiar groups will realize that this is clearly not the case.

3.2. DOUBLE INTERACTING AND PECULIAR COMPANIONS

It has been reported (Arp, 1972b) that double interacting galaxies, as a simpler and more common case of multiple interacting galaxies, also occur characteristically in the neighbourhoods of large, low-redshift galaxies. The interpretation given in the above reference is that compact bodies have been fairly recently ejected from the parent body and are in the process of evolving, as a function of time, from peculiar, disturbed objects with high intrinsic redshift to older, more relaxed objects with a lesser component of intrinsic redshift. The early stages of evolution are apparently characterized by secondary ejection, fission, and disruption. Therefore the double and multiple interacting systems fit very well into this evolutionary scheme as young objects.

It is consequently of great interest to note the recent results of Heidmann and Kalloghlian (1973) in a paper titled 'Evidence for the recent Production of Markarian Galaxies', results undoubtedly related to the doubleness of compact galaxies earlier reported by Bertola *et al.* (1971). From the number of pairs of Markarian galaxies which are observed, it is clear that many Markarian galaxies exist as pairs at the same

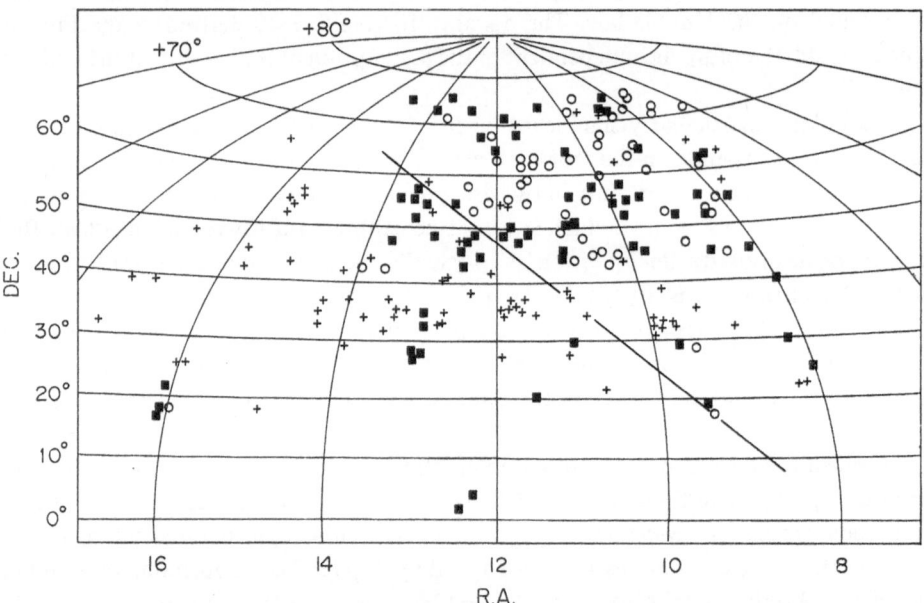

Fig. 9. Distribution of bright Markarian non-Seyfert galaxies ($\leqslant 15.5$ mag) of differing redshifts: Open circles represent Markarian objects with $0.008 \leqslant z \leqslant 0.014$, and solid squares other objects with known redshifts not in the above range. Plus signs represent those objects for which a redshift is unavailable. The line is the same as used by Rubin *et al.* (1973) to divide the areas of their Sc I galaxies of 0.013 to 0.018 redshift (upper area) from their 0.020 to 0.025 redshifts (lower area).

distance from the observer. It is also clear that many pairs have positive energy and are flying apart on time scales of the order of 10^8 to 10^9 yr. On the idea that these Markarian objects are related to the interacting double companions discussed in the previous paragraph, their surroundings have been studied by Arp and Sulentic (unpublished). Preliminary results indicate that, within the area where Markarian galaxies have been most searched for, they are associated with galaxies brighter than 12 mag, the probability of which is estimated to be 0.09 of being accidental. Although the point remains to be formally investigated, presumably most of these Markarian galaxies have considerably higher redshifts than the larger galaxies with which they are associated.

In a further investigation of the distribution of Markarian galaxies on the sky, Arp and Sulentic have noted that Markarian objects of redshift between 0.008 and 0.014 occur predominantly in the area of the sky shown in Figure 9 as occupied by open circles. Curiously, this is the same area in which Rubin *et al.* (1973) found their ScI galaxies of redshift 0.013 to 0.018 to fall preferentially.

The implication, from the close distance for NGC 7319, the association of the companion ScI with NGC 4151, and the general correlation of youth with small luminosity and high redshift, is that the ScI's may have some component of intrinsic redshift, lower luminosities, and be associated with more nearby classes of galaxies.

3.3. Redshifts of companion galaxies

To complete the discussion of anomalous redshifts in galaxies, two important and recent results need to be described. The first is a study by Bottinelli and Gouguenheim (1973) of groups of galaxies identified by de Vaucouleurs in which the main galaxy is at least 50% more luminous than the others. The result is that the companions show an average residual redshift of $+90$ km s^{-1}. The conclusion that there is some residual redshift for these companions is significant at about the 1% level. The Bottinelli and Gouguenheim study repeats the original Arp analysis, which derived a mean residual of 80 km s^{-1}, but has a much larger number of companion galaxies (52), more accurate redshifts, and more stringent inclusion criteria. Therefore this exceedingly surprising and important result appears to be on a much sounder basis at present.

The second important result is a recent study by Jaakkola (1973). As in his previous studies (Jaakkola 1971), he investigates the associations of galaxies in clusters, groups, and pairs, dealing in all with about a thousand galaxies. This time he shows that the residual redshifts are higher for elliptical galaxies with high surface brightness, galaxies with high surface brightness and steep edge gradient of intensity, and galaxies compact in the sense of Zwicky's definition, as well as several other categories of galaxies. The significance of his result lies at the 98 to 99% significance level. The result is important because it independently cross-checks the higher intrinsic redshifts for companion galaxies since it is just the kind of galaxies mentioned above to which the companions tend morphologically. The result also pegs a strong observational confirmation to the hypothesis that the intrinsic redshifts are correlated with the youth of the object, since compactness and high surface brightness are nonstable stages preceding older morphological forms.

The evolutionary age of the system will be an important consideration in the following section where we discuss some analyses of clusters of galaxies.

3.4. REDSHIFT-MORPHOLOGY RELATIONSHIPS IN CLUSTERS OF GALAXIES

We have just considered the neighbourhoods of bright galaxies, typically about a degree radius around a bright Shapley-Ames spiral galaxy, and have reviewed evidence that these companion galaxies are smaller, have higher surface brightness, tend to be multiple and interacting, tend to be emission objects and also be radio sources or be associated in the vicinity of radio sources. All of these characteristics can be subsumed under the loosely descriptive term of young, and most of these objects have redshift excesses of varying amounts with respect to the central dominant galaxy.

The only other independent set of data by which we can test these conclusions is from rich clusters of galaxies. Because the clusters are rich there is greater statistical confidence in demonstrating that the various galaxies are at the same distance. However, it has been known since Hubble's (1936) studies that rich clusters of galaxies tend to be made up of older types of galaxies (E's and S0's) and to lack spirals which are more spread throughout the field. The reason for this may become clearer as we examine the nature of the rich clusters but let us look first at the classical case of a high-density cluster of glaxies.

Tifft (1972 and 1973a) shows that, if the previously established morphological types in the Coma Cluster of galaxies (principally E and non-E classification) are examined as a function of their listed redshifts, the non-E galaxies have a mean redshift about 700 km s^{-1} greater than the E galaxies. Pineau des Forêts and Schneider (1973) confirm that this situation pertains in the core of the Coma Cluster but that the difference diminishes further out in the cluster. The reason for this behaviour becomes clear if one looks at another recent paper by Tifft and Gregory (1973). There he shows that as one goes to the outer regions of the Coma Cluster a greater and greater percentage of blue ellipticals, faint galaxies, and emission galaxies are encountered. This situation was already clear from the work of Sastry (1970) who showed that blue galaxies, and therefore galaxies presumably related to spirals, occurred in a rough ring outside the central regions of the Coma Cluster. (It may be remarked in passing that this kind of structure for a cluster of galaxies implies that the large old galaxies are in the centre and the smaller, younger galaxies around the edge. This explains the previously mentioned tendency for spiral galaxies to be field galaxies, and at the same time is strong evidence for ejection or cascading ejection as the origin of galaxies and clusters of galaxies.)

Recently Tifft (1972) has claimed that the redshift vs. apparent-nuclear-magnitude diagram for rich clusters of galaxies like Coma shows a banded structure. Power spectrum analysis yields significant bands at a given slope. Tifft (this symposium

p. 243) has now analyzed a third cluster in addition to Coma and Virgo, namely A2199. He finds significant banded structure at the same slope of the redshift in these bands, structure which would, of course, be vitiated by even small dispersions in real velocities.

Ignoring the reality or nonreality of the bands for a moment, it is clearly noted in Tifft's work that as one passes towards fainter magnitude and later Hubble type that one encounters increasing amounts of intrinsic redshift. In this respect his results on clusters give direct confirmation of my earlier results for groups and associations of galaxies. But if one accepts the band structure, the confirmation of increasing redshifts with fainter intrinsic magnitudes and younger morphologically type is much stronger. The previous results on correlation of excess redshift with various properties of galaxies are then supported in detail.

If, in fact, the apparent-magnitude vs. intrinsic-redshift plane is banded, this would imply some quantization of the properties of galaxies in different bands and might help selection between the various physical explanations briefly reviewed in Section 5.

4. Frequency Distribution of QSR Redshifts. Periodicity and Grouping

Considerable debate has taken place in the astronomical literature as to whether or not there exists periodicity or clumping at certain redshifts in the total sample of known quasar redshifts. Recently Tifft (1973b) has examined the known quasars in the redshift vs. apparent-magnitude plane and has claimed that the groupings in the distribution become more conspicuous – and significant – when the distribution is projected along a certain angle in the $\log z - m_v$ plane. That angle corresponds to just the angle of the redshift-apparent magnitude bands which he has found in his analyses of clusters of galaxies. He concludes that the major component of quasar redshift is intrinsic and is quantized. Véron and Véron (1973) have criticized Tifft's work and in redoing his analysis have excluded many quasars as photometrically unreliable, and have corrected for galactic absorption. Their results do not confirm the quasar bands.

Without taking any stand as to the reality or non-reality of the quasar bands, however, it is again possible to point out an interesting feature of Tifft's analysis. That feature is illustrated if we assume for the moment that some real grouping actually does take place in the frequency distribution of redshifts. Then we would expect quasars of the same redshift group to have some kind of relation between intrinsic luminosity and z, and some spread in distance (i.e. they would not form horizontal lines in the z-m diagram.) In that case there would be some angle in the z-m plane at which the grouping would appear maximally conspicuous. Projected against the z axis, as has been done heretofore, we would still expect to find some evidence of groupings but weaker, and more subject to contention as to its reality.

Therefore regardless of the actual reality of the Tifft bands, this approach is interesting from the standpoint of the reality of the groupings in redshift, and if it produced a significant improvement in the groupings would be an empirical proof of quantization in the quasar redshifts and therefore compelling evidence for non-velocity redshifts in quasars.

5. Theoretical Explanations of Non-Velocity Redshifts

Until the results on anomalous redshifts became more observationally established and accepted, the author has considered theoretical explanations to be premature. But several groups have now started working on the explanation of the effects reviewed here and for the sake of completeness I will briefly summarize these theories:

(1) *Gravitational redshifts.* Although it was not possible to build plausible models for quasars in which the redshift was gravitational, some investigators still believe in the possibility of such redshifts. There are even a few who believe that anomalous redshifts in galaxies could be due to high gravitational fields, although the fact that redshifts are not observed to vary across resolved regions in galaxies is strong evidence against this latter view. A variant of the gravitational redshift mechanism is the gravitational lens argument – advanced most strongly by Barnothy and Barnothy – that high redshift objects are merely distant objects seen through a relatively nearby gravitational lens.

(2) There are at least two investigators who claim that the difference between co-ordinate time and proper time will give anomalous redshifts at large light-time distances in the universe. Greenberger argues this directly (1970a, b; 1973), and Segal (1974) through a development of special relativity called 'Chronogeometry'. I cannot offer an opinion as to whether these separate authors' claim is true that their results are a natural consequence of currently accepted physics.

(3) *Photon-photon scattering.* Attempts by Sistero, Kierein, the author and others to account for redshifts by means of photons scattering off electrons usually foundered on, among other things, the prohibitive mass of electrons needed. Pecker *et al.* (1973), however, postulate emergent photons scattering off a sea of photons which are in-volved in the objects which have anomalous redshifts. They have impressively extended their predictions to account for redshift anomalies in binary stars in our own Galaxy (Kuhi *et al.*, 1974) and to a claimed anomalous redshift observed in a Pioneer-6 passage behind the sun (Merat *et al.*, 1973). Their type of anomalous redshift increases generally with temperature as T^3. A criticism of this explanation would therefore be that we would expect anomalous redshifts to vary according to the temperature in various parts of a resolved galaxy.

The most sensible comment on this whole proposal from the observational side of the anomalous redshifts, as reviewed here, is that it is very difficult to distinguish between galaxies which have a high temperature ambience and those which are very young. We would expect young galaxies to be generally composed of hotter objects and hence it is phenomenologically difficult to distinguish between the photon-photon scattering explanation and that below, which depends on differences in the fundamental structure of matter at different places in the Universe and at different epochs of creation (i.e. upon the ages of the galaxies concerned).

(4) The final explanation I mention is that put forward by Hoyle and Narlikar (1971). They postulated that matter of different ages and at different places in the Universe could be made of particles of lower mass. The lower mass of the electron in the Rydberg constant would therefore give lower frequency (redshifted) photons from the usual astrophysical radiative transitions. This explanation agrees with my earlier conclusion that the anomalous redshift was a function of the youth of the Galaxy, but it would introduce new physics in other places and times in the Universe.

References

Adams, T. F. and Weymann, R. J.: 1972, *Astrophys. Letters* **12**, 143.

Allen, R. J.: 1970, *Astron. Astrophys.* **7**, 330.

Arp, H.: 1966, *Science* **151**, 1214.

Arp, H.: 1967, *Astrophys. J.* **148**, 321.

Arp, H.: 1970, *Astron. J.* **75**, 1.

Arp, H.: 1971a, *Science* **174**, 1189.

Arp, H.: 1971b, *Astrophys. Letters* **9**, 1.

Arp, H.: 1972a, in D. S. Evans (ed.), 'External Galaxies and Quasi-Stellar Objects', *IAU Symp.* **44**, 380.

Arp, H.: 1972b, *Bull. Am. Astron. Soc.* **4**, 397.

Arp, H.: 1973a, *Astrophys. J.* **183**, 411.

Arp, H.: 1973b, in G. Field (ed.), *The Redshift Controversy*, Addison-Wesley Publishing Co., Reading, Mass.

Arp, H., Burbidge, E. M., Mackay, C. D., and Strittmatter, P. A.: 1972, *Astrophys. J. Letters* **171**, L41.

Bahcall, J. N., McKee, C. F., and Bahcall, N. A.: 1972, *Astrophys. Letters* **10**, 147.

Bahcall, N. A., Bahcall, J. N., and Schmidt, M.: 1973, *Astrophys. J.* **183**, 777.

Balkowski, C.: 1972, Thèse de 3ème cycle, Université de Paris.

Balkowski, C., Bottinelli, L., Chamaraux, P., Gouguenheim, L., and Heidmann, J.: 1973, *Astron. Astrophys.* **25**, 319.

Barbieri, C., Battistini, P., and Nasi, E.: 1967, *Publ. Oss. Astron. Padova*, No. 141.

Bertola, F., and Nasi, E.: 1971, *Mem. Soc. Astron. Ital*, **42**, 517.

Bogart, R, S. and Wagoner, R. B.: 1973, *Astrophys. J.* **181**, 609.

Bottinelli, L. and Gouguenheim, J.: 1973, *Astron. Astrophys.* **26**, 85.

Browne, I. W. A. and McEwan, N. J.: 1973, *Monthly Notices Roy. Astron. Soc.* **162**, 21p.

Burbidge, E. M., Burbidge, G. R., Solomon, P. M., and Strittmatter, P. A.: 1971, *Astrophys. J.* **170**, 233.

Burbidge, G. R. and O'Dell, S. L.: 1973, *Astrophys. J. Letters* **182**, L47.

Burbidge, G. R., O'Dell, S. L., and Strittmatter, P. A.: 1972, *Astrophys. J.* **175**, 601.

de Vaucouleurs, G. and de Vaucouleurs, A.: 1964, *Reference Catalogue of Bright Galaxies*, University of Texas Press, Austin.

Ford, H. C. and Epps, H. W.: 1972, *Astrophys. Letters* **12**, 139.

Greenberger, D. M.: 1970a, *J. Math. Phys.* **11**, 2329.

Greenberger, D. M.: 1970b, *J. Math. Phys.* **11**, 2341.

Greenberger, D. M.: 1973, Preprint on theories of variable mass particles.

Hazard, C. and Sanitt, N.: 1972, *Astrophys. Letters* **11**, 77.

Heidmann, J. and Kalloghlian, A. T.: 1973, *Astrofizika* **9**, 71.

Hoyle, F. and Narlikar, J. V.: 1971, *Nature* **233**, 41.

Hubble, E.: 1936, *Realm of the Nebulae*, Yale University Press, New Haven.

Jaakkola, T.: 1971, *Nature* **234**, 534.

Jaakkola, T.: 1973, *Astron. Astrophys.* **27**, 449.

Jennison, R. C. and Das Gupta, M. K.: 1953, *Nature* **172**, 996.

Kuhi, L. V., Pecker, J. C., and Vigier, J. P.: 1974, *Astron. Astrophys.* **32**, 11.

Lynds, R. and Millikan, A. G.: 1972, *Astrophys. J. Letters* **176**, L5.

Maltby, P., Matthews, T. A., and Moffet, A. T.: 1963, *Astrophys. J.* **137**, 153.

Matthews, T. A., Morgan, W. W., and Schmidt, M.: 1964, *Astrophys J.* **140**, 35.

Merat, P., Pecker, J.-C., and Vigier, J.-P.: 1974, *Astron. Astrophys.* **30**, 167.

Pecker, J.-C., Roberts, A., Tait, W., and Vigier, J. P.: 1973, *Nature* **241**, 338.

Pineau des Forêts, G. and Schneider, J.: 1973, *Astron. Astrophys.* **26**, 397.

Peterson, B. A. and Bolton, J. G.: 1972, *Astrophys. Letters* **10**, 105.

Roberts, M. S.: 1969, *Astron. J.* **74**, 859.

Rubin, V. C., Ford, W. K., and Rubin, J. S.: 1973, *Astrophys. J.* **183**, L111.

Sastry, G. N. 1970, *Publ. Astron. Soc. Pacific* **82**, 1051.

Segal, I. E.: 1974, *Proc. Nat. Acad. Sci.* **71**, 765.

Shapley, H. and Ames, A.: 1932, *Ann. Harv. Coll. Obs.* **88**, No. 2.

Shostak, G. S.: 1974, *Astrophys. J.* **187**, 19.

Strittmatter, P. A.: 1971, *Astrophys. J.* **170**, 233.

Tifft, W. G.: 1972, *Astrophys. J.* **175**, 29.

Tifft, W. G.: 1973a, *Astrophys. J.* **179**, 29.

Tifft, W. G.: 1973b, *Astrophys. J.* **181**, 305.

Tifft, W. G. and Gregory, S. A.: 1973, *Astrophys. J.* **181**, 15.

Van der Kruit, P. C.: 1971, *Astron. Astrophys.* **15**, 110.

Verón, P. and Verón, M. P.: 1974, *Astron. Astrophys.* **30**, 155.

Weedman, D. W.: 1970, *Astrophys. J. Letters* **161**, L113.

Zwicky, F., Herzog, E., and Wild, P.: 1961, *Catalogue of Galaxies and Clusters of Galaxies*, vol. 1, California Institute of Technology, Pasadena.

Zwicky, F., Karpowicz, M., and Kowal, C. T.: 1965, *Catalogue of Galaxies and Cluster of Galaxies*, vol. 5, California Institute of Technology, Pasadena.

DISCUSSION

Heidmann: Arp (H.) showed us a very nice picture of Stephan's quintet in red light. But I know of a still nicer one (Figure A), in full colour. It is by Arp (J.) and is in the Tate Gallery, London, with a caption "Constellation of forms according to the laws of chance". Although in 1930 Arp (J.) did not know about the work of Arp (H.), you see that he made the quintet to be a sextet by including NGC 7320 C. I think the high redshift members are even painted red.

From the caption it appears that Arp (J.) knows about astronomy and statistics but that he (Arp, J.) is not completely convinced about the interpretation by Arp (H.)

E. M. Burbidge: Two comments and one question:

(i) Although the faint galaxy between the QSO PHL 1226 and IC 1746 looks compact, it has no central concentration and is hard to observe. I initially took conventional spectrograms and saw only a bluish continuum. Wampler and I then tried scans; it has no emission lines, so we looked for the H and K break where you have to integrate for long times to beat the noise. Preliminary indications were for a redshift about $z = 0.15$, but we must get more observations.

(ii) One of the troubles facing the cosmological redshift interpretation has been, since 1968, the knowledge of multiple absorption redshifts in QSOs. It's been clear for some time that it's highly unlikely that the absorptions can arise in intervening matter (G. Burbidge discussed this, p. 93), and we have to produce a theory to account for them. The radiation-pressure-driven outflow hypothesis has its own problems, and you actually predict a non-negligible (though fairly small) component of gravitational redshift in the emission lines. We started looking again at the possibility of a gravitational redshift model for the absorption lines in regions of different gravitational potential, but no satisfactory model has emerged.

Regarding the absorption line in Wampler *et al.*'s object B in the double QSO, it will be very interesting when he resolves its doublet structure. It coincides with Mg II em in object A, but one is

Fig. A. Jean Arp: Constellation according to the laws of chance. *Circa* 1930, The Tate Gallery, London. © *by ADAGP, Paris*, 1974.

constantly being surprised by QSOs and I wouldn't like to predict whether it should be C IV λ 1549 absorption or Mg II λ 2798!

(iii) Chip Arp mentioned Rubin and Ford's discovery of a velocity anisotropy in the Sc I redshifts over the sky. I'd like to ask Wal Sargent whether Sandage has considered this effect in the new calibration of the Hubble plot and what does it do?

Sargent: Sandage's view is that the observation by Rubin and Ford results from a very large scale inhomogeneity in the distribution of Sc I galaxies. Since only a 1 magnitude interval was involved in their work, it does not alter the recent determination of the Hubble constant based on Sc I galaxies.

Arp: A difficulty with this interpretation of the Rubin and Ford results is the sharpness of the velocity peaks.

Abell: One observational datum of possible significance that you discussed is the comparison of the frequency distributions of angular sizes of H II regions in NGC 7318 and NGC 7320. Both seemed similar on the slide you showed, and were peaked around one or two arc sec with tails extending to about 4 or 5″. Now for $H = 50$ km s^{-1} Mpc, at the implied cosmological distances of both galaxies, H II regions of diameter 100 pc would subtend angles of only 1″ or less. Seeing at Palomar is seldom that good, and with photographic effects combined with seeing, it seems to me one could expect to find measured diameters of up to several arc sec. In other words, can the frequency distributions you showed actually be describing the frequency distribution of seeing at Palomar?

Arp: I can answer that question quite definitely. I find H II regions in both the high and low redshift systems ranging from apparent diameters of 1″ (the seeing limit) to 5″, the diameter of the largest observed H II regions. Some of the large, high redshift H II regions are of low surface brightness and the observed apparent diameters are clearly the real apparent diameters.

Allen: Two comments:

(i) With respect to the relative sizes of the H II regions in NGC 7318 and NGC 7320, one of the things we are learning from the extensive work by Sandage and Tammann is that the H II regions, or rather complexes, are larger in giant luminous spirals than in other galaxies. Some of the complexes in M101 attain dimensions of order 1 kpc quite easily, on the new distance scale. What this clearly means is that, before one can use the relative size distribution of H II complexes as a distance indicator, the Van den Bergh luminosity class of the galaxies must be estimated.

(ii) My second remark concerns the apparent absence of major non-circular motions in the velocity field of NGC 7320. Dr Sullivan and I have recently obtained H I observations of NGC 7320 with the Westerbork Telescope at a resolution of 25''. The optical size of NGC 7320 is about 2'. Our results confirm the initial detection of H I in this galaxy by Allen (*Astron. Astrophys.* **7**, 330, 1970) and the more recent observations by Balkowski *et al.* (p. 237) and by Shostak (*Astrophys. J.* **187**, 19, 1974). In addition, we have some information on the distribution and motions of the H I in this galaxy. Compared to similar observations of other spiral galaxies, we find no abnormal features either in the H I distribution or in the velocity field. The pattern of velocities resembles that of circular, weak differential rotation to within about 20 km s^{-1}. An upper limit can therefore be placed on the tidal forces exerted on the outer parts of NGC 7320, and thus a lower limit to the separation between this galaxy and the other four members of the group. Although these calculations are quite sensitive to the geometry and the interaction times, we can say that if the separation distance remains roughly constant during one rotation period of NGC 7320, then that distance must exceed ten times the radius of NGC 7320.

Seielstad: I contest the initial philosophical viewpoint taken by Sargent reflected in labelling one viewpoint as orthodox and the other as unorthodox. The point is important because the answer will ultimately be determined observationally, and observers should be careful to avoid bias. In Sargent's parable, the challenge is not to prove that some apples fall up; we need more evidence that apples normally fall down. Specifically, when Arp finds adjacent objects with discrepant redshifts, 'orthodoxy' decrees that the two are chance alignments, even when material apparently connects them, whereas when Gunn finds objects with redshifts similar to those of neighbouring objects, 'orthodoxy' accepts these as coincident in all three dimensions. Had Arp's talk been labelled orthodox and Sargent's unorthodox, the chance coincidence arguments could be reversed.

Let us measure which way apples fall, with no initial prejudgment.

G. Burbidge: The situation is one that is summarized in England by calling it 'a lack of fair play'.

Sargent: Two comments: first on the size of H II regions. The orthodox view has to be that H II regions can be larger in interacting systems than in normal systems, and I think there is evidence for this from the Arp Atlas of Peculiar Galaxies where you often find resolved H II regions in galaxies with redshifts far too big for them to be resolved in normal galaxies if the upper limit to the size is 600 pc as Sandage believes. In the Hercules cluster, IC 4182 is an interacting system with H II regions 2 kpc across.

On the philosophical point, the redshift-distance law is not in dispute. What is in dispute is whether there are exceptions to it, and the exceptions have to be proved.

Seielstad: There seems to me to be a circular argument here, in that the redshift-distance law depends on the existence of 'standard candles', but if an object does not fit the law it is described as 'non-standard'.

G. de Vaucouleurs: We have re-analyzed the data regarding velocity residuals in the Virgo cluster (*Astron. Astrophys.* **28**, 109, 1973) and find that in the *true* Virgo I cluster, i.e. the 12° diameter classical cluster as defined by Hubble and Shapley, the mean velocity of the E, L-cloud is +1000 km s^{-1}, that of the S-cloud is +1300 to +1400 km s^{-1} (depending on whether spirals are subject to systematic corrections or not) and, that within each group, the mean velocity is independent of magnitude over a 4–5 mag interval, as it should be in a bona fide cluster. Objects outside the 12° cluster follow the usual *V-m* relation and should not be included since they are evidently not cluster members. Either the two clouds are at different distances and accidentally aligned or – if there is only one Virgo cluster – there are systematic non-cosmological velocity components.

Lewis: I can add a few comments to the anomalous redshift case, which are appropriate when evidence is to be adduced from small groups (cf. Bottinelli and Gouguenheim, *Astron. Astrophys.* **26**, 85, 1973) or from pairs and large clusters (Jaakkola, T., *Nature* **234**, 534, 1971). All optical velocities appear to be subject to substantial errors, which are demonstrable for all galaxy types and most velocity ranges (Lewis, *Observatory* **94**, 9, 1974). Perhaps the best independent estimate of errors is to compare the 21-cm velocities with the optical velocities, as the 21-cm line frequency is known and the measurement procedure is constant from one galaxy type to another. Using the sample of galaxies available to M.S. Roberts (*IAU Symp.* **44**, 12, 1972), and averaging *all* sources of optical velocities to get V_{opt} and all independent 21-cm measurements to get V_{21}, I obtain the results of the following table, in which all the units are km s^{-1}.

Type	S0/Sa	S(ab to c)	S(cd to m)	Im
$\langle V_{21} - V_{opt} \rangle$	− 64	− 22	+2	− 33
Standard error	± 29	± 15	± 8	± 11
No. of objects	12	36	58	8
Dispersion	99	60	64	30

The most interesting features of this comparison are the large overestimates of velocity which result for both the Im and the S0/Sa types. For the Im type the result is a surprise as the velocities of these galaxies are observed from numerous emission lines. The deviation, if due to mistaken estimates of their centres, would be expected to scatter about the true value, while in practice most of the objects show a negative residual $V_{21} - V_{optical}$, for all of the independent observers.

With the S0/Sa class, the discrepancy is liable to be in part a reflection of an increasing dependence on absorption lines. But the existence of such large corrections for S0/Sa makes it likely that elliptical galaxies require similar correction, though this is much harder to estimate numerically.

G. de Vaucouleurs: Systematic errors in optical velocities depend not only on galaxy type, but on velocity, type of spectrograph, kind of lines measured (absorption or emission), adopted rest wavelengths of blends ($\lambda 3727$, G band) and other effects (e.g. confusion by superimposed night sky spectrum). Several earlier studies (Holmberg, *Arkiv Astron.* **2**, 559, 1961; G. and A. de Vaucouleurs, *Astron. J.* **68**, 96, 1963), based on optical data alone, supplement the radio-optical comparisons, in particular for types earlier than S0/a.

B. Peterson: With regard to QSOs in the fields of clusters of galaxies, John Bolton and I have found two QSOs, out of about 100 identifications we have made, that are within $\frac{1}{2}'$ of 15 mag galaxies in groups or small clusters. In the case of 2020-370, the 17^m5 QSO is 21″ from an S-type galaxy (Figure B). The QSO has a redshift of 1.05 while the galaxy has a redshift of 0.029. The projected distance of the QSO is 18 kpc from the nucleus of the galaxy. In the case of 1953-325, the QSO is 32″ from an E-type galaxy (Figure C). We do not have any redshifts for this system. The QSO appears to be double, or near a faint star, or to have a jet. The ultraviolet excess of this object is entirely due to the brighter component.

The probability of finding a 15^m (or brighter) galaxy within one square arc min of a given position is $\sim 10^{-4}$. In ~ 100 QSO fields we have found two cases when 10^{-2} would be expected. If the area is increased to 100 square arc min, the number of QSO fields containing galaxies $<15^m$ increases from two to about six.

Ekers: I have used the Westerbork telescope to obtain an upper limit of 5×10^{-29} W m^{-2} Hz^{-1} on the 1415 MHz continuum radiation from any galaxy in the VV 172 chain. This observation was part of an unsuccessful search for other chains of the 3C31 (NGC 383) type.

Arp: It is interesting to note that the QSO 2020–320 shown by Peterson and Bolton lies in a chain of galaxies.

G. de Vaucouleurs: Would Arp care to comment on Holmberg's statistics of companion galaxies?

Arp: Holmberg concluded that the ejection from the nucleus is isotropic but that material in the plane prevents companions from escaping in this direction, hence causing the observed concentration of companions along the minor axis. If this is so, then one might expect to find objects in the plane which are trying to get out and I suggest that these are the companions one finds on the ends of spiral arms. It all fits together very well.

G. de Vaucouleurs: This could be checked by a spectroscopic survey to see if there are systematic differences in redshift between companions near the plane and companions near the poles.

Arp: Yes, I've been doing that every time I get an observing night – every once in a while – and am finding that I get much higher redshifts for the companions, for instance, the 6 companions of

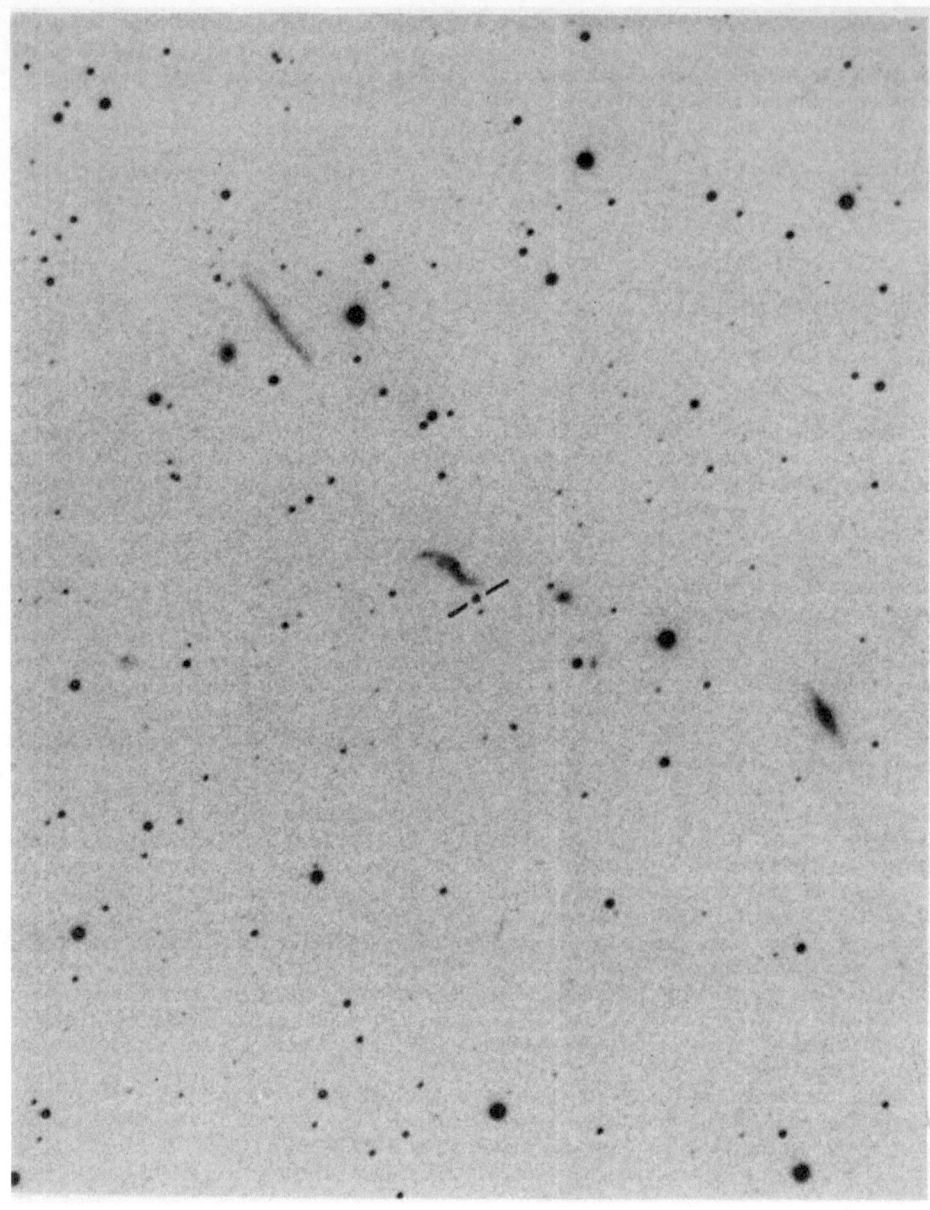

Fig. B.

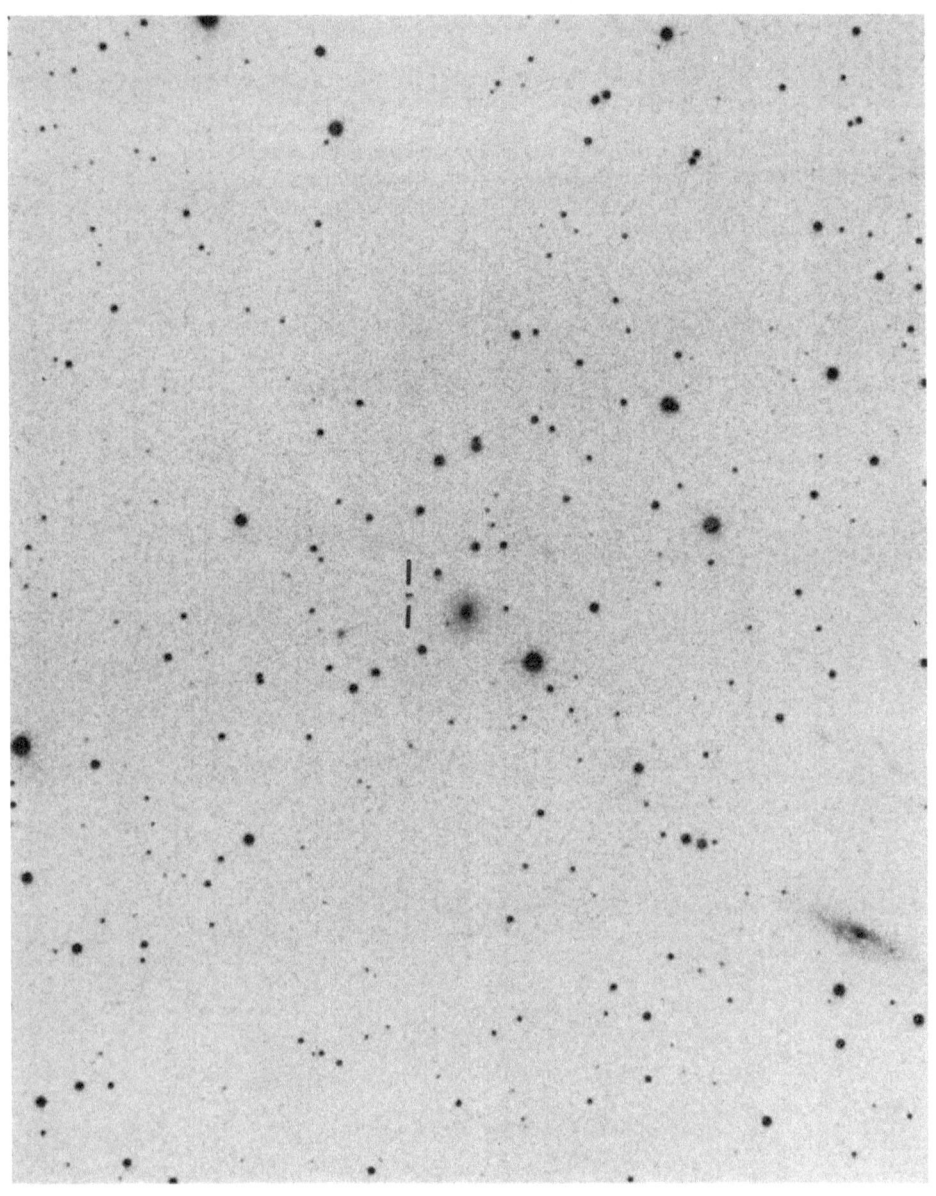

Fig. C.

NGC 2403. But it will require much greater statistics before the generally smaller ejection redshifts can be separated from the intrinsic redshifts.

Rickard: If compact galaxies are objects of mass $10^{6-7} M_\odot$ ejected from the nuclei of galaxies, why don't we observe much more disruption of the parent galaxies?

Arp: If the ejected body is to emerge from the nucleus it must be initially quite small. If it comes out in the plane, where it burrows through gas, then you get spiral arms.

If it is ejected out of the plane then, since it is compact and moving fairly fast, it will not perturb the bulge stars very much. Or the observations may ultimately require that the mass increases with time, as in the Hoyle-Narlikar theory.

THE EXTRAGALACTIC RELATIVE DISTANCE SCALE

G. O. ABELL

Dept. of Astronomy, University of California, Los Angeles, Calif., U.S.A.

Abstract. Arguments are given to support the hypothesis that a characteristic point on a cluster luminosity function, m^*, corresponds to the same absolute magnitude in all clusters, and thus is an appropriate standard candle for determination of relative cluster distances. It is shown, however, that if m^* *is* a good standard candle, then the first and tenth brightest cluster galaxies, $m(1)$ and $m(10)$, respectively, are not, for $m^* - m(1)$ and $m^* - m(10)$ both differ significantly from cluster to cluster. Moreover, $m(1)$ and especially $m(10)$ depend on cluster richness. Distances derived from them for remote clusters chosen for analysis are thus subject to selection effects that depend on cluster richness and on Bautz-Morgan type. Finally, a procedure suggested by Bautz and Abell is described whereby m^* can be estimated from observed properties of distant clusters.

1. Introduction

It is well established (Sandage, 1968, 1972) that a useful criterion of distance for a cluster of galaxies is the apparent magnitude of its brightest member, its third or tenth brightest member, or some combination thereof. The possibility is ever-present, however, that selection effects which depend on distance may affect the observed magnitudes of the brightest cluster galaxies. Years ago, for example, Scott (1957) showed that, with various 'reasonable' assumptions about the luminosity functions of galaxies in clusters, the choice of apparent magnitudes of the first, third- or tenth-brightest cluster galaxies as distance indicators could lead to the derivation of values of the deceleration parameter, q_0, supporting almost any cosmological model. Essentially, the so-called *Scott effect* is that the very distant clusters of galaxies chosen for investigation are selected or discovered because they are richer in membership than average, and thus may have brighter than average brightest member galaxies.

That such selection effects are indeed possible has been shown by the writer (Abell, 1960) from a study of the luminosity functions of several clusters. In particular, Matthews *et al.* (1964), Morgan and Lesh (1965) and Bautz and Morgan (1970) have called attention to the existence of certain clusters (Bautz-Morgan type I and II clusters) which contain superluminous brightest galaxies – the Morgan class cD galaxies. Bautz and Abell (1973) have found evidence for a weak correlation between cluster richness and the presence or absence of a cD galaxy in a cluster. Sandage (1973), while not confirming the correlation between Bautz-Morgan class and cluster richness, nevertheless finds that the absolute magnitude of the brightest cluster galaxy is correlated with the cluster Bautz-Morgan class.

Thus there are several reasons why selection effects can play an important role. There may be a correlation between the presence of a cD galaxy and cluster richness; hence richer clusters, those most easily recognized at large distances, may have brighter than average first brightest members. In any case, a Bautz-Morgan type I

John R. Shakeshaft (ed.), The Formation and Dynamics of Galaxies. 225–235. *All Rights Reserved.*

or II cluster is more likely to be noticed at a very remote distance, simply because of the presence of the cD galaxy. If, in addition, clusters are selected because they contain radio sources, they are especially likely to be type I clusters with cD galaxies, for the latter are often radio sources. All of these considerations cast suspicion on the validity of the use of the first-brightest cluster galaxy as a criterion of distance. We shall present evidence below that the tenth-brightest cluster member is, in fact, very strongly correlated with cluster richness.

2. Cluster Luminosity Functions

Another approach to finding the relative distances of clusters is by comparison of their luminosity functions. The determination of a cluster luminosity function is a painstaking investigation, for it requires photometry of many individual galaxies in a cluster. Nevertheless, the data are rapidly accumulating. Elliptical galaxies, which comprise the dominant membership of most rich clusters, are found to display a very characteristic frequency distribution of magnitudes. The logarithmic integrated

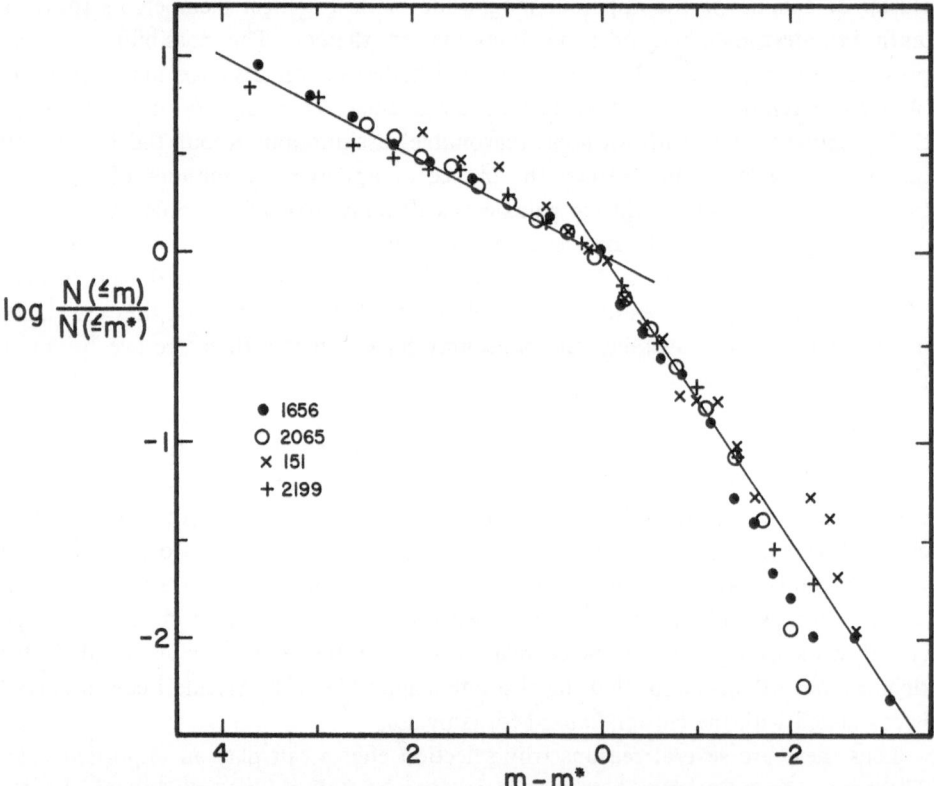

Fig. 1.   Superimposed logarithmic integrated luminosity functions for four rich clusters of galaxies. Clusters are identified by their catalogue numbers in Abell (1958).

luminosity function of elliptical galaxies in the Coma cluster (Abell, 1965), for example, is typical of that of elliptical galaxies in every other rich cluster so far investigated. Data for several clusters are shown in the composite plot of Figure 1. High quality photometry with zero-point calibration is available for only eight clusters at present, but less precise data for many other clusters show that they, too, have similar luminosity functions for their elliptical membership. Photometry by Oemler (1973), for example, while not yet on a standard photometric system, shows cluster luminosity functions like that in Figure 1. Also noteworthy are data by Krupp (1973), who has done crude 'flyspanker' photometry on Palomar Sky Survey prints of all galaxies brighter than $m_v = 16.75$ in Abell clusters of richness class 1 or greater and distance class 3 or less. The data for 27 of the 43 clusters studied by Krupp, for which the observations cover a great enough magnitude range to determine the form of the luminosity functions, are exhibited in Figure 2.

As can be seen from Figures 1 and 2, the integrated logarithmic luminosity functions of clusters can be made to coincide by vertical and horizontal shifts. The vertical shifts necessary for proper fits are measures of relative cluster richnesses, and the horizontal shifts indicate relative cluster distances if the magnitudes are all on the same system. A convenient point on the magnitude scale for the luminosity functions of the clusters can be defined as m^*, the intersection of straight lines fitted respective-

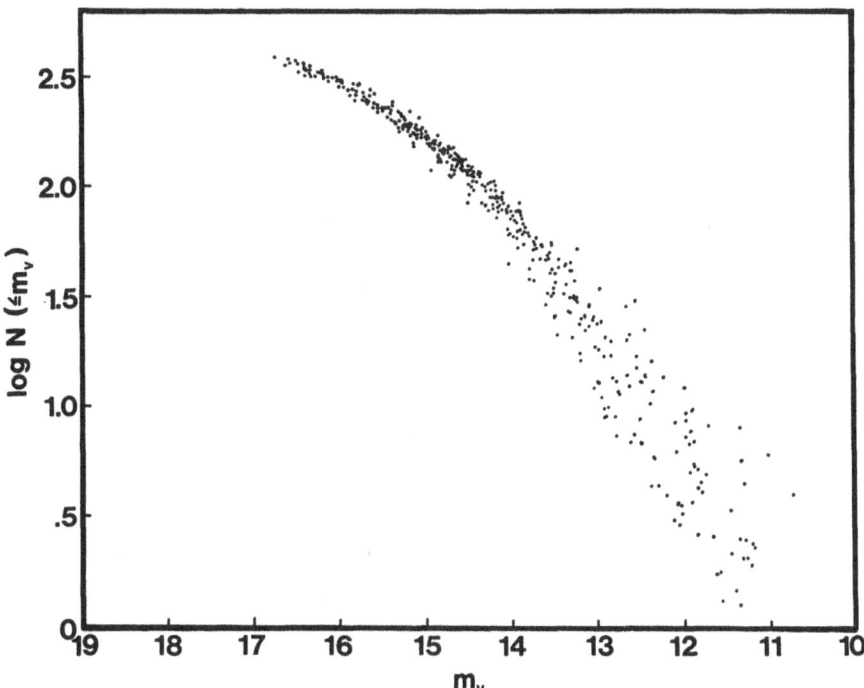

Fig. 2.   Integrated logarithmic luminosity functions of the E-S0 galaxies in 27 rich clusters. Log$N$ and m_v have been scaled to the values for the Coma cluster (from Krupp, 1973).

ly to the bright, steep, and to the fainter, less steep portions of plots such as those
in Figure 1 or 2 (that is, m^* is the magnitude at which the luminosity function changes
slope). If the magnitude determinations in a cluster are properly calibrated to a
consistent zero point, a plot of log cz against m^* should provide a Hubble diagram,
based not on the brightest cluster members, but on the form of the luminosity functions. Figure 3 is such a plot for 8 clusters for which appropriate photometric data
exist (Bautz and Abell, 1973). The scatter, $\sigma \approx 0.1$ mag, is about that expected from
the uncertainty of the interstellar absorption corrections alone (a plane-parallel
absorbing model of half-thickness $m_{pv} = 0.16$ mag is assumed); thus there is no
evidence for any intrinsic scatter at all in the $\log cz - m^*$ plot.

Krupp's step-scale photometry on Palomar Atlas prints can not be expected to
have the same zero-point calibration for different clusters observed on different prints,
because of the variations in quality of the original Sky Survey plates, and different
sky background densities. These zero-point errors doubtless account for most, if
not all, of the scatter ($\sigma \approx 0.5$ mag) in the similar plot of $\log cz$ vs m^* for the clusters
observed by Krupp. Krupp (1973) has shown, however, that this scatter is nevertheless
significantly less than those in plots of $\log cz$ vs $m_v(1)$ or $m_v(10)$, the magnitudes of

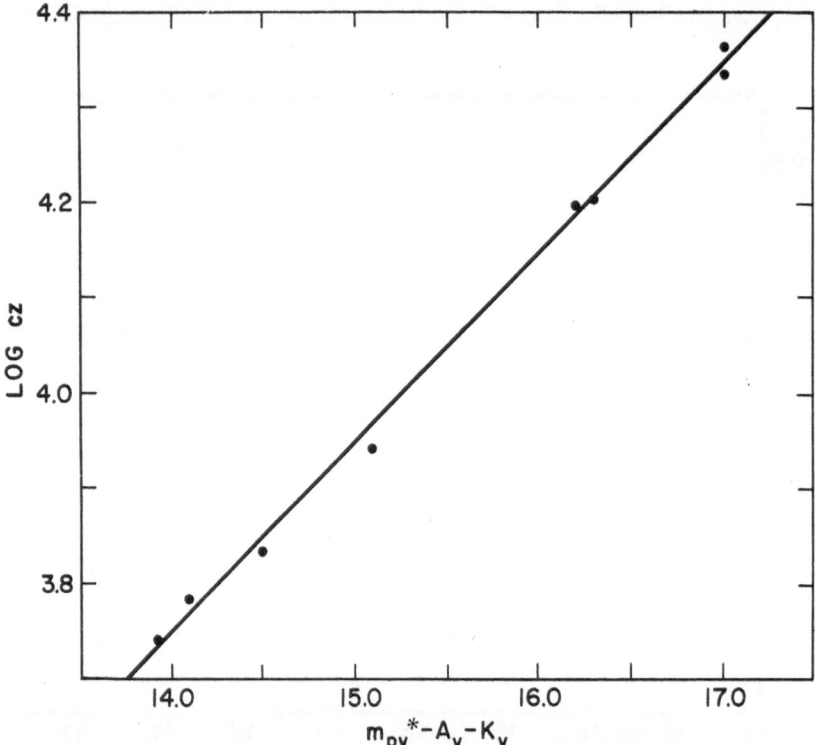

Fig. 3.   Hubble diagram for m^* for eight clusters with well-determined luminosity functions. Ordinates are redshifts, corrected for galactic rotation. Abcissas are m^* corrected for K-dimming and
galactic absorption.

the first and tenth brightest cluster galaxies, respectively. The comparison of the scatters on the different plots does not depend on the zero-point differences referred to above, because all magnitude estimates in a single cluster are, of course, made on the same print.

The evidence suggests, therefore, that comparison of luminosity functions of clusters of galaxies can provide relative distances for the clusters that may be more reliable than comparison of the brightest cluster members. If the characteristic change of slope in the integrated logarithmic luminosity functions near m^* results from some fundamental property of cluster and galaxy formation, it may well be independent of cluster type and richness. At least available data indicate that m^* has a smaller intrinsic variation from cluster to cluster (indeed, perhaps none at all) than does $m(1)$ or $m(10)$. It may, of course, still depend on evolutionary effects.

3. Evidence for the Scott Effect in Clusters of Galaxies

If it is assumed that m^* does, indeed, correspond to the same absolute magnitude in every cluster, and also that $N(\leqslant m^*)$, the number of cluster elliptical galaxies brighter than m^*, is a measure of cluster richness, then Krupp's data provide an excellent test for the Scott effect. All clusters investigated by Krupp are relatively nearby, so evolutionary effects should not be present. In Figure 4, $\log N(\leqslant m^*)$ for each cluster is plotted against $m_v^* - m_v(10)$, the difference between the photovisual magnitudes m^* and of the tenth brightest cluster elliptical. Zero-point differences in the magnitude systems from print to print do not enter into Figure 4 because the magnitude differences plotted are always obtained from one print. Krupp's 'flyspanker' magnitude estimates on a single print are consistent to about $\frac{1}{4}$ mag; the differences in $m^* - m(10)$ between the different clusters are thus very large compared to expected errors in magnitude estimates.

Because m^* is assumed to correspond to the same absolute magnitude in all clusters, Figure 4 shows that the absolute magnitudes of the tenth brightest cluster members differ significantly from cluster to cluster, and moreover that the absolute magnitude of the tenth brightest elliptical cluster galaxy is strongly correlated with the cluster richness. This, of course, is just what is predicted by the Scott effect, and shows clearly that the tenth, or in general, the nth brightest cluster galaxy is suspect as a distance indicator.

A similar plot (Figure 5) of $\log N(\leqslant m^*)$ against $m_v^* - m_v(1)$, on the other hand, shows an entirely different situation. Whereas $m^* - m(1)$ varies from cluster to cluster even more widely than does $m^* - m(10)$, it is correlated, if at all, only very weakly with $N(\leqslant m^*)$. The data show (if m^* corresponds to a constant absolute magnitude) that the absolute magnitude of the brightest elliptical galaxy, $M(1)$, varies greatly from cluster to cluster, but evidently not in a manner that depends strongly on the cluster richness. In fact, both Bautz and Abell (1973) and Sandage and Hardy (1973) find $M(1)$ to be strongly correlated with the Bautz-Morgan (1970) form-type classification of the cluster. (The form-types of the clusters are shown as

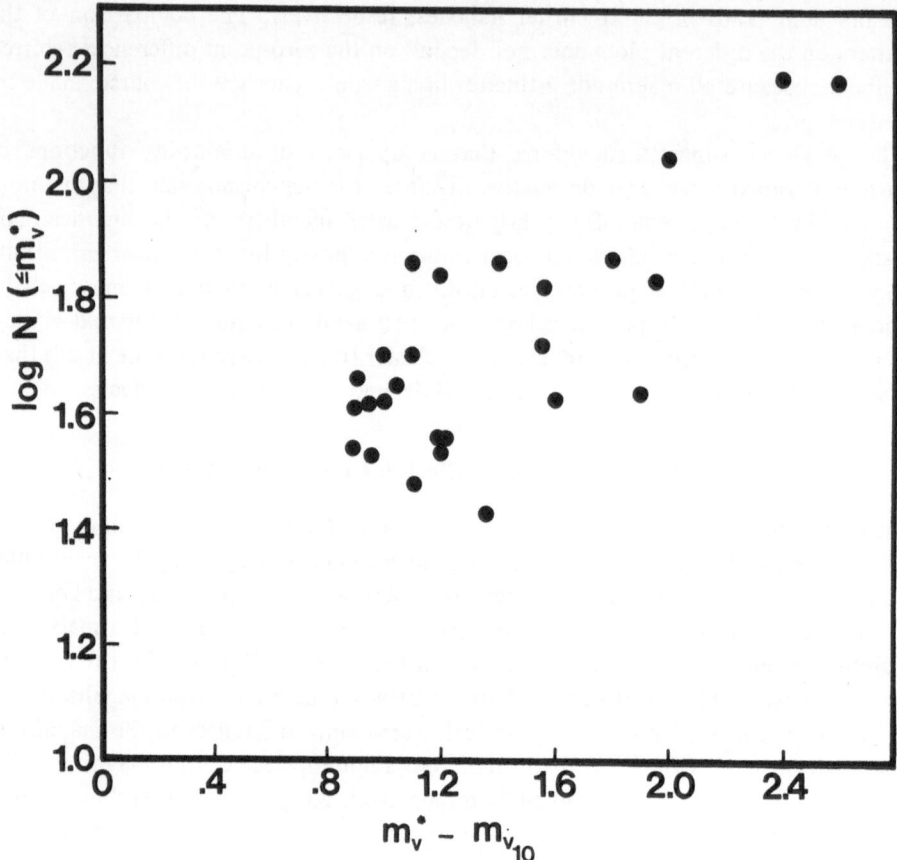

Fig. 4.   Cluster richness, $\log N(\leqslant m_v{}^*)$, vs $m_v{}^* - m_v(10)$ (from Krupp, 1973).

Roman numerals beside the data points in Figure 5). Bautz and Abell find that there is also a weak correlation between $m^* - m(1)$ and cluster richness, and hence that both Bautz-Morgan type and cluster richness are needed to predict $m^* - m(1)$. This conclusion is consistent with the Krupp data in Figure 5, which suggest a possible weak correlation of $m^* - m(1)$ with richness, and is not inconsistent with data of Sandage and Hardy (1973), although they themselves express the opinion that there is no significant correlation between $M(1)$ and cluster richness. At any rate, the Sandage-Hardy, Krupp, and Bautz-Abell data are not inconsistent.

The principal conclusion to be drawn is that selection effects which can bias the determination of the relative distances of clusters are possible. If remote clusters are selected for study because they are preferentially richer than average, use of the tenth brightest cluster galaxy as a standard candle can result in an underestimate of the distances of those clusters, and an overestimate of the deceleration parameter, q_0. If remote clusters are selected because they contain radio sources (for example, the cluster containing 3C 295) or because they contain outstanding galaxies bright

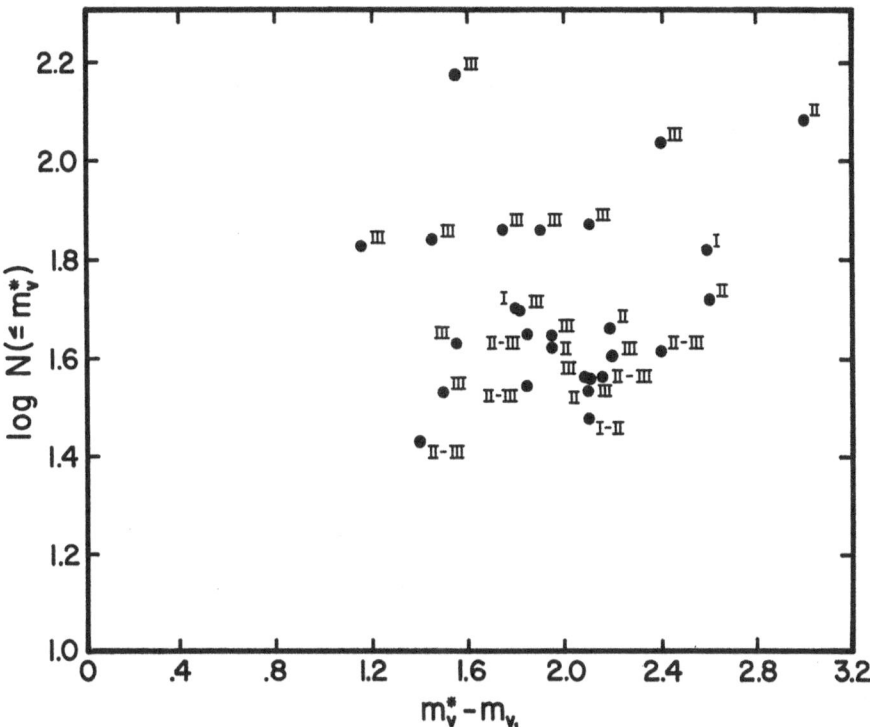

Fig. 5.   Cluster richness, $\log N(\leqslant m_v{}^*)$, vs $m_v{}^* - m_v(1)$. The Bautz-Morgan form types for clusters are shown. (From Krupp 1973.)

enough for convenient redshift measurement (possibly cD galaxies), the use of $m(1)$ for a standard candle can produce the same bias.

4. A Procedure for Correcting for Selection Effects

In the writer's view, the evidence strongly suggests that some representative point on the luminosity function, such as m^*, is a better standard candle for determination of relative cluster distances than is the apparent magnitude of the first, tenth, or nth brightest galaxy. Unfortunately, luminosity functions are tedious to determine at best, and are difficult or even impossible to obtain for those clusters with cosmologically interesting distances. For the latter, in fact, photometry is sometimes reliable only for the brightest member galaxy. Even for these remote clusters, however, classifications can be made of the Bautz-Morgan form type, and rough estimates of cluster richness can be made. Bautz and Abell (1973) have published a provisional calibration of $m^* - m(1)$ in terms of form type and richness class (as defined in the Abell (1958) catalogue of rich clusters), based on those nearby clusters for which the most complete data are available. From this calibration, the form type and estimate of richness class for a remote cluster allow a prediction of how much fainter m^* would be than

$m(1)$, could m^* actually be observed. Estimates of richness classes of remote clusters are, admittedly, rather uncertain, but fortunately the quantity $m^* - m(1)$ is far less sensitive to cluster richness than it is to form type. The provisional calibration of Bautz and Abell is reproduced in Table I. Comparison of Table I with Figure 5 shows that the Krupp data are entirely compatible with the calibration and lend considerable support to its validity.

TABLE I

A provisional calibration of $m^* - m(1)$ in terms of Bautz-Morgan form type and richness classes for clusters of galaxies

Richness class	0	1 and 2	3
Form type			
I and II	> 2	+ 2.9	+ 3.3
II–III and III	+ 1.5	+ 1.9	+ 2.3

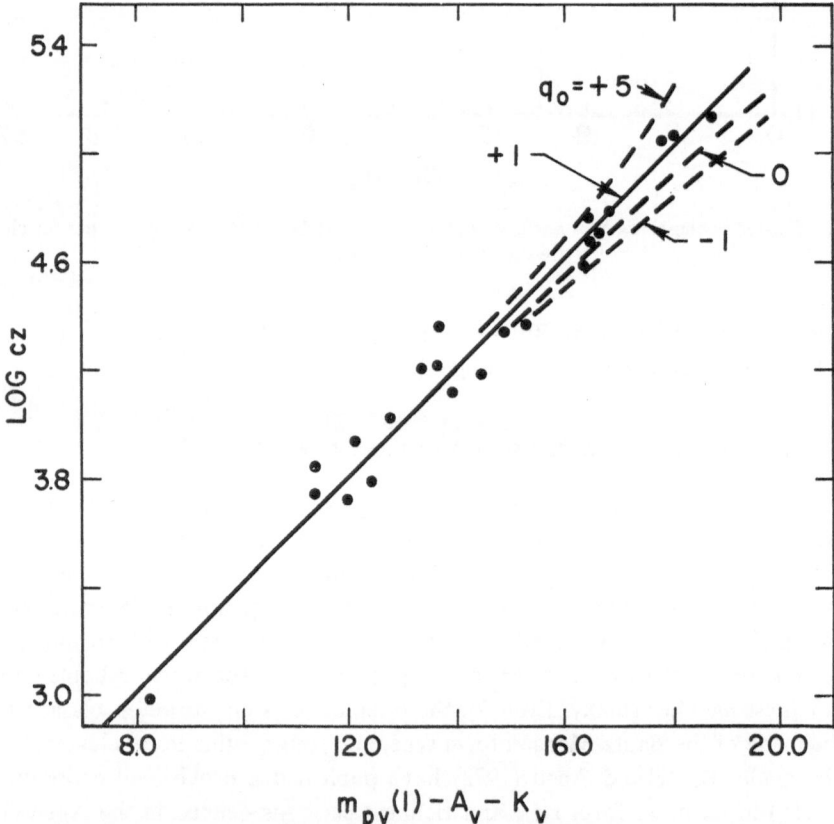

Fig. 6.   Hubble diagram for first ranked cluster galaxies. Ordinates are redshifts, corrected for galactic rotation. Abcissas are $m_v(1)$, corrected for K-dimming and galactic absorption
(from Bautz and Abell, 1973).

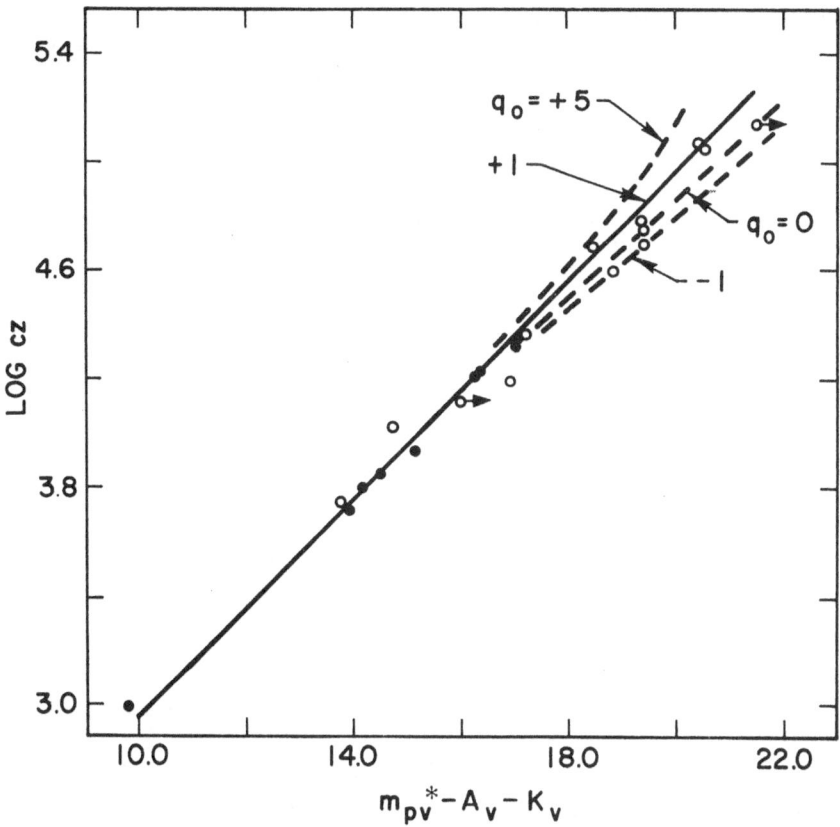

Fig. 7.   Same as Figure 6, except abcissae are m_v*. *Filled circles*: clusters with known m_v* from observed luminosity functions. *Open circles*: clusters for which m_v* has been predicted from observed
$m_v(1)$ and the calibration of Table I (from Bautz and Abell, 1973).

Figures 6 and 7, reproduced from Bautz and Abell (1973) show how a plot of
$\log cz$ vs $m(1)$, in this case based on data from Humason *et al.* (1956), can be converted to a plot of $\log cz$ vs $m*$ with the use of the calibration of Table I.

5. Conclusions

Taken at face value, the use of $m(1)$ as a standard candle, as in Figure 6, leads to a
Hubble diagram which is compatible with a value of $q_0 = +1$. When the $m(1)$ values
are converted to $m*$, however, the resulting Hubble diagram (Figure 7) is more
compatible with $q_0 = 0$. Obviously, the calibration in Table I is too provisional, and
the data are too crude (as seen from the significant scatter) to attribute any significance
to a determination of q_0 by this procedure at present. Moreover, it must be emphasized
that evolutionary effects have not been considered, and they almost certainly also
affect the determination of q_0 from the Hubble diagram (e.g., Tinsley, 1972). The

purpose here is to demonstrate the probable existence of selection effects that must be understood and taken into account in interpreting observations in terms of cosmological models, and also to review a suggested procedure by which such effects might be reduced.

References

Abell, G. O.: 1958, *Astrophys. J. Suppl.* **3**, 211.
Abell, G. O.: 1960, *Publ. Astron. Soc. Pacific* **72**, 459.
Abell, G. O.: 1965, *Ann. Rev. Astron. Astrophys.* **3**, 1.
Bautz, L. P. and Abell, G. O.: 1973, *Astrophys. J.* **184**, 709.
Bautz, L. P. and Morgan, W. W.: 1970, *Astrophys. J. Letters* **162**, L149.
Humason, M. L., Mayall, N. U., and Sandage, A. R.: 1956, *Astron. J.* **61**, 97.
Krupp, E. C.: 1973, *Thesis*, University of California, Los Angeles.
Matthews, T. A., Morgan, W. W., and Schmidt, M.: 1964, *Astrophys. J.* **140**, 35.
Morgan, W. W. and Lesh, J. R.: 1965, *Astrophys. J.* **142**, 1364.
Oemler, A.: 1973, *Astrophys. J.* **180**, 11.
Sandage, A. R.: 1968, *Astrophys. J. Letters* **152**, L149.
Sandage, A. R.: 1972, *Astrophys. J.* **173**, 485.
Sandage, A. R. and Hardy, E.: 1973, *Astrophys. J.* **183**, 743.
Scott, E. L.: 1957, *Astron. J.* **62**, 248.
Tinsley, B. M.: 1972, *Astrophys. J. Letters* **173**, L93.

DISCUSSION

Nandy: Is there any possibility that the members of the fainter clusters of galaxies could be reddened? Put another way, can intergalactic absorption affect the curvature of the redshift-distance relation?

Abell: Some years ago, Zwicky (Herzog, E., Wild, P., and Zwicky, F.: *Publ. Astron. Soc. Pacific* **69**, 409, 1957) suggested that intergalactic absorption caused by dust associated with nearby galaxies hides remote galaxies, accounting for an apparent anti-correlation between positions of nearby and remote galaxies. The absorption required, however, is far greater than what is allowed by the observed slope of the redshift-magnitude relation, and I suspect that Zwicky's results were due to a confusion factor, i.e. the difficulty of recognizing distant galaxies through nearby groups.

Otherwise, I know of no evidence for any intergalactic absorption. The $\log cz$ vs m relation (Hubble diagram) is linear for all cosmological models out to about $z = 0.1$. The data to $z = 0.1$ are good enough for us to place an upper limit on the total absorption to that distance. There could not, for example, be as much as 0.3 to 0.5 mag absorption to $z = 0.1$.

On the other hand, I think we could not rule out a small amount of dust producing, say, absorption of as much as 0.1 mag. Now because of the expansion of the Universe, a given mass of dust would produce greater optical depth per unit redshift at greater redshift. Thus, we could not rule out absorption that may amount to significantly more than 1.0 mag at $z = 1.0$. It would be very hard to detect this amount of absorption – and separate effects it would produce from cosmological effects – especially if, say, the dust were something like hydrogen snow which might be gray and produce no reddening.

Thus, intergalactic absorption could conceivably be present and is another uncertainty in the interpretation of the Hubble diagram. I think that this is an important point, one well taken.

Irwin: Has Sandage changed his mind? Originally he believed that the small scatter in the redshift diagram indicated essentially identical absolute magnitudes of all the *first* brightest galaxies in clusters of galaxies.

Abell: Yes, he now recognizes (Sandage, A. R. and Hardy, E.: *Astrophys. J.* **183**, 734, 1973) that at least part of the scatter in the Hubble diagram of first brightest members of clusters is due to the presence of cD galaxies in some (Bautz-Morgan type I and II) clusters and not in others (type III clusters).

Irwin: Do your results on the Virgo cluster suggest a change in the Hubble constant?

Abell: The effects I have discussed today do not depend on the Hubble constant. Nevertheless, if we do tentatively accept the fit of the luminosity function of the Virgo cluster to those of more remote clusters, we can find the relative distances of the Virgo and these other clusters, the redshifts of the latter being certainly cosmological. Then if we have a modulus for the Virgo cluster from independent methods (say photometry of globular clusters in M87), we can find the Hubble constant. Present estimates of the modulus of the Virgo cluster lead to a value of H of at most about 50 km s^{-1} Mpc^{-1}. This value is in reasonable agreement with the recent estimates of H from other methods.

OBSERVATIONAL EVIDENCE FOR NON-VELOCITY REDSHIFT IN STEPHAN'S QUINTET

C. BALKOWSKI, L. BOTTINELLI, P. CHAMARAUX, L. GOUGUENHEIM,
and J. HEIDMANN

Observatoire de Paris, Meudon, France

Abstract. Distance determinations of the high-redshift galaxy NGC 7319 in Stephan's Quintet, based on 21-cm line observations, yield a value much smaller than the cosmological distance. At the 98.7% level, a part of the redshift of this galaxy which could amount to 5000 km s^{-1} is anomalous.

Stephan's Quintet is a tight group of five galaxies (Figure 1). Four of them have a redshift $cz \sim 6000$ km s^{-1} while one, NGC 7320, has a much smaller $cz \sim 800$ km s^{-1}. To find out whether the galaxies in the Quintet are at the same distance from us, which would imply that they have anomalous redshifts, we observed two of them, the low redshift NGC 7320 and the high redshift NGC 7319, with the 21-cm line receiver at Nançay so as to be able to apply distance criteria entirely independent from redshifts (Balkowski *et al.*, 1973).

Figure 2 shows the line profiles obtained. From them one derives the systemic redshifts (which agree with the optical determinations), the 21-cm line flux densities F_H and the total widths of the lines W.

Combining them with optical data for the apparent luminosities l and apparent photometric diameters a, one may derive the values of four integral parameters:
 – the intrinsic luminosities: $L \propto l D^2$, where D is the distance;
 – the linear diameters: $A \propto a D$;
 – the neutral hydrogen masses: $M_H \propto F_H D^2$; and
 – the *indicative* total masses: $M_i \propto a D W^2$,
to which one may add the morphological type T as a fifth parameter.

We used a, l and F_H values corrected for inclination and absorption effects in galaxies following Heidmann *et al.* (1972). Our statistical studies (Balkowski, 1973) lead to various relations or correlations, of which those where D enters are:

$$M_i/L \, (\propto D^{-1}) \text{ vs } T, \tag{1}$$

$$M_H/M_i \, (\propto D) \text{ vs } \sigma_H, \tag{2}$$

where σ_H is the mean neutral hydrogen surface density ($\propto F_H/a^2$),

$$M_H/M_i \, (\propto D) \text{ vs } T, \tag{3}$$

$$\sigma_H/A \, (\propto D^{-1}, \text{ the quasi-volumic density}) \text{ vs } T, \tag{4}$$

$$L/A^{2.6} \, (\propto D^{-0.6}) \text{ a constant.} \tag{5}$$

These five relations are distance indicators. For example, observations of a, W and l lead to

$$M_i/L = \text{a numerical factor} \times a W^2 l^{-1} D^{-1}$$

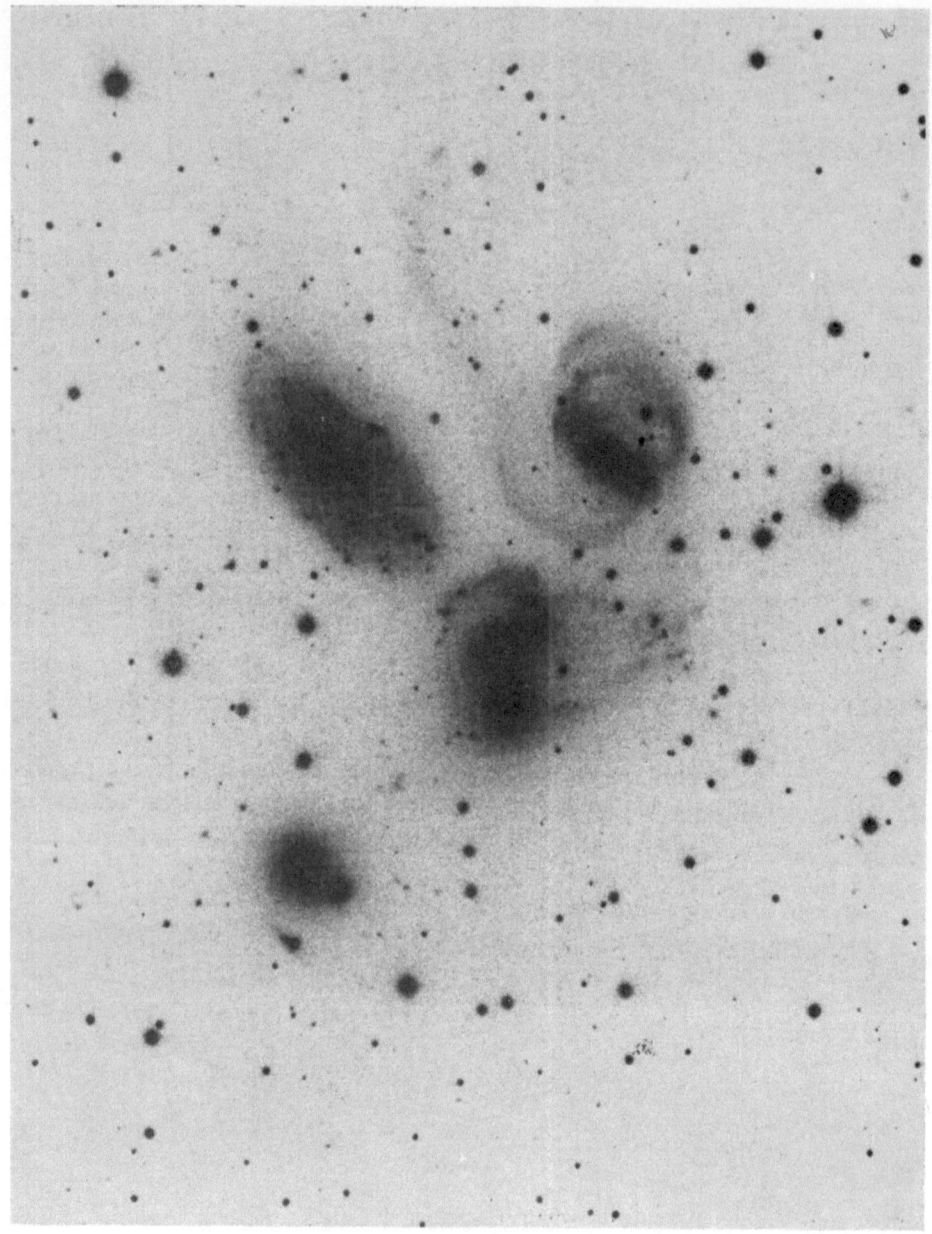

Fig. 1. Stephan's Quintet. NGC 7319 is at top right and NGC 7320 at top left. Mount Palomar 200-in. photograph in red light by Arp (private communication).

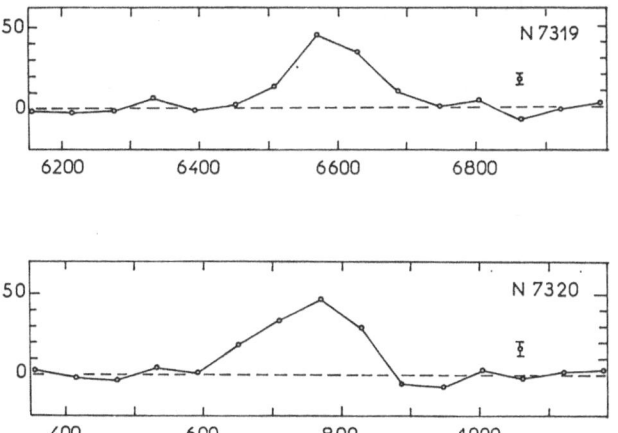

Fig. 2.   21-cm line profiles of NGC 7319 and 7320. Ordinates are flux densities (10^{-29} W m^{-2} Hz^{-1}); abscissae are redshifts cz with respect to the Sun (km s^{-1}). The error bars correspond to ± 1 rms deviation.

and relation (1) gives the value of M_i/L corresponding to the type T of the galaxy, hence D.

The D values obtained by the five methods are shown in Figure 3, together with those obtained in the same way for the neighbouring large spiral NGC 7331. Our studies have shown that there are only three independent D determinations among these five. For NGC 7319 the mean logarithmic value of the first three is

$$\bar{D} = (22^{+15}_{-9} \text{ m.e.}) \text{ Mpc.}$$

Our determinations are all in agreement with a common value ~ 20 Mpc for the three

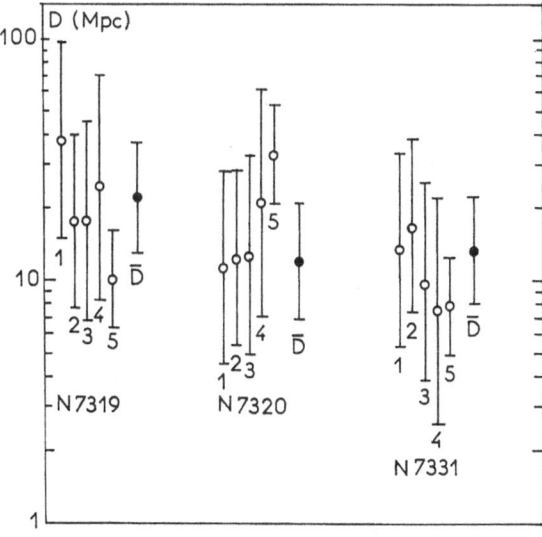

Fig. 3.   Distance determinations in Mpc for NGC 7319, 7320 and 7331 with their mean errors from the five distance criteria. $\bar{D}$ is the mean logarithmic value for the first three criteria.

galaxies NGC 7319, 7320 and 7331, and thus support Arp's (1973) views about the Quintet.

From the redshift and a Hubble constant $H = (80 \pm 20)$ km s^{-1} Mpc^{-1} (Heidmann, 1973), the cosmological distance of NGC 7319 would be $D_{cosmo} = (84^{+28}_{-17})$ Mpc. Taking into account our error on $\bar{D}$ and the error on H, D_{cosmo} differs from $\bar{D}$ by 2.2 rms deviations, *i.e.* there is only a 1.3% chance that NGC 7319 is at D_{cosmo}. Then, at the 98.7% level, a part of the redshift of this galaxy which could amount to 5000 km s^{-1} is anomalous.

References

Arp, H. A.: 1973, *Astrophys. J.* **183**, 411.

Balkowski, C.: 1973, *Astron. Astrophys.* **29**, 43.

Balkowski, C., Bottinelli, L., Chamaraux, P., Gouguenheim. L., and Heidmann, J.: 1973, *Astron. Astrophys.* **25**, 319.

Heidmann, J.: 1973, in G. Cayrel de Strobel and A. M. Delplace (eds.), 'L'âge des Etoiles', *IAU Colloq.* **17**, XXXVII-1.

Heidmann, J., Heidmann, N., and de Vaucouleurs, G.: 1972, *Mem. Roy. Astron. Soc.* **75**, 85, 105 and 121.

DISCUSSION

Allen: Could Dr Heidmann comment on the recent observations of H I in the Quintet made by Shostak (*Astrophys. J.* **187**, 19, 1974; **189**, L1, 1974) with the 300-ft NRAO telescope? Although Shostak observes line profiles for NGC 7319 and NGC 7320 similar to those of Dr Heidmann, he reaches the opposite conclusion, viz. that the two galaxies are most probably widely separated from each other.

Heidmann: I think, in fact, that Shostak's results are in good agreement with ours. When you plot his distance values for NGC 7319, 7320 and 7331 on our diagram you see that they could all agree with a common distance ~20 Mpc. Shostak's statement in his preprint according to which NGC 7319 is unlikely to be at the same distance as NGC 7320 is misleading, as the difference between his two distances is only one standard deviation.

Ekers: Could you remind us what the Hubble distances are for these galaxies and how your conclusions are affected by the value of the Hubble constant?

Heidmann: The statistics we used for evaluating the distances in Stephan's Quintet are completely independent of the Hubble constant; they do not use any distance deduced from redshift (see Balkowski, C. *et al.*: *Astron. Astrophys.* **25**, 319, 1973).

Ekers: I find it puzzling that your distance determination can be independent of the Hubble constant, since that seems to imply that in other situations you could apply your procedure in reverse to *determine* the Hubble constant.

Heidmann: This is exactly one of our current programs.

Longair: In view of the complexity of the relations between the observed quantities, and their large dispersions, it would seem that the best way of estimating the errors in the correlations is by Monte Carlo techniques. Have you done this?

Heidmann: No, it was not necessary; we have statistics large enough to show that the distributions of the values are gaussian, so we used the standard procedure.

B. Peterson: The ratio you find for the radial velocities is about 8. Is this right outside the range covered by the error bars?

Heidmann: Yes – the error bars correspond to a factor 1.7, much smaller than 8.

Lewis: Your use of average values of the various ratios implies that the objects are typical members of their classes. Have you any independent evidence to show that this is a valid assumption, e.g. that they have typical values of colour index, $B - V$?

Heidmann: Yes, we checked that point. We have three integral properties which are independent of distance: hydrogen mass-to-light ratio, mean surface hydrogen density and maximum rotational velocity; all of them have values falling inside the ranges of normal galaxies. For example, $M_H/L_0 = = 0.37$ (solar units) for NGC 7319 which, as you see from the figure (next page), is quite normal.

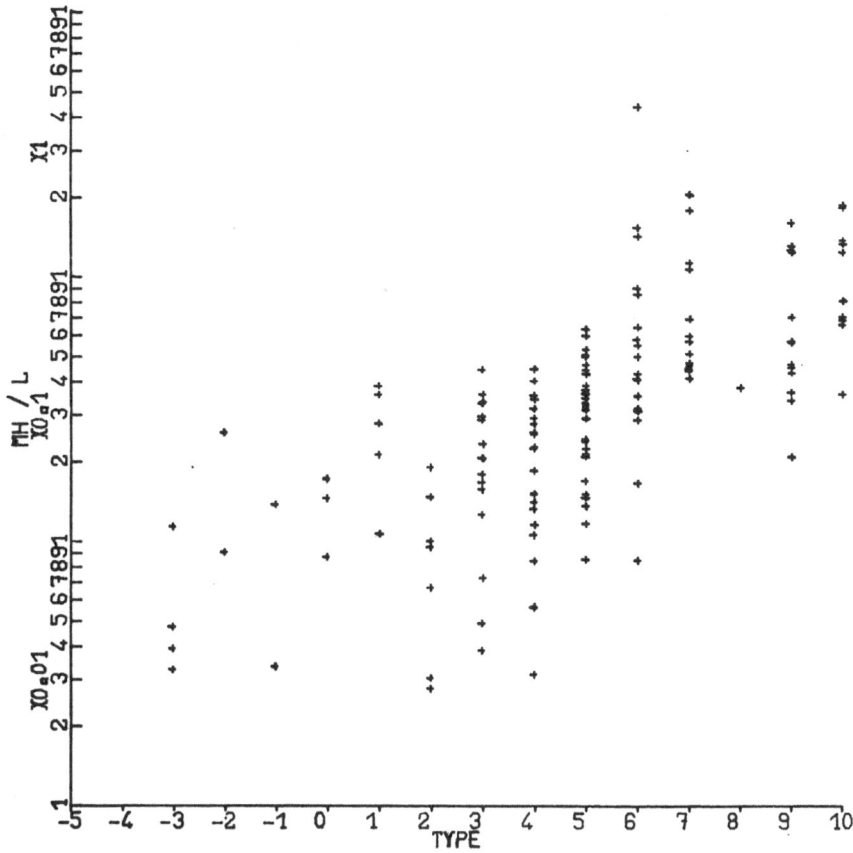

M_H/L_0 ratio in solar units versus morphological type T.

Sargent: I was planning to discuss Dr Heidmann's observations in my talk on the redshift question. My opinion is that the data as presented are consistent with either distance.

Arp: Two comments:

(i) The redshift distance of NGC 7319, using a Hubble constant of 50 km s^{-1} Mpc^{-1}, is $D = 120$ Mpc.

(ii) The Heidmann *et al.* distance for NGC 7319 is obtained from a comparison with data on field galaxies obtained by Balkowski and is not dependent on redshift distances. The Shostak distance is obtained by reference to Morton Roberts' data, some of which are based on a Hubble constant of 100 km s^{-1} Mpc^{-1}.

Oort: The fact that the galaxies in Stephan's Quintet lie in such a close group may well make them abnormal, so that the relations found for normal galaxies cannot safely be applied.

Heidmann: Yes, this may be so but, as I said, the three integral properties M_H/L_0, σ_H and V_m are normal, so we have nothing to compel us to assume they are abnormal. Even the photometric diameter of NGC 7319 appears normal, as is discussed in our paper.

Oort: On the other hand, the result you get is so improbable that I would try any way of escape first.

G. de Vaucouleurs: Has Dr van den Bergh derived luminosity classes for any member of the Quintet?

van den Bergh: The fact that the galaxies in Stephan's Quintet (?) are interacting makes it very difficult to assign unambiguous luminosity classifications to them.

Allen: Of course, the interesting thing to know is: just where is the gas located at the two different redshifts, and what is its dynamics? Dr W. Sullivan and I (in preparation) have mapped the distribution and velocity field of H I in NGC 7320 with the Westerbork telescope. Our conclusion is that the distribution and motions of the H I are quite reasonable for a dwarf galaxy in circular differential rotation. We put an upper limit of 20 km s^{-1} on any peculiar gas velocities caused by tidal effects of the other four Quintet members on the outer parts of NGC 7320. The minimum separation which one can infer from this observation is quite model dependent, but we can say that if we assume the five galaxies have been close to each other for at least one rotation period of NGC 7320 then the minimum separation between the latter galaxy and the other four members would be of order ten times the radius of NGC 7320.

Peebles: I have a philosophical question. If you assume that some galaxies have a non-Doppler redshift, does it make sense to assume that other physical parameters, such as transition rates, are the same as in normal galaxies, so that there is just this one aberration?

Toomre: Philosophical questions cannot be answered!

FINE STRUCTURE WITHIN THE
REDSHIFT-MAGNITUDE CORRELATION FOR GALAXIES

W. G. TIFFT

Steward Observatory, University of Arizona, Tucson, Ariz. 85721, U.S.A.

Abstract. Previous work on the redshift-magnitude banding correlation is briefly reviewed. New tests of the concept are applied successfully to a second cluster (A2199) and the outer portions of the Coma cluster. Using more than 200 redshifts in Coma, Perseus, and A2199 the presence of a distinct band-related-periodicity in redshift is indicated. Finally, a new sample of accurate redshifts of bright Coma galaxies on a single band is presented, which shows a strong redshift periodicity of 220 km s⁻¹. An upper limit of 20 km s⁻¹ is placed on the internal Doppler component of motion in the Coma cluster.

Redshift-magnitude bands are, therefore, now recognized to consist of discrete 'spin states' organized into 'spin groups' which show strong morphological associations. Bands are probably in turn organized into band systems. The individual spin states are suggested to represent distinct configurations of matter at the nuclear or fundamental particle structure level. Transitions between states in a time systematic sense of decreasing redshift (increasing internal binding) are suggested to occur in galaxy nuclei. Energy released in such transitions may be the driving energy in radio sources and related objects.

Investigations of the Coma cluster have played and continue to play a major role in the development of intrinsic redshift concepts which I shall present here. Figure 1 is a reminder of the original Coma cluster observations (Tifft, 1972) which introduced two of the basic phenomena, the nuclear-magnitude-redshift bands and the morphological separation of galaxies along the band direction, both correlations being statistically quite significant.

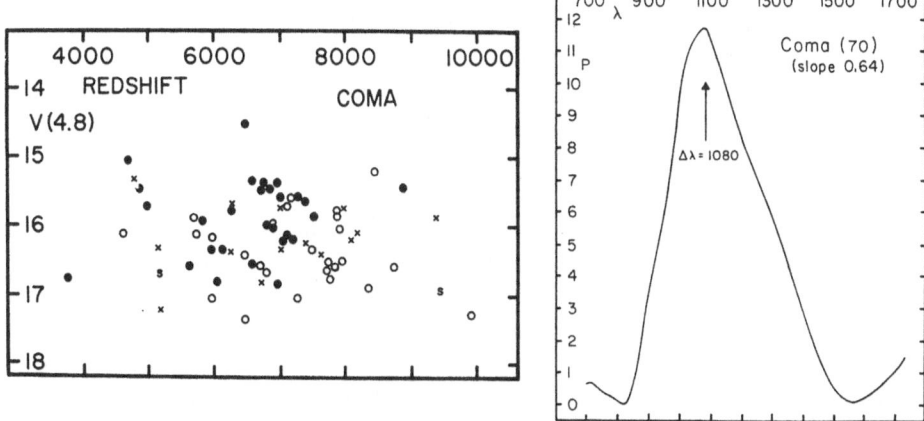

Fig. 1. (left) The original Coma cluster core redshift-magnitude diagram (Tifft, 1972) showing banding and morphological separation along the bands. Symbols denote morphology: filled circles = E, open circles = S0, X = SB0, and S = spiral. (right) Power spectrum of the redshift distribution of the core sample of 70 galaxies projected to a uniform magnitude along a slope of 0.64 mag per 1000 km s⁻¹. The projection slope corresponds to the direction of maximum morphological separation. The probability of finding such a peak by accident in a power spectrum is about 0.0005.

John R. Shakeshaft (ed.), The Formation and Dynamics of Galaxies. 243–256. *All Rights Reserved.*
Copyright © 1974 by the IAU.

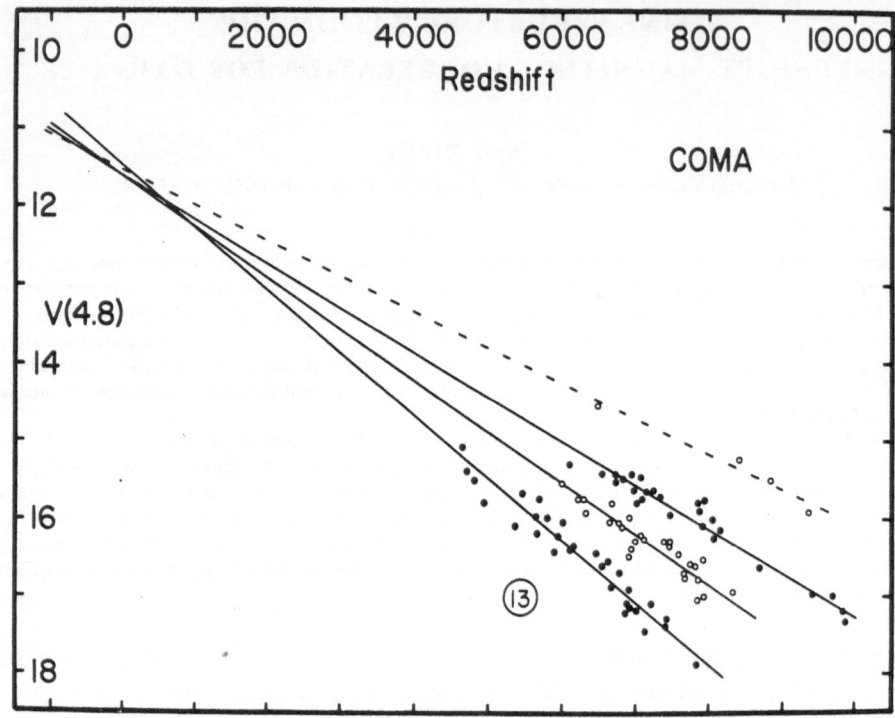

Fig. 2.   The extended Coma cluster core redshift-magnitude diagram (Tifft, 1973a) showing convergence of the bands toward zero redshift. Alternate bands are shown with alternate symbols and 13 points below the main bands are omitted. The convergence toward zero suggests that the bands are scaled intrinsic phenomena with no significant Doppler component.

With the extension of the Coma cluster sample (Tifft, 1973a) came one of the first direct evidences that the entire redshift phenomena could be an intrinsic property of matter and completely or nearly completely non-Doppler. Figure 2 illustrates the convergence of the bands toward zero redshift, permitting their interpretation as scaled versions of one another with no major Doppler component. This was further strengthened by the investigation of QSS objects (Tifft, 1973b). Figure 3 illustrates the concept of bands organized into convergent band systems which can be visible in an all-sky sample only if the redshift is virtually entirely non-Doppler, and matter and galaxies evolve systematically in cosmic time.

On the basis of the band properties, slope and spacing observed in Coma, and the universal band concept, specific predictions concerning bands in other clusters can be made. Figure 4 shows such a test for the A2199 cluster (Tifft, 1974). Power spectral analysis at the predicted slope and spacing confirms the bands with a probability of 10^{-3}. The cluster also shows morphological separation as in Coma.

One of the areas which has caused some discussion relates to the difficulty in seeing band effects in total magnitudes. For example, Figure 5 shows a sample of all brighter galaxies with known redshifts within 6° of the Coma cluster centre (Tifft,

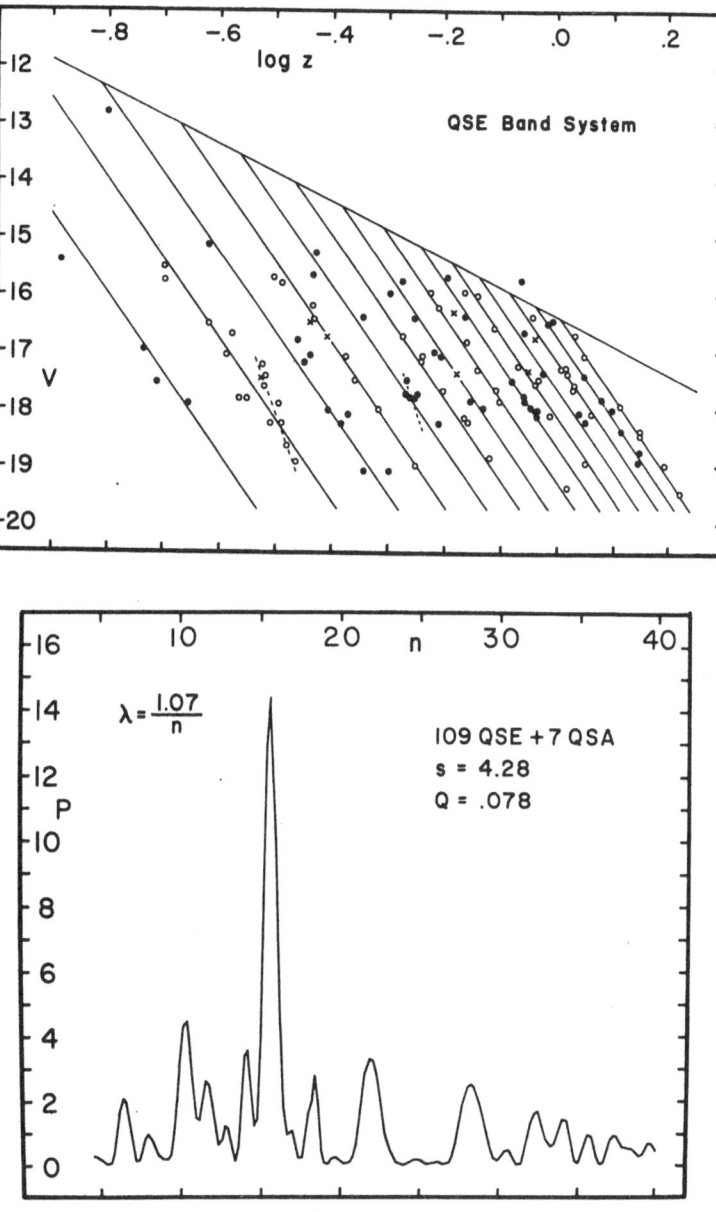

Fig. 3. (upper) Suggested organization of QSS redshifts and magnitudes into a system of bands (Tifft, 1973b) as an entirely intrinsic phenomenon. (lower) Power spectrum of the QSS distribution illustrated after projection to a uniform magnitude along a slope $s=4.28$ mag per factor of two in redshift and 'linearized' by a quadratic stretching factor Q (i.e. $z''=z'+Q(z')^2$ where z' is the value of z projected to $V=16.2$). Although the details of the higher bands are probably doubtful, the lower z bands appear distinct. (QSE refers to quasi-stellar emission-line objects and QSA to quasi-stellar absorption-line objects).

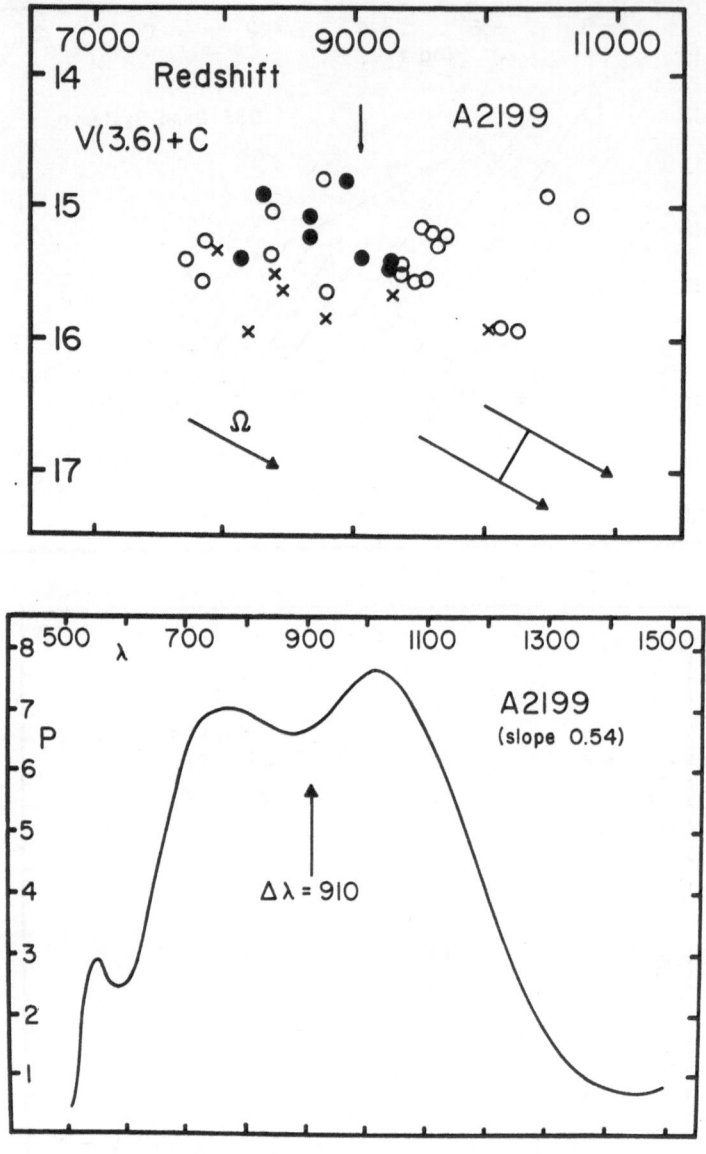

Fig. 4.   (upper) Redshift-nuclear magnitude diagram for galaxies in the core of the A2199 cluster
(Tifft, 1974). Magnitudes refer to the central 3.″6 region of each galaxy and are uncertain within a
constant. The band direction and spacing predicted from Coma are shown with the coupled vectors.
The single vector labelled Ω denotes the direction of morphological separation in the sample (filled
circles = E galaxies, open circles = non-E galaxies, X = unclassified) which is identical to the band
direction as seen in Coma. The downward pointing vector marks the mean cluster redshift. (lower)
Power spectrum of the redshift distribution in the A2199 sample projected at the predicted slope and
wavelength. The probability of finding a power of 7 at the predicted wavelength in a random
sample is 10^{-3}.

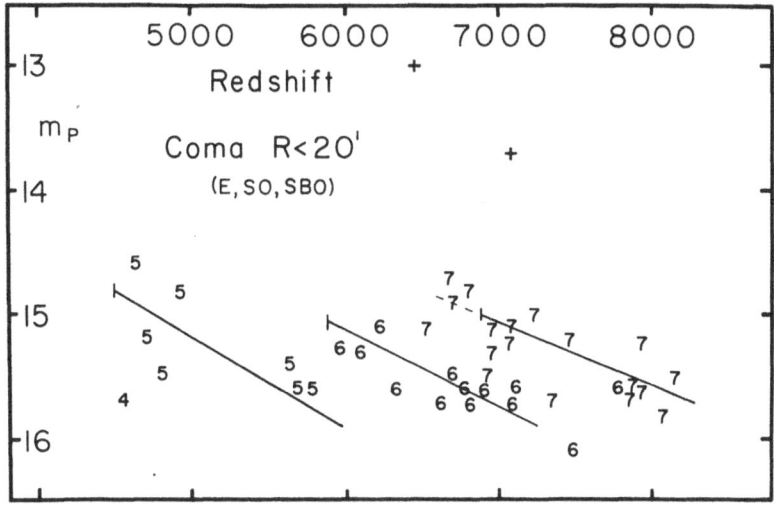

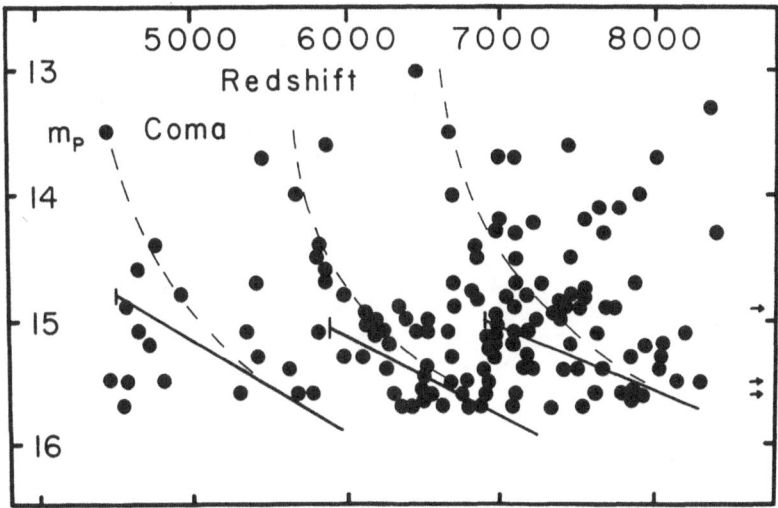

Fig. 5. (lower) Redshift-magnitude (m_p) diagram for all Coma galaxies within 6° of the cluster centre which have known redshifts and published m_p magnitudes (see Tifft, 1974). (upper) Redshift-magnitude (m_p) diagram for all E, S0, or SB0 galaxies in the above mentioned sample which are in common with central region studies and hence have a band assignment, indicated by the numerical symbol. The two supergiant galaxies are shown with crosses. The overlap objects along with known band spacing and convergence permit the mapping of mean band location lines into the diagram. These mean lines and suggested curving extensions are repeated on the lower figure. No obvious banding appears in the lower figure, although association of galaxies with the curved extensions is suggested.

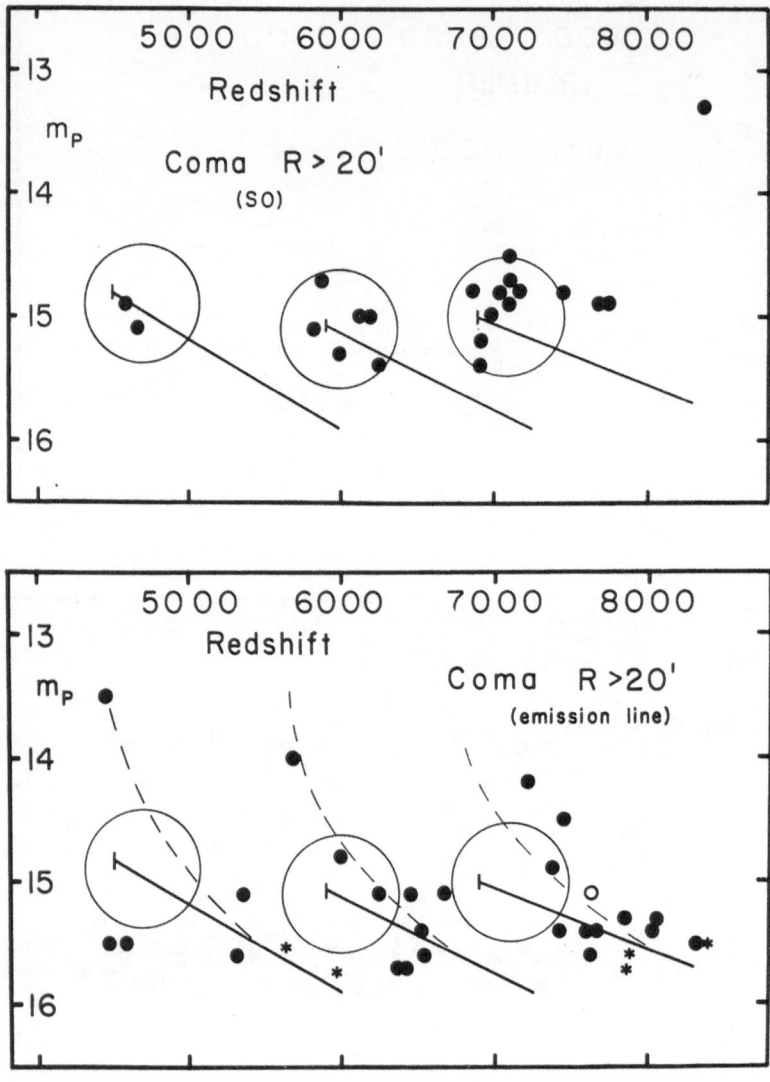

Fig. 6. (upper) Non-emission S0 galaxies in the outer part of Coma contained in the sample of objects shown in the lower part of Figure 5. The band location lines from Figure 5 are also shown. The S0 galaxies concentrate into three band-associated regions indicated by the reference circles. (lower) Emission line galaxies in the extended Coma sample (filled circles = outer Coma region, open circle = inner Coma region, * = additional new outer region data derived by S. A. Gregory at Steward Observatory, July 1973). Emission galaxies associate with the bands as do the S0 types, but are systematically shifted from the S0 regions along each band. When observed with morphological subsets, the band pattern becomes visible and a distinct band-related morphological separation in the same sense as the inner Coma investigations appears.

1974). The upper part of the figure illustrates the location of the bands as mapped by using objects in common with the cluster core studies. Because of the relatively low accuracy of m_p magnitudes, secondary effects in galaxy luminosity, and in-homogeneities in the sample, no band effects are visible by simple inspection of the total sample. If, however, we recall that there is morphological separation along the bands, and confine our attention to specific morphological types, the bands can be made visible as shown in Figure 6. Non-emission line S0 galaxies concentrate in specific band-associated regions as do emission line galaxies. Power spectra of the redshift distributions show that the morphologic alsamples possess the known band-spacing periodicity. Band-related-morphological separation is in the same sense as seen before, later types toward higher redshift along each band. Note that one can have strong band-related-morphological separation but little or no separation in the *total* sample. Redshift-morphology studies which do not allow for multiple bands are likely to be negative or inconclusive.

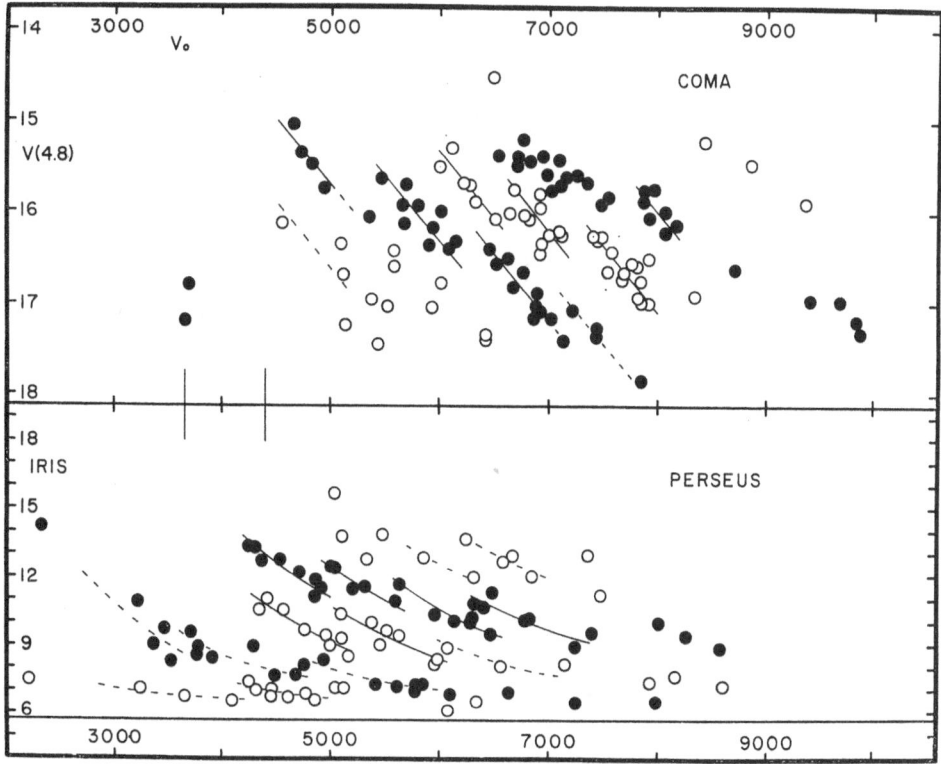

Fig. 7.   'Spin grouping' in the Coma and Perseus redshift-magnitude diagrams. The Perseus data are shown with nuclear region iris readings since magnitudes are not yet available. There is a general convergence toward the background level at 5.7. Alternate bands are indicated with different symbols. The sawtooth spin groups are shown with sloping lines, solid lines for those which appear clearly and dashed lines for other possible cases. Two vertical lines near 4000 km s^{-1} indicate the possible corre-spondence of band terminations in both clusters and the concentration of points on a lower band seen in Perseus but barely reached in Coma.

The first evidence that there might be substructure within individual bands came from the extended Coma studies where a sloping sawtooth 'spin group' structuring was suggested. Figure 7 illustrates this structure for Coma and in new Perseus cluster data. Magnitudes are not yet available in Perseus, and nuclear iris photometry has been substituted. Such readings correlate closely with nuclear magnitudes (Tifft, 1973c). Since the iris index converges to the plate limit, the Perseus banding is curved and cannot yet be examined statistically as in Coma or A2199. There appears very little doubt of its reality, however, and the band spacing in redshift is greater than Coma as expected from the lower mean redshift.

It has always appeared somewhat inconsistent to me that there were discrete bands but that the distribution along the bands was continuous, except, perhaps, for the sawtooth spin grouping. A fully quantized picture with discrete states on discrete bands seemed more logical. To determine if such discrete redshift states might exist at or near the resolution limit of the data, power spectra of the redshift distributions of individual bands were considered. Unfortunately, individual bands generally contain too few points for a definitive study and some model for combining bands

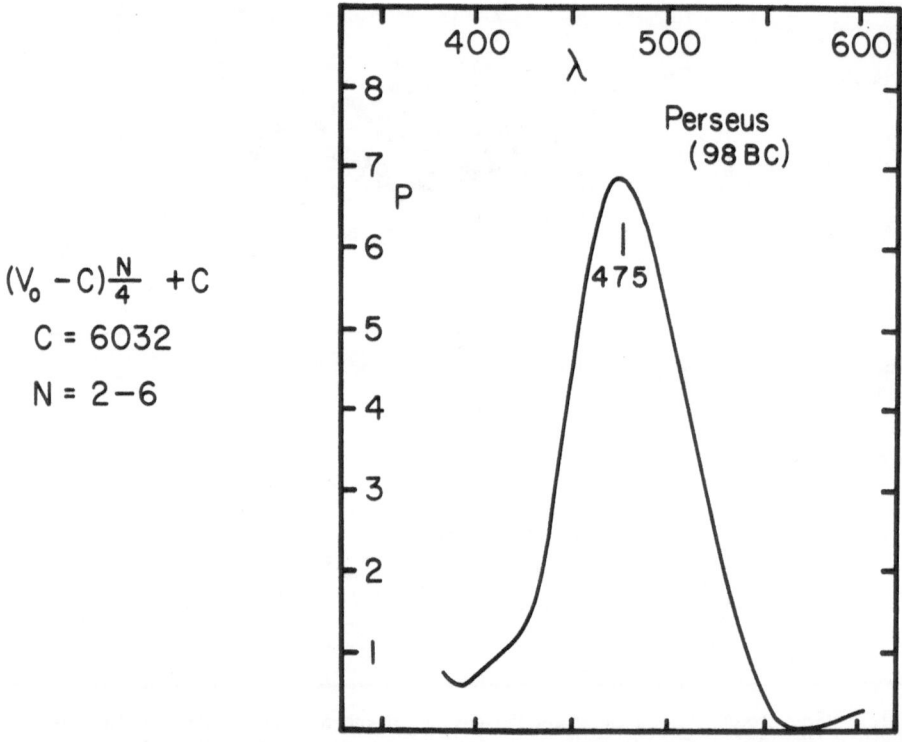

Fig. 8. Power spectrum of 98 Perseus redshifts from Chincarini and Rood (1971) = B or Tifft (1973c) = C. Redshifts have been combined after transformation by the expression given. C in the expression (which is unrelated to the redshift reference C above) is a common zero point and N is a sequential band number. Thus all bands are scaled to refer to band 4. When this combination procedure is followed, a modest power spectrum peak is found for a redshift interval $\lambda \equiv 474$ km s^{-1}.

was required. The first model developed was based upon the fact that superficially the spin group length appeared to vary with band number in such a way that by numbering the bands sequentially toward higher redshift from an appropriate origin, the spin group length times the band number was roughly constant. Figure 8 illustrates what happens when the Perseus bands are combined with such scaling about a common zero point. A modest power peak is produced with some phased contribution from each band.

By itself the Perseus band-related-periodicity has no significance, considering that the power is not exceptionally large and there is choice of the parameter C and origin of n. The interesting test of this concept came by applying the identical relationship to Coma and A2199 as independent samples. Figure 9 shows that both Coma and A2199 have substantial power peaks near the predicted frequency. There is a slight

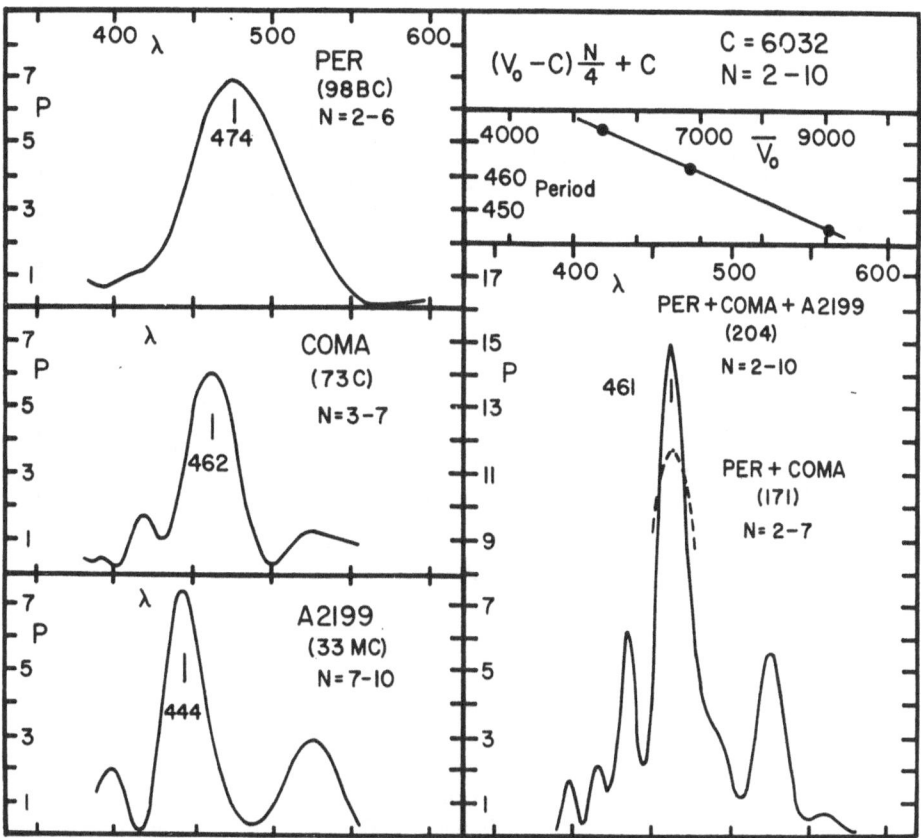

Fig. 9.   Power spectra of Perseus, Coma, and A2199 redshift distributions are shown on the left. Galaxies have been combined according to band number using the expression at top right as discussed in Figure 8 and the text. The individual cluster power spectra show peaks which shift slightly with mean cluster redshift as shown on the right. The combined cluster power spectrum is shown at lower right. In calculating these power spectra, older redshifts tabulated by de Vaucouleurs and de Vaucouleurs (1964) and values from Kintner (1971) have been omitted, since they contain some large scatter.

smooth systematic drift with mean cluster redshift. Finally, when the three cluster samples were combined in a single analysis to test the relative phasing of the periodicity in the individual clusters, the result was a powerfully reinforced peak of power level 15. This reinforcement has, of course, to be present if the concept of a universal intrinsic redshift is valid.

My interpretation of the observed periodicity is not that the model is correct in detail, but that it represents an interference effect of a real band-related-periodicity still not directly visible. To observe a true discrete periodicity, which I was now convinced existed, a precision redshift sample was required. Such a sample was generated

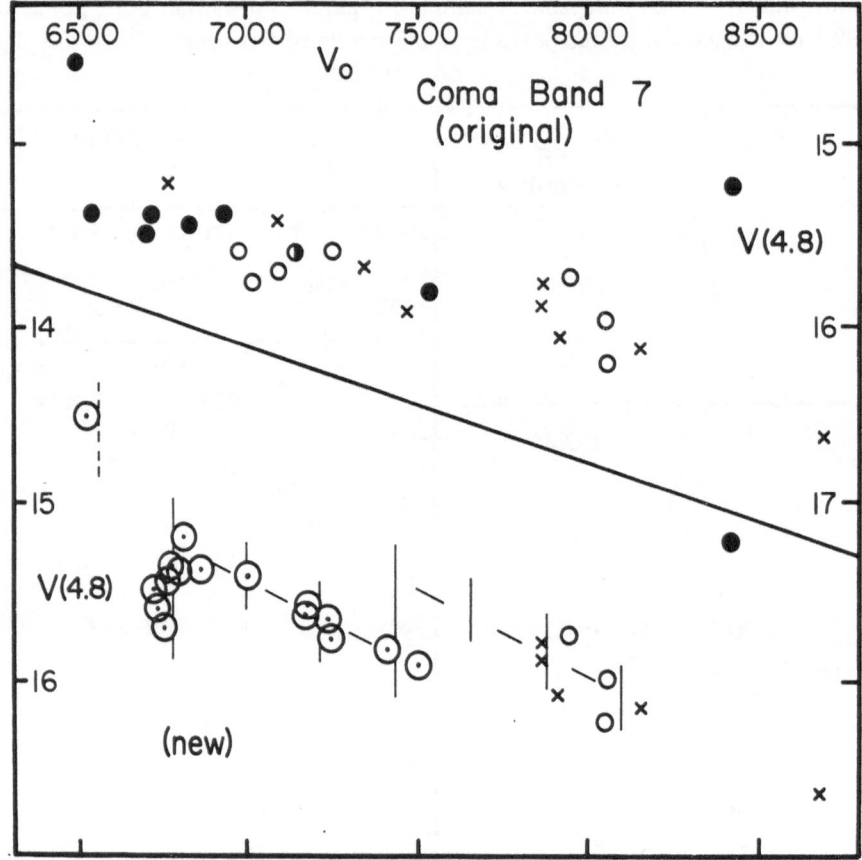

Fig. 10.   Galaxies on and above the brightest Coma cluster band. The upper figure shows the band with the original, previously published redshifts (filled circles = de Vaucouleurs and de Vaucouleurs (1964), open circles = Kintner (1971), X = Tifft (1972, 1973a), composite symbols = means of two sources). In the lower figure the redshifts below 7600 km s^{-1} have been replaced by accurate new mean redshifts from multiple spectrograms. The symbol size indicates the one sigma redshift uncertainty of about 35 km s^{-1}. Reduction is incomplete above 7600 km s^{-1} so the upper symbols are repeated. Some of the older redshifts contained errors in excess of 300 km s^{-1}. The new values show both the 'spin grouping' sawtooth wave on the band (dashed lines) and a 220 km s^{-1} 'spin state' periodicity (vertical lines).

by observing each member of the brightest Coma band three times independently. Most of these galaxies had redshifts from other sources, some of doubtful accuracy. This band also was the only one in Coma not showing clear spin grouping.

Figure 10 shows the original and new accurate Coma band 7 redshift-magnitude diagrams. The data are at present complete only for the 16 bright objects where the

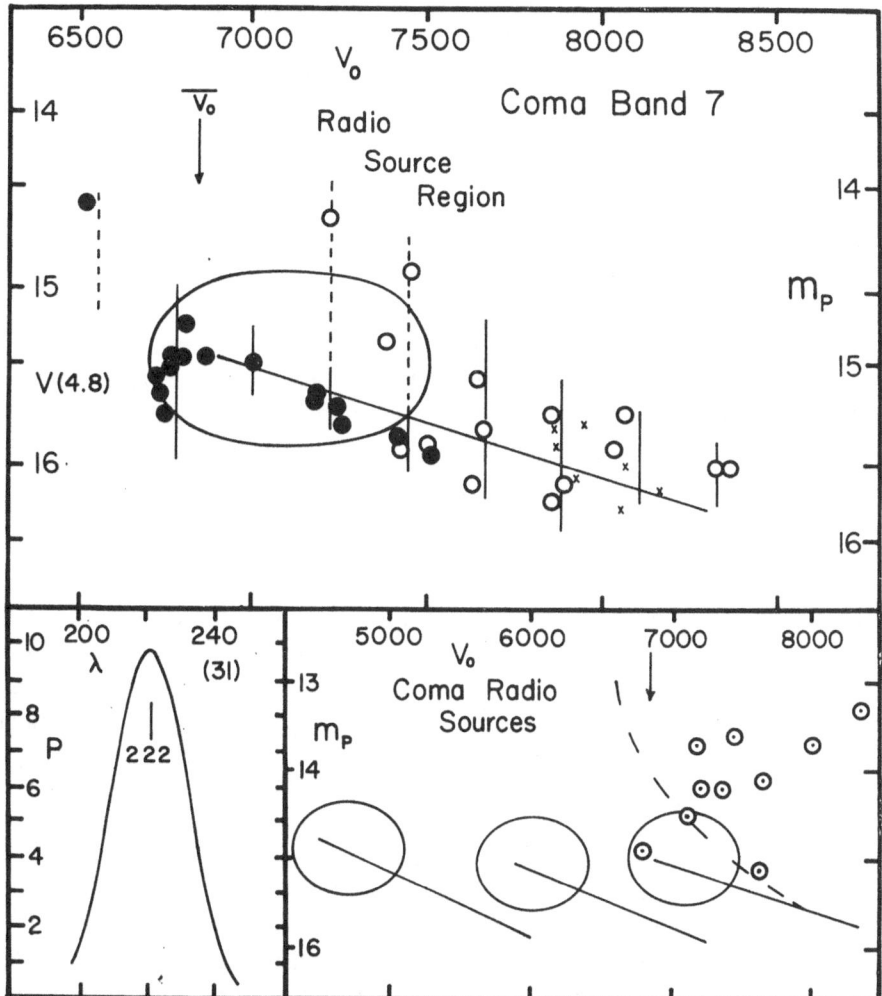

Fig. 11. (upper) Composite Coma band 7 with filled circles for the accurate data from Figure 10 and open circles for emission-line objects from the upper band in the lower part of Figure 6. Some of the lower accuracy data from Figure 10 are shown with small X symbols. The left-hand scale is $V(4.8)$ from Figure 10 and the right-hand scale is m_p from Figure 6 shifted for a rough match. The m_p band line and S0 reference region are repeated for comparison. The 'spin state' periodicity is shown with vertical lines. (lower) The power spectrum of the redshift distribution of the filled and open circle data is shown at the lower left. At the lower right the location of the 5C Coma radio sources is shown (Willson, 1970). They fall directly above the band and well above the mean cluster redshift, shown with a vertical vector.

radii of the circles represent the 35 km s^{-1} mean internal error in the final redshifts. Errors in some of the original redshifts exceeded 300 km s^{-1}. Both the spin grouping sawtooth and a rather clear 220 km s^{-1} discrete 'spin state' periodicity are now visible.

Since 16 objects were not really enough for a good power spectrum test, I added to the sample the Coma emission-line galaxies, previously discussed, which were assigned to the same band. Emission-line redshifts are generally fairly accurate. Figure 11 shows the composite accurate Coma band 7 picture and its power spectral analysis, which gives a peak power of 10 at the 222 km s^{-1} spin state periodicity. The probability of accidentally finding such a peak is 10^{-3}.

Figure 11 contains some intriguing information on band-related morphology and periodicity. Note that the bright group of ellipticals represent one spin grouping and this corresponds perfectly with the location of the outer Coma S0 concentration. The next spin group consists of non-ellipticals in the inner region and is the main concentration region for outer Coma emission line galaxies. Thus morphology appears closely tied to spin grouping. Pervading all is the 220 km s^{-1} spin state periodicity. We now have a concept of discrete 'spin states' organized into 'spin groups' and 'bands', and perhaps bands into band systems – a totally non-Doppler intrinsic redshift picture which I suggest has its origin in multiple substates of matter itself. One number of interest, the redshift dispersion about the set of discrete states is about 40 km s^{-1}, scarcely larger than the internal redshift uncertainty of about 35 km s^{-1}. An upper limit of about 20 km s^{-1} on real Doppler motion of galaxy nuclei within clusters is implied.

In order to produce the apparent observed pattern of redshift increase with distance, we can invoke systematic cosmic time dependent evolution of matter which we then see displayed in lookback time. The focus of this evolution is, I believe, galaxy nuclei. There are, apparently, conditions in nuclei of galaxies which process, create, or somehow operate on matter systematically with cosmic time. Matter may undergo transitions between states being systematically collapsed, possibly by great gravitational compaction, into progressively more tightly bound configurations with the systematic release of positive energy. Galaxies – the visible outflow products of these nuclei – can be systematically removed from the high redshift (low energy) end of a distribution with energy released in the process. The inset in Figure 11 shows how the Coma radio sources all lie above the high redshift band, and I suggest that the energy source of such radio objects arises from matter state transitions.

Because of the remarkable uniformity in redshift-magnitude diagram details, I suspect that nuclei must also be considered as certain 'standard' units in themselves, not simply random lumps of matter. Any explanation of multiple states of matter presumably resides in the realm of nuclear physics and the structure of fundamental particles.

References

Chincarini, G. and Rood, H. J.: 1971, *Astrophys. J.* **168**, 321.
de Vaucouleurs, G. and de Vaucouleurs, A.: 1964, *Reference Catalogue of Bright Galaxies*, University of Texas Press, Austin.

Kintner, E. C.: 1971, *Astron. J.* **76**, 409.
Tifft, W. G.: 1972, *Astrophys. J.* **175**, 613.
Tifft, W. G.: 1973a, *Astrophys. J.* **179**, 29.
Tifft, W. G.: 1973b, *Astrophys. J.* **181**, 305.
Tifft, W. G.: 1973c, *Astron. J.* **78**, 594.
Tifft, W. G.: 1974, *Astrophys. J.* **188**, 221.
Willson, M. A. G.: 1970, *Monthly Notices Roy. Astron. Soc.* **151**, 1.

Acknowledgement

Figures in this paper are reproduced from *The Astrophysical Journal*; Copyright: The American Astronomical Society, by permission of the University of Chicago Press.

DISCUSSION

B. Peterson: Would you describe the method and equipment used to obtain the nuclear magnitudes for the Coma cluster?

Tifft: The Coma nuclear magnitudes are from the Rood and Baum studies (*Astron. J.* **72**, 398 1967; **73**, 442, 1968). These were based upon tracings taken from 200-in. calibrated V photographs and were tied to photoelectric observations for the zero point. Unpublished portions of the Rood-Baum photometry should be available soon, probably in *Astrophysical Journal Supplements*. The m_p magnitudes in my outer Coma studies are from the catalogue of Zwicky and collaborators. All my own redshifts are from 90-in. Steward Observatory image tube spectra at 240 Å mm^{-1} and have an internal rms uncertainty of about 60 km s^{-1} on bright galaxies, and about 100 km s^{-1} on faint objects. There are good comparisons of my redshifts with new Gunn and Sargent data in Coma. The agreement is excellent.

Freeman: If the redshifts you used came from the whole Galaxy rather than just the nucleus, would this affect your argument significantly?

Tifft: If we take seriously the concept that nuclei could undergo 'transitions' in redshift, there might well be a systematic difference between 'nuclear' and 'envelope' redshifts in some cases. In fact the envelope of E type galaxies is virtually impossible to observe for redshift, and all present values are 'nuclear'. It is also quite possible that multiple redshifts (nuclear-envelope) might only be briefly observable. Envelopes might dissipate – to be really far out but not impossible, one might postulate gravitational decoupling of states which would lead to rapid dissipation of envelopes and a complete replacement of the galaxy in a 'new' state. Seriously, the important problem is to proceed with good solid data and tests, and minimize the interpretations.

Freeman: Three bands in Coma converge at approximately zero redshift and $V = 12$. Has this any particular significance?

Tifft: Not to my knowledge. The QSS bands seem to converge to a similar value.

Lewis: Comparison of the 21-cm velocities with the optical systemic velocities gives a reasonable test for differences between the disk and the nucleus. The 21-cm velocity is an average velocity from gas over the whole disk, while the optical velocity generally represents the velocity of the nucleus only. Roberts showed (*IAU Symp.* **44**, 12, 1972) that the regression V_{21} against $V_{optical}$ for about 130 galaxies had no systematic residual.

Following Tifft's (*Astrophys. J.* **175**, 613, 1972) result on the core of the Coma cluster, I examined the magnitude and velocity data for the *whole* Coma cluster, to look for evidence of non-Doppler redshifts. There are 197 galaxies with velocities satisfying $3000 < V < 10000$ km s^{-1} and with types quoted in Rood *et al.* (*Astrophys. J.* **175**, 627, 1972), Tifft (*Astrophys. J.* **179**, 29, 1973), and Tifft and Gregory (*Astrophys. J.* **181**, 15, 1973). Self-consistent magnitudes m_p are adopted from these papers or derived from Figure 5 of Tifft (1972) after an upward adjustment of 0.7 mag. Table I lists the mean data for all 197 galaxies in rows (1) and (5), as well as the means for Mayall's (*Ann. Astrophys.* **23**, 344, 1960) statistically uniform selection of bright galaxies in rows (2) and (6) and the non-Mayall sample means in rows (3) and (7). These figures suggest that

   (i) the largest velocity difference between E and non-E or S0 galaxies is (221 ± 222) km s^{-1} (row 1).

(ii) a 200 km s^{-1} uncertainty exists in the mean velocity of any type. This appears in the differences listed in row (4) between the Mayall and non-Mayall samples.

(iii) Mayall's sample is 1.3 to 1.5 mag brighter than the non-Mayall sample (row 8).

(iv) with these magnitude differences, Tifft's redshift-magnitude relations predicts the velocity differences in row (9). These disagree with the observed values in row (4) by a factor of ten.

(v) $\bar{m}_{S0} - \bar{m}_E = (0.48 \pm 0.22)$ mag from row (5). If the brighter ends of all Tifft's bands are preferentially populated by elliptical galaxies, Tifft's redshift-magnitude relation should be applicable. It

TABLE I

Average systemic velocity and magnitude of Coma cluster galaxies

Sample	Datum	E	S0	Non-E	Units
(1) Total sample	(a)	6728 ± 98	6949 ± 124	6880 ± 110	km s^{-1}
	N	78	95	118	
(2) Mayall's sample	(a)	6828 ± 138	6786 ± 191	6662 ± 155	km s^{-1}
	N	33	23	39	
(3) Non-Mayall sample	(a)	6652 ± 135	7001 ± 149	6952 ± 142	km s^{-1}
	N	45	72	79	
(4) row (3) – row (2)	$\overline{\Delta V}$	-176 ± 273	$+215 \pm 340$	$+290 \pm 297$	km s^{-1}
(5) Total sample	(b)	15.25 ± 0.12	15.73 ± 0.10	15.55 ± 0.10	m_{pg}
	N	78	95	118	
(6) Mayall's sample	(b)	14.49 ± 0.10	14.63 ± 0.12	14.55 ± 0.09	m_{pg}
	N	33	23	39	
(7) Non-Mayall sample	(b)	15.81 ± 0.11	16.09 ± 0.09	16.05 ± 0.09	m_{pg}
	N	45	72	79	
(8) row (7) – row (6)	$\overline{\Delta m}$	1.32 ± 0.21	1.46 ± 0.21	1.50 ± 0.18	m_{pg}
(9) (row (3) – row (2)) calc.	ΔV	2200 ± 350	2400 ± 350	2500 ± 300	km s^{-1}

(a) $\langle V \rangle \pm \sigma(V)/\sqrt{N}$;   (b) $\langle m \rangle \pm \sigma(m)/\sqrt{N}$.

predicts a velocity difference of $(+800 \pm 370)$ km s^{-1}.

These data do not confirm Tifft's fundamental hypotheses. The data for the whole Coma cluster suggest that $V_{S0} - V_E \lesssim +200$ km s^{-1}.

Tifft: The critical question is the association of morphology with redshift within *individual* bands compared with the total sample. Since several bands are present and they overlap in redshift, it is not clear that morphological separation should appear in mean overall redshift even if it were very strong in individual bands. This is quite obvious in my figure comparing S0 and emission line galaxies. A distinct shift in morphology occurs *on each band* but is insignificant in the overall average. It may be accidental that the total core sample shows the effect so well. Individual bands show it much more clearly and at *all* radii.

A second important factor concerns changes in the sample for different radial regions about the cluster. There are definite changes in average morphology with radius. Outlying galaxies may be poor objects to examine in a search for an *identical* correlation as seen in the core region. For example, the outlying E galaxies include 'blue' and active (emission, radio) objects to a higher degree than inner region E types. It is dangerous to consider them to be comparable for mean redshift or magnitude tests. The outer sample is also incomplete, while the core region is complete to some specified magnitude.

RADIO FREQUENCY OBSERVATIONS OF THE
NUCLEI OF GALAXIES

R. D. EKERS

Kapteyn Astronomical Institute, University of Groningen, The Netherlands

Abstract. For spiral galaxies the nuclear radio emission is usually dominated by a complex distribution of emission with median diameter of 200 pc and median power of 10^{19} W Hz^{-1} sr^{-1} at 1400 MHz. There is a large range in both power and diameter. The power is independent of morphological type for the normal spirals but is correlated with the absolute optical magnitude and with the infrared emission. For Seyfert galaxies the emission is generally stronger, in some cases by several orders of magnitude (e.g. NGC 1275, 3C 120).

Elliptical galaxies have been found with very compact radio sources, some less than a parsec in diameter. These are as powerful as the strongest spirals ($\sim 10^{21}$ W Hz^{-1} sr^{-1}). Even stronger compact nuclear sources are now being found in the nuclei of those elliptical galaxies which also have extended radio sources (the radio galaxies). The presence of nuclear sources of this strength is so highly correlated with the presence of extended sources that this suggests a continuing involvement of the nucleus.

1. Introduction

The properties of strong radio sources have been reviewed by Kellermann and Pauliny-Toth (1968) and Kellermann (1971, 1972). They discuss the angular sizes, radio spectra, temporal variations and the physical conditions required to explain these observations. In this report I will concentrate more on the radio nuclei of normal spiral and elliptical galaxies.

2. Spiral Galaxies

This is the most common type of galaxy and consequently the closest systems are mainly of this type. Even so, our knowledge of the radio properties of the nuclei of these galaxies is still very fragmentary except, of course, for our own Galactic centre.

2.1. THE GALAXY

The radio frequency emission from the centre of our Galaxy has three components: an extended source about $1°0 \times 0°4\,(180 \times 70$ pc) elongated along the galactic plane, a complex of giant H II regions in a similar area, and a non-thermal source, Sgr A, about $3'0 \times 2'4\,(9 \times 7$ pc) very close to the centre of the Galaxy. All these components can be seen in the 408 MHz map from Little (1974) reproduced in Figure 1. The extended component is of interest since it is such a component as would dominate our observations of extragalactic systems. It used to be thought that this component had a fairly flat flux density spectrum and could thus be an optically thin thermal source (Downes and Maxwell, 1966), but more recent observations (e.g. see Gordon, 1974) indicate that it is mainly non-thermal. Detailed reviews of the centre of our Galaxy have been prepared by Oort (1974) and by Gordon (1974).

John R. Shakeshaft (ed.), The Formation and Dynamics of Galaxies. 257–277. All Rights Reserved.

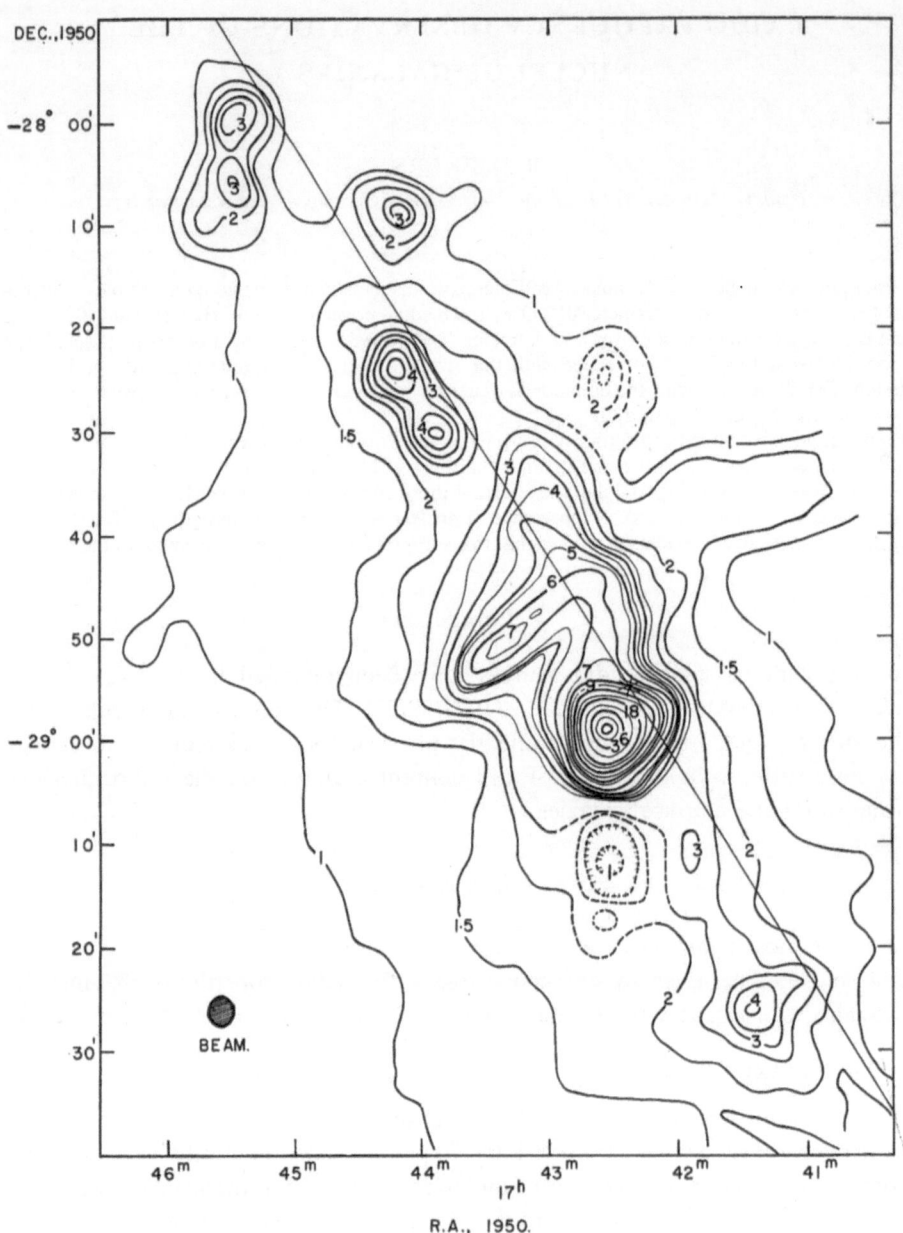

Fig. 1.   Galactic Centre at 408 MHz. The contour interval is 680 K and the unit contour is at 1075 K.
The dashed contours are sidelobe effects from Sgr A. This map obtained with the University of Sydney's
Mills Cross radio telescope has been kindly supplied by Mr A. G. Little.

2.2. M31

The nuclear emission from M31 was observed by Pooley (1969a) at 408 and 1407 MHz and, with higher sensitivity, by van der Kruit (1972) at 1415 MHz with 23″ resolution. A contour map based on the latter data is shown in Figure 2. The nuclear source has a similar scale to, but is an order of magnitude weaker than, the extended component in our galactic centre. It also is resolved into a complex of sources, although these do not appear to lie in the plane of the galaxy. Observations at 2695 MHz with 8″ resolution (Spencer, 1973) failed to detect any of these components, indicating that they have steep non-thermal spectra and that they are resolved.

2.3. M51

The nuclear source in M51 is clearly seen in the map by Mathewson *et al.* (1972). It is 16″ (700 pc) in extent and 10 times stronger than our galactic centre. The higher resolution observations by Spencer (1973) and de Bruyn (private communication) show that this source also breaks up into at least two components.

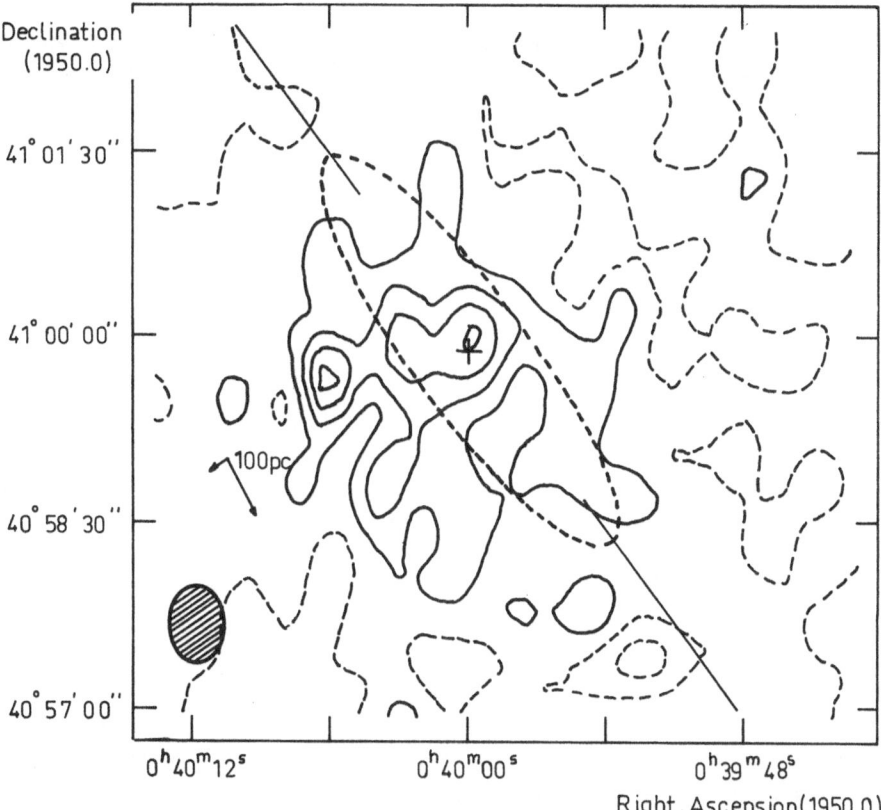

Fig. 2.   1.4 GHz continuum map of the nucleus of M31 from the data of van der Kruit (1972). The contour interval is 1.6 K and the zero and negative contours are dashed. The arrows indicate the directions of the major and minor axes and the apparent length of 100 pc in the plane of M31.

2.4. OTHER SPIRAL GALAXIES

In order to separate clearly the nuclear radio sources from the sometimes complex emission of the galactic disks, I have used only high resolution observations with good spatial coverage. Van der Kruit (1973c) discusses the properties of a sample of 45 galaxies observed in detail with the Westerbork array. I have enlarged this sample of galaxies by adding observations from the Cambridge and Greenbank arrays and some more recent observations from Westerbork. The resulting sample is not complete to a given optical magnitude since different selection criteria have been used in the various programmes, and in such a heterogeneous sample it is especially important to avoid erroneous correlations arising because of differing sensitivities and differing distances for various types of galaxies. For each galaxy I have used its distance and the sensitivity of the observation to calculate the absolute radio power level at which its nucleus would still be detectable. All the galaxies for which a nuclear source could be detected if it were stronger than the median for the final sample are listed in Table I. Hence, for galaxies in this table, there will be no bias with radio power above the median value of 8×10^{18} W Hz^{-1} sr^{-1} at 1410 MHz. There may still be bias in the sample with respect to other properties, e.g. the numbers of each optical type, but these have less serious consequences. Nuclei have only been included if they are known to be less than 1 kpc in extent. Most galactic nuclei for which we have linear dimensions are considerably smaller than this, but the resolution of many of the observations is such that the sample would be too much reduced by use of a smaller size criterion.

Galaxies which have distinct nuclear radio sources but which do not form part of the complete sample are listed in Table II. In all cases the most recent distance determinations available and a Hubble constant of 55 km s^{-1} Mpc^{-1} (Sandage, 1972) have been used.

2.5. SEYFERT GALAXIES

Wade (1968) and van der Kruit (1971) have shown that the radio emission from the nuclei is stronger for Seyfert galaxies. In some cases, e.g. NGC 1275 (3C 84), it is more than a thousand times stronger than the normal spiral nuclei, while other Seyfert galaxies have radio powers in the upper range of the distribution of power from normal spirals.

For NGC 1275 there are two recent results of considerable interest. Detailed interferometer observations between Algonquin Park and Chilbolton by Legg et al. (1973) at 2.8 cm wavelength indicate a structure for the smallest nuclear component which is elongated by 3×10^{-3} arc sec (2 pc) in position angle $-6°$ and consists of at least four components less than or about 10^{-3} arc sec (0.5 pc) in diameter. A similar distribution is required to fit the 2 cm observations between Greenbank, Owens Valley and Fort Davis (Kellermann, private communication). The second result is the detection by De Young et al. (1973) of the 21-cm neutral hydrogen line in absorption at a redshift corresponding to $+2820$ km s^{-1} with respect to the systemic velocity for NGC 1275. This line is only 6 km s^{-1} wide.

TABLE I

Spiral galaxies – selected sample

Name NGC	Other	Type	Dist. Mpc	Total $(Mo)_p{}^a$	Total $S_{1415}{}^b$	Nucleus $S_{1415}{}^b$	Nucleus $\log(P_{1415}{}^c)$	Ref.[d]
224	M31	Sb	0.65	− 23.1	8.0	0.10	17.6	1, 15
253		Sc	4.3	− 21.6	5.0	2.2	20.6	2
598	M33	Sc	0.72	− 20.0	3.1	< 0.003	< 16.3	3
628	M74	Sc	14	− 21.4	0.1	< 0.004	< 18.9	4
891		Sb	7	− 20.7	0.7	0.025	19.1	5, 17
1003		Sc	13	− 19.9		< 0.005	< 18.9	6
1058		Sc	4	− 19.2		< 0.005	< 18.9	6
2403		SABcd	3.5	− 20.5	0.3	< 0.003	< 17.6	3
2681		Sa	14	− 19.8		< 0.003	< 18.8	7
2685		S0	17	− 19.8		< 0.003	< 18.9	7
2841		Sb	11	− 21.1	< 0.2	< 0.005	< 18.7	6
2903		SABbc	13	− 21.9	0.4	0.07	20.1	3
3031	M81	Sab	3.5	− 21.6	0.5	0.08	19.0	3, 8
3034	M82	I0	3.5	− 21.0	8.0	8.0	21.0	9, 10
3077		Ir?-Sef	6	− 18.8	< 0.3	0.021	18.9	7
3184		Sc	8	− 19.5	< 0.2	< 0.005	< 18.5	6
3359		SBc	18	− 20.9	< 0.2	< 0.003	< 19.0	7
3432		SBm	8	− 19.2	0.11	< 0.01	< 18.8	4
3953		Sc	19	− 21.2	< 0.2	< 0.003	< 19.0	7
4051		Sb-Sef	12	− 20.1	0.05	0.021	19.5	7, 8
4156		SBb	12	− 18.5		< 0.004	< 18.8	7
4258	M106	Sbc	8	− 21.2	0.9	< 0.025	< 19.1	11
4425		Sa	20	− 19.5		< 0.003	< 19.1	12
4438		Sap	20	− 21.4	0.12	0.090	20.6	12
4485			8			< 0.005	< 18.5	13
4490		SBdp	8	− 21.1	0.8	0.02	19.1	13
4631		SBd	8	− 20.9	1.3	0.025	19.2	3, 14
4736	M94	Sab	8	− 20.8	0.3	0.013	18.9	7
4826	M64	Sab	7	− 20.4	0.09	< 0.052	< 19.4	3
5055	M63	Sbc	8	− 21.0	0.6	0.020	19.1	3
5194	M51	Sbc	9.5	− 21.5	1.5	0.080	19.9	16, 19
5457	M101	Sc	7	− 21.7	0.5	0.004	18.3	17
5907		Sc	8	− 19.9		< 0.005	< 18.5	6
6946		Sc	10	− 21.2	1.3	0.088	19.9	18
7331		Sbc	14	− 22.0	0.6	< 0.005	< 19.0	4
7640		SBc	14	− 21.1		< 0.005	< 19.0	4
--	IC342	SABcd	6		2.0	0.13	19.6	4
−	our Galaxy	–	0.01			4600.	18.7	

[a] Absolute photographic magnitude corrected for inclination and galactic absorption according to Holmberg (1958).

[b] 10^{-26} W m^{-2} Hz^{-1}.

[c] W Hz^{-1} sr^{-1},

[d] 1. van der Kruit (1972)
2. Becklin et al. (1973)
3. van der Kruit (1973a)
4. van der Kruit (1973b)
5. Baldwin and Pooley (1973)
6. de Bruyn, private communication
7. van der Kruit (1971)
8. Wade (1968)
9. Macdonald et al. (1968)
10. Kronberg et al. (1972)

11. van der Kruit et al. (1972)
12. Ekers, in preparation
13. Allen et al. (1973)
14. Pooley (1969b)
15. Pooley (1969a)
16. Mathewson et al. (1972)
17. Allen, private communication
18. Rots, private communication
19. Spencer and Burke (1972), Spencer (1973).

TABLE II

Other spiral galaxies with nuclear radio emission

NGC	936	2655	3504	4151	4945	7469
	1068	2782	3516	4321	5005	7552
	1097	2798	3521	4569	5035	
	1275	3079	3623	4579	5383	
	1569	3227	3627	4594	5548	
	1808	3310	3628	4670	5866	
	2445	3351	3888	4676A	6052	

2.6. STATISTICS FOR THE SPIRAL GALAXIES

Figure 3 shows the distribution of linear diameters for those spiral galaxies in which the nucleus is resolved or for which the upper limit is less than 400 pc. The median diameter for the sample is 200 pc, which is similar to the diameter of the most extended component of the Galactic centre. Some radio nuclei are very much smaller than this and the resolved nuclei may also contain components of smaller diameter.

The distribution of radio spectral indices between 1400 and 5000 MHz is shown in Figure 4. There is a broad spread around $\alpha = -0.7$ containing all the resolved nuclei, but a few galaxies have inverted spectral indices, including the Seyfert galaxy NGC 1275 which contains the very compact variable radio source 3C84 discussed in the

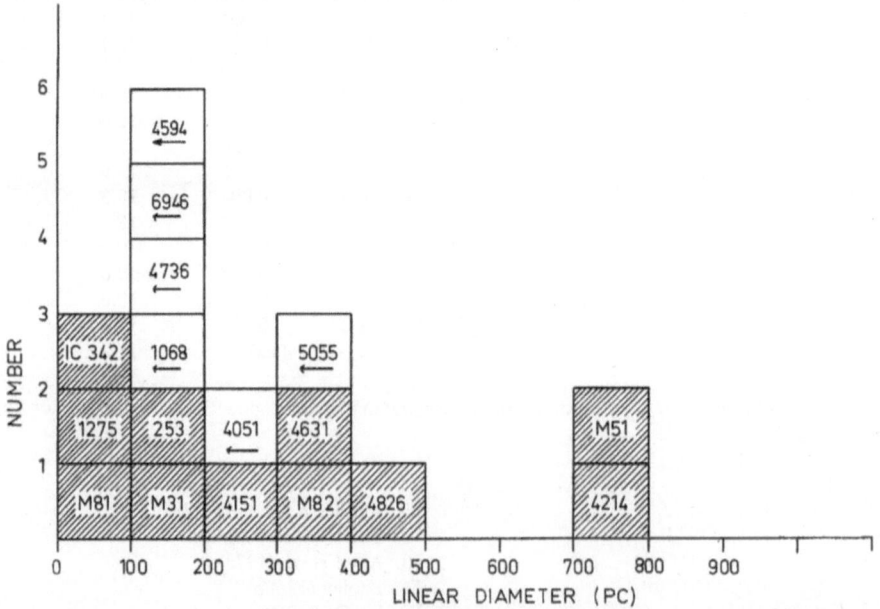

Fig. 3.   Distribution of the major axis diameters of the nuclear radio sources in spiral galaxies. All limits for diameters are included if they are < 500 pc.

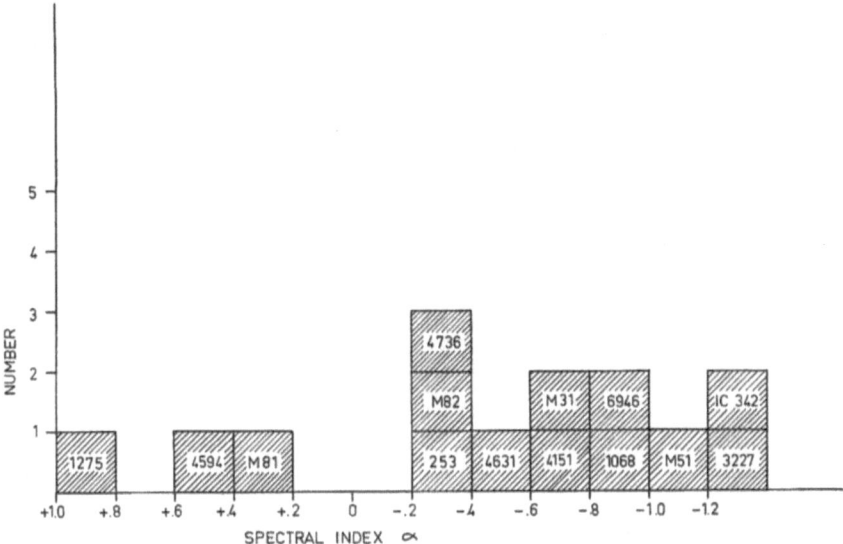

Fig. 4.   Distribution of the radio spectral index, α, for the nuclei of spiral galaxies, α is defined by $S \propto \nu^{\alpha}$.

last section, and M81 which has diameter $<2''$ $(<35\,\mathrm{pc})$ and is the smallest of the normal spiral nuclei known (Wade 1968). From a sample of observations obtained with somewhat lower resolution, Le Squéren and Crovisier (1974) have found a mean spectral index of -0.7 for the nuclei and -1.1 for the disk emission from spiral galaxies.

Figure 5 shows that there is correlation between the radio power of the nucleus and the absolute optical magnitude. Since only the sample of Table I is included, this cannot result from the selection effects. There is a weak correlation between the strength of the nuclear source and the strength of the disk radio emission, and the disk radio emission is also correlated with the absolute optical magnitude, so it is not clear which pairs of parameters are causally related. Van der Kruit (1973c) also compares the nuclear power with the brightness of the disk and finds an absence of galaxies with bright disks and nuclear power below $10^{19}\,\mathrm{W\,Hz^{-1}\,sr^{-1}}$. Although this may still be correct for this sample of galaxies, it is weakened by the presence of two galaxies, NGC 7331 and NGC 4490, which have brighter disks but no distinct nuclear sources. Hence, at present we cannot make a strong case for the nucleus providing a significant supply of radiating electrons for the disk emission.

Essentially no correlation is seen between the radio power of the nucleus and the optical type for the normal spirals plotted in Figure 6. This and other correlation diagrams are discussed by van der Kruit (1973c). He also concludes that for normal spirals neither the radio power of the nucleus nor its ratio to the total power is dependent on optical type or the presence of a bar. There is a weak correlation of radio power with Byurakan type but this is only statistically significant for the Seyferts.

2.7. CORRELATION WITH INFRARED EMISSION

Van der Kruit (1971) showed that there was a correlation between the 10μ emission
and the nuclear radio emission. Although this correlation is maintained in the larger
sample of galaxies in van der Kruit (1973c), he comments that for the normal spirals
the sample is very incomplete and that the correlation could arise from the selection

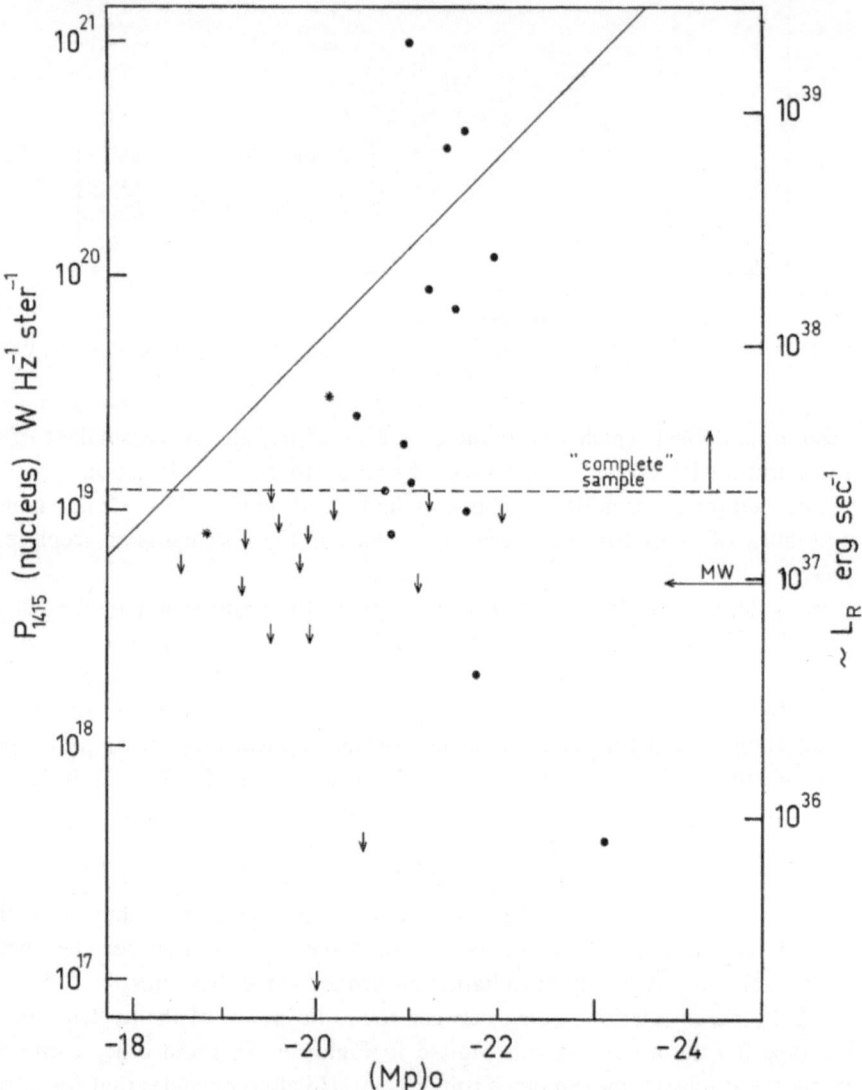

Fig. 5.   Plot showing the correlation between the monochromatic radio power at 1.4 GHz and the
absolute optical luminosity of spiral galaxies. The two Seyfert galaxies in the selected sample are
indicated by *. A radio luminosity scale, L_R, is shown for a power law spectrum of index -0.7
integrated from 10^7 to 10^{11} Hz.

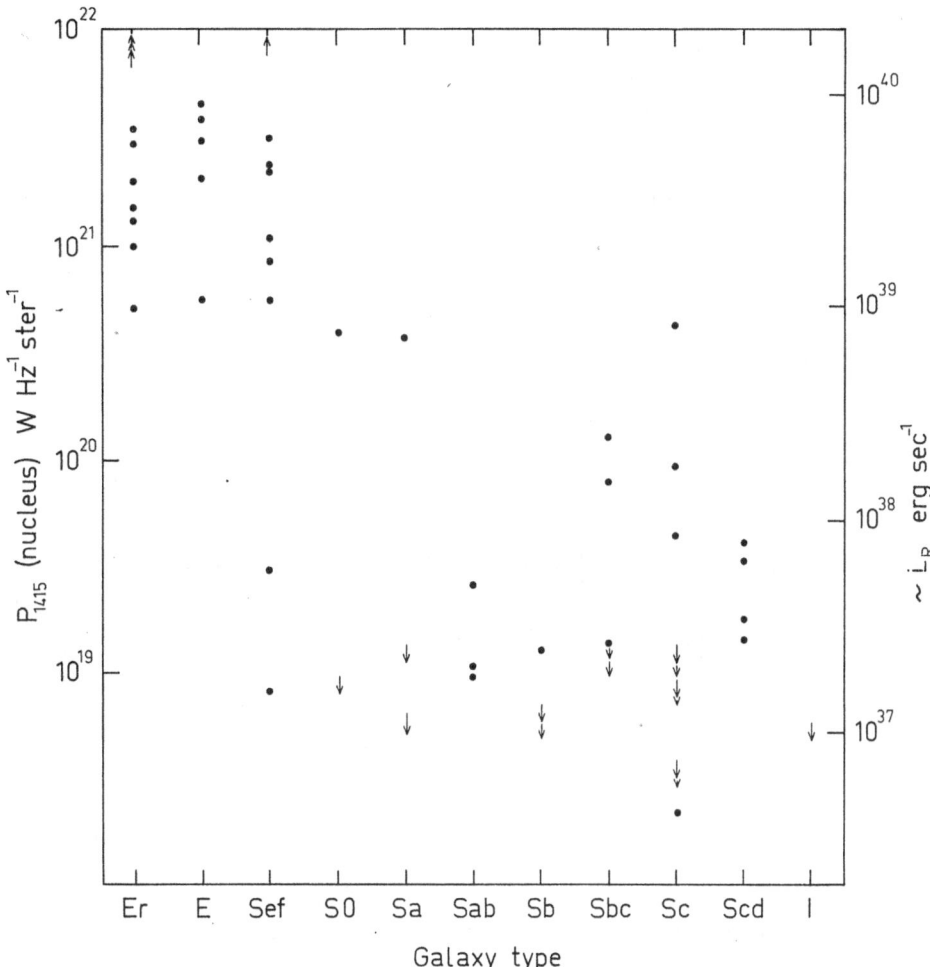

Fig. 6   Plot of the monochromatic radio power at 1.4 GHz for different Hubble types and for the Seyfert galaxies, Sef, and the radio galaxies, E_R. The radio luminosity scale, L_R, is defined in Figure 5.

effects. The Seyfert galaxies do constitute a complete sample and for these there is also a clear correlation.

For M82 (Klienmann and Low, 1970) and NGC 253 (Becklin *et al.*, 1973), both the radio and infrared nuclei are extended and have comparable dimensions. In M82 the radio and the infrared structure have been resolved into bright knots (Kronberg *et al.*, 1972) but there is no detailed correlation between the positions of the infrared and radio knots.

2.8. OTHER RADIO EMISSION POSSIBLY ASSOCIATED WITH NUCLEAR ACTIVITY

In a few spirals there is evidence for radio emission which, although outside the nuclear region, could be interpreted as resulting directly from nuclear activity.

In NGC 4258 van der Kruit *et al.* (1972) observed a pair of radio arms and proposed

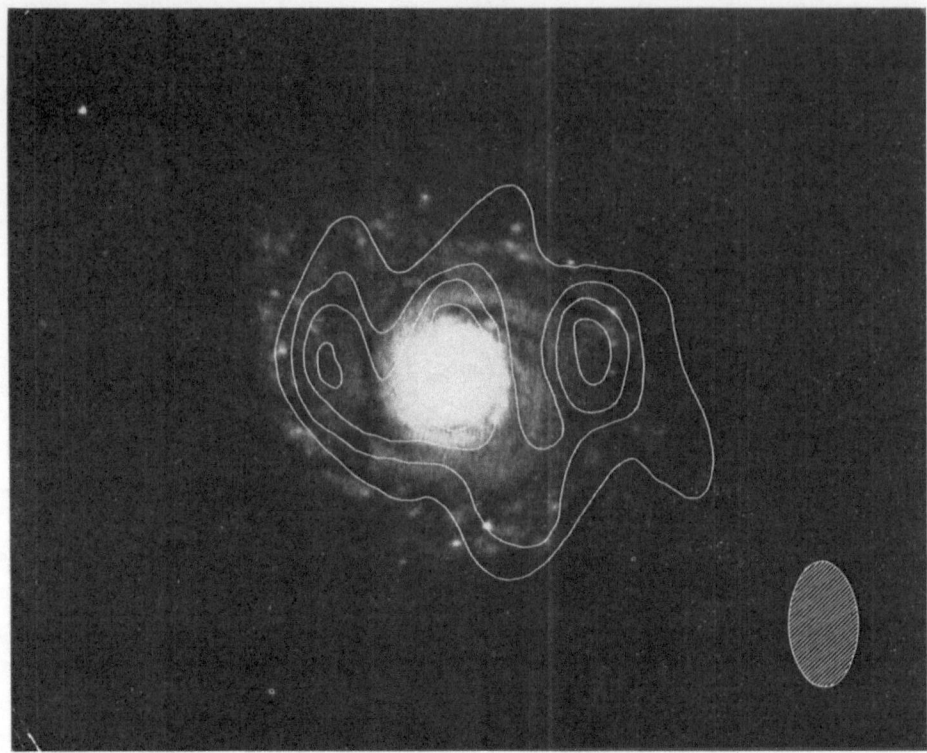

Fig. 7. 1.4 GHz continuum observations of NGC 4736 from van der Kruit (1971) superimposed on an optical photograph from the Mt. Wilson 100-inch telescope. Contour values are 2, 4, 6, 8 and 12 K. This composite was kindly supplied by Prof. J. H. Oort.

a model involving the violent expulsion of material from the nucleus of this galaxy. There is no striking optical or radio activity present in the nucleus now, although the radio arms do continue smoothly into the nuclear region. Further discussions of this galaxy are given by Oort (p. 375) and by van der Kruit (p. 431) in this symposium.

NGC 4736* and NGC 4631 (Figures 7 and 8) are spiral galaxies with double radio sources which, by their symmetry with respect to the nucleus, are suggestive of direct nuclear involvement. Since NGC 4631 is an edge-on galaxy, it is possible that in this case we are seeing the normal spiral arm emission, as is suggested by Pooley (1969b). NGC 3726 (Figure 9) was thought to be a similar case (van der Kruit, 1971), but the more accurate radio position determined by de Bruyn (private communication) shows that these radio components are not symmetrical about the nucleus. More sensitive observations are required to see whether these components are discrete sources or are two peaks in a complex distribution of emission. At present the sample of sources in spiral galaxies is too small to say whether there is a class of spiral galaxies with the same kind of double radio emission as found in the radio galaxies. However, the

* Note added in proof: Additional observations now indicate that at least part of the western source may be associated with H II regions (A. Bosma, private communication).

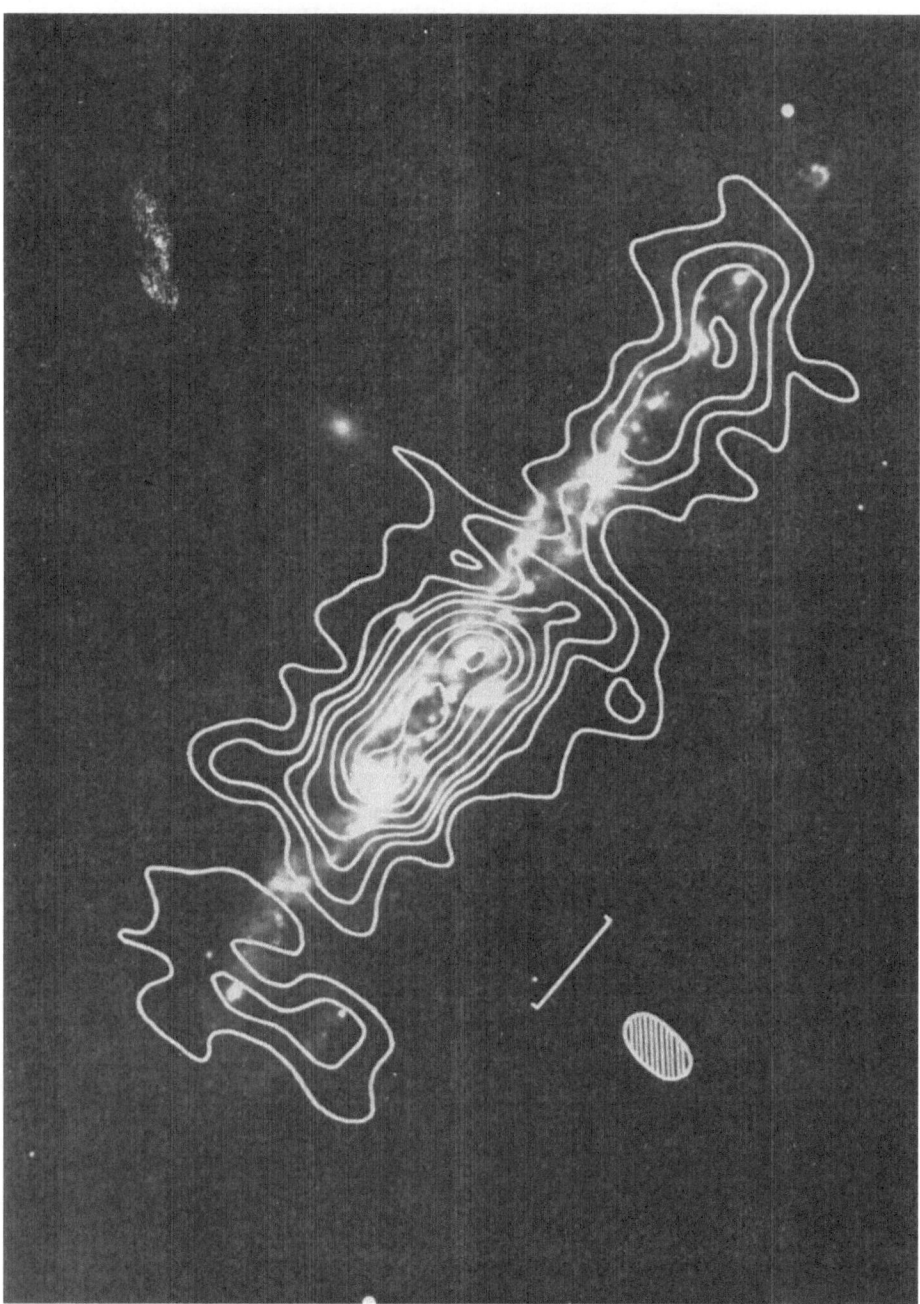

Fig. 8. 1.4 GHz continuum map of NGC 4631 from Pooley (1969). The contour interval is 3 K, and alternate contours above the fourth have been omitted. The line is of length 1′.

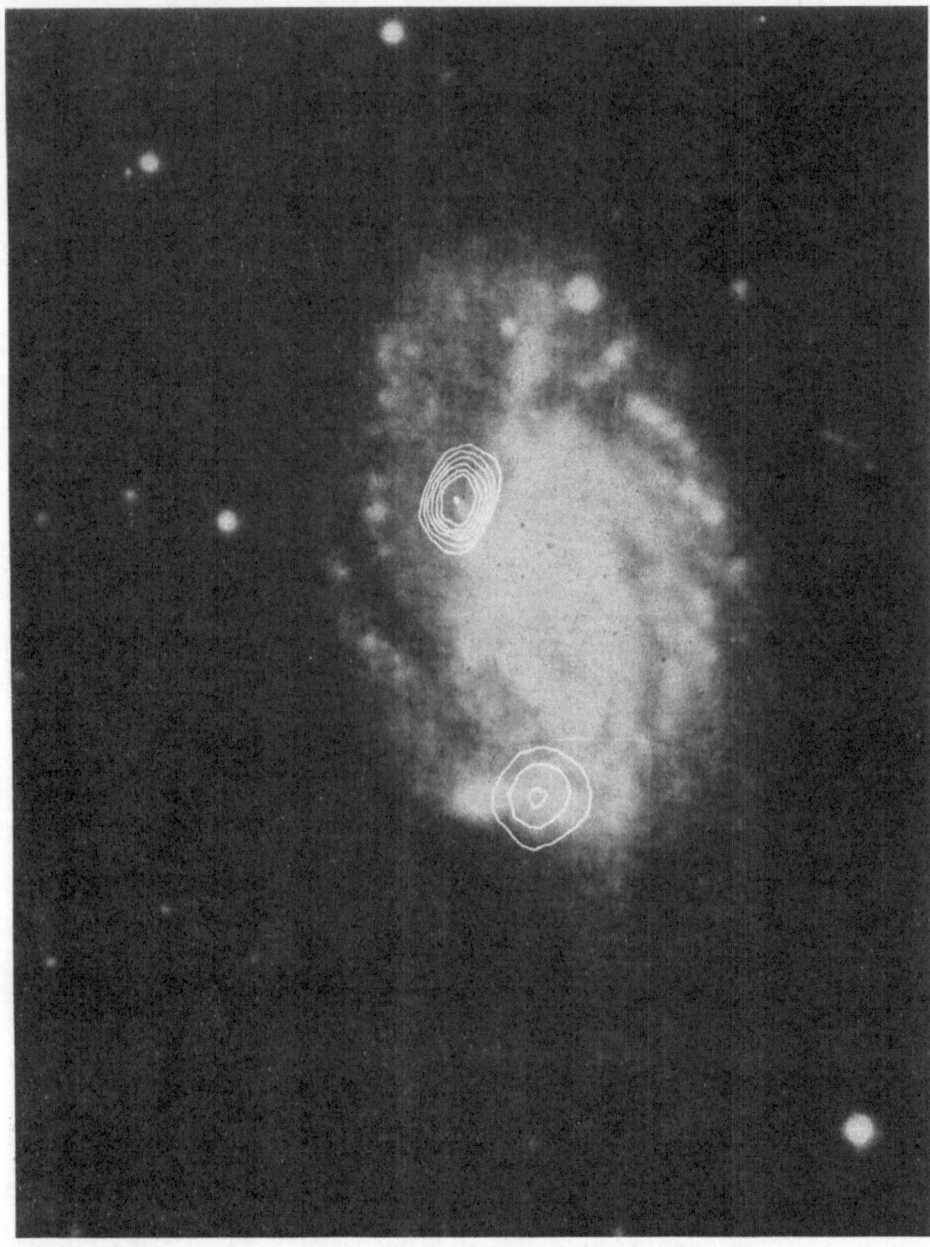

Fig. 9.   1.4 GHz continuum map of NGC 3726 based on the observations of van der Kruit (1971) and superimposed on a print from the Palomar Sky Survey by de Bruyn (private communication) using an improved determination of the position.

range of radio power for the radio galaxies is now known to extend down to values comparable to these spiral galaxies (Ekers, 1974), e.g. the giant elliptical NGC 4472 is an example of a double radio galaxy with a radio power at 1.4 GHz of only 10^{20} W Hz^{-1} sr^{-1} and a separation of 7 Kpc, quite comparable to NGC 4736.

3. Elliptical Galaxies

The radio properties of the elliptical galaxies differ from those of the spiral galaxies in just as marked a way as do the optical properties. The elliptical galaxies are often associated with powerful extended radio sources – the radio galaxies – a phenomena which is rare, if present at all, in the spiral galaxies. The compact radio sources in the nuclei of the elliptical galaxies are also quite different from those of most of the spirals. The nuclear sources in the elliptical galaxies have been found by two different procedures and since these have quite different selection effects I will discuss them separately.

3.1. Surveys of nearby elliptical galaxies

Heeschen (1968) found that two elliptical galaxies NGC 1052 and NGC 4278 contained small diameter sources with radio spectra typical of the spectra of the optically thick variable radio sources found especially in the QSO's. One of these, NGC 1052, was strong enough to be detected subsequently in a very-long-baseline interferometer experiment and was found to have a diameter $<0\rlap{.}''001$, corresponding to a linear diameter of <0.1 pc (Cohen *et al.*, 1971). Further surveys of elliptical galaxies (Heeschen, 1970a; Ekers and Ekers, 1973) yielded the additional sources given in Table III. Because of their greater average distance and the lower sensitivity of the surveys we cannot construct a sample of elliptical galaxies complete to the same luminosity level as used for the spirals. Thus we cannot easily make a direct comparison between the frequency of nuclear radio emission in spiral and elliptical galaxies but it is clear from the data in Table III, plotted in Figure 6, that many elliptical galaxies have more powerful radio nuclei than are found in any of the normal spirals. The lowest absolute monochromatic luminosity for which we have some elliptical galaxies left in the sample is $P_{1415} = 6 \times 10^{20}$ W Hz^{-1} sr^{-1}. At this luminosity Ekers and Ekers estimate a space density of 2.4×10^{-4} Mpc^{-3}, only a few times less than the density of spiral galaxies with this luminosity.

The nuclei of elliptical galaxies generally have complex spectra and show a low frequency cut-off (Heeschen, 1970b). The distribution of radio spectral index for these galaxies (Figure 10) is quite different from that of the spiral galaxies (Figure 4).

Most of the nuclear sources in Table III are known to be less than a few arc sec in extent but only one, Cen A, has probably been resolved. It was not detected in the VLB observation of Broderick *et al.* (1972), implying an angular size $>0\rlap{.}''004$ (0.14 pc), but this was near the sensitivity limit. Wade *et al.* (1971) obtained a limit of $<0\rlap{.}''5$ (<17 pc). Two other galaxies, NGC 1052 (Cohen *et al.*, 1971) and Virgo A (Cohen *et al.*, 1969), are still unresolved with very-long-baseline interferometers,

TABLE III
Radio emission from the nuclei of nearby elliptical galaxies

Name		Type	Dist. Mpc	Total		Nucleus				Ref.[e]
NGC	Other			M_p	S_{1415}[a]	S_{1415}[a]	$\log(P_{1415}$[b]$)$	$\alpha_{1415-5000}$		
1052		E3	25	−20.4	0.35	0.35	21.3	+0.6		1
–	VV6-8-34 4C 39.12	E0	110	−21	1.05	0.45	22.7	−0.5		2
1587		E1	69	−21.3	0.10	0.10	21.7	−0.3		1
2911		S0p	54	−20.2	0.11	0.11	21.5	+0.1		1
3078		E2	40	−21.2	0.25	0.15	21.4	−0.1		1, 2
3998		S0	22	−20.5	0.08	0.08	20.6	+0.3		1
4278	B1217+29	E1	18	−20.1	0.51	0.51	21.2	−0.2		1
4472		E1	20	−21.8	0.18	0.13	20.7	−0.7		2
4486	Vir A	E0	20	−22.0	214	0.85[e]	21.5			3
4552		E0	20	−20.6	0.14	0.14	20.7	+0.6		1
5077		E3	45	−20.8	0.19	0.19	21.6	+0.1		1
5128	Cen A	Ep	7	−22.1	1330	2.40[d]	21.0	+0.2		4

[a] 10^{-26} W m^{-2} Hz^{-1}.
[b] W Hz^{-1} sr^{-1}.
[c] Flux density at 13 cm.
[d] Flux density at 11 cm.
[e] 1. Ekers and Ekers (1973)   3. Cohen *et al.* (1969)
   2. Ekers, in preparation   4. Wade *et al.* (1971).

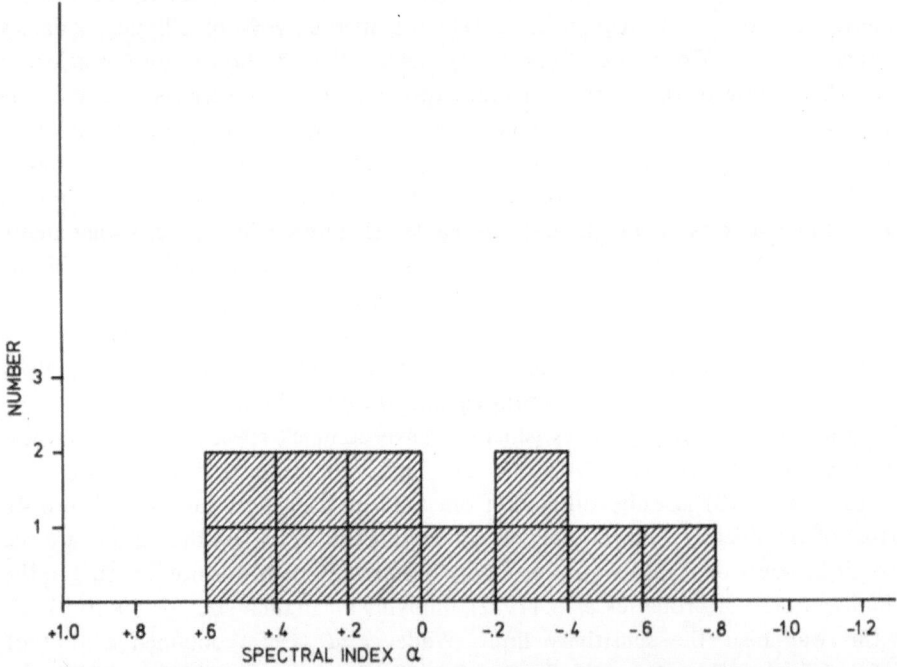

Fig. 10. Distribution of the radio spectral index, α, for the nuclei of elliptical galaxies. α is defined $S_{1.4}/S_{5.0} = (1.4 \text{ GHz}/5.0 \text{ GHz})^\alpha$ but note that in general these sources do not have power law spectra.

indicating components less than 0.2 pc in diameter. If the low frequency cut-off seen in the spectra of other nuclei is due to synchrotron self-absorption then those sources are also likely to have comparably small linear dimensions.

3.2. COMPACT SOURCES IN THE NUCLEI OF RADIO GALAXIES

Two of the galaxies in Table III, Virgo and Centaurus A, also have powerful extended radio components and are well-known members of the class of radio galaxies. Many of the other galaxies in Table III also have extended components, e.g. NGC 4472, and although these are much weaker than the well-known radio galaxies they are presumably the same class of object. This high correlation between the presence of compact and extended components was noted by Ekers (1972). High resolution, high frequency radio observations now show that the presence of a compact radio nucleus is indeed a very common phenomena in the radio galaxies. A list of radio galaxies

TABLE IV

Radio galaxies with nuclear radio emission

Name		Dist.	Total			Nucleus		Ref.[e]
Radio	NGC	[z]	Optical[a]	S_{5000}[b]	S_{5000}[b]	$\log(P_{5000}$[c]$)$	
P0036+03	193	0.0145	14.3 E2	0.57	0.05	21.5	3
3C66		0.0215	12.6 ED2	3.7	0.23	22.5	13
3C83.1 B	1265	0.0181[d]	14. ED3-4	3.5	0.015	21.1	1, 2
3C109		0.306	15.7 N	1.6	0.38	24.9	4
3C129			(17) E, Obsc	2.2	0.07		2, 5
3C264	3862	0.0206	12.7 DE1	2.0			6
3C274	4486	0.0037[d]	8.7 E2	72.1	(0.85)	21.6	7
Vir A							
3C277.3		0.0857	15.9 E2	1.2			6
Cen A	5128	0.0016	7.0 DE3	126	(2.7)	20.8	8
3C310		0.0543	15.2 DE2, 3	1.3	0.35	23.5	14
3C315		0.1086	16.8 DE2, 3	1.3			6
3C338	6166	0.0303	12.5 cD4	0.49	0.11	22.5	10
3C382		0.0586	14.7 D3	2.2	0.23	23.4	12
3C390.3		0.0569	15.4 N	4.5	0.35	23.5	11
3C402			14 D3	0.9	0.15		9
			15		0.15		
3C405		0.0570	15.1 cD3	371	1.1	24.0	6
Cyg A							
3C452		0.0820	16.0 ED	3.3	0.13	22.4	12
3C465	7720	0.0301	13.2 D4	2.8	0.28	22.3	12, 14

[a] Photometry mostly from Sandage (1973).
[b] 10^{-26} W m^{-2} Hz^{-1}.
[c] W Hz^{-1} sr^{-1} assuming $q = -1$, $\alpha = 0$.
[d] Mean z for cluster.
[e] 1. Miley et al., this symposium, p. 109
2. Riley (1973)
3. Ekers, in preparation
4. Branson et al. (1972)
5. Hill and Longair (1971)
6. Longair, private communication
7. Cohen et al. (1969)

8. Wade et al. (1971)
9. Miley, private communication
10. Jaffe, private communication
11. Harris (1972)
12. Riley and Branson (1973).
13. Northover (1973)
14. Miley and van der Laan (1973)

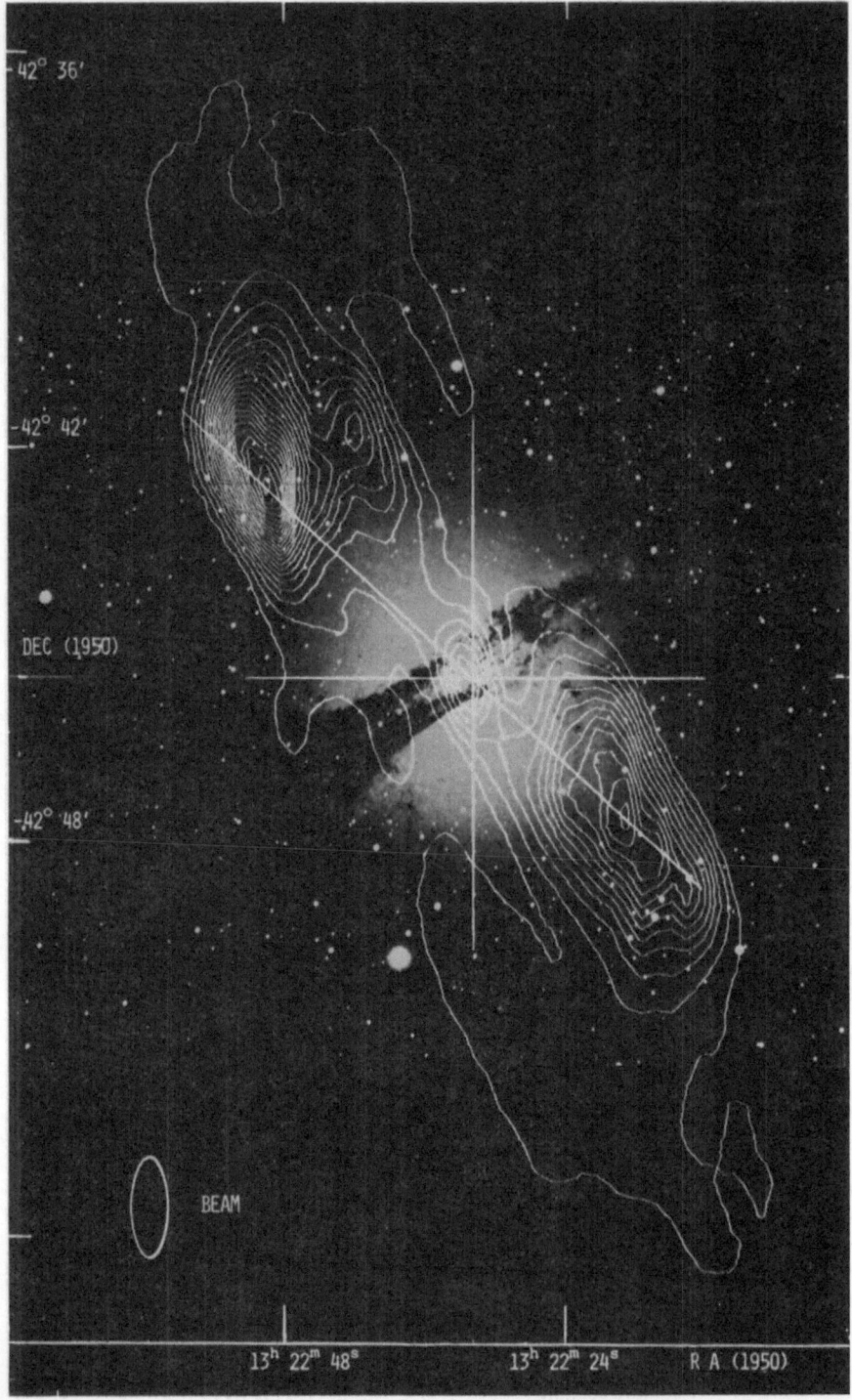

Fig. 11. 1.4 GHz continuum map of the central components of Cen A superimposed on an optical
photograph from the Hubble atlas. This map has been obtained with the Fleurs Synthesis Radiotelescope and has been kindly supplied by Dr R. Frater.

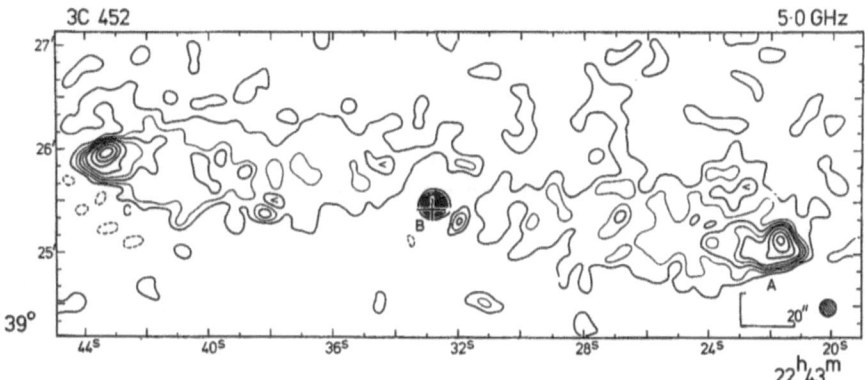

Fig. 12.   5 GHz map of 3C 452 obtained with the One-Mile Radiotelescope at Cambridge (Riley and Branson, 1973).

with nuclear components is given in Table IV and two examples are shown in Figures 11 and 12. These components have eluded detection until now, since their flat spectra and weakness relative to the extended sources have made them inconspicuous in previous lower frequency maps.

Most of the galaxies in Table IV are optically bright, indicating that these are relatively nearby objects and in most cases the compact nuclear sources are near the radio detection limit. This suggests that in the more distant radio galaxies there may be many nuclear sources which are below the present detection limit.

Compact nuclear sources have been detected in 15 out of 25 extended 3C sources identified with galaxies brighter than $m_p = 16$ which have been observed with either the Cambridge or Westerbork synthesis telescopes at 5 GHz. In comparison, out of a sample of 114 E galaxies brighter than $m_p = 16$ *without* extended radio sources, Ekers and Ekers (1973) found no nuclear components as strong as any of those in Table IV. The presence of compact nuclear sources in such a large fraction of the radio galaxies, and only in the radio galaxies, is of considerable interest since it implies that it is usual for the compact and extended components to be present concurrently. This could happen if (i) both types of radio emission are continuous for most of the lifetime of a unique type of elliptical galaxy, or (ii) both types of radio emission are transient but switch on and off together. Both these cases argue that the extended component is maintained by a continuous injection of energy – in (i) because the present energy in relativistic particles in the extended sources is insufficient to enable it to continue radiating for the life of the galaxy and in (ii) because the information about whether the nucleus is on or off has to be transmitted to the extended component and the only plausible way to do this is to switch on or off the energy supply.

4. Quasi-Stellar Objects

Although there is still some disagreement on whether QSO's should be included in a discussion of nuclei of galaxies, I will do so and also make the simplest consistent assumption about the nature of the redshifts, namely that they are cosmological.

The radio properties of the QSO's are similar to those of the elliptical galaxies in that they have both compact and extended components. The extended radio components associated with QSO's are indistinguishable from those associated with elliptical galaxies. The compact radio sources identified with QSO's have very similar radio spectra but are $\sim 10^5$ times brighter than the elliptical galaxy nuclei and 100 times brighter than the brightest Seyfert nuclei. This class of compact radio source has been studied intensively and observations of the radio spectra, intensity variations, angular structure and variations in structure have been reviewed by Kellermann (1972) and others. The observational situation regarding the spectral and intensity variations has not changed significantly since then, but detailed observations with very-long-baseline interferometers show increasingly complex structures involving two or more spatially separated components (Broderick *et al.*, 1972; Clark *et al.*, 1973), some of which exhibit temporal changes. Initially these changes were interpreted as components separating with apparent velocities in excess of the speed of light (Whitney *et al.*, 1971; Cohen *et al.*, 1971), whereas further analysis indicates that the changes may be equally well explained by components of variable intensity (Dent, 1971; Kellermann *et al.*, 1973). An apparent contraction has been observed for the N galaxy P1934–63 by Robertson *et al.* (1973).

5. Summary

Figure 13 shows the positions of the extragalactic radio sources, both compact and extended, on a linear size-luminosity plot. For clarity, many of the upper limits on radio galaxy diameter are not included in the diagram but the centroid of the linear size distribution for all radio galaxies is within the distribution of the galaxies shown. There may be a few objects straddling the gap between the extended and compact sources but a definite separation in the distribution of linear size will remain.

Spiral galaxies clearly produce a different kind of radio nuclei than those observed in elliptical galaxies. They generally have weaker, larger radio nuclei with much steeper spectra than do the elliptical galaxies. Some of the unresolved, flat spectrum, spiral nuclei, such as in M81, could be of the same type as the elliptical galaxy nuclei, but this would be a rare, perhaps intermittent, phenomena. It is also possible that subcomponents of the spiral nuclei will be similar to the elliptical galaxy nuclei in linear size, but they will be many orders of magnitude weaker.

Two of the elliptical galaxies, M87 and NGC 1052, contain the smallest extragalactic radio sources known. Unfortunately, the angular sizes of most of the elliptical galaxy nuclei are not known and they do not appear in Figure 13. If they are just as compact, they would form a continous distribution running all the way up in power to the two Seyfert galaxies 3C120 and NGC 1275. These two Seyfert galaxies have nuclei quite different from the rest of the Seyfert galaxies which lie much closer in radio properties to the ordinary spiral galaxies. It appears that there is a real gap in power between the compact QSO's and the strongest galaxy nuclei, but statistically

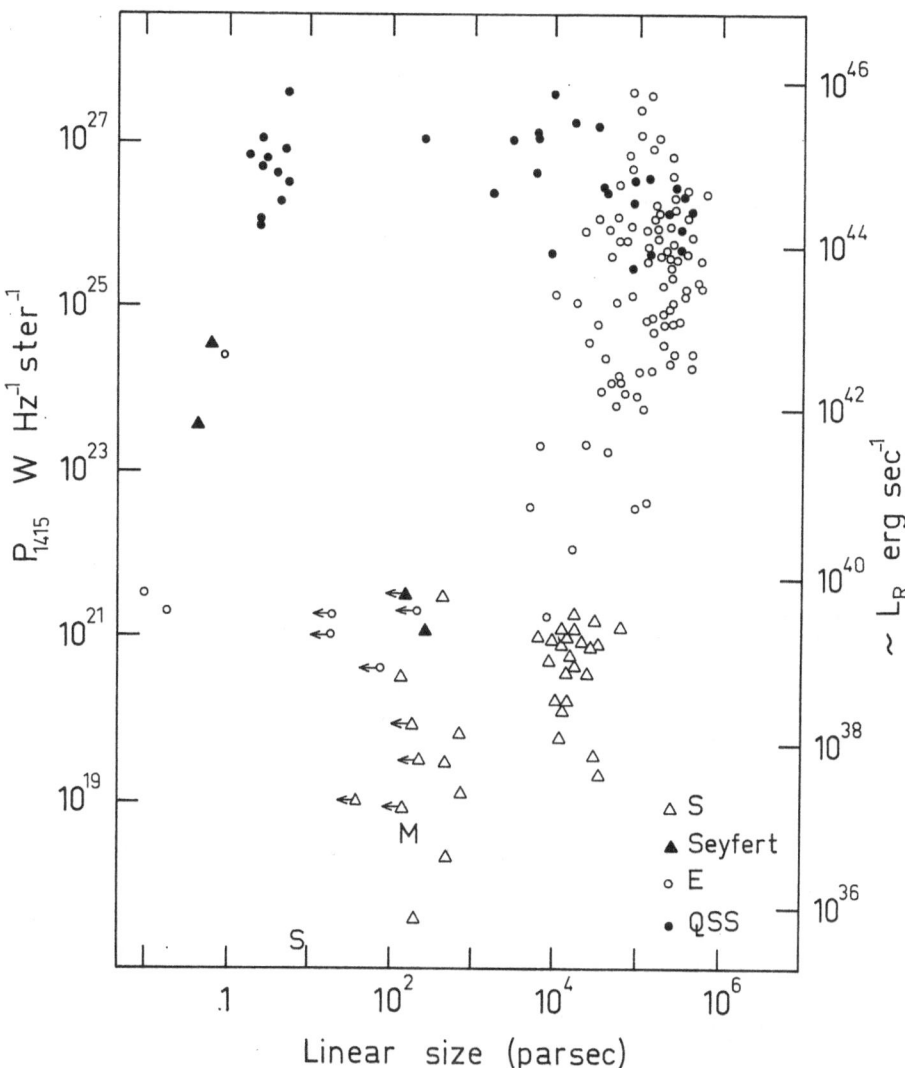

Fig. 13. Plot of the monochromatic radio power at 1.4 GHz against linear diameter for all known classes of extragalactic radio source. For clarity, objects with upper limits to the diameter are not plotted except for the weak radio nuclei. The radio luminosity scale, L_R, is defined in Figure 5. The extended source in our galactic centre is indicated by M and the Sgr A source by S.

complete observations with very-long-baseline interferometers would be necessary to check this further.

The high correlation between the presence of compact and extended sources in elliptical galaxies implies that the nucleus has a continuing involvement in the production of a radio galaxy, but perhaps the much more powerful optical and radio events called QSO's are a transient phase in the life of a galaxy.

References

Allen, R. J., Ekers, R. D., Burke, B. F., and Miley, G. K.: 1973, *Nature* **241**, 260.

Baldwin, J. E. and Pooley, G. G.: 1973, *Monthly Notices Roy. Astron. Soc.* **161**, 127.

Becklin, E. E., Fomalont, E. B., and Neugebauer, G.: 1973, *Astrophys. J. Letters* **181**, L27.

Branson, N. J. B. A., Elsmore, B., Pooley, G. G., and Ryle, M.: 1972, *Monthly Notices Roy. Astron. Soc.* **156**, 377.

Broderick, J. J., Kellermann, K. I., Shaffer, D. B., and Jauncey, D. L.: 1972, *Astrophys. J.* **172**, 299.

Clark, B. G., Kellermann, K. I., Cohen, M. H., Shaffer D. B., Jauncey, D. L., Matveyenko, L. I., and Moiseev, I. G.: 1973, *Astrophys. J. Letters* **182**, L57.

Cohen, M. H., Moffet, A. T., Shaffer, D. B., Clark, B. G., Kellermann, K. I., Jauncey, D. L., and Gulkis, S.: 1969, *Astrophys. J. Letters* **158**, L83.

Cohen, M. H., Cannon, W., Purcell, G. H., Shaffer, D. B., Broderick, J. J., Kellermann, K. I., and Jauncey, D. L.: 1971, *Astrophys. J.* **170**, 207.

Dent, W.: 1971, *Science* **175**, 1105.

De Young, D. S., Roberts, M. S., and Saslaw, W. C.: 1973, *Astrophys. J.* **185**, 809.

Downes, D. and Maxwell, A.: 1966, *Astrophys. J.* **146**, 653.

Ekers, R. D.: 1972, in D. S. Evans (ed.) 'External Galaxies and Quasi-Stellar Objects', *IAU Symp.* **44**, 222.

Ekers, R. D.: 1974, in preparation.

Ekers, R. D. and Ekers, J. A.: 1973, *Astron. Astrophys.* **24**, 247.

Gordon, M. A.: 1974, *IAU Symp.* **60**, in preparation.

Harris, A.: 1972, *Monthly Notices Roy. Astron. Soc.* **158**, 1.

Heeschen, D. S.: 1968, *Astrophys. J. Letters* **151**, L135.

Heeschen, D. S.: 1970a, *Astron. J.* **75**, 523.

Heeschen, D. S.: 1970b, *Astrophys. Letters* **6**, 49.

Hill, J. M. and Longair, M. S.: 1971, *Monthly Notices Roy. Astron. Soc.* **154**, 125.

Holmberg, E.: 1958, *Medd. Lund.* Ser II, No. 136.

Kellermann, K. I.: 1971, in D. J. K. O'Connell (ed.), *Nuclei of Galaxies*, North-Holland Publ. Co., Amsterdam, p. 217.

Kellermann, K. I.: 1972, in D. S. Evans (ed.), 'External Galaxies and Quasi-Stellar Objects', *IAU Symp.* **44**, 190.

Kellermann, K. I. and Pauliny-Toth, I. I. K.: 1968, *Ann. Rev. Astron. Astrophys.* **6**, 417.

Kellermann, K. I., Clark, B. G., Shaffer, D. B., Cohen, M. H., Jauncey, D. L., Broderick, J. J., and Niell, A. E.: 1973, *Astrophys. J. Letters* **183**, L51.

Klienmann, D. E. and Low, F. J.: 1970, *Astrophys. J. Letters* **161**, L203.

Kronberg, P. P., Pritchet, C. J., and van den Bergh, S.: 1972, *Astrophys. J. Letters* **173**, L47.

Legg, T. H., Broten, N. W., Fort, D. N., Yen, J. L., Bale, F. V., Barber, P. C., and Quigley, M. J. S.: 1973, *Nature* **244**, 18.

Le Squéren, A. M. and Crovisier, J.: 1974, *Astron. Astrophys.* **31**, 447.

Little, A. G.: 1974, *IAU Symp.* **60**, in preparation.

Macdonald, G. H., Kenderdine, S., and Neville, Ann C.: 1968, *Monthly Notices Roy. Astron. Soc.* **138**, 259.

Mathewson, D. S., van der Kruit, P. C., and Brouw, W. N.: 1972, *Astron. Astrophys.* **17**, 468.

Miley, G. K. and van der Laan, H.: 1973, *Astron. Astrophys.* **28**, 359.

Northover, K. J. E.: 1973, *Monthly Notices Roy. Astron. Soc.* **165**, 369.

Oort, J. H.: 1974, *IAU Symp.* **60**, in preparation.

Pooley, G. G.: 1969a, *Monthly Notices Roy. Astron. Soc.* **144**, 101.

Pooley, G. G.: 1969b, *Monthly Notices Roy. Astron. Soc.* **144**, 143.

Riley, J. M.: 1973, *Monthly Notices Roy. Astron. Soc.* **161**, 167.

Riley, J. M. and Branson, N. J. B. A.: *Monthly Notices Roy. Astron. Soc.* **164**, 271.

Robertson, D. S., Gubbay, J. S., and Legg. A. T.: 1973, *Proc. Astron. Soc. Australia* **2**, 184.

Sandage, A. R.: 1972, *Quart. J. Roy. Astron. Soc.* **13**, 282.

Sandage, A. R.: 1972, *Astrophys. J.* **178**, 25.

Sandage, A. R.: 1973, *Astrophys. J.* **183**, 731.

Spencer, J. H.: 1973, Ph.D. Thesis, MIT, Cambridge, Mass.

Spencer, J. H. and Burke, B. F.: 1972, *Astrophys. J. Letters* **176**, L101.

van der Kruit, P. C.: 1972, *Astron. Astrophys.* **15**, 110.
van der Kruit, P. C.: 1972, *Astrophys. Letters* **11**, 173.
van der Kruit, P. C.: 1973a, *Astron. Astrophys.* **29**, 231.
van der Kruit, P. C.: 1973b, *Astron. Astrophys.* **29**, 249.
van der Kruit, P. C.: 1973c, *Astron. Astrophys.* **29**, 263.
van der Kruit, P. C., Oort, J. H., and Mathewson, D. S.: 1972, *Astron. Astrophys.* **21**, 169.
Wade, C. M.: 1968, *Astron. J.* **73**, 876.
Wade, C. M., Hjellming, R. M., Kellermann, K. I., and Wardle, J. F. C.: 1971, *Astrophys. J. Letters* **170**, L11.
Whitney, A. R., Shapiro, I. I., Rogers, A. E. E., Robertson, D. S., Knight, C. A., Clark, T. A., Goldstein, R. M., Marandino, G. E., and Vandenberg, N. R.: 1971, *Science* **173**, 225.

DISCUSSION

Toomre: How many components are there in Kellermann's models of 3C 273 and 3C 279?

Kellermann: Three.

Bracewell: The source at the centre of NGC 5128 has been observed at 2.8 cm at Stanford with an 18 arc sec beam. The flux density of 1.5×10^{-26} W m^{-2} Hz^{-1}, taken in conjunction with the values at 3.8 and 21 cm, will help to indicate the spectrum. Do you think that it could represent a new pair of ejecta being cooked up?

Ekers: The energy content of the compact component is too small by many orders of magnitude for it to be able to expand adiabatically to the size of the double source.

Seielstad: You have pointed out that the radio properties of the quasars are continuous with those of the bright radio galaxies. But if you bring them 1000 times closer, so that their radio powers are 10^6 times less, then they are continuous with other types of source.

Ekers: They become continuous with the disks of spiral galaxies, which they look nothing like.

Seielstad: There are ellipticals like NGC 1052.

Ekers: Right, you can try to scale the compact radio components in quasars, but for the QSOs with double radio components I am less sure that a local interpretation would give a class of objects like any other known radio source.

Longair: I can comment on the statistics of nuclear radio components. We have detected central components in about 12 out of 50 sources. In the complex, nearby radio sources associated with Abell's clusters, Julia Riley informs me that all have now been found to contain compact components. For double sources, the statistics are less complete but at least 10% of all doubles have central components. This is a very conservative lower limit and it is quite conceivable that all double sources contain such components.

Miley: We find a compact nuclear component in 3C 310. This is a large relaxed double radio galaxy with a very steep radio spectrum. Since it is presumably old, the detection of a nuclear component reinforces your point that continuous injection may be occurring.

Ekers: All 17 of these are identified with galaxies of magnitude < 16.2. Maybe the fact that they are in clusters is relevant, but the main point is that they are all close, low-luminosity objects and we have detected practically a complete sample out to a certain distance.

OPTICAL OBSERVATIONS OF NUCLEI OF GALAXIES

MARIE-HELENE ULRICH

Dept. of Astronomy, University of Texas at Austin, Austin, Tex. 78712, U.S.A.
and
Observatoire de Paris, 92190 Meudon, France

Abstract. Several problems concerning the nuclei of galaxies are particularly interesting at present, either because new data have become available from advances in observational techniques, or else because recent observations suggest that earlier conclusions should be revised. These problems are discussed in this article.

Spectrophotometry in the near infrared shows that the late-type stars in the nucleus of M31 are mostly late giants; the M/L ratio in the nucleus is approximately 15, a value in agreement with that calculated independently from the velocity dispersion of the stars.

Infrared emission at 10μ has now been detected in 50 galaxies; the Seyfert-type galaxies are the brightest at $10\ \mu$ and there is evidence of variations of the $10\ \mu$ flux in NGC 4151, which rules out re-radiation by dust as the mechanism of infrared emission. In each of the galaxies NGC 253 and NGC 3034 an infrared source of dimension ~ 200 pc is present in the centre which coincides in position and dimension with the central component of the radio source at 1415 MHz; since both galaxies are rich in dust and in early-type stars it is plausible that, in these cases, re-radiation by dust is the origin of the infrared emission. Most of the other galaxies detected at $10\ \mu$ are spirals with intense and narrow emission lines.

Comparison of the optical and radio properties of nuclei of galaxies show that galaxies with compact nuclear radio sources are more likely to have optical spectra with emission lines than are galaxies without central radio sources. This correlation holds for elliptical and spiral galaxies. Its cause is not understood at present.

The relatively accurate profiles now available for the permitted lines of several Seyfert galaxies show that mass motions are the main broadening agent for these lines. Variations of line intensities have been observed in several Seyfert galaxies. The most likely explanation of these variations is that they are due to the variations of the photoionizing flux; in some cases, however, weakening of the permitted lines could be caused by high-velocity mass motions present in the region emitting the broad component of these lines.

From the investigation of the relatively nearby galaxies it is estimated that at least 5 % of galaxies have expansion motions or non-thermal optical fluxes.

1. Introduction

The meaning of the term 'galactic nuclei' is evident, at least in a general descriptive sense, yet attempts at defining precisely the nuclei of galaxies immediately meet with difficulties. In practice the nucleus of a galaxy can be defined as the maximum of the surface brightness of the stellar continuum. The nucleus is usually close to or coincident with the centre of symmetry of spiral and elliptical galaxies, the centre of the rotation curve, and the maximum of star density.

At present, the existence of a central object whose limits can be drawn from the surface brightness distribution is not established. The photographs taken by Stratoscope II give a conservative upper limit of 0″.08 or 4 pc for the half-power width of the nucleus of NGC 4151 (Schwarzschild, 1971) and show the presence, at the centre of M31, of an elliptical structure with a semi-major axis slightly less than 1″ or 3.3 pc at half-intensity (Danielson *et al.*, 1971). It is not known whether this is a common feature in galactic nuclei.

John R. Shakeshaft (ed.), The Formation and Dynamics of Galaxies. 279–304. All Rights Reserved.
Copyright © 1974 by the IAU.

Phenomenologically, nuclei of galaxies have an extremely interesting property: in those galaxies where large amounts of energy are released in the form of relativistic particles, thermal and non-thermal radiation and mass motions, these violent events are occurring in or have originated in the nuclei. This fact has an obvious and important consequence which is that the onset of these violent events is linked – in a way which is still unknown – to some of the properties of the nuclei such as, for example, the high stellar density or the strong gravitational field.

Two types of nuclei of galaxies can be clearly distinguished:

(a) quiescent or passive nuclei, in which the optical radiation is emitted by normal stars and a small quantity of interstellar gas.

Much of our knowledge on normal nuclei is based on observations of the nearby spiral galaxies, especially M31, whose nucleus is the most extensively studied at all wavelengths and with the best angular resolution.

(b) active nuclei, for which some of the optical flux is of non-thermal origin and/or is emitted by large clouds of ionized gas. There are extreme cases of active nuclei where the stellar population of the nucleus is undetected, most of the optical flux being of non-thermal origin; the other signs of extreme activity in these nuclei are variations or polarization of the optical flux, infrared or high-frequency radio emission, and mass motions with velocities larger than, say, $1000 \ km \ s^{-1}$. However, it is important to bear in mind that there exists a number of cases where the level of activity of the nucleus is intermediate between that of the quiescent nuclei and that of extremely active nuclei. These intermediate cases display some of the properties of extremely active nuclei, but the total energy involved is smaller. Evidently, low or medium-level activity in the nucleus of a galaxy is difficult to detect, so there is observational selection against mildly active nuclei as compared with extremely active nuclei.

In this paper we discuss problems which are particularly interesting at present, because new data have just become available and are accumulating rapidly due to the advances of observational techniques, or because recent observations yield compelling evidence that earlier conclusions be revised.

The outline of the article is as follows: stellar populations and abundances are discussed in Section 2. Section 3 deals with gas and star motions and Section 4 with dust, interstellar gas and infrared emission. In Section 5, we discuss the observations of profiles and variations of absolute intensity of the permitted lines, and in Section 6 estimate the fraction of galactic nuclei which show signs of activity in the optical range.

2. Stellar Populations and Abundances

Most of our knowledge about stellar populations and abundances in nuclei of galaxies comes from narrow-band photometry. As observational techniques develop, the wavelength range which can be observed widens toward the infrared and the ultra-violet, providing new information on the populations of very cool and very hot stars.

To understand the recent developments in determination of the stellar populations and abundances in galaxies it is useful to recall the three steps from the narrow-band photometry observations to the synthetic model of a stellar population:

(i) choice of a set of narrow bands centred on lines whose equivalent widths are particularly sensitive to stellar types, luminosities, and abundances.

(ii) choice of the types of stars to be included in the population synthesis; the colour indices of the various synthetic models are then compared to the observed colour indices.

(iii) the third step consists of choosing the criteria according to which some of the models will be rejected and others found acceptable. To be astrophysically acceptable, a model has to have at least the two following properties: firstly, a smooth distribution of stars along the main sequence and, secondly, the proportion of any given type of stars must be consistent with the proportions of the two types immediately following and preceding it in stellar evolution.

2.1. THE NUCLEUS OF M31 AT OPTICAL AND NEAR INFRARED WAVELENGTHS

The nucleus which has been most studied is that of M31, and the extensive observations by Spinrad and Taylor (1971) together with other studies by Wood (1966), McClure and van den Bergh (1968), and McClure (1969) have led to the following results relating to a region 10″ or 33 pc in diameter: (i) the stars are very strong-lined, implying that the abundances of some elements are higher than normal and (ii) there is a strongly dwarf-enriched lower main sequence, indicating that $M/L \approx 44$.

We shall see that recent work confirms the overabundance of some elements but that there is now considerable evidence that the ratio M/L is smaller than 44 and probably close to 15 (However, see the remark on the overabundance in 2.1.1.).

For fitting the computed model indices to those observed, Spinrad and Taylor (1971) employed a trial-and-error method; with this method it is almost unavoidable that some models be overlooked, and Faber (1972) has therefore set up another method based on quadratic programming. The criterion for best fit is that the quantity $\Sigma \sigma^2$, which is the sum of the squares of the residuals, be a minimum. Faber applied this method to the data of Spinrad and Taylor with very little alteration. She found that the model obtained by Spinrad and Taylor was not the one giving the minimum value of $\Sigma \sigma^2$, and that a model including a smaller proportion of late M dwarfs and having $M/L = 15$ gave a smaller value of $\Sigma \sigma^2$ than theirs. The exclusion of super-metal-rich stars from the ingredients of the synthetic model was found to increase $\Sigma \sigma^2$ significantly, confirming the overabundance of some elements in the nucleus of M31.

Two recent observations in the near infrared also indicate that the ratio M dwarfs/M giants must be smaller than was found by Spinrad and Taylor. Whitford (1972 and also this volume p. 169) has observed the nucleus of M31 at the wavelength of the unidentified Wing-Ford band at 9910 Å. This band is a useful discriminant of stellar luminosity for types cooler than late K, being much stronger in M dwarfs than in M giants. From the observed weakness of this band in the nucleus of M31, Whitford concludes that there is no significant contribution from M dwarfs at that wavelength.

The second observation in the infrared is reported by Baldwin *et al.* (1973) who have observed the absorption band of CO at $2.3\,\mu$. The variations of the intensities of this band with stellar types and luminosities have been studied for a number of stars from G0 to M8. The CO band is very sensitive to the luminosity for stars cooler than late K, being much stronger in giants than in dwarfs. The value of the CO index measured in the nucleus of M31 (and also of M81) is not consistent with models rich in late-type dwarf stars and this result agrees well with the conclusion drawn by Whitford from the weakness of the Wing-Ford band.

In summary: (i) the re-analysis by Faber of the data obtained by Spinrad and Taylor, (ii) the weakness of the Wing-Ford band as observed by Whitford and (iii) the strength of the CO band at $2.3\,\mu$, all suggest smaller values of the ratio M dwarf/M giants and of the M/L ratio than in Spinrad and Taylor's best model. The value $M/L = 15$ fits well the observed CO indices in the nuclei of M31 and M81 and agrees with Whitford's observations of M31.

This value of M/L refers to the luminous material only and therefore represents a lower limit to the true value of M/L. In Section 3 we see that a recent determination of the velocity dispersion in the nucleus of M31 also leads to a value of M/L close to 15.

Up to now the CO band has been observed in the nuclei of M31, M81 and NGC 5195 only. If strong CO bands – implying a large proportion of M giants – are also observed in giant elliptical galaxies, the cosmological implications would be important because the change with time of the luminosity of a giant-dominated galaxy population is faster than for a population rich in dwarfs; the true value of q_0 obtained from the magnitude-redshift relation after taking into account a fast rate of luminosity evolution would be smaller than the value which is obtained from the data available now on the stellar content of giant elliptical galaxies (Tinsley, 1973).

2.1.1. *Cautioning Remark on the Overabundance of the Elements*

It has been mentioned above that in the nucleus of M31 the abundance of some elements is higher than normal. When synthesizing populations of galaxies, the abundances of the elements are determined by calculating the number of super-metal-rich stars which must be introduced in order that the indices of the synthetic population match the observed indices. The analyses by Spinrad and Taylor (1971) and by Faber (1972) show that in the nucleus of M31 the number of super-metal-rich K giants is about 1/50 the number of K dwarfs of normal abundances. The expression super-metal-rich stars refers to the specific stars (e.g. μ Leo, ϕ Aur, etc. ....) which are chosen to be included in the mixture. It must be noted that the super-metal-rich stars constitute the ingredient of the stellar mixture which is known with the least accuracy. In general, in these stars, the elements are overabundant by a factor 4 or less than 4. These stars do not form an homogeneous class; the overabundance is not the same for all these stars and in a given star it is not the same for all the elements; the exact amounts by which the various elements are overabundant are still a matter of controversy (Strom *et al.*, 1971; Blanc-Vaziaga *et al.*, 1973) and *it is not clear how much of the apparent overabundance in these stars is due to atmosphere peculiarities.* Sounder evidence for overabundance in nuclei of galaxies is provided by the analysis of emission lines from the interstellar gas.

2.2. The population of hot stars – OAO results

The effect of a small number of hot stars on the energy distribution at $\lambda > 3300$ Å has been investigated by Peimbert (1968), Spinrad and Taylor (1971), Tinsley and Spinrad (1971) and Faber (1972). All these authors agree that the presence of a small number of O and B stars (~ 40 in Spinrad and Taylor's study) is compatible with the narrow-band photometry observations just longward of 3300 Å.

Results on 35 galaxies observed by OAO 2 have been presented by Code *et al.* (1972). The energy distribution shows a minimum near 2400 Å followed by a steep upturn shortward. It does not seem possible to fit the observed curve by any combination of early-type stars since the upturn shortward of 2400 Å is steeper than that produced by the hottest stars. This minimum occurs near the maximum of the galactic interstellar extinction curves.

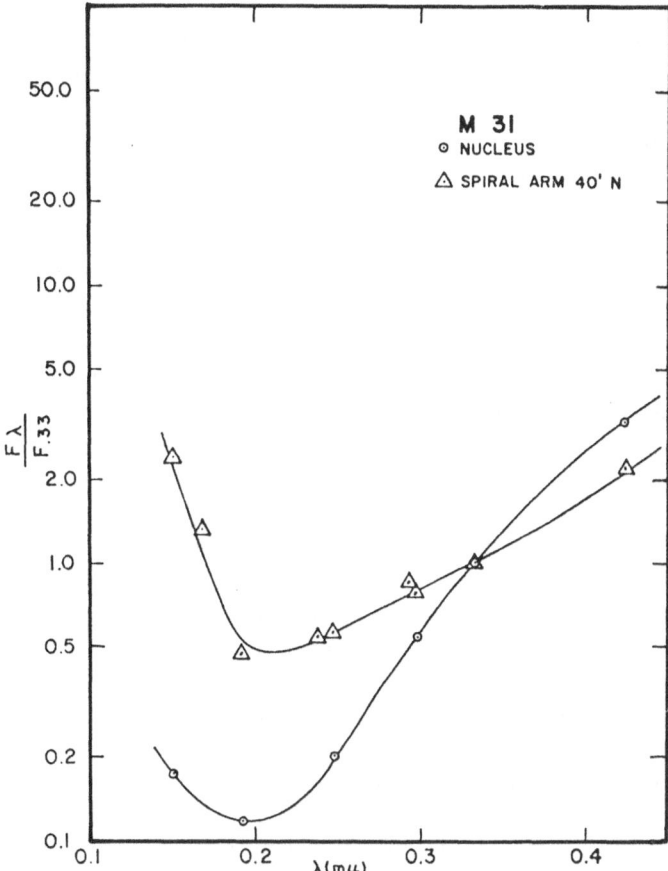

Fig. 1.   Energy distribution between 1500 and 4500 Å, observed through a 10′ aperture at 2 locations in M31. From Code *et al.* (1972)

Figure 1 shows the energy distribution of M31 observed through a 10' aperture (equivalent to 1800 pc in the plane of the galaxy) at two different locations: one centred on the nucleus of M31, the other centred on a bright super-association 40' south of the nucleus. It is suggested by Code *et al.* (1972) that the peculiar shape of the energy distribution between 1500 Å and 3500 Å is due to the presence of early-type stars whose energy distribution is modified by the interstellar dust present in the galaxies. In view of the similarity between the two extinction curves in Figure 1, the dust particles in the central region and in the spiral arms have at least one property in common, a minimum albedo at 2200 Å.

2.3. Variations of stellar content with distance from centre in M31

The variations of the strengths of several absorption lines with increasing distance from the centre of M31 show that the abundances and probably also the stellar population change from the centre outward. The changes of line strengths and colours occurring within 120" along the major axis and with a resolution of 10 Å can be summarized as follows:

– the CN band strength drops sharply with radial distance; the strengths of the Na D-lines drop off slowly, while the Mg I triplet at 5180 Å shows essentially no change (McClure, 1969; Spinrad *et al.*, 1971).

– the $U - B$ index decreases from 0.79 at the centre to 0.60 at 60" away then remains constant across the disk. The other colours $B - V$, $V - R$, and $V - K$ show no change with distance from the centre (Sandage *et al.*, 1969).

The sharp fall of the CN band just outside the nucleus, combined with the fact that the Mg I triplet remains constant, implies that the abundances decrease rapidly and also that the ratio K dwarfs/K giants increases outward. The changes in abundance and luminosity must be such that the weakening of the Mg I triplet due to the decrease of abundance is balanced by the increase of the triplet due to the strong negative luminosity effect (the triplet being much stronger in dwarfs than in giants). The relatively slow weakening of the Na I D-lines is consistent with the fact that these lines are more sensitive to abundance but have a smaller negative luminosity effect than the Mg I triplet.

A complete set of narrow-band photometric data has been obtained at 1' south of the nucleus of M31 (Spinrad *et al.*, 1971) and analyzed following the procedure used by Spinrad and Taylor (1971) for the nucleus itself. The models with only stars having normal abundances work best at 1' from the centre of M31, in contrast to the strong-lined stars required at the nucleus. The best model for the disk has $M/L \approx 45$ as Spinrad and Taylor (1972) had found in the nucleus, but this value may need to be revised when the intensities of the Wing-Ford band and of the CO band at 2.3μ are available for this region of the disk.

2.4. Results from narrow-band photometry in other nearby galaxies

The variations of the CN index with distance from the centre along the major and minor axes of several bright nearby elliptical, S0, Sa and Sb galaxies have been investigated by Spinrad *et al.* (1972) with a 5″ aperture. They found that the CN index either decreases or stays constant with increasing distance from the centre; they also found that the CN gradient is correlated with the degree of flattening of the central regions of the galaxies, the steepest CN gradients being found in galaxies which have the steepest gradients of surface brightness. Flattened systems show a lower CN gradient than do round elliptical galaxies. In NGC 3115, the CN index appears to be constant along the major axis, whereas along the minor axis it decreases rapidly as in elliptical galaxies.

These observations support the view of Freeman (p. 129) who considers that a galaxy is made up of two components which have had different kinematic and chemical histories – a disk of uniform chemical abundance in centrifugal equilibrium and a bulge which is dynamically relaxed and chemically inhomogeneous.

In the nuclei of the dwarf elliptical galaxies M32, NGC 205 and NGC 5195, the hydrogen lines are particularly strong and the fitting of a synthetic population does not require any super-metal-rich stars (Spinrad, 1972).

Finally it must be remembered that the contribution from a quiescent or mildly active nucleus to the integrated light of a galaxy is negligible. For this reason, the fact that in elliptical galaxies the integrated colours and line strengths are closely correlated with absolute magnitudes at all luminosities (Faber, 1973) reflects differences in large fractions of the total populations rather than differences in the populations of the nuclei alone.

2.5. Stellar populations in nuclei of Seyfert galaxies and active galaxies

The stellar populations in nuclei of Seyfert galaxies cannot be analyzed by narrow-band photometry because of the contamination by the non-stellar continuum and by the emission lines. However, a number of lines or bands of stellar origin have been identified in the spectra of the nuclear regions of several Seyfert galaxies, as listed in Table I.

Alloin *et al.* (1971) have investigated the stellar populations in the nuclei of some Seyfert and other active galaxies in an attempt to determine the reddening and/or the contribution from non-stellar continuum: they based model stellar populations upon the observed equivalent widths of about 10 absorption lines measured on spectrograms. For a given galactic nucleus, a unique model cannot be determined, but the various models which fit equally well the observed equivalent widths give rise to the same stellar continuum. This calculated continuum is compared with the observed continuum and the difference is interpreted in terms of reddening and/or non-stellar continuum, i.e. thermal emission from gas or non-thermal radiation. This method applied to NGC 1068 (Andrillat and Souffrin, 1971) shows that the reddening determined from an analysis of the stellar continuum in the nuclear region 7″ in diameter is definitely smaller than the reddening measured by Wampler (1971) in the emission line region, using an aperture of 10″. In the nucleus of NGC 3031, Alloin *et al.* (1971) find an ultra-violet non-stellar source which can be readily accounted for by thermal emission. In NGC 2903, the observed continuum near

TABLE I

Absorption lines and bands of stellar origin observed in the nuclear regions of
some Seyfert galaxies

NGC 1068[a]	H + FeI (3758–3787), FeI + MgI + H (3802–3846), CaII (3933), CH (4300), FeI (4326), MgI (5167–5184), NaI (5890–5896)
NGC 3516[a]	H + FeI (3735–3755), FeI + MgI + H (3802–3846), FeI + H (3874–3895), CaII (3933–3966), Hδ, CaI (4227), CH (4300), NaI (5890–5896).
NGC 4051[a]	H + FeI (3735–3755), H + FeI (3758–3787), H$_{10}$, FeI + MgI + H (3802–3846), FeI + H (3874–3895), CaII (3933–3968), CaI (4227), FeI (4272), CH (4300), FeI (4326), MgI (5167–5173)
NGC 4151[b]	CN (3835), CaII (3933), FeI (4045), CH (4300)
NGC 6814[c]	FeI (3735–3758), FeI + MgI (3812–3846), CaII (3933–3968), MgI (5167–5184), NaI (5890–5896)

[a] Andrillat and Souffrin (1971).
[b] Cromwell and Weymann (1970).
[c] Ulrich (1971).

4000 Å is found to be larger than the sum of the radiation from the F and G stars whose features dominate the spectrum at this wavelength and the thermal brems-strahlung calculated from the emission line intensities; it is concluded that up to ~30% of the light at 4000 Å may be of non-thermal origin (Andrillat *et al.*, 1972).

3. Stars and Gas Motions

The most important quantity provided by observations of stars and gas motions in nuclei of galaxies is the central mass, obtained from the velocity dispersion of the stars or from the rotation curve measured from emission or absorption lines in a small region surrounding the nucleus. However, there are many cases where the emission lines cannot be used for mass determination: in the spectra of elliptical galaxies the lines are usually very weak or absent, and in some active nuclei the ionized gas has expansion motions and therefore the emission lines do not provide a reliable value of the central mass. In these cases, observations of absorption lines are very desirable, but there are further problems: firstly the lines are difficult to observe outside the nucleus, and secondly the interpretation of the observed rotation-curve is not straightforward because the stellar population in elliptical galaxies and in the central regions of spiral galaxies has a three-dimensional distribution. At a given point on the spectrum the observed velocity results from the combined effect of all the stars along the line of sight; these stars are not all at the same distance from the centre so the features of the rotation curve are blurred. In principle, the determination of the true rotation curve in the equatorial plane from the observed rotation curve requires a knowledge of the spatial distribution of the stars.

A most important parameter which is needed to interpret the observed velocities is the orientation in space of the equatorial plane of the galaxy; this is usually determined from the position angle of the ellipse which is the projection on the sky of some morphological feature of the galaxy, e.g. isophotes, 'rings' of H II regions, etc. It is assumed that such features are circular and lie in the equatorial plane of the galaxy, an assumption which is not always valid since different features of the same galaxy sometimes describe ellipses with major axes in different position angles. M31 and NGC 4151 (see below) are examples of galaxies where such ambiguity appears.

As far as non-circular motions are concerned, the observational situation has not changed since the review by Burbidge (1970), except in the case of M31 (see below) and the Seyfert-type galaxies (see Section 5). I summarize it as follows: non-circular motions are known to exist in a dozen of mildly active nuclei; in most cases the data are consistent with expansion motions but are insufficient to enable one to build an accurate model of the geometry and of the velocity field; the mass and physical conditions of the gas involved in the expansion motions are very poorly known.

The major results of recent observations of the motions in the central region of M31 are described below because they illustrate the variety of motions which can be found in the central region of a normal galaxy.

3.1. VELOCITY FIELD IN THE CENTRAL REGION OF M31

M31 is the galaxy for which the motions of the stars and gas have been the most extensively studied. The other observations in the X-ray, optical, infrared and radio ranges of wavelength are consistent with M31 being a normal massive spiral galaxy.

In the region 15″ in radius surrounding the nucleus, the stellar motions have been observed by Lallemand *et al.* (1960) and by Morton and Thuan (1973). The first observations were made in P.A. 37°.7 along the major axis of the whole galaxy, and with very good atmospheric seeing. The absorption lines show that the rotation curve reaches a maximum of 80 km s^{-1} at about 2″ or 6 pc from the centre; beyond 2″.2, the rotation velocity decreases and reaches a minimum of 40 km s^{-1} at 6″ or 18 pc from the nucleus. The rotation curve obtained from the recent observations (Morton and Thuan, 1973) in the EW direction and with a 4″ seeing disk displays qualitatively the same features as those found by Lallemand *et al.*, viz. a maximum of velocity near 3″ from the centre, then a decrease to almost 0 km s^{-1} near 4″. The maximum of velocity is 44 km s^{-1} on the west of the nucleus and 62 km s^{-1} on the east of the nucleus, significantly smaller than the value obtained by Lallemand *et al.* Morton and Thuan note that, according to Johnson (1961), the nucleus within a 3″.3 × 2″.4 isophote has its major axis at P.A. = 52° ± 2°. If this value of the position angle of the major axis is relevant for the small region discussed here, then Lallemand *et al.*'s observations were made 14° away from the major axis of the nuclear region, and Morton and Thuan's observations were made 38° away from the major axis and on the opposite side of the major axis; in these conditions the two sets of observations roughly agree with purely circular motions whereas, if the major axis lies in P.A. 37°.7, the large value of the maximum of the velocity found in the EW direction should imply a significant radial expansion of the stars. The velocity dispersion observed by Morton and Thuan is 120 ± 30 km s^{-1}. The M/L ratio in the region 9 pc in radius, calculated from this value of the velocity dispersion, is 13 with an uncertainty by a factor a little less than 2. This agrees well with the value of 15 discussed in Section 2.

An extensive study of the motions of the interstellar gas within 600 pc of the centre has been made by Rubin and Ford (1971). The major conclusions are as follows: (i) the gas is gathered in a very thin rotating disk about 25 pc thick and has a maximum rotational velocity of 200 km s^{-1} at 400 pc from the centre; (ii) expansion motions with velocities up to 100 km s^{-1} occur in a fairly symmetric pattern in P.A. 68°–128° and P.A. 248°–278°; (iii) the emission lines are also seen in front of the absorbing dust lanes present in the NE quadrant, showing that there is gas outside the galactic plane. In front of the dust, the lines emitted by ionized gas outside the plane are broader and also show velocities up to 60 km s^{-1} in excess of the rotation velocity; this is tentatively interpreted as gas falling towards the plane.

The same spectrographic material was also used to investigate the stellar motions (Rubin *et al.*, 1973). The maximum rotation velocity observed along the major axis is 140 km s^{-1}; this is shown by Rubin *et al.* to be consistent with the bulk of the stellar population having the same rotation motions as the thin layer of gas. Because the galaxy is viewed only 13° from edge-on and also because the distribution of the stars is thicker than that of the gas, stars with distances from the centre differing by several hundred parsecs contribute to the absorption lines measured; this integration along the line of sight has the effect of lowering the value of the observed maximum of the rotation curve. Along the far minor axis, the stars appear to have the same expanding motion as is observed in the gas and it is possible (Rubin *et al.*, 1973) that the motion in the area may be linked with the branching of the spiral pattern.

In summary, the following features are observed within 500 pc of the centre of M31:

(i) an elliptical structure with a semi-major axis slightly less than 1″ or 3.3 pc at half-intensity (Danielson *et al.*, 1971);

(ii) a first maximum of the rotation curve at 2″2 or 7 pc, measured from the absorption lines;

(iii) a second maximum of the rotation curve at 400 pc, measured from the emission lines, the radial velocities of the absorption lines being consistent with this feature;

(iv) expansion motions of the ionized gas in two opposite sectors in the equatorial plane, motions which extend over 400 pc and have velocities ~100 km s^{-1}; and

(v) gas outside the equatorial plane which is perhaps falling back to the plane.

3.2. Velocity dispersions in nuclei of galaxies

Determination of the mass of the central region or of the total mass of spheroidal galaxies is possible through the observations of the stellar velocity dispersion σ, provided several assumptions are made (Poveda, 1958). As observational techniques develop, it is possible to observe σ with an increasingly high accuracy. Recently, new determinations of σ have been published for 5 elliptical or S0 galaxies (Morton and Chevalier, 1973), for M32 (Richstone and Sargent, 1972), and for the central region of M31 (Morton and Thuan, 1973). The results are listed in Table II. The values of σ from the recent observations are smaller than had been found earlier and consequently lead to lower values of M/L. The value of σ in Table II were obtained by fitting the spectra observed in the range 4000–4500 Å with artificially broadened spectra of G8III or K0III stars, which contribute most of the light in that range. However, a large fraction of the mass could be contained in fainter low-mass stars which may not have the same σ and this is the source of the main uncertainty in the calculation of the masses of nuclei. Only in the case of M31 has the new value of σ been used to calculate the mass of the nucleus: within 9 pc from the centre the mass is $(1.8\pm0.8)\times10^8\ M_\odot$ (Morton and Thuan, 1973).

TABLE II

Recent measurements of the velocity dispersions in
7 galactic nuclei

NGC	Galaxy type	Velocity dispersion (km s^{-1})	M_{pg}
1889	E0	110 ± 20[a]	-18.0
3115	E7/S0	215 ± 20[a]	-18.7
4473	E5	160 ± 20[a]	-20.6
4494	E1	160 ± 20[a]	-19.9
7332	S0	160 ± 20[a]	-19.6
221 (M32)	E2	60 ± 8[b]	-15.5
224 (M31)	Central region 8 pc in radius	120 ± 30[c]	-12.1

[a] Morton and Chevalier (1973).
[b] Richstone and Sargent (1972).
[c] Morton and Thuan (1973).

Notwithstanding the difficulties of this problem, it is desirable to measure σ in a larger number of nuclei, in particular in galaxies which are radio sources, and then to compare the values of σ in active and passive galaxies.

4. Gas, Dust and Infrared Emission

Data have recently become available in two areas which are extremely important for understanding the energy production in nuclei of galaxies: (1) correlations between the presence of a small radio source in the nucleus and the presence of emission lines in the optical spectrum, (2) measurement of infrared fluxes from nuclei. These are examined below.

4.1. PRESENCE OF EMISSION LINES IN THE OPTICAL SPECTRA OF NUCLEI WHICH HAVE A SMALL RADIO SOURCE; PHYSICAL CONDITIONS OF THE IONIZED GAS

Comparison of optical and radio properties of galactic nuclei shows that there exists a correlation between the presence of a small radio source in the nucleus and the presence of emission lines in the optical spectrum of the nucleus. This correlation holds both for elliptical galaxies (Disney and Cromwell, 1971; Ekers and Ekers, 1973) and for spiral galaxies (Alloin, 1973).

Tables IIIa and IIIb summarize the data relative to the proportions of nuclei with radio emission and optical line emission for elliptical and spiral galaxies respectively. The galaxies used for these statistics are within 100 Mpc of the Earth (taking $H = 100$ km s^{-1} Mpc^{-1}). The data on the emission lines in the optical spectra are from Humason et $al.$ (1956).

TABLE IIIa

Optical line emission in the nuclei of elliptical galaxies[a]

Radio Characteristics	Total number	Number with [O II] 3727
Galaxies with a radio source less than 6″ in nucleus	12	8
Galaxies with extended sources	7	0
Galaxies not detected	172	37

[a] From Ekers and Ekers (1973).

TABLE IIIb

Optical line emission in the nuclei of spiral galaxies[b]

Type, and % of galaxies of that type in the sample		Nuclei with a small central radio source	Nuclei without small central sources
Sa	17%	37%	63%
Sb	32%	53%	47%
Sc	51%	32%	68%
Presence of emission lines in the spectra of the nuclei, for all types of spirals		93%	67%

[b] From Alloin (1973).

Both tables show that emission lines in the optical spectrum occur more often in nuclei which have a central radio source than in nuclei without a central radio source.

The sample of elliptical galaxies analyzed by Ekers and Ekers is formed from galaxies observed by Humason et $al.$ (1956). The radio observations were all carried out with the same instrument, and the criterion of radio structure (although necessarily arbitrary) according to which a galaxy is classified as having a small radio source is precisely defined: the central source must be smaller than 6″. However, as was noted by Ekers and Ekers, even within 6″ there is a large variety of radio sizes and structures. The sample analyzed by Alloin is made of spiral galaxies observed by Lequeux or by Wade, who used other criteria for defining a small central source (see Alloin, 1973), and as a result this sample of spiral galaxies is less homogeneous than the sample of Ekers and Ekers.

At present, only in a very few galaxies has the gas in the central region been analyzed in detail and these galaxies have radio sources in their nuclei. Therefore it is not possible to compare the physical conditions of the gas in galaxies with small central sources and without central sources.

The major uncertainty in our knowledge of the physical conditions of the gas in nuclei of galaxies comes from the fact that, although there is indirect evidence for the presence of various regions with different electron temperatures and densities and degrees of ionization, these regions cannot be resolved spatially with a ground-based telescope; as a consequence the line intensities which are measured result from the contributions of various regions where the gas has not necessarily the same physical conditions.

The physical conditions and abundances of the gas in the nuclei of M81, M51 and a few other relatively nearby large spiral galaxies with small central radio sources have been analyzed by Peimbert (1968), Alloin (1973) and Warner (1973). The major conclusions of these analyses can be summarized as follows:

(i) Since the intensity ratios of the lines of O I, O II, O III are not compatible with collisional ionization, it is concluded that the ionization must be radiative. The number of main sequence hot stars at 40000 K and $10\,R_\odot$ which would be necessary to maintain the ionization is about 40 in M81 and M51 (Peimbert, 1968); in all the nuclei considered, the contribution of the flux emitted by these hot stars to the total flux of the nucleus would not exceed a few percent at $\lambda \sim 3500$ Å and therefore cannot be detected.

The [O I]/[O II] line intensity ratio is considerably higher than in normal H II regions, therefore it has been suggested that a substantial fraction of the ionization in the nuclei of these galaxies is due to hard UV radiation (Bergeron and Souffrin, 1971). The line intensity ratios calculated by Bergeron and Souffrin are in fair agreement with the observations, the largest disagreement occurring in the case of He I λ 5876/Hα for which the predicted value seems a little too strong as compared with the upper limits observed by Peimbert and Spinrad (1970) in M51 and M81. Other ionization mechanisms like cloud collisions suggested by Peimbert (1973) should be investigated before the problem of the ionization in these nuclei can be considered settled.

(ii) Nitrogen is found to be overabundant by a factor 6 (with an uncertainty by a factor 2). This overabundance of N provides the most likely explanation for the rapid change of the line intensity ratio Hα/[N II] λ6584 from the value ≈ 3 in the spiral arms to a value $\leqslant 1$ in the nucleus.

iii) There are large fluctuations of the electron density. The emitting gas is clumped in small clouds or filaments, the clumpiness factor being of the order of 10^{-4}. There is also a gradual decrease of the electron density from the nucleus outward, on a scale of 5″ (Warner, 1973).

That galaxies with compact nuclear radio sources have optical spectra with emission lines more often than do galaxies without central sources is a fact which is not understood at the present and raises interesting questions on which one can comment only in a very speculative way: Are the stars which are necessary to keep the gas ionized normally present in the nucleus, or has the onset of the radio source been accompanied by a small outburst of star formation? Is the gas which is revealed by the emission lines normally present in the nuclei, or has it gathered or been released at the epoch the radio source started? The situation is further complicated by the fact that there exist a few galaxies with compact radio sources which apparently have only a very small amount of ionized gas and where the presence of non-thermal optical radiation is demonstrated by UV excess, variations, or linear polarization of the optical continuum. An example is B2 1652+39 (4C 39.49) which is an elliptical galaxy with a compact central radio source (Ekers, 1973). The nucleus is very bright in the optical range, and linear polarization of 2% has been measured by Kinman (1973). Emission lines can hardly be seen on the optical spectrum; the redshift measured from the H and K lines is 0.034. Another example of such an elliptical galaxy is 3C 371. Perhaps the objects like BL Lac are extreme cases of such nuclei which emit optical non-thermal radiation but have extremely weak emission lines.

4.2. Infrared emission

Infrared emission at $10\,\mu$ has now been detected in 47 galaxies (Rieke and Low, 1972a). The total span of infrared luminosity at $10\,\mu$ is 10^6. The weakest of all the galaxies detected are M31, Maffei 1 and the galactic centre. The next in increasing luminosity are at least 50 times brighter than these galaxies of the Local Group. Most of the non-Seyfert galaxies detected by Rieke and Low have optical spectra with intense narrow emission lines, like NGC 7714 for example, whose spectrum in the Hα region is shown in Figure 2. The non-Seyfert galaxies are, in general, weaker at $10\,\mu$ than the Seyfert galaxies; yet there are a few exceptions like NGC 1614 (see Figure 2). This galaxy is the fourth brightest galaxy observed at $10\,\mu$ and is 40 times brighter than NGC 4151. The emission lines in NGC 1614 are definitely narrower than in Seyfert-type spectra – even considering the wing of Hα extending down to [N II] $\lambda 6548$, which is very probably caused by gas flowing out from the nucleus (Ulrich, 1972b).

The spectra from $2\,\mu$ to $25\,\mu$ have been measured for several of the galaxies with large apparent infrared luminosities (Rieke and Low, 1972a). The galaxies with Seyfert-type or active nuclei usually have a steep straight spectrum in the $\log f_\nu$ vs $\log \nu$ diagram. Flattening of the spectrum at $\lambda < 3\,\mu$ occurs in several cases and is readily explained by the contribution from cool stars in the vicinity of the nucleus. Some of the galaxies which are intrinsically weak infrared emitters have spectra which differ markedly from a steep straight spectrum, as can be seen in Figure 3.

4.2.1. *NGC 253 and NGC 3034 (M82)*

In each of these two galaxies the region emitting the $10\,\mu$ flux is extended. In NGC 3034, the $10\,\mu$ source is elongated along the equatorial plane with dimensions $25'' \times 8''$ or 400×120 pc, and the radio source at 1420 MHz is also elongated along the plane

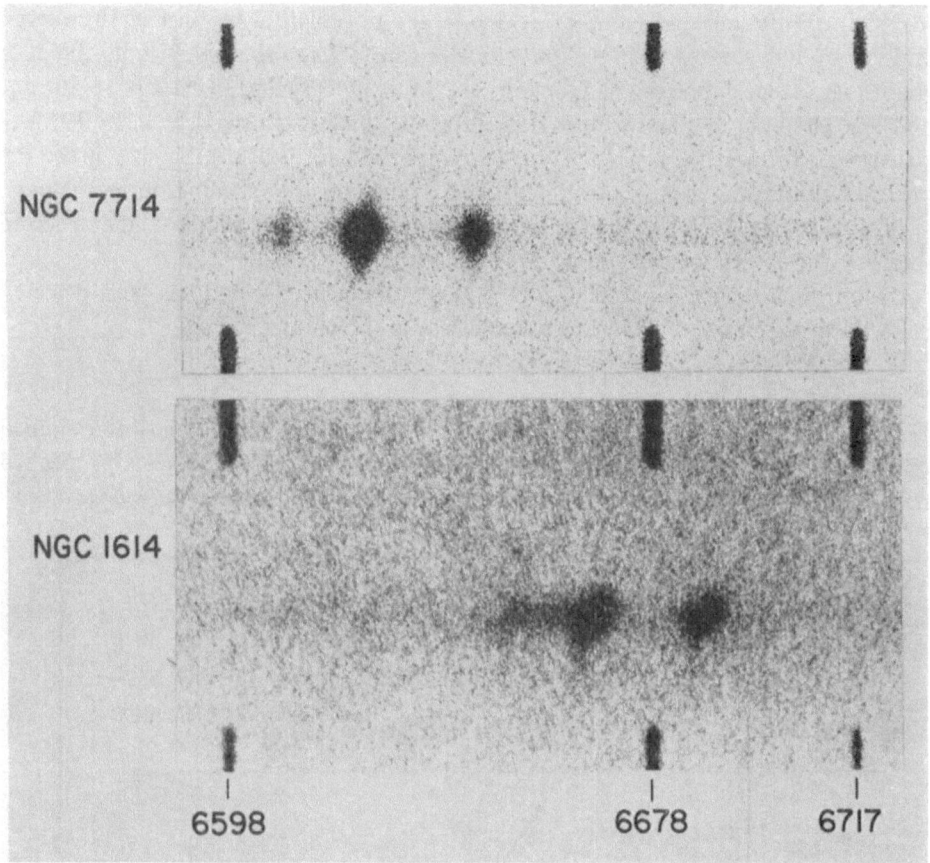

Fig. 2. 28 Å mm⁻¹ spectrograms of NGC 7714 and NGC 1614 in the region of Hα and [N II] λλ 6548, 6584. The 10 μ-fluxes emitted by these galaxies have been measured by Rieke and Low (1972a) and are respectively two times and forty times more intense than the 10 μ-flux emitted by NGC 4151.

with dimensions a little less than $30'' \times 20''$ (see Kleinmann and Low, 1970). In NGC 253, the core of the 10 μ source is about $10''$ or 150 pc in diameter and coincides in position and shape with the central radio source observed at 2695 MHz (Becklin *et al.*, 1973). M82 and NGC 253 have been detected at wavelengths between 27 μ and 125 μ (Harper and Low, 1973) with an airborne telescope 30 cm in diameter; the angular resolution of these observations is $2'$ as compared to $5''$ for $\lambda < 20$ μ and to $1'$ for the ground-based observations at 350 μ. At 350 μ, Rieke *et al.* (1973) detected NGC 253 and set an upper limit on the flux from NGC 3034. These far-infrared data show that the energy distributions of NGC 253 and NGC 3034 have a very pronounced maximum near 150 μ. If the infrared emission in NGC 253 and NGC 3034 originated from a single synchrotron source, the value of the magnetic field causing synchrotron self-absorption longward of 200 μ would be extremely high because of the large size of the source. This type of argument suggests that in these two galaxies the infrared flux is produced by re-radiation from dust particles. On the other hand, if in NGC 253 the infrared emission is due to the re-radiation by dust of the energy emitted by hot stars, and if the far infrared source has the same size as the 10 μ source, i.e. 150 pc diameter, then the total energy emitted by the hot stars is 3×10^9 $L_\odot$ and this assembly of stars is located in a region which has a mass of a few 10^8 $M_\odot$ (Harper and Low, 1973). Thus it appears important to measure the size of the far infrared source and also to look for direct evidence for the re-radiation mechanism, such as detection of structure in the infrared spectrum near 10 μ similar to the structure found in the spectra of galactic infrared sources.

The infrared spectrum of NGC 253, which is the galaxy which has been detected at the largest number of different wavelengths, is shown in Figure 4.

4.2.2. *NGC 1068 and NGC 4151*

Harper and Low (1973) searched for NGC 1068 at 100 μ and 350 μ, and obtained upper limits on the fluxes which indicate that the spectrum of NGC 1068 is less peaked near 200 μ than the spectra of NGC 253 and NGC 3034. From the observed variations

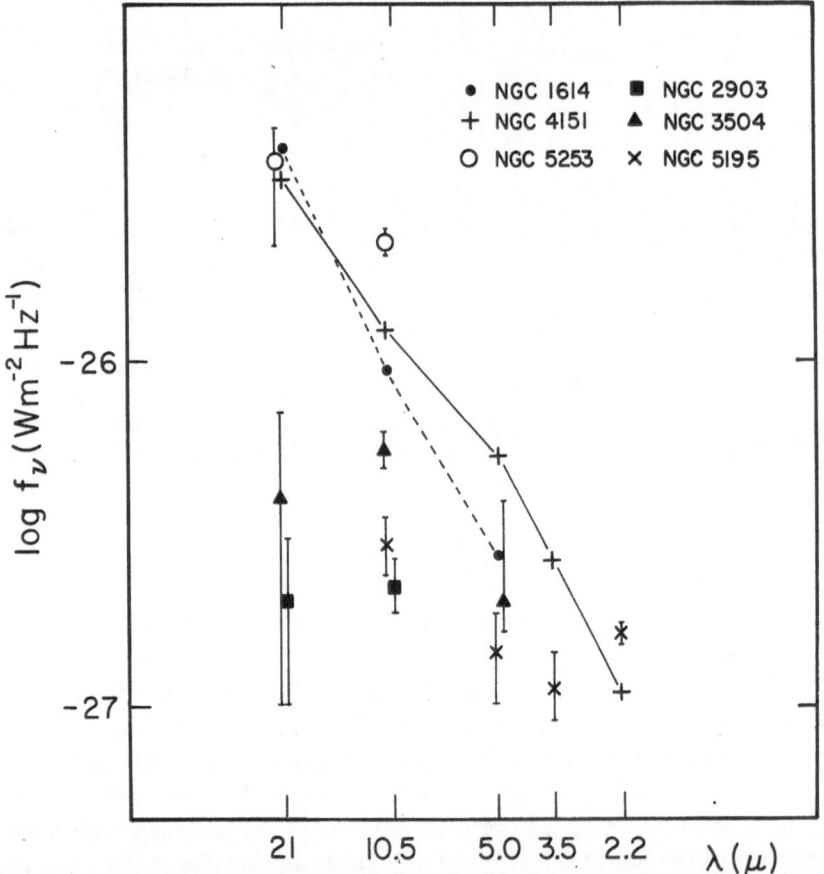

Fig. 3.   Infrared spectra of a few of the galaxies observed by Rieke and Low, 1972a.

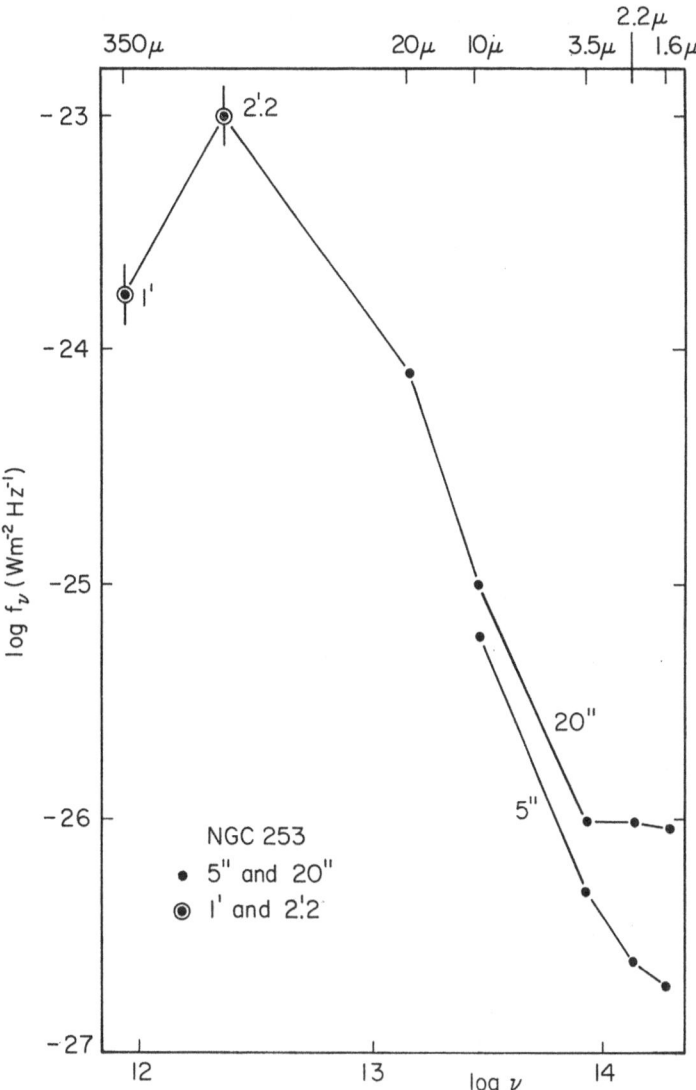

Fig. 4. Infrared spectrum of NGC 253 between 1.6 and 350 μ. Note that the entrance aperture was not the same for all the wavelengths at which the flux was measured. Data are from Becklin *et al.* (1973), from Harper and Low (1973) and from Harper *et al.* (1972).

of the 10 μ flux of NGC 1068, Rieke and Low (1972b) have concluded that in this galaxy the infrared emission is of non-thermal origin. However, the variability of the 10 μ flux is seriously questioned by Stein *et al.* (1974) who monitored NGC 1068 for over three years, occasionally at the same time as Rieke and Low, and it is not clear at present whether the 10 μ flux of NGC 1068 shows variations large enough to rule out its origin as re-radiation by dust.

Stein *et al.* (1974) also observed the spectrum of NGC 1068 between 8 μ and 13 μ

with a resolution $\Delta\lambda/\lambda \sim 0.05$; the spectrum is structureless, in contrast to the spectra observed for galactic sources in which the source of energy is interpreted as thermal re-radiation by grains. The discovery of a structureless spectrum between 8 and 13 μ does not entirely rule out the re-radiation mechanism, although it seems difficult to build a model with dust particles having properties similar to those found in the infrared galactic sources and, at the same time, having an infrared spectrum without structure.

In NGC 4151, Rieke and Low (1972b) and Stein *et al.* (1974) found some evidence for changes of the 10 μ flux by 50% on a time scale of approximately 100 days, and therefore most of the infrared flux is of non-thermal origin.

Our present knowledge on the infrared emission from nuclei of galaxies can be summarized as follows:

(i) Seyfert-type galaxies are the galaxies which have the highest luminosity at 10 μ. In NGC 4151, the variations of the 10 μ flux strongly suggest that the infrared emission is of non-thermal origin.

(ii) In the two nearby galaxies NGC 253 and NGC 3034, which are at a distance of ~ 3 Mpc, the 10 μ source is extended with typical dimensions of 200 pc. The far infrared spectrum has a sharply peaked maximum at 200 μ. Because of the large dimension of the infrared source and of the sharp decrease of the flux longward of 200 μ, emission by a single synchrotron source can be rejected in favour of re-radiation by dust as the source of the infrared luminosity.

(iii) A number of galaxies, mostly spirals, with intense and narrow emission lines have been detected at 10 μ.

5. Active Nuclei

Generally speaking, the nucleus of a galaxy is considered to be active if a detectable fraction of the electromagnetic spectrum cannot be explained by stellar radiation, radiation by interstellar gas at $T_e < 2 \times 10^4$ K and re-radiation by dust particles. Nuclei where expansion motions are present are also considered active: the expansion motions are understood as the result of radiation pressure from a source of ultraviolet continuum on dense gas close to this source, and thus expansion motions are indirect evidence of the presence in the recent past of a source of ultraviolet continuum probably of non-thermal origin. Depending upon the most striking features of their spectra or their morphology, active nuclei have different names, e.g. Seyfert-type nuclei, N-galaxies, but the classification of active galaxies into various groups is not discussed here. It is the view of the author that the current terminology is temporarily useful but not satisfactory.

Recent observations of three properties of active nuclei are discussed below: X-ray emission, profiles of the permitted lines and variations of the line intensities, structure of the forbidden lines.

5.1. X- AND GAMMA-RAY EMISSION

The only nuclear region which has so far been detected with certainty by several groups of observers as a source of gamma radiation is the Galactic Centre. However, it is not yet clear whether the radiation comes from several discrete sources or from one diffuse source (Fazio 1972).

Two other sources of gamma-radiation which may be associated with galaxies are:

(i) a source in the direction of the Seyfert galaxy 3C 120 (Volobuev *et al.*, 1971) and

(ii) a source in the direction of AP Lib (PKS 1514-24) which is an object of BL Lac type (Frye *et al.*, 1971). The angular resolution of the gamma-ray observations is poor, with error boxes of several degrees in size, but 3C 120 and AP Lib seem to be the most likely optical identifications for these two sources. 3C 120 was detected in late 1968 at a time of intense activity at centimetric wavelengths; the source in the direction of AP Lib is variable and was seen in December 1969 but not in February 1969.

In the hard X-ray range, about 50 keV, the only galaxy detected with certainty is NGC 5128. The spectrum between 1 and 180 keV can be fitted with a power law $v^{-\alpha}$ with α between 1.45 and 2.0 (Lampton *et al.*, 1972).

UHURU satellite observations have provided a great wealth of data between 2 and 10 keV (Giacconi *et al.*, 1972). The spectra of the three individual extragalactic objects NGC 4151, NGC 5128 and 3C 273 display a most interesting feature, viz. a definite cut-off for energy lower than ~ 5 keV. Observed upper limits on the angular sizes of the X-ray sources associated with NGC 4151 and NGC 5128 are 15' and 10' respectively, corresponding to linear sizes of 50 and 15 kpc. An indirect argument based on the presence of the cut-off in the spectrum suggests that the X-ray source is substantially smaller than the upper limit on the size obtained by the observations (Giacconi, 1973). In NGC 4151, the large cut-off corresponds to a column density of absorbing material of 10^{23} atom cm^{-2}; if the X-ray source were as large as the observed upper limit on the size, the total mass of the absorbing gas would be $\sim 3.5 \times 10^{11}$ $M_\odot$. Such a large mass of absorbing gas is quite unlikely and a more reasonable configuration is one in which the X-ray source is in the nucleus of the galaxy and surrounded by a relatively small volume of dense gas. Such a model has been discussed in detail by Tucker *et al.* (1973) in the case of NGC 5128.

The X-ray source detected in M31 is comparatively weak and the X-ray spectrum has not yet been measured. The intensity is consistent with the emission being the sum of several discrete sources as is the case in the Magellanic Clouds. The peculiar galaxy M82 has also been detected in the range 2-8 keV. The number of counts per second is about the same as from the X-ray source in the direction of 3C 273, and slightly below the count rate of NGC 4151 (Giacconi *et al.*, 1973).

5.2. PROFILES OF THE PERMITTED LINES; VARIATIONS OF LINE INTENSITIES

In the spectra of many Seyfert nuclei the permitted lines are very broad with widths of 100 Å or more, whereas the forbidden lines have widths less than 20 Å. This fact is regarded as evidence for the existence of at least 2 regions: one with an electron density less than 10^6 cm^{-3} which emits forbidden and permitted lines, and another region with $N_e > 10^6$ cm^{-3} which emits permitted lines only and produces the broad component of these lines. Because of the large value of the electron density in the latter region, electron scattering could be an important broadening agent of the permitted lines. This question has been investigated by Weymann (1970) and by Mathis (1970) who found that electron scattering could in some cases explain the profiles of the permitted lines. In the case of NGC 4151, Weymann showed that a small region with $N_e = 2.5 \times 10^9$ cm^{-3} and a dimension of only 0.005 pc would emit the observed Hα intensity. NGC 4151 is the best known case of a Seyfert-type spectrum where the hydrogen lines have a sharp intense core and smooth symmetrical wings reminiscent of the calculated profile caused by electron scattering.

The observed profiles of the permitted lines available at the time of Weymann's and of Mathis' investigations were not known with an accuracy high enough to allow meaningful comparisons with theoretical profiles produced by electron scattering. Since then, relatively accurate profiles with spectral resolutions of 2 to 5 Å have been published by Anderson (1971) for NGC 5548, by Ulrich (1972a) for NGC 5548 and NGC 3516 and by Shields et al. (1972) for 3C 120. The observed profiles are either asymmetric or show structures or irregularities which rule out electron scattering as the main broadening agent. It is concluded that in these nuclei the profiles of the permitted lines are caused by mass motions. Less detailed observations of other nuclei are also consistent with this conclusion.

In the spectra of some nuclei, the short wavelength side of the hydrogen lines is more intense than the long wavelength side (e.g. NGC 5548, NGC 1614), or the hydrogen lines have a blueshifted component (e.g. 3C 227, 3C 390.3, Markarian 6). This feature can be explained by the presence of dense gas flowing out of the nucleus in two opposite directions, with the gas flowing away from us being partly hidden by material which is optically thick; this material could be the dense gas itself or some other component of the nucleus. It must be noted that if the dense gas itself is optically thick then only a lower limit on its mass can be calculated from the absolute intensities of the hydrogen lines.

The mass necessary to emit an Hβ line of absolute intensity $\sim 10^{41}$ erg s^{-1} – which is typical of a fairly weak Seyfert nucleus – is approximately 15 $M_\odot$ if the electron density is 10^8 cm^{-3}. It is easy to see that if the high velocities indicated by the width of the hydrogen lines represent expansion motions in a single cloud of this mass, observable weakening of the hydrogen lines should occur on a time scale of 3–30 yr. But no such rapid variations are expected if the gas is clumped in small clouds which have high relative velocities but small internal velocity dispersion, a situation reminiscent of the gaseous filaments of the Crab Nebula.

It has been shown that, if the non-thermal optical continuum is extrapolated into the ultraviolet with a power-law spectrum, there is enough energy to accelerate the dense gas to the observed velocities by radiation pressure (Mushotzky et al., 1972; Shields et al., 1972).

We have seen above that mass transfer can produce line intensity variations, but only for permitted lines because of the small dimension of the single cloud from which these lines can in principle originate. Variations of the intensities of the permitted and the forbidden lines can also be caused by variations of the photo-ionizing flux, provided this flux comes from a central source. Bahcall *et al.* (1972) showed that, in some cases, changes in line intensities can occur over times that are small compared with the light travel-time across the nebula.

Rapid variations of line intensities in the nuclei of the Seyfert galaxies NGC 4151, 3516 and 1068 have been reported by Cherepashchuk and Lyutyi (1973). The observations consisted of measuring the fluxes received through a 150 Å pass-band centred on Hα, the fluxes received through the same pass band at ±300 Å from Hα, and the *U* magnitudes.

The amplitudes of variations of the lines reach 34% for NGC 3516. Variations were observed with time scales of 15 days for NGC 4151 and 35 days for NGC 3516. The variations of flux in the band centred on Hα were attributed by the authors to Hα only but this band included the two forbidden lines [N II] λλ6548, 6584.

Cherepashchuk and Lyutyi observed variations of the *U* magnitude which resemble the variations of the Hα line but with a time delay of 15 to 30 days between the variations of the line and the variations of the *U* magnitude, the variations in *U* magnitude always preceding the variations of the line (Figure 5a).

Variations of the profiles of the blended emission lines Hα + [N II] λλ6548, 6584 of NGC 1068 have been reported by Eilek *et al.* (1973) during a time interval which was

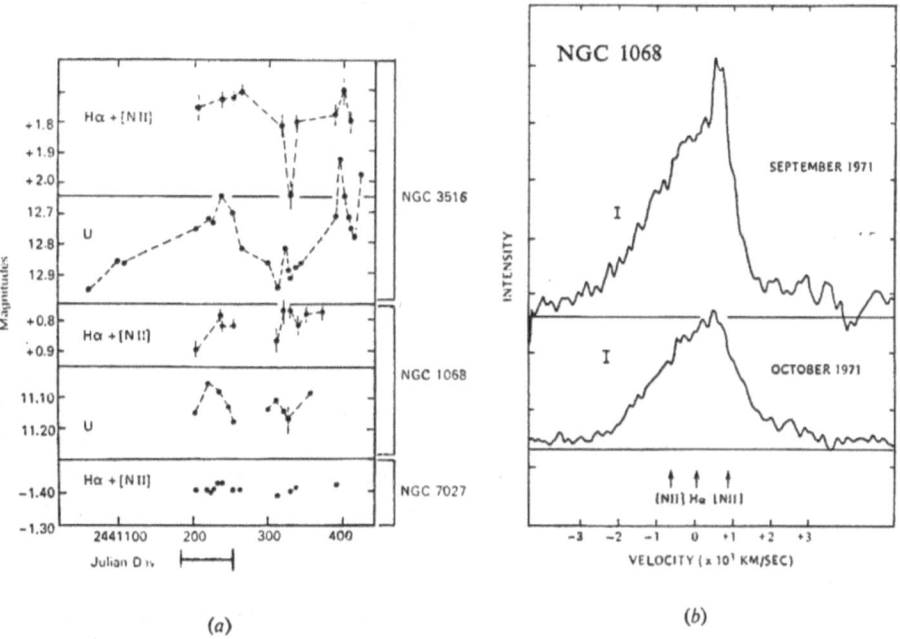

(a) (b)

Fig. 5. Variations of the intensities and of the profiles of emission lines in nearby Seyfert galaxies; Figure 5a is from Cherepashchuk and Lyuti (1973), Figure 5b is from Eilek *et al.* (1973). The horizontal bar below Figure 5a represents the epoch of the observations of NGC 1068 by Eilek *et al.* (1973).

also covered by Cherepashchuk and Lyutyi. The observations were made between August 28 and October 23, 1971 with an isocon television camera. Comparison of the profiles of the blend Hα + [N II] λλ6548, 6584 shows that in September the [N II] λ6584 peak was unusually strong relative to Hα but by October it was back to its level of August (Figure 5b).

Prior to the observations of Cherepashchuk and Lyutyi, and of Eilek *et al.*, variations of the intensity of the hydrogen lines had already been reported to have occurred in NGC 1566 (Pastoriza and Gerola, 1970), NGC 3516 (Collin-Souffrin *et al.*, 1973) and IC 450 (Markarian 6) (Khachikian and Weedman, 1971; Notni *et al.*, 1973). These variations of line intensities can be readily understood as due to the variations of the photoionizing flux. Note, however, the scanner observations of 3C 120 by Shields *et al.* (1972) covering a period of 3 yr, during which the continuum varied by 20% while the emission lines did not vary.

It thus appears that the amplitudes and time scales of the variations of the line intensities triggered by the variations of the continuum flux are different from object to object and probably also from epoch to epoch in a given nucleus.

Observations which are most needed are monitoring of the profiles of the lines (for example Hα and [N II] λλ6548, 6584 or Hβ and [O III] 4959, 5007), and also simultaneous monitorings of the absolute line intensities and of the continuum flux. They will provide data necessary to make progress on the determination of important parameters of the emission line regions such as linear dimensions, clumpiness factors of the gas, etc.

How the high-velocity gas affects the physical conditions of the ambient less-dense gas is a question which has not yet been investigated.

5.3. STRUCTURE OF THE FORBIDDEN LINES

The structures of the forbidden lines in the spectra of NGC 7469, NGC 4151 and Markarian 78 have recently been investigated. In NGC 7469, the motions indicated by the forbidden lines are interpreted by Anderson (1973) as due to rotation. If this interpretation is correct the mass of the central region 400 pc in radius is about $10^9 \, M_\odot$.

In NGC 4151 the four clouds found by Ulrich (1973) are roughly aligned in position angle 40° which is close to the direction of the major axis determined by 21-cm observations (Davies, 1973); therefore the motions of these clouds may plausibly be caused by rotation. The most prominent feature of NGC 4151 on direct plates is an ellipsoidal ring, 4′ × 8′, whose major axis is perpendicular to the major axis determined by Davies. Prior to the 21-cm observations this prominent feature was assumed to be circular and was used to determine the orientation in space of the equatorial plane of the galaxy. If the determination of the major axis from the 21-cm observations is valid, which is quite likely, then the orientation and shape of this ellipsoidal feature becomes very puzzling and it would be of interest to observe its motions.

Clouds of ionized gas have also been found in the inner region of Markarian 78 (Adams, 1973); they are roughly aligned along the major axis. It is not known at the present if their velocity difference is due to the rotation of the galaxy or to expansion motions.

The mass of the nuclear region of an active galaxy, determined from the measurement of intense forbidden lines which show structures or large velocity dispersions, should be regarded with caution because the question arises as to whether the features which are measured truly represent the rotation of the stellar component in which lies most of the mass of the central region.

6. What Fraction of Galactic Nuclei are Active?

We have seen earlier that in the optical range of wavelengths the signs of activity of a galaxy are the presence of expansion motions or emission of non-thermal continuum flux. Evidently it is very difficult to search for these signs in a statistically meaningful fashion. This is in contrast with the situation in the radio range where searches for compact nuclear sources can be made down to well-defined limits of flux and of angular dimension.

The set of galaxies which is the best known in the optical range is that formed by the nearest galaxies. Let us consider the galaxies which were observed by Humason *et al.* (1956), which are not members of the Virgo Cluster nor of the Local Group, and which have $V_{corrected} < 1000$ km s^{-1}. This defines a set of 126 galaxies among which 7 (i.e. 5%) have nuclei which are active* and can be recognized as such by their optical properties alone. The values of 5% is a lower limit to the fraction of active nuclei, because not all the galaxies of this set have been thoroughly searched for signs of activity in the optical range.

Since this set contains very few elliptical galaxies, this discussion of the fraction of 'optically active nuclei' is relevant to spiral galaxies only; the presence of dwarf galaxies in this set is ignored. Finally, it is of interest to note that among this set of 126 galaxies, 16 have small radio sources in their nuclei (Ekers, 1973), and among these 16 galaxies are all the galaxies in which activity has been observed in the optical range.

Acknowledgements

I thank my colleagues for helpful discussions on nuclei of galaxies, in particular R. Ekers and M. Peimbert.

This work was supported in part through grant GP-25220 from the National Science Foundation.

* These 7 galaxies are:

NGC 253 (Demoulin and Burbidge, 1970)
NGC 3034 (Lynds and Sandage, 1963)
NGC 4051 (Seyfert, 1943)
NGC 4151 (Seyfert, 1943)
NGC 4258 (Courtès and Cruvellier, 1961)
NGC 4736 (Chincarini and Walker, 1967)
NGC 5194 (Burbidge and Burbidge, 1964)

References

Adams, T. F.: 1973, *Astrophys. J.* **179**, 417.
Alloin, D.: 1973, *Astron. Astrophys.* **27**, 433.
Alloin, D., Andrillat, Y., and Souffrin, S.: 1971, *Astron. Astrophys.* **10**, 401.
Anderson, K. S.: 1971, *Astrophys. J.* **169**, 449.
Anderson, K. S.: 1973, *Astrophys. J.* **182**, 369.
Andrillat, Y. and Souffrin, S.: 1971, *Astron. Astrophys.* **11**, 286.
Andrillat, Y., Souffrin, S., and Alloin, D.: 1972, *Astron. Astrophys.* **19**, 405.
Bahcall, J. N., Kozlovsky, Ben-Zion, and Salpeter, E. E.: 1972, *Astrophys. J.* **171**, 467.
Baldwin, R. J., Danziger, I. J., Frogel, J. A., and Persson, S. E.: 1973, *Astrophys. Letters* **14**, 1.
Becklin, E. E., Fomalont, E. B., and Neugebauer, G.: 1973, *Astrophys. J. Letters*, **181**, L27.
Bergeron, J. and Souffrin, S.: 1971, *Astron. Astrophys.* **14**, 167.
Blanc-Vaziaga, M-J., Cayrel, G., and Cayrel, R.: 1973, *Astrophys. J.* **180**, 871.
Burbidge, G. R.: 1970, *Ann. Rev. Astron. Astrophys.* **8**, 369.
Burbidge, E. M. and Burbidge, G. R.: 1964, *Astrophys. J.* **140**, 1445.
Cherepashchuk, A. M. and Lyutyi, V. M.: 1973, *Astrophys. Letters* **13**, 165.
Chincarini, G. and Walker, M. F.: 1967, *Astrophys. J.* **147**, 407.
Code, A. D., Welch, G. A., and Page, T. L.: 1972, in A. D. Code (ed.), *Scientific Results from OAO-2*, NASA SP-310, Washington, p. 559.
Collin-Souffrin, S., Alloin, D., and Andrillat, Y.: 1973, *Astron. Astrophys.* **22**, 343.
Courtès, G. and Cruvellier, P.: 1961, *Compt. Rend. Acad. Sci. Paris* **253**, 218.
Cromwell, R. and Weymann, R.: 1970, *Astrophys. J. Letters* **159**, 147.
Danielson, R. E., Light, E. A., Schwarzschild, M., and Tomasko, M. G.: 1971, *Bull. Am. Astron. Soc.* **4**, 230.
Davies, R. D.: 1973, *Monthly Notices Roy. Astron. Soc.* **161**, 25P.
Demoulin, M.-H. and Burbidge, E. M.: 1970, *Astrophys. J.* **159**, 799.
Disney, M. J. and Cromwell, R. H.: 1971, *Astrophys. J. Letters* **164**, 35.
Eilek, J. A., Auman, J. R., Ulrych, T. J., and Walker, G. A. H.: 1973, *Astrophys. J.* **182**, 363.
Ekers, R. D. and Ekers, J. A.: 1973, *Astron. Astrophys.* **24**, 247.
Ekers, R. D.: 1973, private communication.
Faber, S. M.: 1972, *Astron. Astrophys.* **20**, 361.
Faber, S. M.: 1973, *Astrophys. J.* **179**, 731.
Fazio, G. G.: 1972, in H. Bradt and R. Giacconi (eds.), 'X- and Gamma-Ray Astronomy', *IAU Symp.* **55**, 303.
Frye, G. M., Albats, P. A., Zych, A. D., Staib, J. A., Hopper, V. D., Rawlinson, W. R., and Thomas, J. A.: 1971, *Nature* **233**, 466.
Giacconi, R.: 1973, *Phys. Today*, May, p. 38.
Giacconi, R., Murray, S., Gursky, H., Kellogg, E., Schreier, E., and Tananbaum, H.: 1972, *Astrophys. J.* **178**, 281.
Giacconi, R., Murray, S., Gursky, H., Kellogg, E., Schreier, E., Matilsky, T., Koch, D., and Tananbaum, H.: 1973, *Astrophys. J.* **188**, 667.
Harper, D. A. and Low, F. J.: 1973, *Astrophys. J. Letters* **182**, 89.
Harper, D. A., Low, F. J., Rieke, G., and Armstrong, K. R.: 1972, *Astrophys. J. Letters* **177**, 21.
Humason, M. L., Mayall, N. U., and Sandage, A. R.: 1956, *Astron. J.* **61**, 97.
Johnson, H. M.: 1961, *Astrophys. J.* **133**, 309.
Khachikian, E. Ye. and Weedman, D. W.: 1971, *Astrophys. J. Letters* **164**, 109.
Kinman, T. D.: 1973, private communication.
Kleinmann, D. E. and Low, F. J.: 1970, *Astrophys. J. Letters* **161**, 203.
Lallemand, A., Duchesne, M., and Walker, M. F.: 1960, *Publ. Astron. Soc. Pacific* **72**, 76.
Lampton, M., Margon, B., Bowyer, S., Mahoney, W., and Anderson, K.: 1972, *Astrophys. J. Letters* **171**, 45.
Lynds, C. R. and Sandage, A. R.: 1963, *Astrophys. J.* **137**, 1005.
Mathis, J. S.: 1970, *Astrophys. J.* **162**, 761.
McClure, R. D.: 1969, *Astron. J.* **74**, 50.
McClure, R. D. and van den Bergh, S.: 1968, *Astron. J.* **73**, 313.

Morton, D. C. and Chevalier, R. A.: 1973, *Astrophys. J.* **179**, 55.
Morton, D. C. and Thuan, Trinh, X.: 1973, *Astrophys. J.* **180**, 705.
Mushotzky, R. F., Solomon, P. M., and Strittmatter, P. A.: 1972, *Astrophys. J.* **174**, 7.
Notni, P., Khachikian, E. Ye., Butslov, M. M., and Gevorkian, G. T.: 1973, *Astrofizika* **9**, 39.
Pastoriza, M. and Gerola, H.: 1970, *Astrophys. Letters* **6**, 155.
Peimbert, M.: 1968, *Astrophys. J.* **154**, 33.
Peimbert, M.: 1973, Private Communication.
Peimbert, M. and Spinrad, H.: 1970, *Astrophys. J.* **159**, 809.
Poveda, A.: 1958, *Bol. Obs. Tonantzintla Tacubaya* No. 17.
Richstone, D. and Sargent, W. L. W.: 1972, *Astrophys. J.* **176**, 91.
Rieke, G. H. and Low, F. J.: 1972a, *Astrophys. J. Letters* **176**, 95.
Rieke, G. H. and Low, F. J.: 1972b, *Astrophys. J. Letters* **177**, 115.
Rieke, G. H., Harper, D. A., Low, F. J., and Armstrong, K. R.: *Astrophys. J. Letters* **183** L67.
Rubin, V. C. and Ford, W. K.: 1971, *Astrophys. J.* **170**, 25.
Rubin, V. C., Ford, W. K., and Krishna Kumar, C.: 1973, *Astrophys. J.* **181**, 61.
Sandage, A. R., Becklin, E. E., and Neugebauer, G.: 1969, *Astrophys. J.* **157**, 55.
Schwarzschild, M.: 1971, *Bull. Am. Astron. Soc.* **3**, 243.
Seyfert, C. K.: 1943, *Astrophys. J.* **97**, 28.
Shields, G. A., Oke, J. B., and Sargent, W. L. W.: 1972, *Astrophys. J.* **176**, 75.
Spinrad, H.: 1972, *Astrophys. J*, **177**, 285.
Spinrad, H. and Taylor, B. J.: 1971, *Astrophys. J. Suppl.* **2**, 445.
Spinrad, H., Gunn, J. E., Taylor, B. J., McClure, R. D., and Young, J. W.: 1971, *Astrophys. J.* **164**, 11.
Spinrad, H., Smith, H. E., and Taylor, D. J.: 1972, *Astrophys. J.* **175**, 649.
Stein, W. A., Gillett, F. C., and Merrill, K. M.: 1973, *Astrophys. J.* **187**, 213.
Strom, S. E., Strom, K. M., and Carbon, D. F.: 1971, *Astron. Astrophys.* **12**, 177.
Tinsley, B.: 1973, *Astrophys. J. Letters* **184**, 41.
Tinsley, B. M. and Spinrad, H.: 1971, *Astrophys. Space Sci.* **12**, 118.
Tucker, W., Kellogg, E., Gursky, H., Giacconi, R., and Tananbaum, H.: 1973, *Astrophys. J.* **180**, 715.
Ulrich, M.-H.: 1971, *Astrophys. J. Letters*, **165**, 61.
Ulrich, M.-H.: 1972a, *Astrophys. J.* **174**, 483.
Ulrich, M.-H.: 1972b, *Astrophys. J.* **178**, 113.
Ulrich, M.-H.: 1973, *Astrophys. J.* **181**, 51.
Volobuev, S. A., Galper, A. M., Kirillov-Ugryumov, V. G., Luchkov, B. I., and Ozerov, Yu. V.: 1971, *Astron. Zh.* **48**, 1105; *Soviet Astron.* **15**, 879, 1972.
Wampler, E. J.: 1968, *Astrophys. J. Letters* **154**, L53.
Wampler, E. J.: 1971, *Astrophys. J.* **164**, 1.
Warner, J. W.: *Astrophys. J.* **186**, 21.
Weymann, R. J.: 1970, *Astrophys. J.* **160**, 31.
Whitford, A. E.: 1972, *Bull. Am. Astron. Soc.* **4**, 230.
Wood, D. B.: 1966, *Astrophys. J.* **145**, 36.

DISCUSSION

Freeman: Relevant to the last part of this talk are variations observed in NGC 1566 (also noted by the Cordoba group). In 1962 this object had a Seyfert-like nuclear spectrum with Hβ 2000 km s^{-1} wide. In 1968 Hβ was double with two components 800 km s^{-1} apart, one at the same redshift as the other emission lines, and the other to the red. Now there is only one sharp Hβ line and one would not call NGC 1566 a Seyfert system any more.

Danziger: I have a comment which may be relevant to the previous discussion and more particularly the report by Ron Ekers of the presence of a non-thermal radio source in the nucleus of M81. Mrs Jean Goad, a graduate student at Harvard, is studying the dynamics of the nuclear region of M81 and has found evidence for a significant non-circular component in the velocity of the gas which could be interpreted as ejection of matter.

Kerr: Outflow of gas in the central regions has been discussed for our own and other galaxies. Recent work has shown that the 3 kpc arm and other features observed in the Galaxy can be interpreted as a dispersion ring, in which the gas is moving in gravitationally-stable elliptical orbits.

Under these circumstances, there is apparent outward motion but there is no need to assume the existence of energetic explosions. Similar phenomena may occur in other galaxies.

van der Kruit: Fits of 'dispersion rings' to the non-circular motions in central regions of galaxies have been made in particular in our Galaxy (Shane, W. W.: *Astron. Astrophys.* **16**, 118, 1972; Simonson, S. C. and Mader, G. L.: *Bull. Am. Astron. Soc.* **4**, 266, 1973) and in M51 (Tully, R. B.: Univ. of Maryland, Ph.D. Thesis, 1972). In this respect it must be noted that:

(i) the only unambiguous determinations of the direction of non-circular motions (against Sgr A in our Galaxy and along the minor axis in M51) give *expansions*. This is also true for other galaxies such as M31.

(ii) The solutions for the geometry of such dispersion rings are, in both cases, those where the velocities in the line of sight are the same as for a field of expanding motions and thus those where we cannot decide between the two alternatives. This seems rather suspicious.

Davies: I have two comments about NGC 4151 further to my discussion in *Monthly Notices Roy. Astron. Soc.* **161**, 25P, 1973. Firstly, Mr A. Murray has taken photographs for me with the Isaac Newton Telescope, and isodensity tracings by Dr C. Fraser show that the central regions at a mean radius of 5″ have an axial ratio of 1.06, the position angle of the major axis being 35°. This is similar to the H I dynamical major axis at $26° \pm 3°$. Secondly, the velocities found in the inner 6″ by Dr Ulrich are in the same sense as the H I rotational velocities in the outer part of the galaxy, implying that the inner parts of the galaxy are also rotating. The mass inside a radius of 2″(100 pc) would then be $6 \times 10^9 \, M_\odot$, assuming an inclination of about 30°, and gravitational equilibrium.

THEORY OF GALACTIC NUCLEI

WILLIAM C. SASLAW

*Dept. of Astronomy and Center for Advanced Studies, University of Virginia,
National Radio Astronomy Observatory, Charlottesville, Virginia and
Institute of Astronomy, Cambridge Mass., U.S.A.*

> We dance around in a ring and suppose,
> But the Secret sits in the middle and knows.
>
> ROBERT FROST

Abstract. This paper reviews current ideas about the possible constituents of galactic nuclei, and the mechanisms for ejecting gas and massive objects.

We come now to the very innermost regions of a galaxy. What generates the enormous energies that pour from galactic nuclei? How do they evolve? And are new physical laws needed to understand them? Even though I've only asked three questions so far, already the ratio of questions to answers is infinite! At present there is no complete, comprehensive and compelling theory of galactic nuclei. Rather, there are a number of approaches, each emphasizing different major constituents of galaxies. It is likely that several of these constituents, and perhaps some not yet imagined, will play important parts in our eventual understanding of galactic nuclei.

By the nucleus, I mean the region where most of the violent action is. In many galaxies this has a radius less than ~ 10 pc, and it may be very much less. There is no shortage of possible explanations for all the radio and optical activity we've recently heard about. Rather, the main problem is to find definitive tests of these proposals. The difficulty, as we will see, is that galactic nuclei can be very complex objects with a rich array of physical phenomena. Thus we should be careful not to be over-simple in our picture of galactic nuclei. On the other hand, we should also keep in mind the syndrome paraphrased from Ecclesiastes: Of the making of many models there is no end, and too many parameters are a weariness of the flesh.

In this review I will try to describe the current state of our understanding of the basic physics underlying galactic nuclei. Although model-building is certainly important to progress in this field, I won't say much about the virtues or disadvantages of particular models, except as occasional illustrations which are related fairly directly to observations. Many of these observational connections would improve significantly in the next decade if statistical properties of nuclei become better known. Our main problem is that we don't know how typical the present observations are, in any reasonably rigorous sense. These observations are a bit impressionistic. There is a danger in biasing general theories toward a few well-studied cases which may not be very representative of active galactic nuclei. I'll try also to emphasize results since the subject was last reviewed in 1970 at the Vatican Symposium (O'Connell, 1971) and by Burbidge (1970). The general picture of the evolution of galactic nuclei as dense stellar systems is reviewed in detail elsewhere (Saslaw, 1973) so I won't spend much

John R. Shakeshaft (ed.), The Formation and Dynamics of Galaxies. 305–334. *All Rights Reserved.*
Copyright © 1974 by the IAU.

time on it here. The scope of this review is broader, but less detailed, than the one of dense stellar systems.

Let us start by considering, in turn, the major constituents of galactic nuclei. These fall into two classes: Those observed to exist in galactic nuclei, and those observed to exist in the minds of theoreticians. The first class contains:

(1) electromagnetic radiation, (2) gas, (3) dust, and (4) stars;
and the second class contains:

(5) supermassive objects, (6) rotating magnetoids, (7) pulsar systems, (8) singularities in general relativity, (9) accretion disks, (10) anti-matter, (11) gravitational radiation, and (12) new physics.

We discuss first the general physical properties of these objects and their relation to some observational questions such as the spectra, variability, and lifetimes of galactic nuclei. Then we use some of these constituents to review one question of current interest in more detail: How can galactic nuclei eject large amounts of matter?

1. Constituents of Galactic Nuclei

1.1. ELECTROMAGNETIC RADIATION

The optical, radio, and infrared observations reviewed by Drs Ulrich (p. 279) and Ekers (p. 257) provide direct evidence for intense electromagnetic radiation in galactic nuclei. The small sizes of the radio emitting volumes are found from very long baseline measurements with resolution $\sim 4 \times 10^{-4}$ arc sec at $\lambda = 6$ cm (Kellermann *et al.*, 1971). Many sources, both quasars and galaxies, have strong radio components less than several parsecs in diameter. In Centaurus A (NGC 5128) Kunkel and Bradt (1971) have detected an infrared hot spot with a size of about 75×110 pc and a total luminosity of $\sim 2 \times 10^{41}$ erg s^{-1}. They identify this as the nucleus of the galaxy. Within this nucleus is a non-thermal radio source (Wade *et al.*, 1971) whose linear size is measured to be less than 13 pc. If this is synchrotron radiation which is self-absorbed at centimetric wavelengths the size of the emitting region is only about one light year.

Photometry between 2 and 25 μ shows that the infrared luminosity of galaxies ranges from 10^{37} to $\sim 10^{45}$ erg s^{-1}. In many cases, and especially for Seyferts, the infrared losses dominate the total luminosity of the galaxy. Unfortunately the lower limits on the size of the emitting regions are usually much greater than for the radio regions. Thus far, detailed comparisons of the locations of radio, infrared, and optical hot spots in galactic nuclei have been made only for the Milky Way (Downes and Martin, 1971) and the exploded galaxy M82 (Kronberg *et al.*, 1972). In neither case is there any detailed correspondence between the positions of strong infrared and radio regions. However, these are not typical strongly concentrated nuclei, and many more observations of this type will be useful for understanding the radiation mechanisms.

In the optical, the smallest upper limit to the size of a Seyfert nucleus comes from the Stratoscope II photograph of NGC 4151 (Schwarzschild, 1973). The photographs taken on the seventh Stratoscope flight had higher definition than those of the previous flight. From the earlier flight, an upper limit to the half-intensity diameter of the nucleus of NGC 4151 was $\sim 1\overset{''}{.}7$. The new measurement reduced this by about a factor of two. However, in the interval between the two observations, the value of the Hubble constant also decreased by a factor of about two, leaving the upper limit to the linear size of the nucleus much the same. The upper limit for the half-power radius is now 3.5 pc and the luminosity, mostly non-thermal, is $\sim 2 \times 10^{43}$ erg s^{-1}.

There appear to be some relations between the optical, infrared, and radio emission of galactic nuclei. In spiral galaxies, there is a strong tendency for compact radio sources to be present if the nucleus is starlike, and especially if the nucleus is split into two or more unresolved components (Tovmassian, 1972). For elliptical galaxies there is not yet much information. Some ellipticals have sharp central peaks of optical luminosity, while others are smoother. Although the compact active radio ellipticals usually have optical emission lines (Disney and Cromwell, 1971), their correlation with central optical luminosity peaks is not established (Heeschen *et al.*, 1971). One relation which is better established, however, is that the radio luminosity of Seyfert or related galaxies is usually proportional to its infrared luminosity (van der Kruit, 1971; Rieke and Low, 1972a).

Thus there is evidence for strong radiation in galactic nuclei. The intense pressure of this radiation may drive gas out of the system. In the simplest case when the pressure is caused by resonance line scattering in an optically-thin spherically-symmetric medium, the radiative acceleration due to a species i is (Mushotzky *et al.*, 1972)

$$g_r(i) = \frac{F_v}{c} \frac{N_a}{N_H} \frac{N_i}{N_a} \frac{X}{m_H} \frac{e^2}{m_e c} f. \qquad (1)$$

Here, F_v is the continuum flux at the line centre, N_H, N_a, and N_i are respectively the number densities of hydrogen (which has mass fraction X), of the element, and of the relevant ionization state, f is the f-value of the transition, and the other physical constants have their usual meaning. If the scattering is not from a resonance line, the left-hand side of Equation (1) is multiplied by the probability of finding an ion in the relevant excited state. When the only radial dependence in $g(r)$ comes from the flux, or if all the factors fortuitously give $g(r) \sim r^{-2}$, then both radiative and gravitational accelerations of the gas have the same spatial dependence. Thus there will be a critical mass which can produce a static situation where gravity balances radiation pressure. More mass makes the gas fall in, less mass lets it blow out. In general, however, the situation will be too complicated to be in balance and the results will depend on details of the model.

One important aspect of radiation-driven mass loss is less model-dependent. This is a 'line locking' mechanism which can produce narrow velocity ranges for the outflowing gas. Originally suggested nearly 50 yr ago by Milne (1926) to produce solar cosmic rays, 'line locking' is now being revived as a possible explanation for narrow absorption lines in quasars, stars and Seyfert nuclei (Scargle, 1973; Williams, 1972; De Young *et al.*, 1973). The basic idea is that gas in the nucleus produces a continuum with strong absorption lines. Filaments of gas near, or outside the edge of, the nucleus are accelerated by gravity and by radiation pressure in the absorption line. If the balance between these two accelerations is destroyed and the filaments increase their velocity, they will begin to absorb the more intense radiation from the blue wing of the line and eventually, as the velocity increases still further, the filaments absorb from the strong continuum and are driven out of the galaxy. Similarly, infalling filaments would be decelerated.

There are several ways in which these filaments may reach a definite line-locked limiting velocity: (a) if there is only one dominant absorption line, the filament will be accelerated by the continuum and its final velocity will depend on the depth of the line as well as the size and luminosity of the accelerating region, (b) if there are two strong absorption lines, the filament is accelerated along the continuum until its absorption is blueshifted to the second line, at which point the radiation pressure drops and the filament achieves a limiting velocity equivalent to the separation of the two lines, (c) if a strong absorption line is located to the red of the Lyman continuum drop, the filament can similarly be accelerated to a velocity equivalent to the difference between the line and the Lyman drop. For a filament to be accelerated coherently by radiation pressure, photons created within the filament must not be absorbed inside it (Williams, 1972). The simplest condition this implies is that, when the filaments are accelerated, they are optically thin in the absorption line. We will return to this effect of radiation in our discussion of the ejection of matter from galactic nuclei. Radiation pressure can be especially important if the brightness temperature at low frequencies, T_b, exceeds the rest mass of the electron, for then induced Thomson scattering becomes important and may increase the temperature of the electrons to nearly T_b, and distort the radiation spectrum (Levich et al., 1972).

1.2. Gas

Clouds of gas in galactic nuclei have been detected from their radiation in Balmer lines, their emission in the forbidden lines of nitrogen, oxygen and other heavy elements, and in 21-cm absorption. In two galaxies, 3C 390.3 (Burbidge and Burbidge, 1971) and NGC 1275 (Burbidge and Burbidge, 1965) the emission lines show two systems of redshifts, suggesting that gas is being expelled from the system. In other cases, such as the Seyfert I 4329 Å, there is evidence for high-velocity motion of the gas in the nucleus, at velocities up to ~ 13000 km s^{-1} (Disney, 1973).

During the last few years, people have studied a variety of fairly detailed models to account for the spectra of galactic nuclei and quasars. In addition to constraints from the observed spectrum, ranging from radio to γ-rays, the models are strongly constrained by intensity fluctuations observed over periods of days to years (e.g. Pacholczyk, 1970; Penston et al., 1971). Among the physical processes which have been invoked to reproduce the general form of the spectrum are thermal radiation by gas and dust, electron and proton synchrotron emission, inverse Compton scattering, induced Compton scattering, heating of gas by X-ray sources or by intense low frequency radiation from pulsars or spinars, scattering of photons by non-thermal plasmons, non-thermal emission from the surfaces of massive objects, and collisions of large gas clouds. Recent discussions of these processes may be found in Jones and

Kellogg (1972), Davidson (1972), Bergeron and Salpeter (1973), MacAlpine (1972), de Sabbata *et al.* (1972), Ozernoy (1973), Levich and Sunyaev (1971), Arons *et al.* (1974), Daltabuit and Cox (1972).

Various combinations of these models can account, in a fairly consistent manner, for most observed properties of gas and radiation in galactic nuclei. While this may not be too surprising, in view of the large number of possible geometries and physical parameters, it does indicate that there is no compelling need to invoke completely new kinds of physics to explain these particular phenomena.

These models are also important for estimating the basic physical parameters of galactic nuclei, and their uncertainties. For example, a simple model of NGC 1068 (Bergeron and Salpeter, 1973) in which most of the infrared radiation is synchrotron and the X-rays are from inverse Compton scattering on the infrared, gives a radius of the active region between 10^{-2}–10^{-4} pc, a magnetic field between about 10–10^3 G, an electron density of about 10^2–10^6 cm^{-3} and a thermal gas density of $\sim 10^{11}$ cm^{-3}. The ranges in these values are due to the assumed ratio of X-ray to infrared luminosity, between 10^{-1} and 10. Bergeron and Salpeter have also constructed a simple model of this Seyfert galaxy in which the infrared is produced by thermal grains, and the X-rays by inverse Compton scattering of the infrared. For this the emitting region must be much larger, between 10^2–5×10^4 pc, the electron density much less, between 6×10^{-4}–1 cm^{-3}, the number of grains between 10^{-14}–3×10^{-10} and the total mass of grains between 5×10^4–2×10^9 $M_\odot$. Again this range is for the same assumed ratio of X-ray to infrared luminosities. The large size of the infrared emitting region in dust models makes it difficult for them to account for rapid luminosity fluctuations.

The stability of gas in the nucleus is an unsettled fundamental problem. There are two main possibilities: either the gas may be unstable to the formation of stars or massive objects near the centre, or it may be blown out in a galactic wind. Some time ago Arny (1970) argued that the gas in galactic nuclei would be thermally unstable and form clouds. The size of these clouds depends strongly on the manner of heating the gas, for which only the supersonic passage of stars through the gas and the deposition of energy from stellar winds into the surrounding gas were considered. Heating by X-rays, ultraviolet, or low energy cosmic rays could change the picture considerably, and it would be useful to know how they affect the size and mass of the clouds.

Gravity may cause these clouds to collapse and fragment. It is likely that both turbulence in the cloud (Arny, 1971) and the presence of the background gravitational field of stars in the nucleus (Mathews, 1972) encourage the formation of more massive stars. Objects of several hundred solar masses could possibly form, although this is highly uncertain.

If the gas which stars shed in the nucleus is further heated to temperatures $\gtrsim 10^6$ K by shocks from supernovae explosions and photo-ionization, it may be expelled from the nucleus as a galactic wind (Burke, 1968; Johnson and Axford, 1971; Mathews and Baker, 1971; Wolfe, 1974). Unlike the situation in the solar wind, the sources of the gas and the gravitational field are distributed throughout the region where the wind forms instead of being a single compact body. As with the solar wind, the early solutions were for idealized spherically-symmetric steady flow of a perfect gas. The later solutions added enough details of heating and cooling to produce unsteady flows and possible thermal instabilities.

The most detailed models are those of Mathews and Baker. As an example, they consider a hot wind of 2×10^6 K, fed by a stellar mass-loss rate of 3.4 $M_\odot$ yr^{-1} in an elliptical galaxy with total mass 9×10^{10} $M_\odot$ and a stellar density of 440 $M_\odot$ pc^{-3} at its centre. Starting with no initial gas, a steady-state wind establishes itself in $\sim 5 \times 10^7$ yr. This is basically the time it takes gas to flow from the centre to the edge of the galaxy (14 kpc). The mass involved in the steady wind is $\sim 8 \times 10^7$ $M_\odot$. It starts moving subsonically at ~ 100 km s^{-1} in the nucleus, and becomes supersonic in the outer parts of the galaxy, reaching ~ 1000 km s^{-1} at ~ 15 kpc. Such a wind would not be observable directly, though it could carry radiating dust along as a marker. On the other hand, if the gas is cooler ($\lesssim 10^4$ K), the wind is unsteady. This may occur when the central gas density rises and cooling becomes more effective, or if the supernova rate is small. Now the gas falls in toward the centre of the galaxy, perhaps to form new stars or massive objects related to the strong radio sources associated with $\sim 10\%$ of giant elliptical galaxies.

Wolfe has shown that a hot wind flowing out from the nucleus can be thermally unstable. If so, it may form a two-phase system in which cool dense clouds are in pressure equilibrium with the hotter, more tenuous intercloud gas. These clouds would then be borne outward by the wind. If observed between us and a central continuum source, their spectra could show the narrow lines seen in some quasars and Seyfert galaxies. It is, however, difficult to use nuclear winds to produce clouds moving out from quasars at very high velocities or, in Seyferts, to produce large central clouds which do not contain a continuum source. Details of all these models are tested by computing line strengths and shapes for highly ionized atoms, especially iron. Usually it is possible to find a particular structure of the gas and radiation field which is reasonably consistent with observations.

Magnetic fields are often associated with the ionized gas in galactic nuclei. Relativistic electrons interacting with these fields produce synchrotron radio and possibly infrared and optical emission. A tell-tale sign of synchrotron radiation is its circular polarization. For an isotropic array of pitch angles, the fractional polarization is approximately $10^{-2} (\lambda B)^{1/2}$ where λ is the wavelength in centimetres and B the magnetic field in gauss (Sciama and Rees, 1967). Several radio sources have measured circular polarizations of a few tenths of one percent in the radio, indicating magnetic fields of $\lesssim 1$ G. The degree of circular polarization can be reduced, however, if the synchrotron radiation is Compton scattered to higher energies by the same electrons that made it (Bonometto and Saggion, 1973). Thus the lack of circular polarization in the observed non-thermal optical emission (Landstreet and Angel, 1972) does not preclude the optical being produced from synchrotron radiation by inverse Compton scattering.

1.3. DUST

Dust is often seen in reddened Seyfert and other nuclei. Another observational reason for introducing dust into galactic nuclei was to account for the high infrared luminosity observed in many Seyferts and quasars (Rees *et al.*, 1969). While thermal emission from grains at different temperatures can explain the steep spectrum observed between 2.2 and 22 μ, it runs into difficulty with the rapid time fluctuations observed in some sources. 3C 273 fluctuates especially violently, changing its 10 μ luminosity by 10^{12} $L_\odot$ – nearly a factor of 2 – in two months, and there is also evidence for substantial 10 μ variability in NGC 1068 and NGC 4151 on time scales of weeks to years (Rieke and Low, 1972b).

To obtain thermal infrared emission by irradiating grains from a central UV source requires dust temperatures between about 10–1000 K. At these temperatures, the sizes of dusty disks needed to produce the high infrared luminosities are greater than the light travel distances during the period of fluctuation. Thus large amplitude fluctuations would not be produced. While it may be possible to diminish this incompatibility by clever modelling, the variations found recently by Rieke and Low (1972b) are inconsistent with straightforward models (Kaneko *et al.*, 1972). However, grains could be an important contribution to the weaker ($L \sim 10^{42}$ erg s^{-1}) infrared emitted in the centre of our own Galaxy (Bergeron and Salpeter, 1973; Krishna Swamy, 1971), since this comes from a relatively large region of ~ 175 pc.

1.4. STARS

Although the *direct* evidence for stars in galactic nuclei is not as compelling as the evidence for radiation and gas, there seems to be little doubt that they are present. The difficulty in determining stellar content arises mainly because the angular sizes of galactic nuclei are so small and because the non-stellar continuum and light emission lines would overpower any stellar absorption lines in the spectra. In the nuclei of two Seyferts, NGC 1068 and NGC 4051, there is some evidence for a stellar population like that of ordinary galactic nuclei (Andrillat and Souffrin, 1971). Possibly the highest stellar density observed in a Seyfert nucleus comes from the Stratoscope II photographs of NGC 4151. Assuming that most of the mass of its nucleus is in stars of ~ 0.2 $M_\odot$ ($M/L \approx 20$) which occupy about the same region as the non-thermal radiation, the properties of this nucleus are approximately (Schwarzschild, 1973) mass$=$ $=4 \times 10^9$ $M_\odot$, L_{vis} (nonthermal)$=6 \times 10^9$ $L_\odot$, L_{vis} (stars)$=2 \times 10^8$ $L_\odot$, number of stars$=2 \times 10^{10}$, diameter of nucleus <7 pc, velocity of stars >1500 km s^{-1}, star collision rate >2 per year.

Since I have recently reviewed the properties and dynamical evolution of dense stellar systems elsewhere (Saslaw, 1973), they will be discussed only very briefly here. I shall try to sketch, in a very schematic way, the highlights of the evolution of a dense stellar system, but the previous review should be consulted for details, caveats, and references.

Dense stellar systems may form in a variety of ways. It could be that the initial conditions are very favourable. For example, in a cosmology with local inhomogeneities of large amplitude (represented by turbulence, lagging cores, great density perturbations, etc.) dense agglomerations may grow after matter and radiation decouple in the standard big-bang picture. These could then form the nuclei of galaxies, accreting the rest of the galaxy from initially more uniform surroundings.

But what if the relevant initial conditions (as determined by their other observational implications, if any) do not turn out to be so favourable? Suppose initial perturbations first grow into a gaseous, rather more uniform cloud. Such a cloud will normally have a density which decreases outward from its centre. Characteristically, in contracting, the central density rises more rapidly than the density in the outer parts, and the inhomogeneity becomes exacerbated. This situation is probably unstable, and the centre splits into stars (Larson, 1969).

Other processes can further increase the density of the nucleus. As stars lose mass during their natural evolution, some of the lost mass may devolve to the centre and there form new stars which in turn lose mass which falls further to the centre and there forms new stars which in turn... As an illustration (Spitzer, 1971), if 10% of the entire stellar mass becomes gas which falls to the centre and forms new stars in a region with radius 5% of the original system, there is a density enhancement of $\sim 10^3$. Starting with 10^{11} $M_\odot$ in a 10 kpc radius, three such steps create a core with 10^8 $M_\odot$ in 1 pc. A system of stars with the solar luminosity function would evolve about 20% of its stellar mass into gas (of which about half is assumed to fall to the centre) after $\sim 10^8$ yr. Thus a dense core could be formed rather quickly in the evolution of a galaxy. This process depends on the gas having little or no angular momentum, and on fairly normal stars being formed at each stage.

Purely stellar dynamical effects will also tend to produce a dense core. If the system is nearly in equilibrium, the stars will perturb one another's orbits slightly. Some stars will slowly gain energy until they escape, and form a halo of high energy stars. Since the remaining stars have lower average energy (i.e. their total energy is more negative), they form a more compact core. The time scale for stars to evaporate is about

$$T_{\text{evap}} \approx 10^2 \, T_R,$$

where

$$T_R = 8.3 \times 10^5 \, \frac{N^{1/2} R_{(\text{pc})}^{3/2}}{(M/M_\odot)^{1/2} (\log N - 0.3)} \, \text{yr}$$

is the relaxation time for stellar orbits to be deflected substantially from their original motion. R is the rms radius of the system having N stars each of mass M. The core which remains behind becomes highly condensed after about 20 initial relaxation times T_R. The evaporation time scale can be quite long. For example, use of the values for NGC 4151 given earlier with $R < 7$ pc, gives $T_{\text{evap}} \lesssim 4 \times 10^{13}$ yr. Unless this nucleus is much smaller than 7 pc, evaporation of stars will be negligible.

Formation of a dense core may be faster if the system relaxes violently. To do this it must start far from equilibrium. Then stars will scatter not just against small fluctuations in the mean gravitational field, but against large collective oscillations of the whole stellar system. If the gravitational field is inhomogeneous in space and varies rapidly in time, the time needed for relaxation may be decreased compared to T_R by a factor of as much as $N/\log N$, for the limiting case when the whole system oscillates with period $\sim (G\varrho)^{-1/2}$. Violent relaxation is quickly damped, however, by phase mixing and Landau damping, and the net results of this relaxation in a realistic system are not yet well understood.

Another effect which produces dense cores is the lack of energy equipartition among stars of different masses. If there were equipartition, the massive stars with lower velocities would sink to the centre of the system, and there form a quasi-stationary subsystem. Stars would evaporate so slowly compared to the crossing time (which measures collective gravitational response) that the system could still be considered to satisfy a quasi-stellar virial equilibrium. However, in nearly all systems there can be no equipartition. As a result, the more massive stars will lose kinetic energy to the lighter ones, fall toward the centre, continue to lose the kinetic energy they gain by falling, and continue to fall. The tendency of energy exchange to produce equipartition never succeeds, and ultimately a dense core forms at the centre. Although the dynamical evolution is not fully understood, numerical experiments suggest that the lack of equipartition may produce a core about an order of magnitude faster than in a comparable system with equal mass stars.

When the core becomes sufficiently dense, stars will collide bodily. For a typical star, the mean time between collisions is

$$ t_c = \frac{1}{n\sigma v} = \frac{9.7 \times 10^{21} R_{(pc)}^{7/2}}{N^{3/2} (m^{1/2} r^2 / M_\odot r_\odot^2)(1 + 8.8 \times 10^7 R_{(pc)} r_\odot / Nr)} \, \text{yr}, $$

where r is the radius of the star. In deriving this relation, the virial theorem has been used and the geometrical cross-section has been increased by a factor $(1 + 2\, Gm/rV^2)$ to account for the gravitational attraction, ignoring tidal deformation. As examples, for 10^8 stars and $R_{(pc)} = (0.1, 1, 10, 100)$, we have $T_{R(yr)} = (3.4 \times 10^7, 1.1 \times 10^9, 3.4 \times 10^{10}, 1.1 \times 10^{12})$ and $t_{c(yr)} = (2.8 \times 10^6, 5.2 \times 10^9, 3.1 \times 10^{12}, 1.1 \times 10^{15})$.

If two similar stars collide at relative velocities exceeding several hundred kilometres per second, most of the gas will interact supersonically relative to the local sound speed, and shocks will convert much of the kinetic energy of stellar motion into thermal energy which is then radiated. Thus the collision is highly inelastic. If, moreover, the total binding energy of the two stars substantially exceeds the total kinetic energy of their orbital motion, most of the stars' material will coalesce. The distended, newly formed object pulsates for some time, then settles down as a well-defined star. Eventually it may become a supernova.

If the velocities of the stars are so great that their orbital kinetic energy exceeds their gravitational energy, the collision will be mostly disruptive. For two sun-like stars this implies a relative velocity $V \gtrsim 1500 \, \text{km s}^{-1}$, approximately. Using simple models, a number of calculations of colliding stars were made some years ago (Spitzer and Saslaw, 1966; Mathis, 1967; Colgate, 1967; De Young, 1968; Sanders, 1970; Seidle and Cameron, 1972). These have greatly clarified, but not really solved, many basic questions such as the amount of mass loss, the conditions for coalescence, the growth of very massive stars, the role of thermonuclear reactions, the generation of relativistic particles. Other questions such as the structure and evolution of a coalesced star, the most massive stars that can form by coalescence, and the fraction of stars in the nucleus that survive the coalescence phase are still quite open.

Geometric collisions begin to dominate the evolution of the core when the collision time t_c becomes less than T_R. This happens when $R \approx Nr$, so that if the whole system were put into one dimension with the stars just touching one another, the length of this line would be about the radius of the three-dimensional system. Initially most collisions occur at low velocities and the stars coalesce. At first, the main sequence lifetime of the coalesced stars is less than the collision time, so the newly combined stars have a chance to evolve off their main sequence, perhaps becoming supernova. Collisions are not yet so rapid that they add fuel to stars and mix their cores faster than the hydrogen is burned. The general evolution of a cluster during this period has been simulated by Monte Carlo calculations (Sanders, 1970). Although the results are quite model-dependent, it is easy to find plausible conditions which produce the approximate luminosities and lifetimes of Seyfert galaxies and quasars.

As the system contracts still further, coalescing collisions become more frequent until the collision time becomes less than the time for the coalesced star to evolve into a supernova. If enough time is spent in this regime (before disruptive collisions take over), stars of extreme mass may form. At first the mass of a typical star is built up by coalescence with smaller stars. Every addition of hydrogen with mixing is assumed to be so effective that it sets the star's evolutionary clock back to zero (an important question for further calculation). In this way stars of $\sim 500\ M_\odot$ may form (Sanders, 1970). These in turn sink to the centre and, coalescing one with another, accelerate the building of even more massive stars. Eventually the cluster becomes so compact and the velocities so high that it enters a phase dominated by the disruption of the objects which have already grown.

In systems where a large number of low mass stars survive the coalescence phase (what detailed conditions are necessary for this?), the subsequent evolution will be even more complex and dramatic. Collisions at velocities $\gtrsim 2000\ \mathrm{km\ s^{-1}}$ liberate appreciable amounts of gas from the stars. This gas cools by free-free emission and radiation by heavy ions, especially carbon, nitrogen and oxygen. A simple homogeneous model (Spitzer and Saslaw, 1966) suggests that when enough gas is liberated, its cooling time can become substantially less than the collision time of the stars. As the temperature falls the cloud begins to shrink, its density rises, and the cooling processes accelerate rapidly. As a result, the pressure gradient no longer supports the gas, and the cloud goes into free-fall. Beginning with negligible interstellar matter, this sequence of the building, cooling, contraction, and collapse of a gas cloud takes ~ 5000 yr and involves $\sim 200\ M_\odot$ in a nucleus containing 10^9 stars within a 1 pc radius (about 5×10^3 ions $\mathrm{cm^3}$). It takes ~ 200 yr and involves $\sim 4000\ M_\odot$ for a nucleus with 10^{10} stars in 1 pc and it takes $\sim 2 \times 10^7$ yr and involves $\sim 50\ M_\odot$ for 10^9 stars in 10 pc. At this stage of evolution the cooling of the gas is so rapid that the average mass of gas in the nucleus is too small to influence its dynamics, except at the very centre.

The gas which falls to the centre forms a flattened disk whose radius is determined by its angular momentum. Since stars with opposite orbital angular momentum have a somewhat greater probability of colliding than stars moving in the same direction, it is likely that the specific angular momentum of the liberated gas will be less than that of the stars. The rms height of the gas above the disk, on the assumption of hydrostatic equilibrium, is approximately (Spitzer, 1942) $Z_g \approx R V_{gas}/V_{stars}$, where R is the rms radius of the stellar system, V_{gas} and V_{stars} are the rms velocities of these components of the system. Since V_{stars} must be at least ~ 2000 km s^{-1}, and the gas may cool below 10^4 K ($V_{gas} \lesssim 10$ km s^{-1}), the disk is very thin compared to the stellar system.

At this stage (and neglecting the more exotic possibilities to be discussed later), there are four main sources of luminosity in the nucleus: the general background of old stars including coalesced objects, the radiation of gas liberated by collisions during the collision and later as it falls toward the centre, new O stars or supermassive stars formed in the disk, and supernovae. If we assume that an energy $\frac{1}{2} V_{stars}^2$ per unit mass of gas liberated is radiated as the gas cools, and an equal amount is lost as the gas falls into the disk radiating the potential energy released, then the luminosity of the gas $L \approx V_{stars}^2 \xi M/2t_c$. Here ξ is the fraction of the total mass lost per collision. Substituting the virial expression for V_{stars}, and the previous equation for t_c, gives

$$ L \approx 7 \times 10^{10} \xi N^{7/2} (m/M_\odot)^{5/2} (r/r_\odot)^2 R_{(pc)}^{-9/2} \text{ erg s}^{-1}. $$

For velocities of 1500–2000 km s^{-1}, $\xi \approx 0.06$ and, as an illustration, if there are 10^8 suns in ~ 0.05 pc radius, then $L \approx 10^{43}$ erg s^{-1}. The contribution of the other three sources could exceed this luminosity, depending on details of the model.

The late evolution of the central disk depends strongly on its gravitational interaction with the rest of the nucleus. If the gas forms stars of fairly normal mass (which is by no means guaranteed), the degree of mixing between old and new stars is governed by the ratio of the relaxation time, T_R, to the timescale t_c/ξ for collisions to liberate a substantial fraction of the entire stellar mass. As long as this ratio remains small, distant encounters between the low-velocity new stars with the old high-velocity ones are able to transfer enough kinetic energy to enable the newly formed stars to rise from the disk. At the same time, the orbits of old stars contract further and the average density increases. It is difficult to determine the actual amount of energy exchange, and models have been constructed (Spitzer and Stone, 1967) for the two extremes of zero exchange and of the maximum exchange possible. The luminosity of the liberated gas depends strongly on the angular momentum of the gaseous disk, and is greatest in cases of low angular momentum. Plausible models give luminosities and lifetimes characteristic of Seyfert nuclei and quasars during this phase.

When the core becomes so tight that $\xi T_R/t_c > 1$, collisions destroy most of the older stars before they have time to impart their kinetic energy to the younger ones. Gradually the stellar system becomes a flattened disk, the random stellar velocities are greatly reduced, and the luminosity produced by the colliding stars begins to die away. But even this state does not end the career of the nucleus.

Now, if not before, a new menagerie of objects may grow to affect the further evolution of the nucleus. These are the more speculative members of the second list. But before describing their roles, let us pause to face the basic question: Is there one main source of energy in the nucleus, or are there many?

Evidence on this question is scanty, and comes mainly from the nature of intensity variations in the objects. The longest and most detailed optical variations are for the quasar 3C 273, for which there is some suggestion of a ~ 10 yr cycle. This has given rise to an inconclusive quasi-periodic debate in the literature as to whether the light curve can be represented by a superposition of randomly occurring outbursts. For the latest instalment, with references to previous debate, see Chertoprud *et al.* (1973). In the radio, the most extensively observed variations are for the Seyfert-like galaxy 3C 120 (Dent, 1972a) and for the quasar 3C 279 (Dent, 1972b). Both these objects vary substantially on a time scale of weeks. If they are coherent objects, they give the impression of expanding faster than the speed of light. The alternative explanations of this faster-than-light expansion are that the objects are really much closer than the Hubble relation applied to their redshift indicates, that the expansion is a phase velocity effect, or that short-lived outbursts occur fairly randomly in various parts of the object. Very few astronomers at this time feel that the evidence for the breakdown of the Hubble law is compelling. The second explanation requires rather artfully contrived, but still possible, models. The third explanation seems to be the simplest. A possible problem with this random event model – sometimes likened to lights blinking on and off on a Christmas tree – is that in some cases successive brightenings tend to occur along the same line on the sky. This would require that we see a disk nearly edge-on. However, there are not yet enough of these objects to determine whether the required geometry is statistically improbable.

In addition to evidence from individual objects, there is some information for the statistical properties of radio spectra in compact extragalactic objects (De Young, 1971). Nearly 50 sources, including both quasars and radiogalaxies, have more than one intensity peak in their spectrum. If these peaks are due to synchrotron self-absorption of radiation from an outburst, then the size and magnetic field strength of the outburst can be found from the frequency and flux density at the peak, if an average brightness temperature (which is not very critical) is assumed. The distribution of these peaks seems to be more consistent with multiple outbursts occurring in physically separate regions than with successive outbursts from a central source into a pre-existing or regenerated magnetic field.

Although the balance of evidence seems to favour the multiple source model, it is not at all decisive. There is no reason why both types of source could not be present in the same object. The discovery of a galaxy or quasar with unambiguous periodic fluctuations would go a long way to showing that this could be the case.

With this background, we consider the forms that single sources might take. In olden days, cartographers drawing maps of unexplored regions of space often populated them with myriads of strange phantastical creatures. So also today, at the edge of understanding, we try to imagine the denizens of galactic nuclei.

1.5. SUPERMASSIVE STARS AND DISKS

Historically, hot supermassive stars (Hoyle and Fowler, 1963a, 1963b) were the first single sources proposed for the energy of strong radio sources and quasars. If these objects of $\sim 10^5$–10^{10} $M_\odot$ were stable, they could burn thermonuclear energy with a luminosity of $\sim 10^{45}$ erg s^{-1} for $\sim 10^6$ yr. However, it has become apparent that non-magnetic, non-turbulent, non-rotating supermassive stars are difficult to form and stabilize, and recently this has been computed in detail (Appenzeller and Fricke, 1972; Fricke, 1973). As a supermassive star forms (either from a gas cloud or from coalescence of less massive stars) and begins to contract, there are four possible ways it may develop. If $M \lesssim 4 \times 10^5$ $M_\odot$, it settles down into thermonuclear equilibrium. But if $M \gtrsim 4 \times 10^5$ $M_\odot$, it can explode or collapse into a black hole. Explosion occurs, for a given mass, if the initial heavy element is great enough to produce rapid thermonuclear burning. For example, if $M = 10^6$ $M_\odot$, explosion occurs if $z > 0.04$. When z is too small for a given M, or M is too great for a given z, thermonuclear energy cannot halt the gravitational collapse. The fourth possibility is that relaxation oscillations occur in which the radius, luminosity, and rate of energy production change periodically in the pulsating star. But this does not seem to occur unless the rate of burning is arbitrarily damped during the explosive phase.

It is worth noting that thermal radiation from the surface of stable supermassive stars will not generally have a Planck spectrum (Illarionov and Sunyaev, 1972). When electron scattering is more important than free-free processes in the outer atmosphere, the radiation spectrum can be quite complex, depending on the run of temperature and density in the atmosphere. For a simple power-law dependence of plasma temperature on surface depth, the radiation may follow a power-law spectrum with negative index, thereby mimicking the spectrum of synchrotron radiation over a range of frequencies.

To overcome some of the stability problems of supermassive stars, one can consider differentially rotating supermassive disks (e.g. Salpeter and Wagoner, 1971; Quirk and McKee, 1971; Scharlemann and Wagoner, 1972). Unlike stars in which gravity is opposed by thermal gas and radiation pressure, these disks are maintained by the balance of gravitational and centrifugal forces. If the matter is completely cold, the disk will be infinitesimally thin. In the more realistic case where the pressure, total energy density, and redshift at the centre of the disk are P_c, E_c, and z_c respectively, the ratio of the half-width to the equatorial radius is approximately (Salpeter and Wagoner, 1971)

$$\frac{W}{R} \approx \frac{P_c}{E_c} \frac{1 + z_c}{z_c},$$

for an idealized uniformly rotating disk, and the period of rotation of the disk is approximately

$$\tau \approx 8 \times 10^{-5} \, (M/M_\odot) \left(\frac{cJ}{GM^2}\right)^3 \, \text{s},$$

where J is the angular momentum. To within a factor of order unity, $z_c/(1+z_c)$ is the ratio of the maximum binding energy to the rest mass energy. For a hot disk,

$$\frac{L}{L_\odot} \approx 3 \times 10^4 M/M_\odot.$$

An important difference between disks and stars is that when a pressure-supported star loses only angular momentum, its radius does not change significantly and therefore its rotation slows down. But a centrifugally-supported disk contracts and rotates faster when it loses angular momentum. The disk cannot rotate too rapidly, or it will begin shedding mass. And if it rotates too slowly, it collapses. The exact regime of stability depends on the detailed rotation curve.

Several results on the stability of these disks are described by Quirk and McKee (1971). For a maximum global stability against collapse, and local stability against fragmentation, one wants a rotating, very centrally condensed, disk supported either by radiation pressure or a tangled magnetic field (particle pressure does not give sufficient internal energy for stabilization). The minimum stable ratio of thickness to radius for a uniformly rotating Newtonian disk is about 1/20; differential rotation can reduce this by a factor of ~ 2, but relativistic effects may increase it by a factor of 1–2 for $R/R_{\text{Schwarzschild}} \lesssim \frac{1}{2}$. These results are based on analytic idealizations, and numerical experiments suggest they may be off by about 50%, to give some idea of the uncertainty in these stability calculations. No realistic stability calculations have yet been done for highly-relativistic, differentially-rotating disks supported by magnetic pressure.

For $R/R_{\text{Sch}} \approx \frac{1}{2}$, the disk can release $\sim 6\%$ of its rest mass energy during contraction to this radius, and possibly more if it accretes surrounding gas and dust. Disks supported by radiation pressure radiate thermally at the rate (Quirk and McKee, 1971)

$$L \approx 1.3 \times 10^{38} \frac{M}{M_\odot} (1 - R/R_{\text{Sch}}) \text{ erg s}^{-1}.$$

Therefore the time for them to lose all their pressure support and become infinitesimally thin, if no processes other than thermal radiation are involved, is

$$E_{\text{binding}}/L \approx 10^8 \frac{R_{\text{Sch}}}{R} \left(1 - \frac{R_{\text{Sch}}}{R}\right)^{-1} \text{yr}.$$

But with only thermal radiation losses the disk would contract to an unstable axial ratio of $\sim \frac{1}{20}$ in about $3 \times 10^7 \, R/R_{\text{Sch}}$ yr. So if the disk is to avoid fragmenting, it must contract even faster, which means that non-thermal energy losses (e.g. magnetic dipole radiation) must be more important than thermal losses.

Fragmentation is not a particularly great evil, however; it is just difficult to understand in detail. If the disk fragments, its early evolution will depend on the mass and size of the fragments relative to the central condensation in the disk. If the outer part of the disk (which may be more unstable if its differential rotation is less) fragments into many small objects, these might orbit the centre as a reasonably stable satellite system, perhaps surrounding a black hole. On the other hand, if a few massive fragments form, the system would probably be quickly unstable: some pieces might be ejected and others would coalesce as they radiated gravitational waves. The effects of tidal forces make the picture more uncertain. No one knows whether the fragments would be neutron stars, smaller disks, or black holes. We should also remember that the instability may be set off not by an infinitesimal perturbation, but by a collision of the disk with a massive star or supernova, or by the eruption of a magnetic flare, and this could change the nature and growth rate of fragments considerably. These problems are very difficult, and our lack of understanding of Newtonian fragmentation and star formation just increases the uncertainties (cf. Arny and Weissman, 1973). Moreover, among all these questions lurks the possibility that the equations of general relativity are not applicable to strong gravitational fields – where they have never been tested.

1.6. ROTATING MAGNETOIDS

In many massive rotating objects, a magnetic field may be important. These configurations have been called magnetic rotators, magnetoids, or spinars. They come in two main forms: hot and cold. Hot magnetoids, with high entropy per baryon, are the more spherical in shape and are supported by radiation pressure as well as by the magnetic field. There are a variety of possible structures, depending upon whether the rotation is uniform or differential, and upon the poloidal and toroidal components of the magnetic field (Ozernoy and Usov, 1971). A poloidal field, however, will tend to smooth differential rotation in a time short compared to the evolutionary contraction time of the magnetoid. To account for the activity in galactic nuclei (and radiogalaxies) these objects must have a lifetime of $\gtrsim 10^5$ yr. There are two kinds of hot magnetoids which seem to have this property. Either they have differential rotation and a very weak poloidal field, or they rotate uniformly and have a strong poloidal field. In the first case, however, the field is too small to give appreciable magnetic dipole radiation compared with thermal radiation, so this configuration is unstable to fragmentation.

The thermal luminosity of hot magnetoids is determined approximately by the Eddington limit at which radiation pressure balances gravity:

$$L_{\text{thermal}} \approx 10^{46} \, (M/10^8 \, M_\odot) \, \text{erg s}^{-1},$$

although, as mentioned previously, the emerging Planck spectrum could be substantially modified by electron scattering. The magnetic dipole radiation has been considered by Cavaliere *et al.* (1971) and in more detail by Ozernoy and Usov (1973a, b). It is appreciable for at least $\sim 10^5$ yr in a uniformly rotating object with a strong poloidal field. At the poles, this magnetic field has the value

$$B_{\text{p}} \approx 10^6 \, \xi^{1/2} \, (M/10^8 \, M_\odot) \, (R/10^{16} \, \text{cm})^{-2} \, \text{G},$$

where ξ is now the absolute value of the ratio of total magnetic energy to gravitational energy. If the rotation is as fast as it can be without disrupting the star, i.e. $\Omega \approx (GM/R^3)^{1/2}$, this field gives the luminosity of magnetic dipole radiation as

$$L_{md} \approx 10^{58} \, \xi \sin^2 \chi \, (R_{Sch}/R)^4 \, \text{erg s}^{-1},$$

where χ is the angle between magnetic and rotational axis. This angle may change in a quite complicated manner which depends on how angular momentum is removed from the object. In particular, the rate of loss of angular momentum depends sensitively on the temperature of the object's corona (Anand and Shara, 1972).

The spectrum of this luminosity has not been worked out rigorously for a realistic situation. But the general idea is that intense low frequency ($\nu \approx 10^{-6}-10^{-8}$ Hz for very massive rotators) radiation accelerates the surrounding plasma to relativistic velocities. These high energy particles then radiate both by synchrotron radiation with the dipole field, and by Compton scattering off the low frequency waves which accelerated them (Blandford and Rees, 1972). Simple models suggest that this is a plausible mechanism for producing radio and infrared emission.

As the hot magnetoids radiate, they cool and any remaining support must come from rotation and an internal magnetic field. Thus they flatten and come to resemble the disks of the previous section. The structure of a thin, magnetic, uniformly-rotating disk has been determined by Scharlemann and Wagoner (1972), but little is about known its detailed stability.

Perhaps the most unequivocal observational test for these objects would be the discovery of a definite dominant periodicity in the luminosity of a quasar or active galaxy. So far, this does not seem to have been found, and the other evidence that they exist in quasars is not compelling (Sturrock, 1971). However, they cannot be ruled out.

1.7. Pulsars

Since pulsars can produce moderate amounts of radio and optical radiation, it seems attractive to gather $\sim 10^7$ of them together to explain the activity in galactic nuclei and quasars (Kardashev, 1970; Rees, 1971). The main problem is to account for the 1–100 supernovae per year over $\sim 10^7$ yr in the nucleus needed to produce these pulsars (if supernovae make them with high efficiency). Pressent models for supernova generation during the formation of galaxies are phenomenological, so one simply assumes the parameters required to fit the observations. (Un)fortunately for these models, there are always more parameters than observations. The most detailed model of galactic nuclei as pulsar clusters is by Arons et al. (1974). In particular, they consider collective effects such as the acceleration of particles scattered by moving pulsars in analogy to the Fermi mechanism, and the efficient resonant acceleration of particles by the superposed wave fields of the pulsars, especially when two pulsars have the same frequency in the Doppler-shifted particle frame. These models can be made to agree reasonably well with the observed luminosities, fluctuations, and optical lines. However, their necessary complexity inhibits any definitive tests.

1.8. SINGULARITIES IN GENERAL RELATIVITY

When the ratio, GM/rc^2, of the gravitational binding energy of an uncharged, non-rotating object to its rest mass energy becomes as large as unity, an event horizon forms around the object and prevents radiation from escaping from the Schwarzschild radius $R_{Sch} = 2GM/c^2$. For rotating or charged objects this criterion is modified somewhat, but still a black hole forms. It is an unsolved question as to whether all types of objects have event horizons when their self-gravitational field is sufficiently strong, or whether there can exist 'naked singularities' without such an horizon.

How could we tell whether there is a black hole in the nucleus of a galaxy? As the hole forms, it is unlikely to retain any evidence of a magnetic field if the net charge is zero (Scharlemann and Wagoner, 1972). Thus the black hole itself probably doesn't produce any electromagnetic radiation, although surrounding gas flowing into the hole may drag a magnetic field with it and radiate. The possibility of detecting a black hole when it accretes will be discussed briefly in the next section.

A black hole might be detected, in principle, by its modification of the light from a surrounding cloud of stars. Gerlach (1971) has examined the idealized case of a collisionless system of stars moving in a spherical shell around a black hole. Each star radiates at the same frequency in its own rest frame, and the photons move through the Schwarzschild geometry without scattering or absorption, but the random Doppler motions of the stars broaden the spectrum. The effect of the black hole is to swallow photons emitted toward it with impact distances less than $\sqrt{27} R_{Sch}/2$. This first of all depletes the redshifted photons (as would be seen by a distant observer) relative to blueshifted photons, below what the flux would be without the black hole. The amount of depletion depends mainly on the distance of the stars from the hole, but even for the closest stable orbit it is not more than 40%. Secondly, the total photon intensity, integrated over all frequencies, is also depleted by about 50% (relative to the case of no black hole) at these small impact distances. The system would appear to be a faint central disk surrounded by a ring whose inner edge is about twice as bright as the disk, and whose outer edge is about three times as bright. Unfortunately, to observe even the largest black hole ($\sim 10^{11} M_\odot$) in the nearest large galaxy (M31) in this way would require a resolution $\lesssim 10^{-2}$ arc sec. Moreover, a realistic distribution of stars, and any gas or dust present in the nucleus, will probably smooth this picture beyond recognition.

The gravitational effect of a black hole on the stellar distribution in the nucleus might be detectable under extreme conditions, but the details of this effect are uncertain. If one assumes that the stellar distribution around a black hole reaches a steady state in which the distribution function, $f(r, v) \propto E^p$, is a power of the total energy $E = (GM/r) - \frac{1}{2}v^2$, then the star density $\varrho \sim r^{-9/4}$ and the velocity dispersion $v \sim r^{-1/2}$ (Peebles, 1972) where r is radial distance from the hole. To within observational uncertainties this density distribution cannot be distinguished from an isothermal sphere, $\varrho \sim r^{-2}$. The theoretical uncertainties in the form of $f(r, v)$ are probably at least as large, especially if the stars have a range of masses and do not reach a steady state.

Wolfe and Burbidge (1970) discuss in some detail the possibility that galactic nuclei contain black holes. They also assume that the stars in the nucleus are relaxed but in the form of an isothermal sphere, which is at least quasi-stationary. This is consistent with observations showing a smooth luminosity distribution into the innermost $1''$ radius, at which point the seeing disk smears the image. If the presence of a black hole at the centre is not to change the isothermal nature of the stellar distribution significantly (a problem deserving more study), then a simple model gives upper limits to the possible mass of a central black hole from observations of the luminosity and velocity distributions. For most giant elliptical galaxies, this upper limit is $\sim 10^{10} M_\odot$. A central black hole cannot, therefore, explain the high visual mass/luminosity ratios of these galaxies, but it could account for explosive phenomena in some nuclei. It is possible, however, for there to be a large number of holes, with the same mass spectrum as the stars, spatially distributed like stars throughout these galaxies.

Another point of view is to regard the very existence of galaxies as evidence for black holes in their nuclei (Ryan, 1972). As a way out of the well-known problems of forming galaxies from statistical fluctuations in an initially homogeneous universe, one can start with a finite amplitude perturbation after matter and radiation decouple, and have the perturbation accrete a galaxy (never mind the origin of this perturbation itself, according to this view). This idea was first suggested in the context of steady-state cosmology (Roxburgh and Saffman, 1965), but is more broadly applicable. A perturbation of $\sim 1\%$ the mass of the final galaxy works very well in most cosmological models, but there is no real necessity for it to be a black hole.

1.9. ACCRETING OBJECTS

This category is not concerned so much with a particular type of object, as with ways in which objects may swallow their surroundings. The objects may be black holes, neutron stars, or relativistic disks. The basic physics of accretion is reviewed in Zeldovich and Novikov's (1971) book, and more recently in Novikov and Thorne (1972), so I shall make very few comments here. The subject has been revived in recent years, mainly as a possible explanation for X-ray stars.

Radiation produced by accreted matter ultimately comes from the gravitational potential energy of the objects it falls onto. This energy is greatest in the case of a black hole, even though it is redshifted substantially as it escapes from regions close to the event horizon (the Schwarzschild radius in the case of an uncharged, non-rotating hole). Orbits of stars and other point masses moving around a Schwarzschild singularity are unstable if (Hansen, 1971) $f \leqslant (3+e) R_{Sch}$ where f is the semi-latus rectum and e is the eccentricity. Since real stars have a finite size, they will be broken up by tidal forces on reaching the unstable orbits. Viscous dissipation (possibly magnetic) causes the resulting gas and dust to radiate and fall into the unstable region where it is swallowed for ever. The manner in which stars surrounding a black hole interact to populate the unstable orbits is an important problem on which Wolfe and Burbidge (1970) have made a start. The motion of gas and stars into rotating Kerr black holes is a more involved problem which is discussed by Wilson (1972), and Bardeen *et al.* (1972).

A fundamental question is how much of the energy of a black hole can be extracted under astrophysically plausible conditions. In principle $\sim 42\%$ of the accreted rest-mass energy could be radiated in a Kerr metric (Bardeen, 1970), but this may not be realized in practice. Even for spherically symmetric, non-magnetic, steady-state accretion of pure hydrogen onto a Schwarzschild black hole, the luminosity and spectrum depend very sensitively on the mass of the hole, and the temperature and density of accreted gas (Shapiro, 1973). Models of accretion onto black holes (Lynden-Bell and Rees, 1971; Norman and ter Haar, 1973) in active galactic nuclei suggest that masses $\sim 10^8 \, M_\odot$ are needed.

Accreting neutron stars may also serve as an energy source for galactic nuclei (Bisnovatni-Kogan and Sunyaev, 1972). As with pulsars, one must account for the high concentration of neutron stars in the galactic nucleus. Also, only the dying neutron stars which are not energetic pulsars would be able to accrete substantially. Recently much work has been done on the properties of neutron star accretion in connection with X-ray sources, but that is beyond the scope of this review.

1.10. ANTI-MATTER

If the infrared radiation of Seyfert nuclei is produced by matter-anti-matter annihilation, the minimum flux of μ-neutrinos which is also produced can be predicted for straightforward models (Steigman and Strittmatter, 1971). The original observations of high infrared flux were inconsistent with the low observed upper limits of the μ-neutrino flux. More recent estimates which lower the IR flux are consistent. However, in view of the present difficulties with solar e-neutrino expectations, any conclusions about anti-matter are very tentative.

1.11. GRAVITATIONAL RADIATION

In galactic nuclei, gravitational radiation can be generated by collisions of stars or black holes, or during the accretion process itself. Very little of the radiation is absorbed by other matter; most escapes to transport energy and angular momentum out of the galaxy. This could remove much of the mass from galaxies. Recently Press and Thorne (1972) have reviewed the properties, production, and detection of gravitational waves.

1.12. NEW PHYSICS

There have been a number of proposals, starting with Jeans (1929) and including Ambartsumian (1965), that entirely new physics is needed to explain the violent activity in galactic nuclei. Ambartsumian, particularly, has advocated the view that the galaxies themselves form by the splitting and expansion of very condensed matter. Variants of this view consider lagging cores from the early big-bang Universe (Novikov, 1965; Ne'eman and Tauber, 1967; Harrison, 1971) or pockets of creation in the steady-state cosmology (Hoyle and Narlikar, 1966). This is in contrast to the more general view that galactic nuclei form from the contraction of regions of diffuse gas. A definitive observational test of whether galaxies form from the inside out or from the outside in would be a great step forward. So far, it seems that the features of galactic nuclei can be explained reasonably well by applying conventional physics (including general relativity), and there is no necessity to modify any basic laws. However, this question deserves an open mind.

2. The Ejection of Matter from Galactic Nuclei

Thus far we have mainly tried to learn about galactic nuclei by understanding their manner of emitting radiation. There is substantial evidence that many galaxies also eject large quantities of gas, and perhaps massive compact objects as well. The observations were reviewed by Burbidge (1970) and at the Vatican Symposium (O'Connell, 1971), as well as at this Symposium. In addition to all the evidence for the explosion of some galaxies (e.g. M82, NGC 1275) and the ejection of radio sources from many others, there is an increasing number of anomalies such as the compact components of the nucleus of NGC 1808 which may be dynamically unstable (Arp and Bertola, 1970), the blue condensations associated with elliptical and S0 galaxies (Stockton 1972), and the broad (~ 13000 km s^{-1}) Hα lines in the Seyfert I4329 A which may indicate high outflow velocities (Disney, 1973). Such phenomena may contain important clues to the inner workings of galactic nuclei.

What sets off these vast explosions? In each case, the ultimate source of the energy is gravitational, but the acceleration may be caused directly by magnetic fields, radiation pressure, or hydrodynamic explosions, as well as by gravity. Let us consider each of these possibilities in turn.

Magnetic sources of acceleration arise through the rotational winding up of magnetic fields until they become unstable. One process, numerically calculated by Le Blanc and Wilson for stars (see Wheeler, 1971), may also apply to more massive objects. They consider a collapsing, rotating star of $\sim 7\ M_\odot$ and initial central density $\sim 10^8$ g cm^{-3}. It starts with a small poloidal magnetic field ($E_{magnetic} = -2.5 \times \times 10^{-4} E_{grav}$, $E_{rot} = -2.5 \times 10^{-3} E_{grav}$) which is wound up by differential rotation as the star contracts. The field becomes mostly azimuthal and builds up great pressure along the axis of the star. In a very short time, when the central density reaches $\sim 10^{11}$ g cm^{-3}, this field bursts out along the axis producing a double-sided jet. Although the material in the jet goes off with high velocity, ~ 0.1 c, only $\sim 1.5 \times 10^{-3}$ of the star's mass is ejected, so the process is not very efficient. Whether it becomes more efficient in more massive stars is an open question.

Related models involve magnetic flares in galaxies and large differentially rotating gas clouds. Ozernoy and Somov (1971) consider such a contracting cloud, and are especially interested in the way rotation twists the initially poloidal magnetic field to form neutral lines and the field becomes quasi-radial. With continued twisting the field lines become unstable and a flare may occur, accelerating particles along the polar axis to relativistic velocities. This action would repeat as the cloud contracts further and rotation continues to tighten the residual field lines. Using a different geometry consisting of an annulus and a core rotating about the same axis but with different velocities, Sturrock and Barnes (1972) have indicated how a metastable force-free field can form. This may be explosively unstable and produce a double jet. Since these models are so complex, it is impossible to compute much MHD detail, but they make plausible the ejection of large clouds of relativistic gas approximately along the rotation axis.

Shklovsky (1970) has proposed, in general terms, a variant of magnetic ejection in which the relativistic particles and synchrotron radiation are ejected anisotropically from the magnetoid. Conservation of linear momentum implies that the magnetoid recoils and might be ejected from the galactic nucleus. The frequent similarity between double components on opposite sides of a radiogalaxy is not naturally explained by this approach, however, since one component would be a magnetoid and the other a cloud of gas.

Ejection driven by radiation pressure, and line-locking in particular, was discussed in Section 1.1. There may be some evidence that this occurs in the exploding Seyfert NGC 1275, alias 3C 84, alias Perseus A. Photographs and spectra (Burbidge and Burbidge, 1965; Lynds, 1970) of NGC 1275 show a central amorphous region with medusa-like filaments (rather like the Crab Nebula) coming out. The recession velocity of this region and its filaments is 5270 km s^{-1}. On the northwest side of this central region there is a sector of $\sim 110°$ in size containing a number of emission-line filaments with recession velocities of 8220 km s^{-1}, or $+2950$ km s^{-1} with respect to the systemic velocity. The remarkable thing is that these filaments extend throughout ~ 10 kpc from the centre (modulo unknown projection factors) and their velocities are all in the narrow range 8090–8372 km s^{-1}, i.e. they cluster around the relative velocity of ~ 3000 km s^{-1} to within $\pm 5\%$. Dividing the distance of the farthest filament from the nucleus by 3000 km s^{-1} gives an age of $\sim 5 \times 10^6$ yr. It seems that ejection at the same velocity is still going on at the present time (De Young *et al.*, 1974). Within 10 pc from the nucleus is a cloud of neutral hydrogen, seen at 21 cm absorbed against one of the complex central radio sources, moving outward from the nucleus at 2850 km s^{-1}. The width of the 21-cm line is only ~ 10 km s^{-1}. Thus the cloud has a very low temperature and little turbulent motion.

There is no unique explanation of this cloud. For example, it could be a chance coincidence in position and velocity space, or it could be a small colliding galaxy. If these explanations don't apply, then the only method people have suggested which might accelerate gas to a very narrow velocity range for a long period of time is line-locking. Explosions or galactic winds won't do because they have too many parameters, and not enough known constraints. Now the second and third methods of line-locking discussed in Section 1.1., i.e. between two absorption lines, or between the Lyman continuum drop and an absorption line to its red are likely to the most persistent over long times. For the second possibility, one could consider strong Lyα absorption at 1215 Å and the strong N v multiplet at 1238 Å. The velocity difference of these two absorption lines is ~ 6000 km s^{-1}, giving a velocity projection factor of ~ 2. Alternatively, one might consider the difference between the Lyman continuum drop and absorption multiplets of O IV (922 Å) and N IV (923 Å), giving a velocity of ~ 3000

km s^{-1}, and a projection factor ~ 1. However this latter possibility has the problem that the O IV and N IV lines are not resonance lines, and are therefore less likely *a priori* to give strong absorption. Thus it is possible that a line-locking mechanism is working. On the other hand, the combined hydrodynamic and radiative transfer problems have never been calculated, and it is not clear that this method will accelerate clouds coherently. Clearly, exciting things are happening here, and they are close enough to study in detail.

Moving toward hydrodynamic theories of ejection, we first encounter the buoyant bubble model (Gull and Northover, 1973). In this picture, relativistic particles in the nucleus are supposed to form a bubble of hot gas which bifurcates into two bubbles. These new bubbles are wafted out of the galaxy by gentle intergalactic winds and by their buoyancy in the cooler surrounding gas. The high external gas densities ($\sim 10^{-26}$ g cm^{-3}) needed to confine the bubble and provide sufficient buoyancy are likely to occur only in clusters of galaxies, if anywhere. Moreover, this gas must be very homogeneous and have very small turbulent motion if the bubble is to remain coherent. This model needs some detailed hydrodynamic calculations. For example, under what conditions does a bubble form rather than a galactic wind?

Galactic winds were discussed in Section I.2. While there does not seem to be any observational evidence for their existence, the general idea is certainly plausible. Violent hydrodynamic explosions could produce the phenomena in M82, for example. Such an explosion might be set off by the nuclear reactions in a supermassive star (Appenzeller and Fricke, 1972). In a non-magnetic non-rotating spheroid with an exponential density distribution, a point explosion at the centre tends to break out first along the main axes (Sakashita, 1971). However, this tendency is not pronounced unless the eccentricity of the spheroid is $\gtrsim 0.7$. Strong, global magnetic fields might focus the explosion into a jet, but such problems have not been calculated in detail.

Direct gravitational ejection can be more efficient than methods which first convert gravity to magnetic or rotational forces. Wheeler (1971) has described one such mechanism, the 'tube of toothpaste' instability. This may occur as a star falls into the ergosphere of a massive rotating black hole. The star becomes tidally elongated and disrupts. Some of its material manages to extract energy from the ergosphere and be ejected, perhaps in a manner resembling a jet. Simple models of this process (Mashhoon, 1973) suggest that, although the escaping pieces may leave with speeds ~ 0.2 c, the efficiency of energy extraction is only $\sim 10^{-3}$ in physically reasonable cases.

There are two ways of producing gravitational ejection. One is by imparting enough kinetic energy to a particle so that it escapes. The other is by removing enough binding energy so that a particle can escape with its original kinetic energy. Processes which may cause a loss of binding energy are gravitational radiation, or the explosive expulsion of intergalactic gas. The effects of a sudden loss of binding energy in a cluster of particles are discussed by Field and Saslaw (1971), Aarseth and Saslaw (1972) and Case (1972). This mass release does not generally produce a preferred direction of ejection, so it is probably not relevant to most galactic nuclei, although it may be related to the virial mass discrepancy in clusters of galaxies.

Perhaps the simplest method of gravitational ejection, indeed the simplest ejection process of any type, is the gravitational slingshot. This idea harkens back to Lagrange (1783) who noticed that the Newtonian 3-body problem is generally unstable. Consider three mass points interacting gravitationally with negative total energy. If the masses of two of these particles are much less than that of the third, the two will become relatively stable satellites of the most massive body, and nothing very interesting from the point of view of ejection is likely to happen. However, if the three bodies have masses and separations of the same order of magnitude, the system is rapidly unstable. Two of the objects can give enough kinetic energy to the third so that it escapes. Energy conservation requires that the two remaining objects from a compact binary, and linear momentum conservation requires the binary to move in the opposite direction to the ejected object.

Clearly then, if three or more massive objects can form (e.g. by fragmentation of gas clouds or by coalescence of colliding stars) in a galactic nucleus, the gravitational slingshot may eject them from the galaxy. In systems with many objects, numerical computations indicate that after several crossing-times the three most massive objects will tend to interact strongly at the centre in this way. To explore the gravitational slingshot in detail, Saslaw *et al.* (1974) computed some 25000 cases of three-body scattering and several hundred cases of four-body interaction, for a wide range of initial orbital parameters and mass ratios.

The main features of ejection by the gravitational slingshot are:

(1) the double configuration characteristic of many extragalactic radio sources follows from conservation of momentum, with one component of the source a single massive object and the other component a binary, provided only that the ejection velocities exceed the escape velocity from the parent galaxy or quasar.

(2) the numerical experiments indicate that the ejection velocities are likely to be relatively low – a few thousand km s^{-1} rather than relativistic. There is some evidence for non-relativistic velocities (Mackay, 1973).

(3) the massive objects can account for compact components observed in many radio sources, without the need to invoke additional confinement mechanisms.

(4) the massive objects are nearly always ejected close ($\lesssim 20°$) to the plane of their total angular momentum. Thus if this plane is also approximately the plane of the galaxy (because the massive objects formed from material whose angular momentum was typical of the galaxy), the objects would be ejected roughly in the plane of the galaxy. This feature is especially important since the other uncontrived ejection mechanisms calculated in detail predict that matter escapes perpendicular to the galactic plane, approximately parallel to the polar axis along the path of least resistance. This can be checked observationally. Present evidence suggests that double radio sources tend to be ejected in the plane of their parent galaxy (Mackay, 1971; Bridle and Brandie, 1973). Some consequences of the passage of such massive objects through the interstellar medium of a galaxy are discussed by Saslaw and De Young (1972).

The problems of the gravitational slingshot approach to the structure of extra-galactic radio sources are mainly to understand the formation and stability of the

WILLIAM C. SASLAW

TABLE I

Properties of ejection processes

Ejection mechanism	Primary ejecta	Jet[a]	Natural consequences		Ejection relative to polar axis	Much quantitative calculation
			Double symmetry	Compact radio components[a]		
Magnetic wind up	gas	yes	yes	no	‖	yes
Magnetic flares	gas	yes	yes	no	‖	yes
Recoiling magnetoid	magnetoid	yes	?	yes	?	no
Radiation pressure	gas	no	no	no	...	yes
Buoyant bubble	gas	?	?	no	‖	no
Galactic wind	gas	no	no	no	...	yes
Hydrodynamic explosion	gas	no	?	no	‖	yes
Tube of toothpaste	gas, star bits	yes	?	?	?	no
Mass release	compact objects, gas	no	no	yes	...	yes
Gravitational slingshot	compact massive objects	yes	yes	yes	⊥	yes

[a] without additional confinement mechanisms

massive objects, and their detailed radiation mechanisms. These properties are more complex than the ejection process, and they need careful investigation.

Table I summarizes some important aspects of these ejection mechanisms, including a rather subjective impression of whether the ejection process has been calculated in detail. From the observations it is clear that no one ejection process can operate universally. From the theories of the last few years we now know several possible ways to expel matter from the nuclei of galaxies. But the secret still sits in the middle. Perhaps by the next General Assembly of the IAU, the more likely mechanisms will be sorted out, and this ring of possibilities will start to change into an inward spiral.

Part of this review was written at the Aspen Center for Physics during the summer of 1973. It was completed during a visit to St Andrew's College of the University of Sydney, and I am happy to thank the Principal and Fellows of St Andrew's College for their friendly hospitality.

References

Aarseth, S. J. and Saslaw, W. C.: 1972, *Astrophys. J.* **172**, 17.
Ambartsumian, V. A.: 1965, in *The Structure and Evolution of Galaxies*, Wiley Interscience, New York, p. 1.
Anand, S. P. S. and Shara, M. M.: 1972, *Astrophys. Space Sci.* **16**, 171.
Andrillat, Y. and Souffrin, S.: 1971, *Astron. Astrophys.* **11**, 286.
Appenzeller, I. and Fricke, K. J.: 1972, *Astron. Astrophys.* **21**, 285.
Arny, T. T.: 1970, *Monthly Notices Roy. Astron. Soc.* **148**, 63.
Arny, T. T.: 1971, *Astrophys. J.* **169**, 289.
Arny, T. T. and Weissman, P.: 1973, *Astron. J.* **78**, 309.
Arons, J., Gunn, J. E., Kulsrud, R. M., and Ostriker, J. P.: 1974, in preparation.
Arp, H. and Bertola, F.: 1970, *Astrophys. Letters* **6**, 65.

Bardeen, J. M.: 1970, *Nature* **226**, 64.
Bardeen, J. M., Press, W. H., and Teukolsky, S. A.: 1972, *Astrophys. J.* **178**, 347.
Bergeron, J. and Salpeter, E. E.: 1973, *Astron. Astrophys.* **22**, 385.
Bisnovatni-Kogan, G. S. and Sunyaev, R. A.: 1972, *Soviet Astron.* **15**, 697.
Blandford, R. D. and Rees, M. J.: 1972, *Astrophys. Letters* **10**, 77.
Bonometto, S. A. and Saggion, A.: 1973, *Astrophys. Letters* **13**, 193.
Bridle, A. H. and Brandie, G. W.: 1973, *Astrophys. Letters* **15**, 21.
Burbidge, G. R.: 1970, *Ann. Rev. Astr. Astrophys.* **8**, 369.
Burbidge, E. M. and Burbidge, G. R.: 1965, *Astrophys. J.* **142**, 1351.
Burbidge, E. M. and Burbidge, G. R.: 1971, *Astrophys. J. Letters* **163**, L21.
Burke, J. A.: 1968, *Monthly Notices Roy. Astron. Soc.* **140**, 241.
Case, L. A.: 1972, *Astrophys. J.* **173**, 665.
Cavaliere, A., Morrison, P., and Wood, K.: 1971, *Astrophys. J.* **170**, 223.
Chertoprud, V. E., Gudzenko, L. I., and Ozernoy, L. M.: 1973, *Astrophys. J. Letters* **182**, L53.
Colgate, S. A.: 1967, *Astrophys. J.* **150**, 163.
Daltabuit, E. and Cox, D.: 1972, *Astrophys. J. Letters* **173**, L13.
Davidson, K.: 1972, *Astrophys. J.* **171**, 213.
Dent, W. A.: 1972a, *Astrophys. J. Letters* **175**, L55.
Dent, W. A.: 1972b, *Science* **175**, 1105.
de Sabbata, V., Fortini, P., and Gualdi, C.: 1972, *Astrophys. Letters* **12**, 87.
De Young, D. S.: 1968, *Astrophys. J.* **153**, 633.
De Young, D. S.: 1971, *Astrophys. Letters* **9**, 43.
De Young, D. S., Roberts, M. S., and Saslaw, W. C.: 1974, *Astrophys. J.* **185**, 809.
Disney, M.: 1973, *Astrophys. J. Letters* **181**, L55.
Disney, M. J. and Cromwell, R. H.: 1971, *Astrophys. J. Letters* **164**, L35.
Downes, D. and Martin, A. H. M.: 1971, *Nature* **233**, 112.
Field, G. B. and Saslaw, W. C.: 1971, *Astrophys. J.* **170**, 199.
Fricke, K. J.: 1973, *Astrophys. J.* **183**, 941.
Frost, R.: 1951, 'The Secret Sits' in *Complete Poems*, Jonathan Cape, London, p. 394.
Gerlach, U. H.: 1971, *Astrophys. J.* **168**, 481.
Gull, S. F. and Northover, K. J. E.: 1973, *Nature* **244**, 80.
Hansen, R. O.: 1971, *Astrophys. J.* **170**, 557.
Harrison, E. R.: 1971, *Monthly Notices Roy. Astron. Soc.* **154**, 167.
Heeschen, D. S., Morgan, W. W., and Walborn, N. R.: 1971, *Astrophys. J. Letters* **165**, L65.
Hoyle, F. and Fowler, W. A.: 1963a, *Monthly Notices Roy. Astron. Soc.* **125**, 169.
Hoyle, F. and Fowler, W. A.: 1963b, *Nature* **197**, 533.
Hoyle, F. and Narlikar, J. V.: 1966, *Proc. Roy. Soc. A.* **290**, 177.
Illarionov, A. F. and Sunyaev, R. A.: 1972, *Astrophys Space Sci.* **19**, 61.
Jeans, J. H.: 1929, *Astronomy and Cosmogony*, Cambridge University Press, p. 352.
Johnson, H. E. and Axford, W. I.: 1971, *Astrophys. J.* **165**, 381.
Jones, T. W. and Kellogg, P. J.: 1972, *Astrophys. J.* **172**, 283.
Kaneko, N., Toyama, K., and Nichimuna, M.: 1972, *Astrophys. Space Sci.* **18**, 121.
Kardashev, N. S.: 1970, *Sov. Astron.* **14**, 375.
Kellermann, K. I., Jauncey, D. L., Cohen, M. H., Shaffer, B. B., Clark, B. G., Broderick, J., Rönnäng, B., Rydbeck, O. E. H., Matveyenko, L., Moiseyev, I., Vitkevitch, V. V., Cooper, B. F. C., and Batchelor, R.: 1971, *Astrophys. J.* **169**, 1.
Krishna Swamy, K. S.: 1971, *Astrophys. J.* **167**, 63.
Kronberg, P. P., Pritchet, C. J., and van den Bergh, S.: 1972, *Astrophys. J. Letters* **173**, L47.
Kunkel, W. E. and Bradt, H. V.: 1971, *Astrophys. J. Letters* **170**, L7.
Lagrange, J.: 1873, in M. J. A. Serret (ed.), *Oeuvres*, vol. 6, Gauthier-Villars, Paris.
Landstreet, J. P. and Angel, J. R. P.: 1972, *Astrophys. J. Letters* **174**, L127.
Larson, R. B.: 1969, *Monthly Notices Roy. Astron. Soc.* **145**, 405.
Levich, E. V. and Sunyaev, R. A.: 1971, *Soviet Astron.* **15**, 363.
Levich, E. V., Sunyaev, R. A., and Zeldovich, Ya. B.: 1972, *Astron. Astrophys.* **19**, 135.
Lynden-Bell, D. and Rees, M. J.: 1971, *Monthly Notices Roy. Astron. Soc.* **152**, 461.
Lynds, C. R.: 1970, *Astrophys. J. Letters* **159**, L151.
MacAlpine, G. A.: 1972, *Astrophys. J.* **175**, 11.

Mackay, C. D.: 1971, *Monthly Notices Roy. Astron. Soc.* **151**, 421.
Mackay, C. D.: 1973, *Monthly Notices Roy. Astron. Soc.* **162**, 1.
Mashhoon, B.: 1973, *Astrophys. J. Letters* **181**, L65.
Mathews, W. G.: 1972, *Astrophys. J.* **174**, 101.
Mathews, W. G. and Baker, J. C.: 1971, *Astrophys. J.* **170**, 241.
Mathis, J. S.: 1967, *Astrophys. J.* **147**, 1050.
Milne, E. A.: 1926, *Monthly Notices Roy. Astron. Soc.* **86**, 459.
Mushotzky, R. F., Solomon, P. M., and Strittmatter, P. A.: 1972, *Astrophys. J.* **174**, 7.
Ne'eman, Y. and Tauber, G.: 1967, *Astrophys. J.* **150**, 755.
Norman, C. A. and ter Haar, D.: 1973, *Astron. Astrophys.* **24**, 121.
Novikov, I. D.: 1965, *Soviet Astron.* **8**, 857.
Novikov, I. D. and Thorne, K. S.: 1972, Les Houches Lectures.
O'Connell, D. J. K. (ed.): 1971, *Nuclei of Galaxies*, North-Holland Publ. Co., Amsterdam.
Ozernoy, L. M.: 1973, *Soviet Astron.* **16**, 916.
Ozernoy, L. M. and Somov, B. V.: 1971, *Astrophys. Space Sci.* **11**, 264.
Ozernoy, L. M. and Usov, V. V.: 1971, *Astrophys. Space Sci.* **13**, 3.
Ozernoy, L. M. and Usov, V. V.: 1973a, *Astrophys. Letters* **13**, 209.
Ozernoy, L. M. and Usov, V. V.: 1973b, *Astrophys. Space Sci.* **25**, 149.
Pacholczyk, A. G.: 1970, *Astrophys. J. Letters* **161**, L207.
Peebles, P. J. E.: 1972, *Astrophys. J.* **178**, 371.
Penston, M. V., Penston, M. J., Neugebauer, G., Tritton, K. P., Becklin, E. E., and Visvanathan, N.:
    1971, *Monthly Notices Roy. Astron. Soc.* **153**, 29.
Press, W. H. and Thorne, K. S.: 1972, *Ann. Rev. Astron. Astrophys.* **10**, 335.
Quirk, W. J. and McKee, C. F.: 1971, *Astrophys. J.* **169**, 119.
Rees, M. J.: 1971, *Nature* **229**, 312.
Rees, M. J., Silk, J. I., Werner, M. M., and Wickramasinghe, N. C.: 1969, *Nature* **223**, 788.
Rieke, G. H. and Low, F. J.: 1972a, *Astrophys. J. Letters* **176**, L95.
Rieke, G. H. and Low, F. J.: 1972b, *Astrophys. J. Letters* **177**, L115.
Roxburgh, I. W. and Saffman, P. G.: 1965, *Monthly Notices Roy. Astron. Soc.* **129**, 181.
Ryan, M. P.: 1972, *Astrophys. J. Letters* **177**, L79.
Sakashita, S.: 1971, *Astrophys. Space Sci.* **14**, 431.
Salpeter, E. E. and Wagoner, R. V.: 1971, *Astrophys. J.* **164**, 557.
Sanders, R. H.: 1970, *Astrophys. J.* **162**, 791.
Saslaw, W. C.: 1973, *Publ. Astron. Soc. Pacific* **85**, 5.
Saslaw, W. C. and De Young, D. S.: 1972, *Astrophys. Letters* **11**, 87.
Saslaw, W. C., Valtonen, M. J., and Aarseth, S. J.: 1974, *Astrophys. J.* **190**, 253.
Scargle, J. D.: 1973, *Astrophys. J.* **179**, 705.
Scharlemann, E. T. and Wagoner, R. V.: 1972, *Astrophys. J.* **171**, 107.
Schwarzschild, M.: 1973, *Astrophys. J.* **182**, 357.
Sciama, D. W. and Rees, M. J.: 1967, *Nature* **216**, 147.
Seidle, F. G. P. and Cameron, A. G. W.: 1972, *Astrophys. Space Sci.* **15**, 44.
Shapiro, S. L.: 1973, *Astrophys. J.* **180**, 531
Shklovsky, J.: 1970, *Nature* **228**, 1174.
Spitzer, L.: 1942, *Astrophys. J.* **95**, 329.
Spitzer, L.: 1971, in D. J. K. O'Connell (ed.), *Nuclei of Galaxies*, North-Holland Publ. Co., Amster-
    dam, p. 443.
Spitzer, L. and Saslaw, W. C.: 1966, *Astrophys. J.* **143**, 400.
Spitzer, L. and Stone, M. E.: 1967, *Astrophys. J.* **147**, 519.
Steigman, G. and Strittmatter, P. A.: 1971, *Astron. Astrophys.* **11**, 279.
Stockton, A.: 1972, *Astrophys. J.* **173**, 247.
Sturrock, P. A.: 1971, *Astrophys. J.* **170**, 85.
Sturrock, P. A. and Barnes, C.: 1972, *Astrophys. J.* **176**, 31.
Tovmassian, H. M.: 1972, *Astrophys. J. Letters* **178**, L47.
van der Kruit, P. C.: 1971, *Astron. Astrophys.* **15**, 110.
Wade, C. M. *et al.*: 1971, *Astrophys. J. Letters* **170**, L11.
Wheeler, J. A.: 1971, in D. J. K. O'Connell (ed.), *Nuclei of Galaxies*, North-Holland Publ. Co.,
    Amsterdam, p. 560.

Williams, R. E.: 1972, *Astrophys. J.* **178**, 105.
Wilson, J. R.: 1972, *Astrophys. J.* **173**, 431.
Wolfe, A. M.: 1974, *Astrophys. J.* **188**, 243.
Wolfe, A. M. and Burbidge, G. R.: 1970, *Astrophys. J.* **161**, 419.
Zel'dovich, Ya. B. and Novikov, I. D.: 1971, *Relativistic Astrophysics*, Univ. of Chicago Press.

DISCUSSION

G. de Vaucouleurs: I am surprised that you have not discussed the effects of capture of infalling material by galactic nuclei.

Saslaw: I did not do so because I considered that such matter would already have formed the main constituents of the nuclei, but I agree that it's an important process to be considered in the detailed evolution of nuclei since it may alter the chemical composition.

E. M. Burbidge: I have a comment concerning radiation-driven gas outflow from nuclei of galaxies and QSOs and the way in which radiation pressure outwards can balance gravity acting inwards. Such a balance depends on constancy of light, but in NGC 4151 the He I absorption lines are variable (the triple structure observed at Lick charged in a time ~ 1 yr) and its nuclear light is also variable. None of the QSOs with multiple absorption line redshifts are known to be variable, and the absorption lines have remained visible in some for a few years. It would be interesting to know to what level one can say their light is constant: smallish variations should destroy the balance and the absorption systems should change.

Saslaw: Your first point about the association between constant luminosity and constant velocity in the absorption lines is an important one, and NGC 1275 would be an especially interesting case to study. However, we should keep in mind that the luminosity of these objects may well come from several sources in the nucleus, and not all these sources need have constant luminosity. Thus while a positive correlation between constant luminosity and multiple absorption line systems would be significant, lack of correlation would not provide a strong test if there are multiple sources of luminosity and some of them vary.

E. M. Burbidge: You seemed to suggest that galactic winds flowing outward are a necessary consequence of stellar evolution in the central region. Wouldn't this conflict with the infall of gas postulated to form new stars in the nuclei in dense configurations?

Saslaw: The conditions necessary for galactic winds to occur depend on the temperature and density of the interstellar gas and there wasn't time to describe this in detail. Roughly, if the temperature is high enough ($\gtrsim 10^6$ K), a wind may flow out of the nucleus. For a low temperature ($\lesssim 10^4$ K), pressure support is insufficient and the gas flows into the nucleus. At intermediate temperatures the situation is especially complex since the gas is prone to thermal and hydrodynamic instabilities. The exact dependence on temperature and density, of course, has to be found from the detailed models, but there seems to be a regime of inflow and a regime of outflow.

Arp: With respect to the gravitational sling-shot model, it is interesting to note that the companions on the ends of spiral arms – which I argued had been ejected from the nucleus in the plane – that some of these companions are on the end of curiously *doubled* spiral arms. From what you just said the sling-shot model would give a double ejection in one of the two opposite directions.

Saslaw: The separation of the double would be extremely small, perhaps only 1/100 pc or less.

Arp: Yes, but it would probably be unstable and come apart.

G. Burbidge: Suppose strange things are happening (such as Arp was suggesting this morning) and I'll take money on that, how would your ideas be affected?

Saslaw: If it were indeed the case that nuclei are ejecting other, very massive, compact objects then a knowledge of the velocities would be important. I think the sling-shot idea is in this case the only feasible one which has been looked at in detail. Newtonian calculations indicate velocities of a few thousand kilometres per second, and I don't know whether detailed relativistic calculations would predict relativistic velocities or not. There may then be the problem of blue-shifts, not yet observed. That's one I'll leave to you.

G. Burbidge: That's an almost insuperable difficulty.

G. de Vaucouleurs: I notice that most of your proposed mechanisms imply ejection of matter parallel to the rotation axis. This brings to mind the explosion in M82 and also a study by Holmberg (*Arkiv Astron.* **5**, 305. 1969) of the angular distribution of companions of galaxies seen on edge, where he found a significant excess of companions near the minor axis.

Saslaw: We should bear in mind that several of these mechanisms may be operating and it may be dangerous to try to account for everything in terms of one mechanism; history has shown that the Universe is complex. One of the interesting questions relating to this is whether the components of double radio sources associated with elliptical galaxies tend to lie along the major or the minor axes of the galaxies. There have been two studies, one by Mackay (*Monthly Notices Roy. Astron. Soc.* **151**, 421, 1971) and one by Bridle and Brandie (*Astrophys. Letters* **15**, 21, 1973). The statistics are small and it seems to be too early to draw firm conclusions from their results, but the indication is that the components are preferentially aligned with the major axes, i.e. with the planes of the galaxies rather than the rotation axes. It is most important to improve these statistics, since it is possible in this way to decide between the different mechanisms of ejection.

Oort: In considering nuclei of galaxies it is useful to look at the nucleus of our own Galaxy, which is one that we can observe in considerably greater detail. The infrared observations by Rieke and Low show a structure more complex than has been considered in Dr Saslaw's communication. There are five or more distinct patches of diameter $\frac{1}{4}$ pc or less, all contained in a region of about 1 pc diameter. The observations of the CO emission around Sgr A and Sgr B2 indicate something else that is unexpected, viz. a 'wind' with a velocity of about 50–80 km s^{-1} apparently blowing past these two sources and having a rather steep velocity gradient.

G. de Vaucouleurs: In barred spirals such as NGC 1365 one also gets the impression of a flow through the nucleus.

Sargent: One mechanism you mentioned for forming quasars is to have stellar collisions in very dense stellar systems in the nuclei of galaxies. Now the densest stellar systems we know of, such as the nucleus of M31, have star densities of around 10^5 stars pc^{-3} which is far short of the values required by the stellar collision hypotheses. Have you given any thought to whether systems which are nearly dense enough to become active could be detected, for example in the compact galaxies?

Saslaw: In principle it should be possible to detect such systems. The main problem (apart from getting the necessary resolution) is to look for systems where most of the nuclear radiation is still thermal so that the presence of stars could be established more directly.

Rickard: One has to be careful about observations of concentrations in the nuclei of galaxies. I've taken photographs in Chile, where the seeing is occasionally very good, of some of the 'hot spots' pointed out by Dr de Vaucouleurs in southern galaxies, and resolved them into chains of H II regions. So there is nothing unusual about these particular irregular concentrations.

van den Bergh: Inspection of plates of NGC 4151 shows that the Seyfert nucleus is embedded in a large disk with a rather pathological structure. The fact that this structure extends to a large distance from the nucleus suggests that the nuclear disturbance has been going on for a considerable time.

Saslaw: Is there any evidence that it might have been going on in spurts but not continuously?

G. Burbidge: How many galaxies have been looked at in such detail as NGC 4151, so that one can be sure it really is pathological?

van den Bergh: I think NGC 4151 is quite unique.

G. Burbidge: This contradicts to some extent what was said earlier, i.e. that Seyfert galaxies are fairly normal systems outside their nuclei.

van den Bergh: NGC 4151 is the only case I know of where the outer structure is so peculiar.

G. de Vaucouleurs: It would be very important to know whether Seyfert galaxies are entirely normal outside the nucleus or whether there are features outside the nucleus which may be associated with the nuclear activity. Some authors have recently drawn attention to the outer ring in NGC 1068 as a possible example. Now, of the 12 classical Seyfert galaxies, 3 (NGC 1068, 3516, 7469) have outer ring structures. This may seem at first sight possibly significant until one realises that Seyferts are confined to a rather narrow range in the Hubble sequence, from intermediate lenticulars to Sbc, and that about one-quarter of all galaxies in this range have an outer ring structure.

Heidmann: From the point of view of neutral hydrogen radiation, Seyfert galaxies are not different from normal galaxies, either when looking at the relative amount of neutral hydrogen or its velocity distribution (ref. *Astron. Astrophys.* **6**, 453, 1970).

Disney: I would like to ask Dr Ulrich if there is a real continuity between Seyfert and elliptical nuclear spectra. I have the impression that you do not find strong optical nuclei in ellipticals and that the line widths are nowhere near so broad.

Ulrich: I agree.

Ekers: Since the observations of extended radio sources associated with QSOs argue strongly for QSOs being in the nuclei of E galaxies, it seems surprising that the optical spectra of the QSOs resemble those of the nuclei of Seyfert galaxies, which are spirals, more than they do those of ellipticals.

Ulrich: It does not seem that there is any tight correlation between the amounts of energy released in various ways: for example, the infrared luminosity is not correlated with the X-ray luminosity or with the width of the hydrogen lines.

van Woerden: If Seyferts are cores of spirals and quasars are cores of ellipticals, can there be a transition between Seyferts and quasars, just as there is a transition between spirals and ellipticals?

G. Burbidge: The answer to that is clearly yes.

G. de Vaucouleurs: (i) Seyferts *are* spirals or lenticulars, not just cores.

(ii) There have been suggestions that Seyfert *nuclei* may be related to quasars.

(iii) There is no evidence of a transition between spirals and ellipticals, except in a classificatory sense, lenticulars being placed between ellipticals and spirals; no evolution along the sequence is implied and there seems to be a discontinuity between lenticulars and ellipticals.

Heidmann: Yes, I would like to emphasize this point: the relative amount of neutral hydrogen in ellipticals is much smaller than one would expect from an extrapolation of its variation along the Hubble sequence from irregulars down to lenticulars; there is a break in the sequence between ellipticals and lenticulars (*Astron. Astrophys.* **25**, 451, 1973).

Freeman: Ekers points out how double radio sources occur in elliptical galaxies only. Many lenticulars appear mainly elliptical in the sense that they have dominant bulge components. It would be interesting to observe some lenticulars in the continuum, to see if they look more like ellipticals than spirals (assuming they radiate at all in the continuum).

Larson: I'd like to make a comment in favour of open-mindedness in the interpretation of anomalies in the outer structure of galaxies with active nuclei. People have generally tried to interpret such anomalies as being a result of the activity in the nucleus, but it is also possible that in some cases the anomalous outer structure and the nuclear activity may have a common cause connected with the overall dynamics or evolution of the galaxy. For example, if nuclear activity has anything to do with condensation of matter into the nucleus, it may be that the structural anomalies could be interpreted as manifestations of the same condensation process on a larger scale. Either the infall of intergalactic gas or the continuing condensation of a remnant protogalactic envelope could be a possible cause of such a situation.

G. Burbidge: Are you thinking in terms of accretion from outside, or processes which conserve mass in a galaxy?

Larson: Perhaps just condensation of gas that's already in a galaxy or near it, such as the neutral hydrogen halos round M81 and M82.

G. Burbidge: There is no evidence for H I near ellipticals, which are sometimes the seat of violent activity, and it's hard to argue in these cases that the activity is due to gas. It must be attributed to stars in the nucleus or to something else.

Osmer: At Cerro Tololo, M. Smith, D. Weedman and I have been making spectrophotometric observations of the southern galaxies with peculiar nuclei which were found by Sersic and Pastoriza, as well as the Seyferts NGC 1566 and 3783. The former galaxies have luminosities in $H\beta$ that in many cases are as large as are found in Seyfert galaxies, although the gaseous regions are also much larger in physical extent. NGC 3783 is a barred spiral with an apparently normal Seyfert-type nucleus, while NGC 1566 has the lowest $H\beta$ luminosity yet known, its spectrum otherwise being similar to that of NGC 1068. The Balmer lines had widths of several thousand km s^{-1} in January 1973.

Freeman: We have a collection of spectra of the nucleus of the ex-Seyfert galaxy NGC 1566. In 1956 $H\beta$ was about 3000 km s^{-1} wide (de Vaucouleurs). In 1968 $H\beta$ was clearly double – one component had the same redshift as the forbidden lines and the other was about 800 km s^{-1} to the red. By March 1972 $H\beta$ was very sharp, rather weaker, and single, so the spectrum could not be described as Seyfert-like.

Cox: Infall of material in a galaxy can proceed at low densities and high temperature until a rapid cooling occurs near the centre, bringing about a high density contrast. It seems to me that a very modest infall over a long period of time can cause a large accumulation of matter in the nucleus. Thus we might easily end up with any of several of the dramatic types of objects Saslaw has out-lined.

G. Burbidge: Shklovsky (*Astron. Zh.* **39**, 591, 1962) considered this possibility for the radio source M87 some years ago, but the necessary accretion rates were very high.

Cox: I do not suggest that such a modest infall could sustain the required power output in a steady state, but could slowly generate an object which would then be dramatic for a limited period of time.

Larson: I don't think that the suggested role of infall or condensation processes as a contributing cause of nuclear activity is necessarily inconsistent with the absence of any clear evidence of this process in many galaxies showing nuclear activity. One should distinguish two stages in time: an earlier stage when gas inflow proceeds vigorously and is the dominant hydrodynamic process, and a later stage when a sufficient amount of mass has gone into pulsars (or other energy sources) to produce intense quasar activity. By this later time the infall process may have diminished in intensity, and in any case it is likely that the quasar activity itself might tend to reverse the inflow and disperse much of the residual gas in and around the galaxy. Thus by the time a spectacular object (e.g. quasar) has been produced, the dominant hydrodynamic process may be outflow rather than inflow. Perhaps there is an analogy with the formation of a massive O star: by the time a newborn O star has become visible, the initial collapse process has been terminated and one observes predominantly the gas outflows associated with the development of an expanding H II region.

Tifft: In Virgo, nuclear peculiarities seem to concentrate in the centre of the cluster – *perhaps* where the concentration of intergalactic material is higher. This is in interesting contrast with Coma where peculiarities like emission lines are found mainly in the outer parts of the cluster.

Miley: New measurements with the Westerbork telescope of the structure of NGC 1275 show that there is a component of size $\sim 30''$ to $1'$, comparable with the size of the optical galaxy. The flux density amounts to about 2×10^{-26} W m^{-2} Hz^{-1} at 1.4 GHz and I wonder if emission from this source could be partially responsible for the new H I absorption results which you reported.

Saslaw: It may be related.

STRUCTURES OF CENTRAL BULGES AND
NUCLEI OF GALAXIES

G. DE VAUCOULEURS

Dept. of Astronomy, The University of Texas at Austin, Tex., U.S.A.

Abstract. The isophotal surfaces of central bulges of normal spirals, often described as 'spheroidal', are shown to depart from ellipsoids of revolution and to resemble MacLaurin or Jacobi spheroids with zonal distortion; two typical examples, NGC 4565, 5746, are illustrated. Weaker examples are NGC 2683, 4594 and 891. Stronger examples are NGC 7332 and 128.

The true nucleus of an ordinary spiral is at most a very small object (~ 10 pc) resolved only in the nearest galaxies. The apparently larger nuclei seen in more distant spirals and lenticulars, in particular the so-called N-types, are probably artifacts due to insufficient resolution. It is possible that even in M31 the small 'nucleus' is merely the seeing-convolved image of the central peak in the $r^{1/4}$ luminosity distribution of the spheroidal component.

In the nearest barred lenticulars and early spirals each large ellipsoidal 'nucleus' ($\sim 1.0 \times 1.5$ kpc) in the centre of the bar or central lens is found to include a second bar and inner nucleus on a smaller scale, in the approximate ratio of the corresponding Jeans' lengths. This is illustrated by high resolution photographs of NGC 1291; another example is NGC 1326.

1. The Nuclear Bulge

Among early-type spirals (Sa-Sb) a distinction should be made between systems with a dominant spheroidal component or large 'bulge', e.g. NGC 4594, and systems with a minimal nucleus and a small bulge, e.g. NGC 4866. This distinction has been noted in classification work (Sandage, 1961), but has no expression in the Hubble scheme in which it detracts from the usual close correlation between arm structure and bulge/disk ratio. It is easily represented, however, by 2-parameter photometric models including spheroidal and exponential components.

A curious property of the nuclear bulge of spirals is the shape of the isophotal surfaces of the spheroidal component. These are usually described as ellipsoids of revolution. However, careful inspection of photographs of edge-on spirals and lenticulars discloses that the isophotes of the central bulge are often less convex near the minor axis than would be the case if they were true ellipses and are even slightly concave in a few cases. In other words they resemble MacLaurin or perhaps Jacobi spheroids with marked zonal distortions rather than simple ellipsoids of revolution. This is obvious in extreme cases such as NGC 128 whose isophotes are actually peanut-shaped, and to a lesser extent in NGC 7332, both of which are lenticulars described as 'peculiar' with a 'box-shaped central region' in the Hubble Atlas (Sandage, 1961). A similar shape is also easily detectable in several edge-on spirals, such as NGC 891 and 5746 (Figure 1) and it is still perceptible although by no means obvious in many others, for example NGC 2683, 4565, 4594. This shape may well be a general property of the spheroidal component which should be quantitatively accounted for by realistic dynamical models.

John R. Shakeshaft (ed.), The Formation and Dynamics of Galaxies. 335–340. All Rights Reserved.
Copyright © 1974 by the IAU.

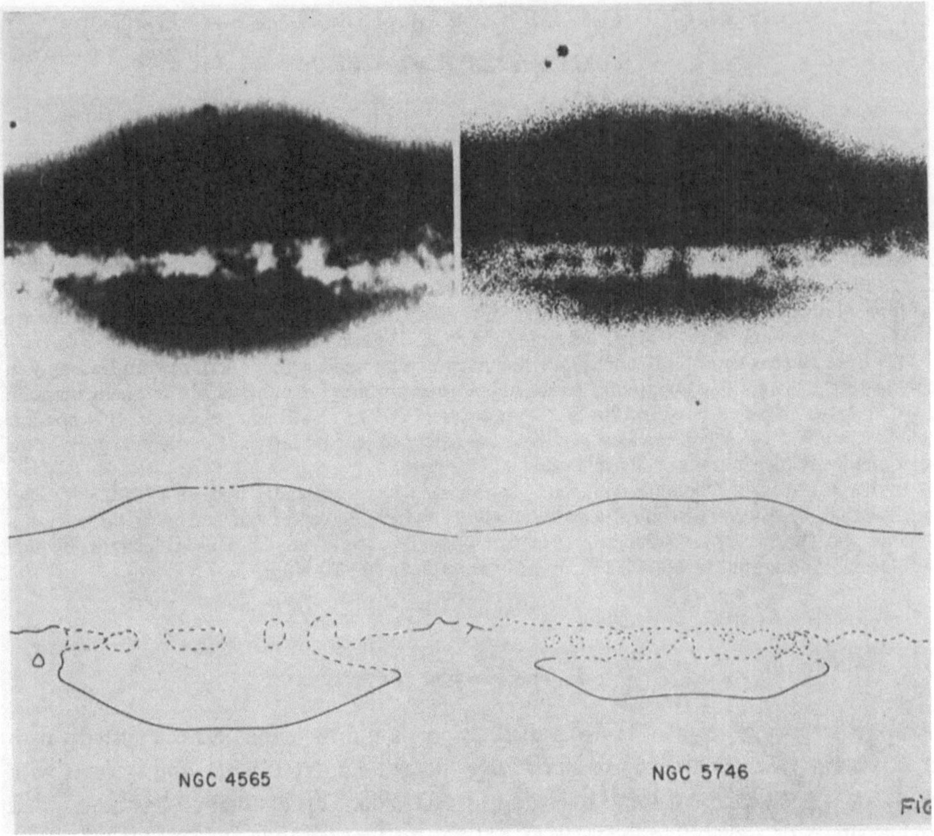

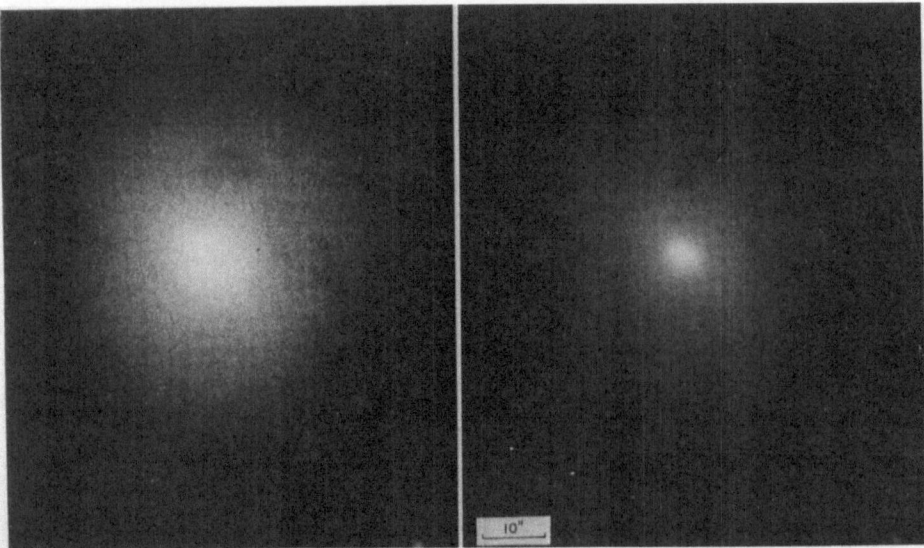

Fig. 1.   The isophotes of the central bulges of NGC 4565 and NGC 5746 depart markedly from ellipses and resemble MacLaurin spheroids with zonal distortions.

Fig. 2.   The very small nucleus of M31 is a typical example of the central peak in the luminosity distribution of the spheroidal component of ordinary spirals. McDonald Observatory, 205-cm Struve reflector, Cassegrain focus (1960 November 15, 103a-0, 10 min; original scale: 7˝5 mm⁻¹).

2. Nuclei of Ordinary Lenticulars and Spirals

In the nearest spirals (M31, M33) the high angular resolution available, relative to the size of the system, allows us to detect a very small ellipsoidal object (about $3\overset{''}{.}3 \times 2\overset{''}{.}4 = 15 \times 10$ pc in M31; Figure 2) rather similar to a globular cluster in size although not in structure or composition (Johnson, 1961; Kinman, 1965).

If this object is a typical example of the true nuclei of ordinary spirals it would not be visible outside the Local Group; at the distance of M81 it would subtend less than 1″, and less than $0\overset{''}{.}1$ at the distance of the Virgo cluster. It is clear that the apparent

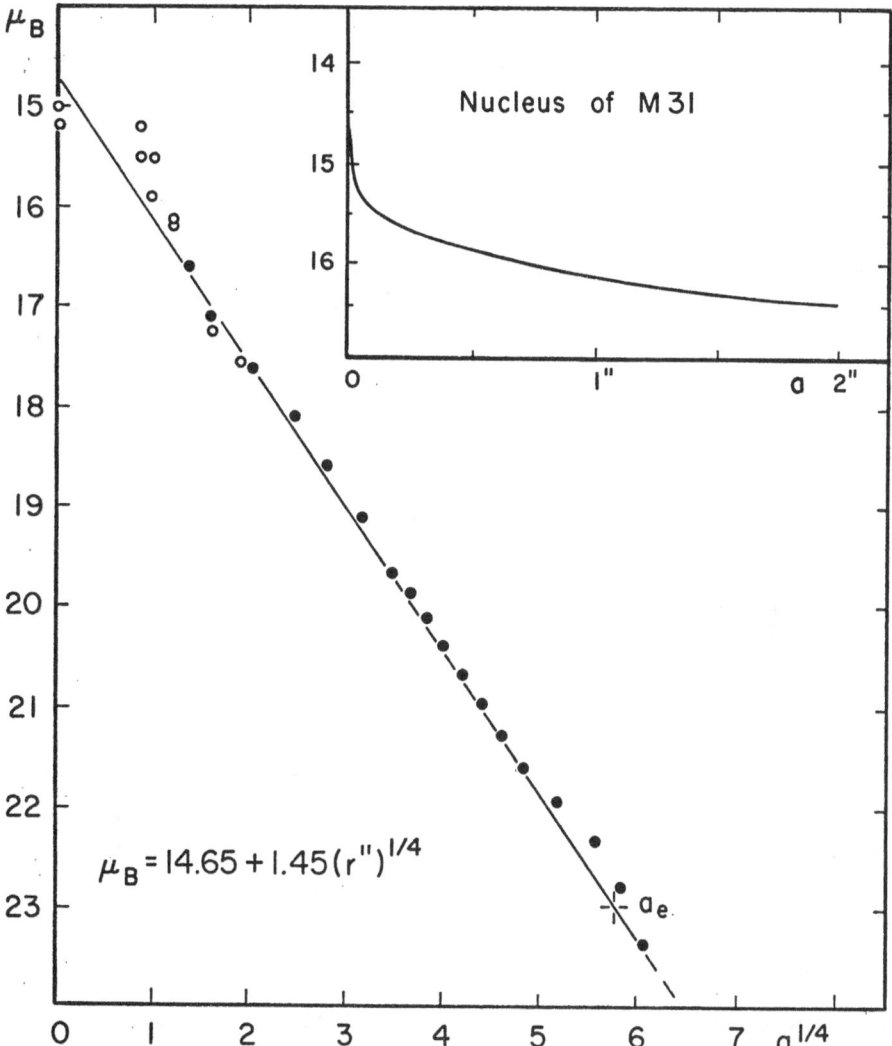

Fig. 3. The luminosity distribution of the spheroidal component of M31 along its major axis follows closely the characteristic $r^{1/4}$ law in the range $2'' < a < 20'$ and, within the errors of roughly decon-volved photographic photometry, even into the 'nucleus'.

'nucleus' seen on photographs of ordinary spirals and lenticulars is to a large extent an artifact due to insufficient resolution; in particular the appearance of many so-called N-type galaxies arises from this effect, especially when a quasi-stellar central source is present. For example, high-resolution photoelectric scans of 3C 120 demonstrate that the luminosity profile of the bright 'nucleus' is not broader than the stellar seeing-disk profiles (de Vaucouleurs and de Vaucouleurs, 1968).

One may ask whether, even in M31, the small photographic nucleus (Figure 2) is really a distinct object and a separate photometric entity, or perhaps is merely the seeing-convolved image of the central peak in the very steep luminosity distribution of the bulge, which is known to follow closely the $r^{1/4}$ law of the spheroidal component at least to $a \simeq 20'$ along the major axis (de Vaucouleurs, 1958).

A plot (Figure 3) of available photoelectric and corrected photographic data (Johnson, 1951; Kinman, 1965) shows that, except perhaps within 2" from the centre where both the original photometry and the (approximate) deconvolution corrections are uncertain, the luminosity distribution follows closely the relation

$$\mu_B = 14.65 + 1.45 \, r^{1/4},$$

where μ_B is the B-system mag (arc sec)$^{-2}$ and r is in arc sec. The corresponding effective radius is $a_e = (3.33 \times 2.5/1.45)^4 = 18'$ where $\mu_B = \mu(0) + 8.32 = 22.97$ mag (arc sec)$^{-2}$. If the relation holds right up to the centre, the central 1.0 mag spike from $\mu_B \simeq 15.65$ at $a \simeq 0''.2$ to $\mu_B \simeq 14.65$ at $a=0$ (Figure 3, inset) might be detectable on the Stratoscope II photographs.

3. Nuclei of Barred Lenticulars and Spirals

In barred spirals the situation is entirely different; here the nucleus is generally a fairly large, distinct ellipsoidal object with typical dimensions of the order of 1.5×1.0 kpc, comparable to the half-width of the bar. This appearance is confirmed by detailed surface photometry, e.g. in NGC 1291, 1433, 1512, etc., which indicates a fairly sharp change of slope in the luminosity profile at the edge of the nucleus. The laws of luminosity distribution in these nuclei are not yet precisely known, because few barred spirals are near enough for high-resolution quantitative analysis. Nevertheless, a remarkable structure, not previously reported, has been noted on high-resolution photographs of the nuclei of two of the largest, nearby, barred systems, NGC 1291 and 1326, which are both classified as (R) SB(s) 0/a, i.e. transition types between late lenticulars and early barred spirals with conspicuous outer rings. Inside the elliptical nucleus at the centre of the bar crossing the primary lens, a second, smaller-scale bar structure is present along the major axis of the large nucleus, or secondary lens, in the centre of which is a second, much smaller nucleus. In other words the lens-bar-nucleus pattern is repeated on two different scales, one inside the other (Figure 4). This repetitive pattern suggests that the nucleus is so much denser than the main bar and lens that its dynamics is almost independent of the extra-nuclear mass distribution, and that the same mechanisms are at work to produce similar structures with different

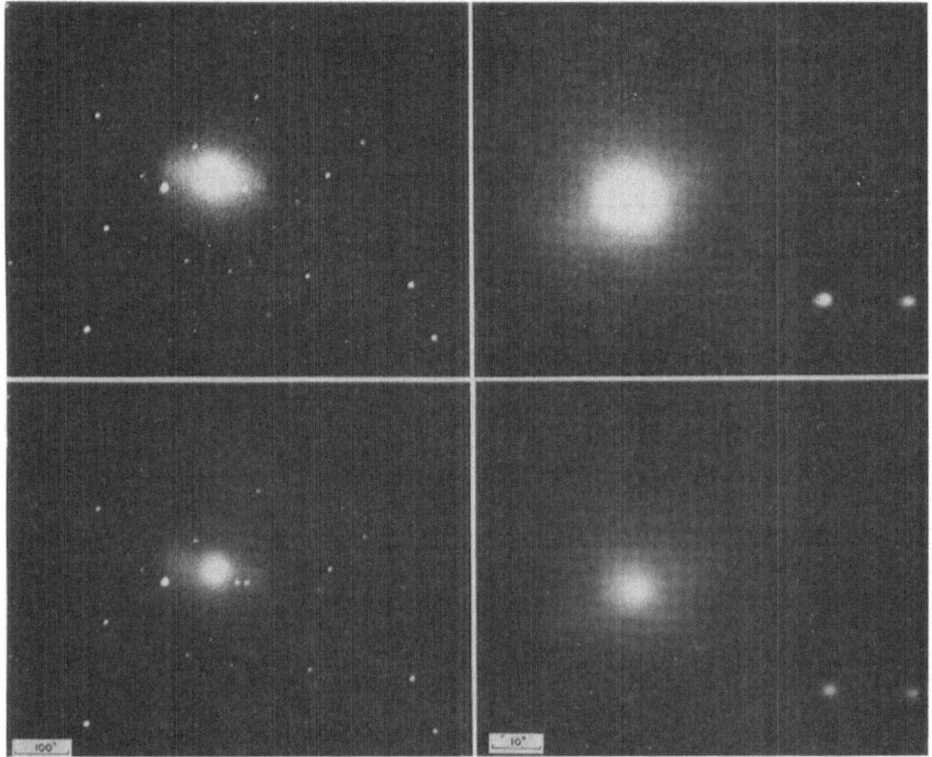

Fig. 4. The lens-bar-nucleus pattern is repeated on two different scales in the large southern barred galaxy NGC 1291, revised type (R) SB(s) 0/a.

At left: the primary lens, bar and 'nucleus' are visible on this small scale photograph. Newtonian focus of Mt. Stromlo Observatory 75-cm Reynolds reflector (1952 November 19, 103a-0, 60 min; original scale: 67″ mm^{-1}).

At right: the secondary lens, bar and inner nucleus are shown on this large scale photograph of the internal structure of the primary 'nucleus'. Cassegrain focus of McDonald Observatory 205-cm Struve reflector (1960 November 14, 103a-0, 65 min; original scale: 7″.5 mm^{-1}). Elongated star images are caused by atmospheric dispersion at $z \simeq 73°$.

scale factors measured, e.g. by the Jeans' length $L = (\sigma_v^2/G\varrho)^{1/2}$. If $\sigma_v = \text{const.}$, $L \propto \varrho^{-1/2}$. Since the mean surface brightness, and presumably the mean density ϱ, in the nucleus of NGC 1291 is about 40 times higher than in the lens, the Jeans' length is 6 to 7 times smaller in the nucleus than in the lens, in approximate agreement with the scale ratio of the two bar structures.

References

de Vaucouleurs, G.: 1958, *Astrophys. J.* **128**, 465.
de Vaucouleurs, G. and de Vaucouleurs, A.: 1968, *Astron. J.* **73**, S174.
Johnson, H. M.: 1961, *Astrophys. J.* **133**, 309.
Kinman, T. D.: 1965, *Astrophys. J.* **142**, 1376.
Sandage, A.: 1961, *The Hubble Atlas of Galaxies*, Carnegie Inst., Washington, D. C.

DISCUSSION

Carrick: I am working on this very subject, the light distributions of the bulges of edge-on systems. Of the six disk galaxies for which I have plates, two definitely show this box-like shape. It appears to be a general phenomenon.

G. de Vaucouleurs: Yes, you can see it plainly on photographs of edge-on spirals such as NGC 891, and Sandage has pointed out in the Hubble Atlas a similar phenomenon in lenticulars, particularly in NGC 7332. So it must be a general phenomenon and theoreticians should explain it.

Mark: There is a possible analogy here with rotating stars. If stars are rotating differentially then one can get isophotes which are 'box-like', though not if they are rotating uniformly. It is possible that, by using the moment equations for stellar dynamics instead of the fluid equations of stellar structure, differential rotation in galactic bulges could give rise to box-like isophotes.

Butcher: May I ask Dr de Vaucouleurs to comment on some observations by Zasov in Moscow, which suggest that the luminosity gradients of the more extreme Seyfert galaxies are steeper than in normal spirals of the same types.

G. de Vaucouleurs: I have not seen this paper, but we have recently completed a detailed study of all classical Seyferts which does not seem to support this view. When the luminosity of the nuclear point source is correctly subtracted, Seyfert galaxies appear normal for their morphological types with respect to luminosity gradients and colours; in one or two cases only is there perhaps a very slight negative colour excess in the innermost regions. This is difficult to assess exactly because of the variable amounts of line emission in the disk.

VARIABILITY OF THE EMISSION LINE SPECTRUM OF THE NUCLEUS OF SEYFERT GALAXY NGC 1275

I. PRONIK

Crimean Astrophysical Observatory, U.S.S.R. Academy of Sciences, Crimea, U.S.S.R.

Abstract. Recent spectroscopic observations of the nucleus of NGC 1275, together with results from earlier workers, confirm the variability of the emission lines and indicate corresponding variations in the electron temperatures and densities of different zones. These changes may be associated with the microwave outbursts from this source.

The nucleus of NGC 1275 is known to be very unstable, exhibiting radio variability (e.g. Kellermann, 1972), UBV variability (Lyutyi, 1972) and infrared variability (Khozov, 1971). We have studied the variability of emission lines from the regions of ionized hydrogen in the nucleus of this galaxy. Such H II regions in Seyfert galaxies were discussed by Dibay and Pronik (1967) who concluded that they can be considered as being divided into several zones having different temperatures and densities. The highest electron temperature T_e and electron density $n_e \sim 10^7$ cm^{-3} correspond to the zone emitting wide hydrogen wings; [O III] and [Ne III] lines are emitted in a region with $T_e \approx 16000$ K and $n_e \approx 3 \times 10^6$ cm^{-3} (called the [O III] zone), while [O II] and [S II] lines originate in the most rarefied region with $T_e \sim 12000$ K and $n_e \approx 4 \times 10^3$ cm^{-3} (called the [O II] zone).

The nucleus of NGC 1275 has been observed at the prime focus of the 2.6-m Shajn telescope with V. Pronik's high-speed spectrograph during the period 1971 October 14 to 1973 April 4. 26 spectrograms with dispersion 380 Å mm^{-1} in the region 3700–

TABLE I

Å	Ion	1930 Dec. (i)	1942 (ii)	1964 Dec.– 1966 Feb. (iii)	1966-1967 (iv)	1968-1971 (v)	1971 Oct.– 1972 Oct.	1972 Dec.– 1973 Apr.
3727 + 29	[O II]	3.1	1.4	3.0	2.5	2.0	3.4	2.1
3869	[Ne III]	0.6	0.4	1.0	0.7		1.1	0.6
4069 + 76	[S II]	3.8	0.5		0.4	0.5	0.7	0.5
4102	Hδ	0.6	0.1	0.2	0.2	0.2	0.1	0.4
4340	Hγ	3.1	0.5	0.4	0.5	0.2	0.2	0.2
4363	[O III]	2.5	0.4		0.3		0.6	0.2
4861	Hβ	1.0	1.0	1.0	1.0	1.0	1.0	1.0
4959 + 5007	[O III]	0.6	3.5	2.7	5.4	5.2	7.3	3.6
6300 + 64	[O I]		1.4		1.6	1.7	3.8	1.3
6563	Hα ⎞		7.0	6.0	11.9	12.9	16.6	4.6
6548 + 83	[N II] ⎠							
6716 + 31	[S II]		2.1	1.0	3.4	3.6	3.9	1.2

References for Table I: (i) Humason (1932). (ii) Seyfert (1943). (iii) Dibay and Pronik (1967). (iv) Anderson (1970). (v) Wampler (1971).

John R. Shakeshaft (ed.), The Formation and Dynamics of Galaxies. 341–345. All Rights Reserved.

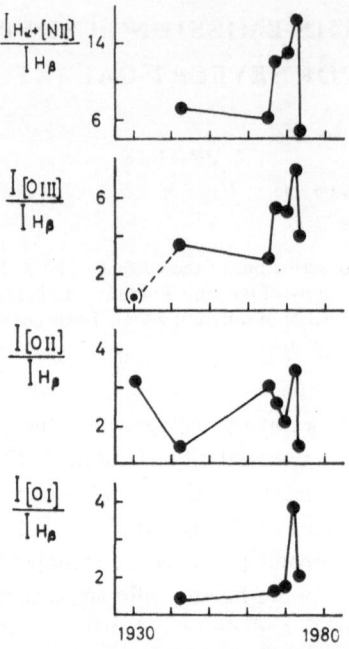

Fig. 1.   Variations of the relative line intensities: [O I] 6300 + 64 Å, [O II] 3727 Å, [O III] 4959 + + 5007 Å, Hα + [N II] 6549 + 83 Å and Hβ, in the gaseous nucleus of NGC 1275.

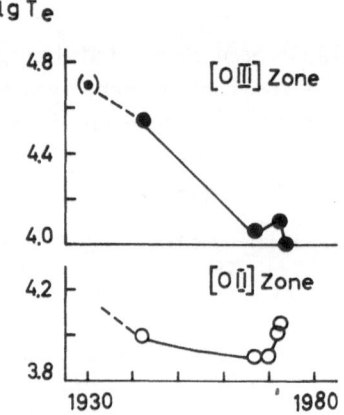

Fig. 2.   Variations of electron temperature T_e in the [O III] (●) and [O II] (○) zones.

6800 Å were obtained, which enabled the relative intensities of the emission lines 3727 [O II], 3869 [Ne III], 4069–76 [S II], Hδ, Hγ, 4363 [O III], Hβ, $(N_1 + N_2)$ [O III], 6300 + 64 [O I], Hα + [N II] and 6716 + 31 [S II] to be measured. The intensities of some of these emission lines were found to vary, the largest variations being between October and December 1972. We have therefore divided all the observational data into two groups: (1) 1971 October to 1972 October, and (2) 1972 December to 1973 April. Our results, combined with others for the years 1930 to 1971, are presented in

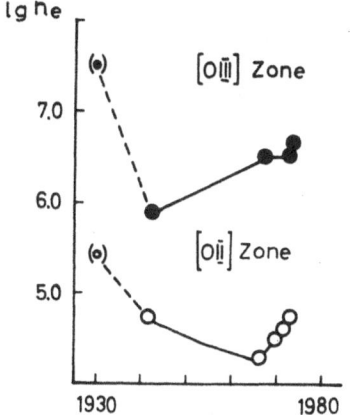

Fig. 3.   Variations of electron density n_e in the [O III] (●) and [O II] (○) zones.

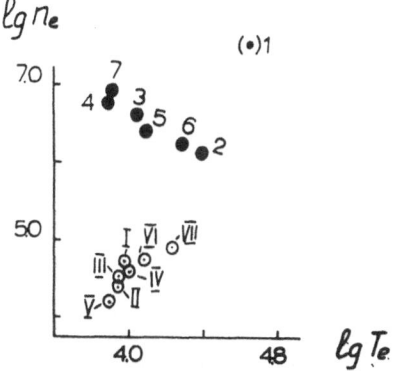

Fig. 4.   The relation $\log T_e$–$\log n_e$ for the [O III] (●) and [O II] (○) zones. The Arabic and Roman numerals refer to dates as follows: 1 – 1930 (Humason, 1932); 2 – 1942 (Seyfert, 1943); 3 – 1966–67 (Anderson, 1970) and author: 4 – 1971, October 12; 5 – 1972 October 11; 6 – 1972 February 6; – – I – 1942 (Seyfert, 1943); II – 1966–67 (Anderson, 1970); III–1968–71 (Wampler, 1971) and author: IV – 1972, April 1; V – 1972, October; VI – 1972, December; VII – 1973, April.

Table I and in Figure 1, both of which confirm the variability of the relative intensities of emission lines in the nucleus of NGC 1275. Line intensities given in the Table have been corrected for interstellar absorption in the nuclei of galaxies according to Wampler's (1968) data. They were used to calculate T_e and n_e in the [O III] and [O II] zones by a graphical method (Boyarchuk *et al.*, 1969), and the results of these calculations are presented in Figures 2 and 3. For the [O III] zone they reveal a systematic decrease of T_e and an increase of n_e during the period under consideration, and for the [O II] zone there are also T_e and n_e variations.

This gradual evolution of the gaseous nucleus of NGC 1275 was interrupted in the first half of 1972 at which time there was a 'flash' in the [O III], [O II] and [O I] zones producing values of T_e higher than in the previous period. The relationship between T_e and n_e for the [O III] and [O II] regions is shown in Figure 4. From points 1 to 4, which

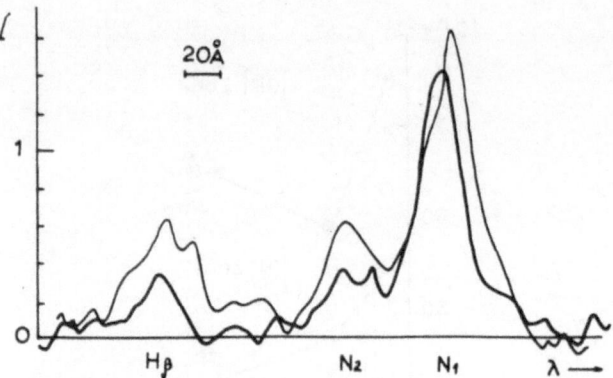

Fig. 5. Mean profile of the emission lines Hβ, N₁ and N₂ of the nucleus of NGC 1275. The thin line corresponds to the quiet state of the [O III] zone, 1972 December 1–3 and 1973 April 4 (10 spectrograms). The thick line corresponds to the period of the [O III] zone flash 1971 October to 1972 October (4 spectrograms).

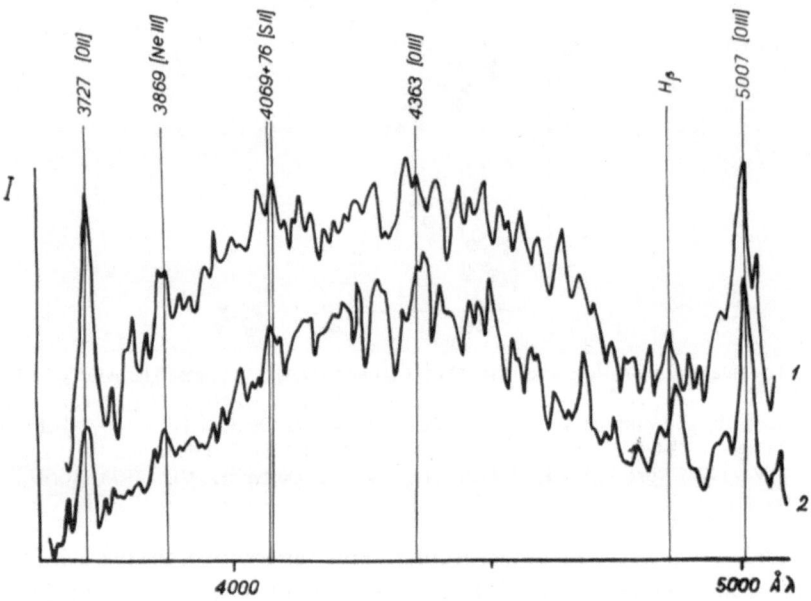

Fig. 6.   Intensities of the blue region of NGC 1275 nuclear spectra (O III zone); 1 – 1972, February 6 (time of flash); 2 – 1972, December 2 (quiet state).

refer to the period from 1930 to 1971, there is seen to be cooling and a density increase in the [O III] zone during this time. Points 5 and 6 refer to the time of the 'flash' (1971 October–1972 October) in the [O III], [O II] and [O I] zones, and point 7 to the quiet state after the flash. One can see that during the flash the physical conditions in the [O III] zone approached those observed in 1942.

Careful measurements of our spectrograms show that the profiles of some forbidden lines have also varied from the flash to the quiet state. One can see from Figure 5 that during the flash the violet wing of the N_1 line was stronger than the red one, while during the quiet state it was *vice versa*. Moreover, during the flash, the [Ne III] 3869, [S II] 4069 + 76, and perhaps the [O III] 4363, lines in the [O III] zone had violet satellites, which became weaker or disappeared during the quiet state (Figure 6). So our material shows that the [O III] zone is not homogeneous. We suppose that its lower layers are denser and are observed during the quiet state, while its higher layers are more rarefied and are observed during the flash.

According to Kellermann (1972) the radionucleus of NGC 1275 is a double one. It consists of a highly variable source of diameter ~ 0.3 pc which was born about 30 yr ago, and another more stable source about 5 pc in diameter and 3×10^4 yr old. Bursts of the small source are observed once or twice a year. Our calculations show that about 40 yr ago, when the small radiosource was born, the [O III] and [O II] zones were very hot and dense; according to Humason (1932), the lines $(N_1 + N_2)$ were very weak and the ratio of line intensities I_{4363}/I_{3869} was very high. Then the [O III] region cooled till the present time. Outbursts of the small radiosource may supply relativistic electrons which excite the flashes in the [O III], [O II] and H II zones. It will be very interesting to investigate the connections between the microwave flashes, flashes in the H II zones and flashes in the [O III] zones of the NGC 1275 nucleus.

Andrillat and Souffrin (1968) observed variability of emission line spectra like that of NGC 1275 in the nucleus of the Seyfert galaxy NGC 3516. They supposed that bursts of relativistic electrons were produced in the nucleus of that galaxy. Variability of the ratio $I_{N_1+N_2}/I_{H\beta}$ in the nucleus of the Seyfert type galaxy Markarian 6 also has been observed by Pronik and Chuvaev (1972), so it is very probable that variability of the emission-line spectrum is a characteristic of all Seyfert nuclei.

The detailed results will be published in the *Astronomiceskij Zhurnal*.

Acknowledgements

I express my sincere thanks to Drs S. B. Pikelner, V. I. Pronik and R. E. Gershberg for stimulating discussions, and to T. Nikulina and T. Korkina for numerical work and for drawing the figures.

References

Anderson, K.: 1970, *Astrophys. J.* **162**, 743.
Andrillat, Y. and Souffrin, S.: 1968, *Astrophys. Letters* **1**, 111.
Boyarchuk, A., Gershberg, R., Godovnikov, N., and Pronik, V.: 1969, *Izv. Krymsk. Astrofiz. Obs.* **39**, 147.
Dibay, E. and Pronik, V.: 1967, *Astron. Zh.* **44**, 952.
Humason, M.: 1932, *Publ. Astron. Soc. Pacific* **44**, 265.
Kellermann, K. I.: 1972, in D. S. Evans (ed.), 'External Galaxies and Quasi-Stellar Objects', *IAU Symp.* **44**, 190.
Khozov, V.: 1971, *Astron. Tsirk.* No. 607.
Lyutyi, V.: 1972, Ph.D. Dissertation, Moscow.
Pronik, V. and Chuvaev, K.: 1972, *Astrofizika* **8**, 187.
Seyfert, C.: 1943, *Astrophys. J.* **97**, 195.
Wampler, E.: 1968, *Astrophys. J. Letters*, **154**, L53.
Wampler, E.: 1971, *Astrophys. J.* **164**, 1.

GRAVITATIONAL INTERACTIONS BETWEEN GALAXIES

ALAR TOOMRE

Massachusetts Institute of Technology, Cambridge, Mass., U.S.A.

Abstract. Recent theoretical studies of the consequences of fierce tidal interactions between galaxies are reviewed and compared.

1. Introduction

Assuming gravity has not deserted us over modest intergalactic distances, one might be tempted to claim all its effects on any interplay between galaxies to be no more than vastly scaled-up analogues of the familiar celestial mechanics that involves moons, planets and suns. Certainly there is a lot to be said for such intuition – and yet one must not trust it too far. The plain truth is that nothing in our usual experience (not even from theories of contact binary stars) has really prepared us adequately for those rare but by no means unimaginable situations where two or more galaxies either graze each other or actually interpenetrate in their respective orbits. Moreover, there is growing reason to believe that many such encounters take place at speeds only comparable to the mutual speed of escape; then already the simplest estimates suggest that the resulting mechanical damage must indeed be large. And finally, even if we were intimately familiar with the behaviour of lesser celestial bodies under similarly fierce but brief external forces, one could still protest that galaxies are only loose confederations of stars and other material endowed with a variety of separate orbits. Such assemblies may react very differently from the rocky or gaseous objects.

In short, of the many aspects of the dynamics of multiple galaxies, it seems most clearly the *violent tides* arising from very close passages which require further and probably quite extensive analyses. With one notable exception from a decade ago, calculations of such severe distortions have only lately begun to be available – and this review now summarizes what has emerged and what has not. First we focus on that seemingly age-old issue of whether at least some of the pronounced spiral galaxies owe their present appearances to the tidal forces of neighbours. Next we touch on some recent and perhaps more surprising demonstrations that certain narrow 'streamers', 'filaments' and 'tails' of peculiar galaxies may indeed have had such tidal origins. And thirdly we also ruminate whether some yet more bizarre-looking specimens in the sky might not represent actual mergers of galaxies which experienced especially severe tidal friction not very long ago.

2. Tidal Spirals

The idea that gravitational interactions might yield spirals goes back at least to Chamberlin (1901). Already he reasoned that any roughly parabolic close passage of

John R. Shakeshaft (ed.), The Formation and Dynamics of Galaxies. 347–365. *All Rights Reserved.*
Copyright © 1974 *by the IAU.*

two comparable masses would subject either body to a fierce tidal force that would not only be two-sided as usual but also rather sudden or impulsive. Hence in the absence of intrinsic strength, any "concurrent rotation must obviously give rise to a spiral form", and "there should therefore be two chief arms to the resulting spiral", and furthermore "these must be curved in a common direction by the rotation of the mass".

With this emphasis on sudden tidal damage followed by differential rotation of whatever cause, Chamberlin might almost have been commenting on Zwicky's (1956) four sketches of a hypothetical passage of two galaxies, here reproduced as Figure 1.

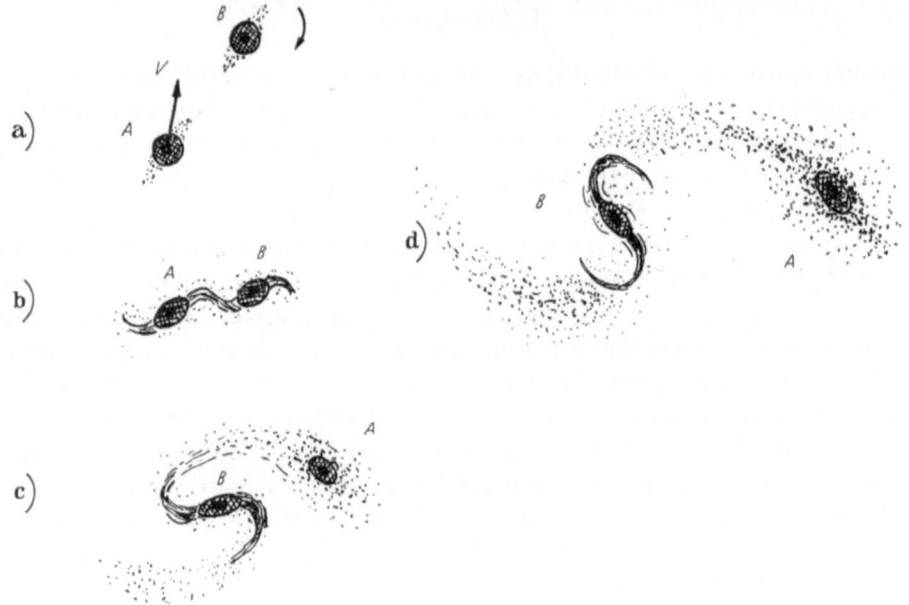

Fig. 1.   "Schematic drawing of the possible formation of an intergalactic bridge between two galaxies passing each other", by Zwicky (1956).

It hardly matters that Chamberlin was in fact thinking only of an encounter between two stars: As presumed examples of such near-calamities, his article offered photographs of no fewer than six spiral nebulae including M51! Yet in retrospect even Zwicky's sketches reveal one startling need for calculations to buttress intuition: unlikely though it may seem today, both the tidal arm and counterarm of the clockwise rotating nebula B were evidently imagined by Zwicky to wrap themselves not in a trailing but a leading direction.

Of course, any serious numerical studies of the tidal spiral-making had to await the electronic computer – even though Holmberg (1941) briefly proved otherwise with his ingenious graphical integrations. Using the new computer, the most significant early work was undoubtedly that of Pfleiderer and Siedentopf (1961) and Pfleiderer (1963). Samples of their results appear in Figure 2.

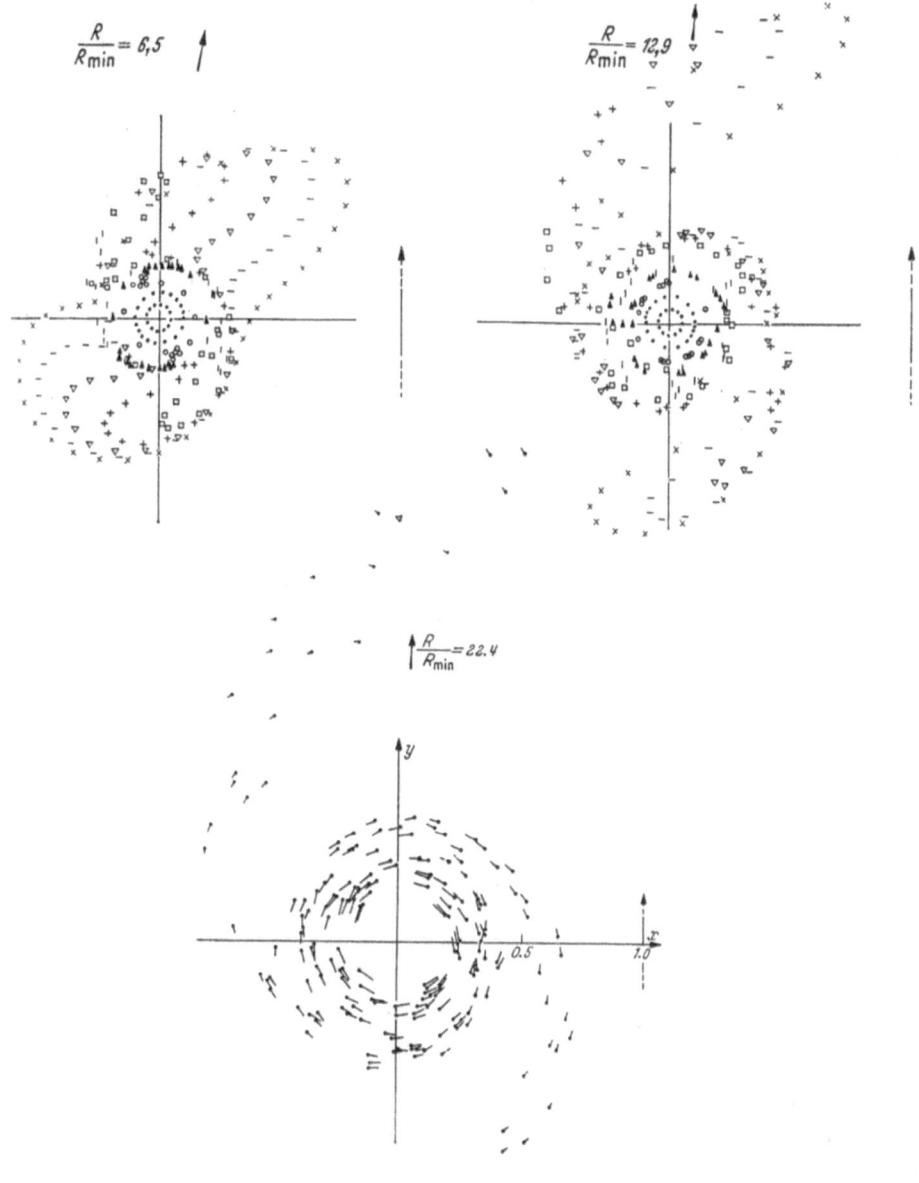

Fig. 2. Evolution of a counterclockwise rotating disk of test particles about a unit mass. The three views shown refer to elapsed times in the ratio 2:4:7 since the instant of closest approach of a three times heavier mass travelling upward on the right at a rate equal to four times its speed of escape. The slanting arrows point toward the present distant locations of that hyperbolic passer-by. The short line-segments in the bottom diagram depict the instantaneous velocity vectors of the various particles.
The first two frames are from Pfleiderer and Siedentopf (1961), and the third from Pfleiderer (1963).

Pfleiderer and Siedentopf pretended boldly that all the mass of a galaxy could for these tidal purposes be thought of as concentrated at its very centre, or at least that its outer disk could be imagined to consist simply of a large number of non-interacting test particles orbiting initially in concentric circles. The main accomplishment of Pfleiderer and Siedentopf was to demonstrate outright that just this original orbital motion of the disk particles – rather than any fierce tidal torques envisaged by Chamberlin and implicitly also by Zwicky – could be relied upon to provide much of the differential rotation needed for the kinematic development (and later dispersal) of some rather impressive two-armed and *trailing* spiral structures after a tidal 'shock' of suitable duration and severity.

Strange to say, this inspired work then lapsed into almost total obscurity for the better part of a decade. Perhaps it was simply misinterpreted as an unpromising attempt to explain all kinds of spirals – even though Pfleiderer himself had expressly disowned that hope on the ground that sufficiently close passages of unbound galaxies seemed statistically much too rare. Or possibly the reason was that, despite Zwicky's and Vorontsov-Velyaminov's efforts especially in the 1950's, the roughly one-in-a-hundred abundance of strange tails or filaments in multiple galaxies, or of companions seeming to lie at the ends of spiral arms, was not yet appreciated nearly as widely as after the publication of Arp's (1966) *Atlas of Peculiar Galaxies*. And besides, Pfleiderer and Siedentopf themselves had offered no satisfactory bridges or tails.

Looking back, one key to producing good connected spirals was of course to slow down the passer-by. This point was first demonstrated in detail by Tashpulatov (1969;

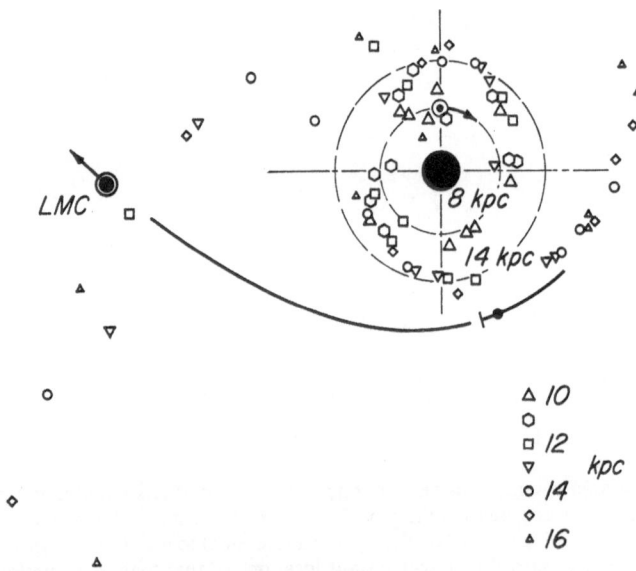

Fig. 3.   Distortion of our Galaxy as the result of an imagined 36°-inclined direct and almost parabolic passage of a 3/16-mass Large Magellanic Cloud (Toomre, 1970). Only the hypothetical relative orbit and a central mass point have here been added for clarity.

see 1970 for some purported tails) and later by Yabushita (1971). But there was one slightly subtler key as well. As Figure 9 will soon attest incidentally, it is now known that passages in a direct sense of partners of roughly *equal* mass at speeds parabolic or slower tend either to accrete the near-side material too efficiently from whichever is to be regarded as the victim disk, or else they leave it much too splattered. Of the two logical alternatives which thus present themselves, the possible merits of using a distinctly more massive passer-by went largely unappreciated until the recent work of Clutton-Brock; more about that shortly. It was the other alternative of using a smaller mass which Toomre (1970) first stumbled upon, almost literally by accident.

To digress personally for a moment, Hunter and Toomre (1969), in their studies of the bending of this Galaxy by the vertical component of tidal force during a possible close passage of the Large Magellanic Cloud, were blissfully unaware not only of Pfleiderer and Siedentopf: they also did not realize the undue sensitivity – which those German authors had already implied – of any such disk to the horizontal components of the same tidal force during a *direct* encounter of low inclination. One can imagine my initial chagrin, therefore, when the kind of direct LMC orbit that had seemed preferable for bending proved, upon further checking with test particles mobile in all three dimensions, to yield the unacceptably spectacular commotion pictured in Figure 3!

Since then, that accidental recipe for making some fairly decent if only transient bridges and counterarms has been corroborated by Wright (1972), whose best example is probably the one in Figure 4. The same formula involving a small companion in a

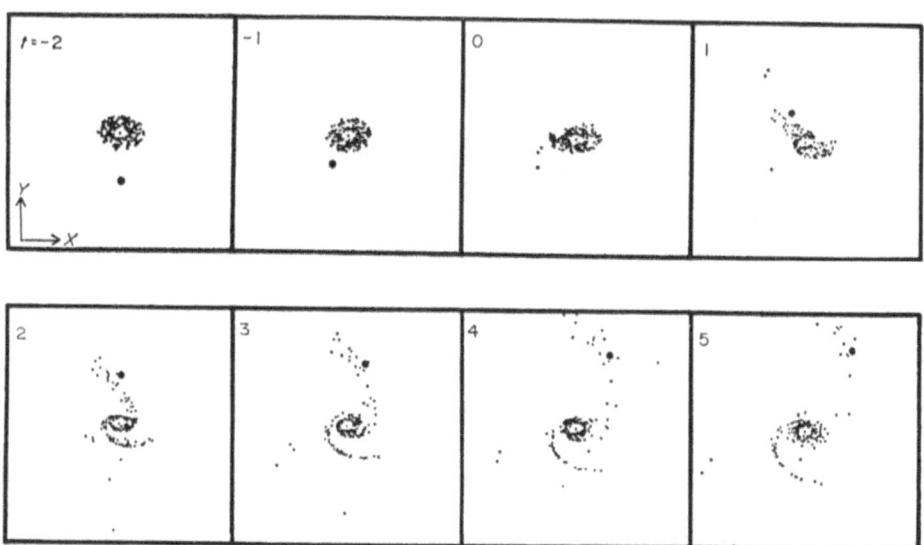

Fig. 4.   Evolution of a disk of test particles during a 45°-inclined parabolic passage of a companion of one-quarter its own central mass. The view in this rearranged copy of Wright's (1972) Figure 7 is exactly normal to the orbit plane; that original diagram also shows simultaneous orthogonal views from two other directions.

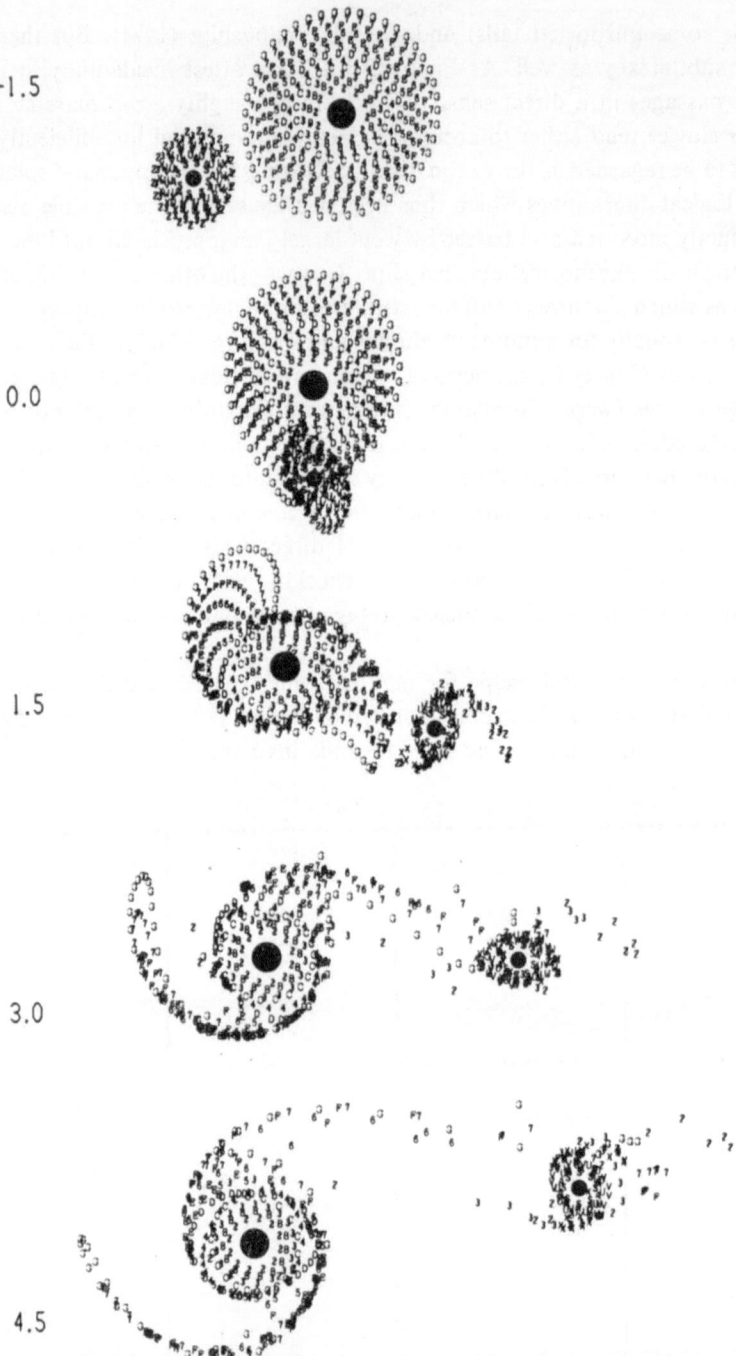

Fig. 5.   Another 45°-inclined parabolic passage of disks of test particles with central masses in the ratio 4:1. Unlike in Figures 3 and 4, the closest approach of those masses here occurs at a position as far as possible above or below the spin plane of either disk; hence the two as-yet-mildly perturbed disks in frame 0.0 are still well separated in depth.

moderately-inclined slow orbit has also been explored extensively by Toomre and Toomre (1972; hereinafter 'TT'). However, rather than duplicate any of the latter material, Figure 5 presents an abbreviated time sequence from a related motion picture which my brother produced in early 1971; these movie frames have the advantage of also conveying the fate of the smaller disk.

Of course one should beware that diagrams like Figures 3–5 are biased toward the most picturesque. There is nothing very sinister about such bias, but it does mean that distinctly uglier and less striking tidal remnants should in reality be numerically predominant by at least an order of magnitude; fortunately, such a preponderance of the unsightly seems evident already in Sky Survey. More seriously, one should also note that those recent slight improvements on the tidally produced spirals of Pfleiderer and Siedentopf still suffer from a total neglect of both the self-gravity and any initial random motions of the particles. And most important, even if such crude computations do prove to have been reasonable, one should still not hasten to conclude that tidal spirals – though almost certainly seen in such examples as NGC 2535/6 and 3808 – constitute any more than some particularly striking exceptions among the answers to what is often abbreviated in the singular as 'the spiral problem'.

Speaking of those exceptions, it is sudden doubtful whether NGC 7752/3 belongs among the select few: Bertola and d'Odorico (1973) report that the line-of-sight velocities of the two members of that apparently connected pair differ by 340 km s^{-1}. However, as regards M51, the only tidal question which in my opinion remains grossly unanswered is whether any of the pronounced inner spiral structure of its larger member could be an *indirect* by-product of the tidal damage that seems undeniable near its periphery (cf. TT; see also Tully, 1972; Weliachew and Gottesman, 1973).

3. Tidal Filaments and Tails

Possibly the most compelling evidence that the two partners in M51 were indeed involved in some recent near-collision is not even the outer shape of NGC 5194. It is rather the general mess near – and in particular the two long and oppositely-directed streamers from – the satellite galaxy NGC 5195. Those two faint streamers may be viewed in the remarkable IIIaJ photograph by van den Bergh (1969) that has been reproduced in Figure 6. Although modelled only crudely in Figure 21 of TT, these filaments seem almost certain to represent the vaguely two-sided tidal damage inflicted upon the smaller galaxy by the larger, somewhat as in the present Figure 5.

Despite this one example that appears so incriminating, the systematic study of geometrical shapes obtainable when a small disk gets ripped fairly suddenly by the non-uniform gravity of a heavy mass during a roughly parabolic fly-by remains even more in its infancy than the reverse. The only important exploration of that nook of parameter space reported so far is one by Clutton-Brock (1972), who concentrated on the totally planar, parabolic encounters with heavier bodies either 8, 64 or 512 times the mass of the small model galaxy. Whatever Clutton-Brock's work may have lacked in quantity was more than offset by his interesting, if only approximate, inclusion of

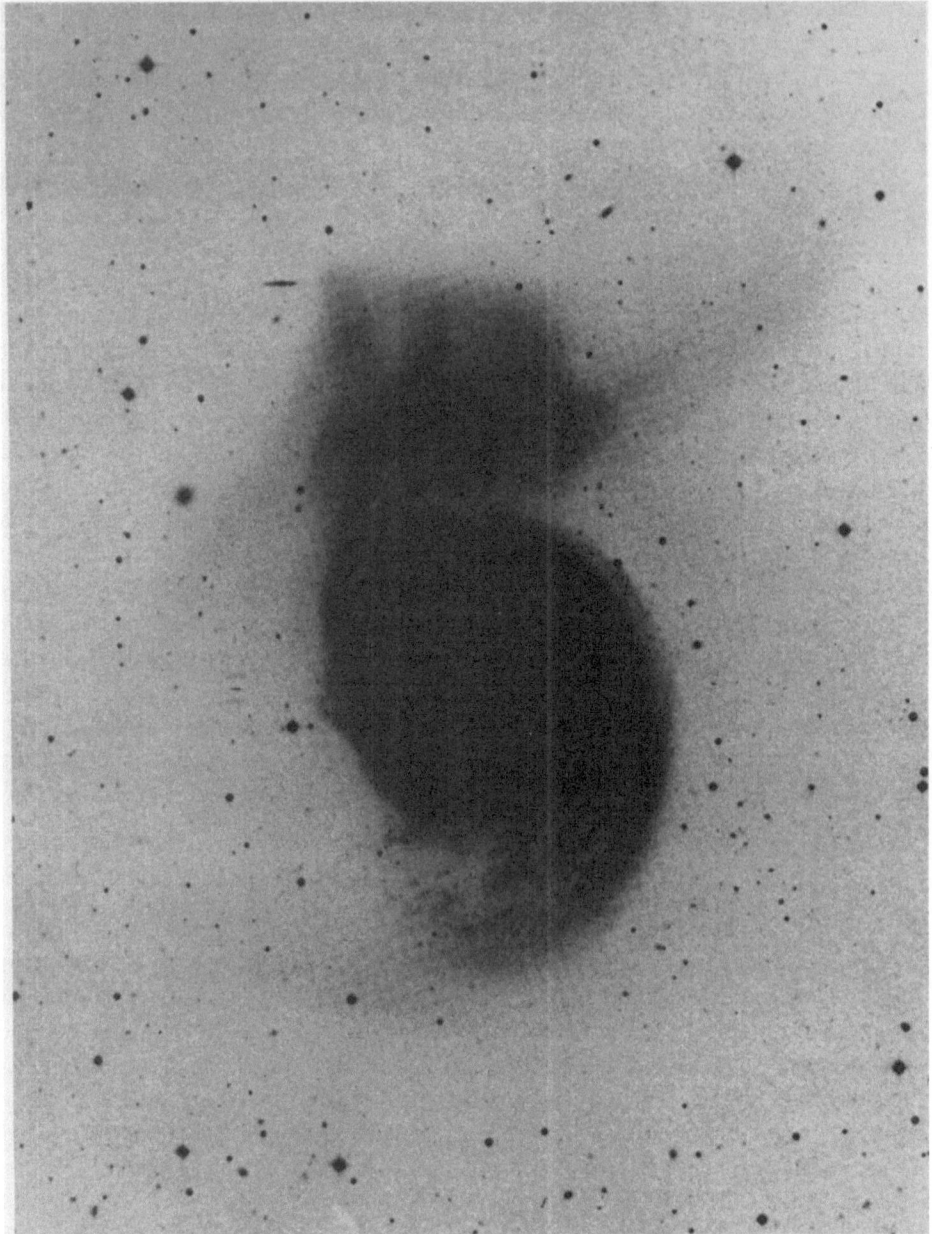

Fig. 6.   Photograph of NGC 5194/5, by van den Bergh (1969).

the self-gravity of the disk stars, and by his assignment to those mass points of significant initial random velocities. He also retained a set of test particles, without such
random velocities, to mimic at the same time the interstellar gas. Figure 7 is typical
of Clutton-Brock's findings; it was traced directly from eight of his separate diagrams,

GAS   STARS

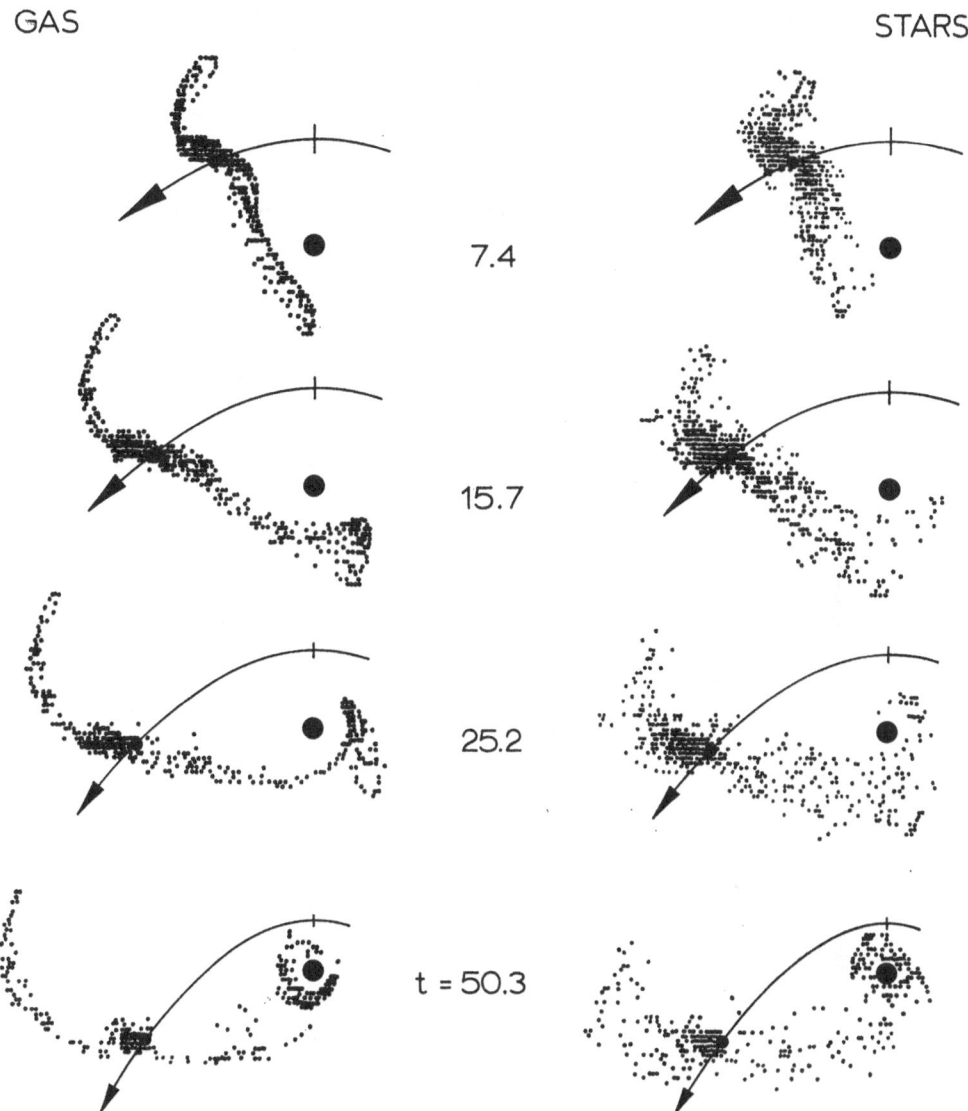

Fig. 7.   Flat parabolic passage of a small, counterclockwise rotating disk past a body eight times as massive, after Clutton-Brock (1972). In each of these four pairs of views, the configurations of 'gas' and 'stars' have been drawn to the same scale; however, as in Clutton-Brock's original figures, that magnification here changes with time.

and only the orbits of the smaller body relative to the large one have been added to lend a sense of proportion.

Figure 7 reports several interesting things, one of which is that – except possibly during a brief early interval in mid-encounter – the small deformed disk simply does not look like a respectable spiral. This is consistent with my own limited experience;

it seems impossible to obtain good two-armed kinematic spirals from these relatively leisurely passages except by choosing the perturbing mass to be small. A second item is that the initially cold medium labelled 'gas' here produces structures – i.e., both the near-side 'bridge' and far-side 'tail', in the usage of TT – that are considerably narrower than those for the initially hot 'stars'. Though not astonishing, it is nice to see this point illustrated so vividly. Thirdly, it is my impression, based on Figures 5 both of TT and of the present article, that the gravity of the stars in Clutton-Brock's example makes these gas features narrower than they would have been in a situation consisting only of test particles. Fourth and most encouragingly, it also seems that the bridge now endures considerably longer.

This praise of the only self-gravitating simulations yet available in the whole subject must be tempered with two cautions. The lesser is that Clutton-Brock's algorithms for approximating that gravity do not yet appear to have been reported or

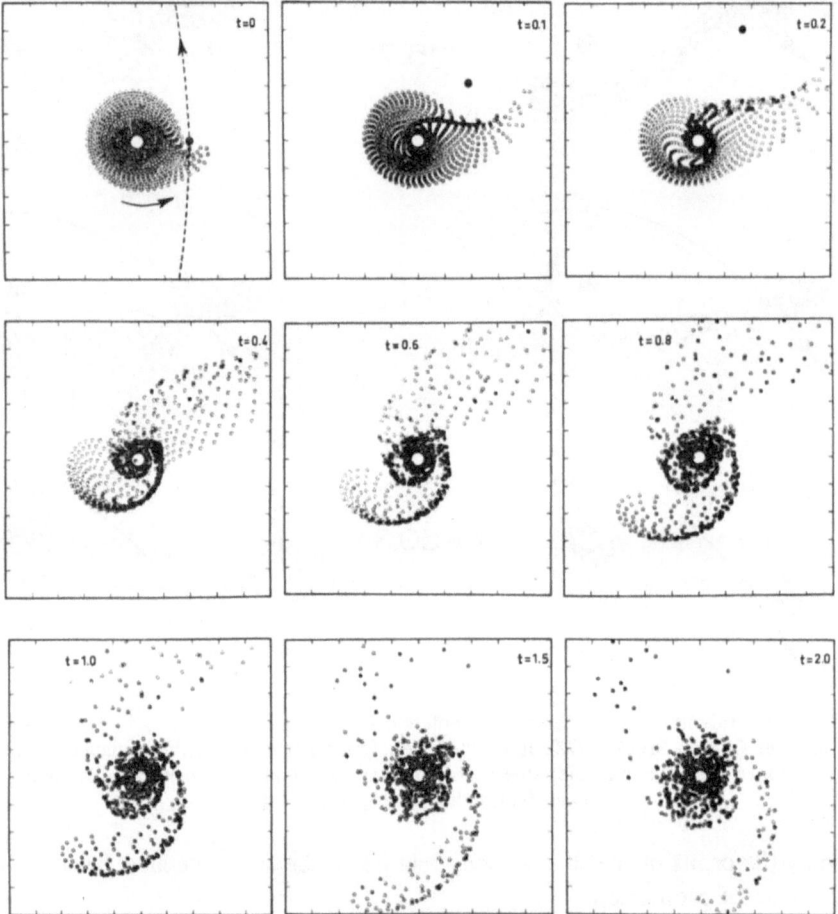

Fig. 8.   Evolution of a disk of test particles following the passage "within the plane in the direction of galactic rotation" of an equal mass at twice the speed of escape, by Eneev *et al.* (1973).

tested in much detail. The graver worry is that also Clutton-Brock "was unable to construct a completely stable disk. ... Sometimes the disk would explode during a tidal simulation". Unfortunately such symptoms are not unheard of; on the contrary, this lack of stable disk models without a great deal of random motion seems a source of fundamental difficulty even to understanding the structures of totally isolated disk galaxies (cf. Ostriker and Peebles, 1973; see also p. 134)

It is, however, not the passages of light or heavy culprits which up to now have yielded tidal results at first most contrary to intuition. That distinction belongs to the close encounters of approximate equals – provided 'close' means very near indeed and also provided the speeds are less than hyperbolic.

To reset the stage, we might recall one major reason why Vorontsov-Velyaminov (1958, 1961) among others grew distrustful of would-be tidal explanations of the various luminous extensions of peculiar galaxies: it was that *tails* of various kinds seem distinctly more common among double galaxies than any bridge-like filaments. How can this fact possibly be reconciled, Vorontsov-Velyaminov asked, with the notorious two-sidedness of tides as we know them? Or even granted an imperfect symmetry, why should the far-side tidal damage now appear so much more prominent than that from the side on which the external pull during a close approach would have been especially severe?

The answers to those questions may perhaps best be approached by reference to the recent Soviet work by Eneev *et al.* (1973; see also Kozlov *et al.* 1972). These authors produced several fascinating displays – and as I understand from Vorontsov-Velyaminov, also some movies – of the evolving tidal damage from the fly-by of an equally massive partner. However, Eneev *et al.* also continued very much in the spirit of Pfleiderer and Siedentopf by requiring those passages to be distinctly hyperbolic; hence our Figure 8 culled from their paper seems ironically most useful as a quick lesson on how *not* to do it.

Somewhat as in Figure 2 – where the perturbing mass had been greater but its passage was yet faster – we observe in Figure 8 that the tidal damage in this moderately rapid fly-by remains not only two-sided but indeed is more pronounced on just the side one would have expected. Yet notice how relatively splattered that near-side damage has already become: if these experiments had been repeated with progressively slower passages, this ugly and extensive spray would only have worsened until speeds about as slow as parabolic had been adopted (cf. Wright's Figure 1, or TT's Figure 2); with orbits of considerable inclination, the problem would indeed have persisted down to some distinctly elliptical encounters (cf. TT's Figure 16). The relief at last comes simply from the fact that a slow enough 'aggressor' actually captures most of the near-side debris, like a sort of gravitational vacuum cleaner. And that in turn explains why the later tidal remnants can then look bizarrely one-sided, with only the counter-arm growing with time into an ever-lengthening tail.

Since the only available theoretical study of tails of that sort seems still to be the one by TT, I may be the wrong man to review it further. However, it is perhaps of interest to view at leisure the development with time of that symmetric model of the 'Antennae',

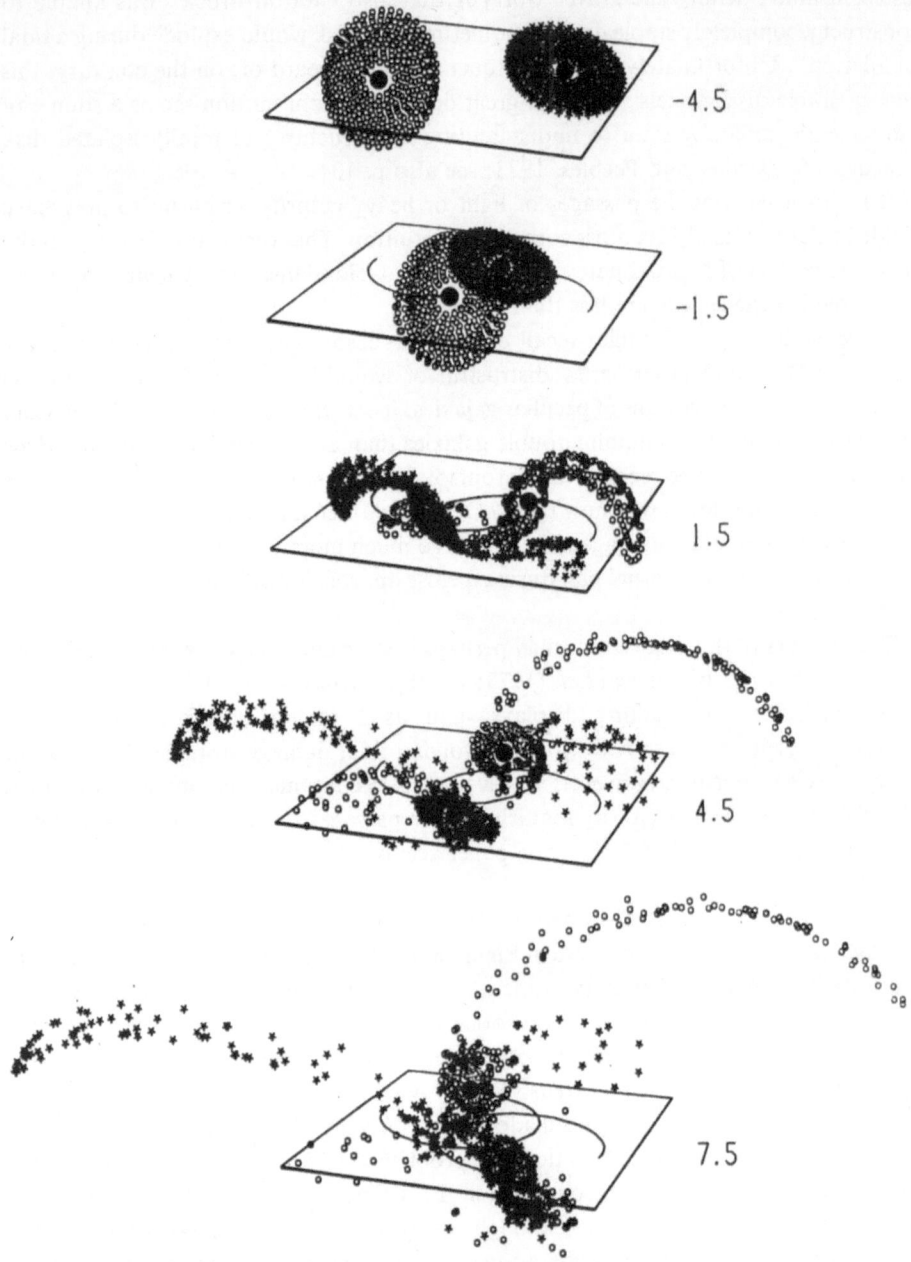

-4.5

-1.5

1.5

4.5

7.5

Fig. 9.   Five views of the symmetric close encounter of two 60°-inclined disks of test particles presumed by TT (§VI.d) to caricature the recent history of NGC 4038/9. These views are equally spaced
in time, with the instant 0 (not shown) meant to represent pericentre. The stereographic projection
used here assumes a vantage point at a distance equal to 16 times that of the closest approach of the
two central masses, or four times the edge length of the square that denotes their common orbit plane.
The elevation of viewing is 20°.

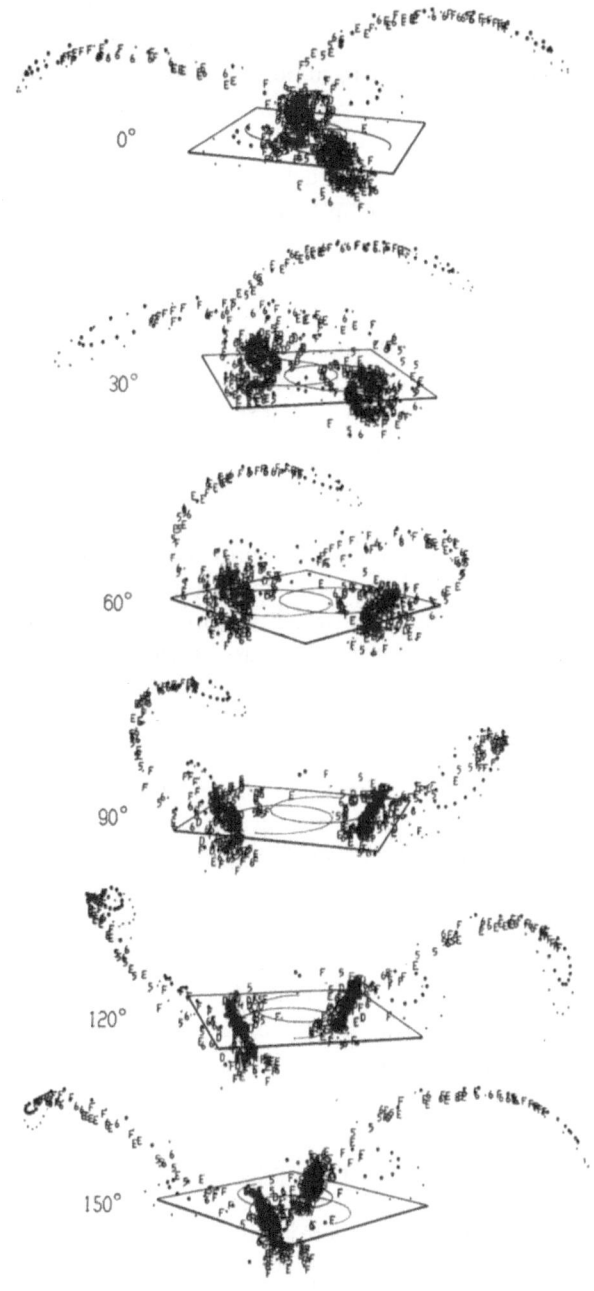

Fig. 10.   End-product of the evolution in Figure 9, viewed at time 8.0 from six equally-spaced longitudes. The viewing elevation here is held constant at 15°.

NGC 4038/9, of which only two orthogonal snapshots could be offered in TT's Figure 23. This evolution is shown in Figure 9, taken directly from a new movie kindly prepared by my brother for this Symposium, which differs only in its perspective viewings from the one described by Toomre and Toomre (1971).

Also from the new movie comes the stack of separate but simultaneous views in Figure 10 depicting the same evolved model at one single instant. That set of pictures has here been included mainly to re-emphasize vividly – as only pictures can – that the real 'Antennae' and many similar peculiars probably have structures which are very three-dimensional. In a motion picture one can obtain an uncannily stereoscopic feeling for these models by pretending one is walking slowly around them!

We close with one sobering reservation about models such as employed in Figure 9, and with a brief, pleasanter report of a tentative success. The criticism once again is that our total neglect of the self-gravity and of the random motions in these tail calculations means at best that they can only mimic the tidal dynamics of *those components* of a real galaxy which (i) stem from regions far enough out to contain no more than, say, one-fifth of the total mass, and (ii) possess initial motions no more eccentric than perhaps the Sun in the case of our Galaxy. Until recently the latter might not have seemed a severe restriction. However, the increasing worries about the large-scale stability of purely disk-like systems to which we have already alluded (and which, of course, arise only when there is self-gravity) raise at least the possibility, discussed by Ostriker and Peebles, that some surprisingly massive but underluminous halos may coexist with most disks. Or else there may need to be a lot of mass in the higher-velocity stars of a disk itself. Neither class of rapidly moving old stars can be expected to form tidal ribbons remotely as narrow as those in NGC 4038/9 – and it would by inference be reassuring indeed to know that at least some such observed narrow arcs are themselves embedded in much thicker (if yet distressingly fainter) stellar crescents, rather as in Clutton-Brock's example.

The pleasanter news concerns the 'Mice', NGC 4676. For them TT had remarked that it is "all but impossible [on a tidal hypothesis] for the centre of mass of component *B* not to be approaching us relative to that of *A*", and also that some decade-old spectroscopic measurements by the Burbidges had indicated the opposite. Recently Stockton (1974) has reobserved NGC 4676. He reports a velocity difference of about 80 km s^{-1}, in the tidally desired sense.

4. Tidal Friction

This last section can be short because there is still not much that can be said definitively on the subject of possible orbital changes as a by-product of the tidal distortions. Of course the very notion that galactic encounters are not elastic has been with us for quite some time. For instance Holmberg (1941) undertook his calculations not so much to study spiral structures as in an attempt to see "whether the loss of energy resulting from the tidal disturbances at a close encounter between two nebulae is large enough to effect a capture". Since then, the capture idea has demonstrably intrigued also Zwicky (1959), Alladin (1965), Lauberts (1973), and several others. And surely today, having just viewed the possible tidal after-effects of some very close encounters, no one can in principle object to the assertion that such violence must cost mechanical energy of some sort.

Yet the real challenge here, it seems to me, concerns not principle as such but chiefly the expected rates. After all, the fractional changes *per encounter* which Alladin had in mind when he wrote that "given long enough, the components of a double galaxy will disrupt one another and give rise to a single loose system", or which Aarseth (1966) envisaged in surmising that "it might be possible that strong interactions between a few closely bound galaxies could in time lead to the formation of one larger common object, containing several galactic nuclei", were presumably rather small.

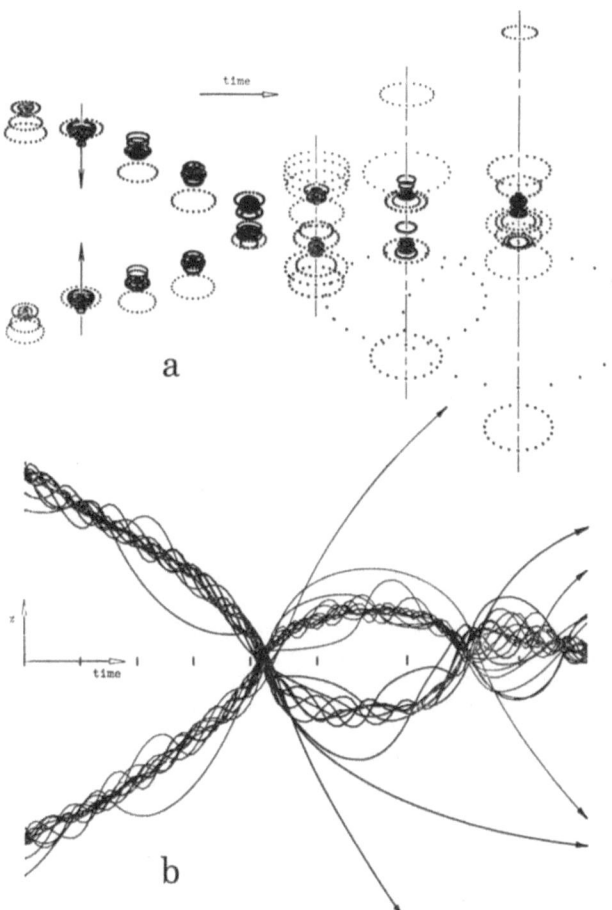

Fig. 11. Head-on encounter between two equal and approximately spherical stellar systems presumed to have started from rest at infinity. (a) Perspective views at eight (not entirely uniformly spaced) instants of the two clusters of twelve equal self-gravitating rings apiece used to simulate those systems. (b) Record of the axial coordinate z of each of the 24 rings, drawn to a 1.8 × taller vertical scale as functions of the time t.

By contrast, as TT discuss, there are perhaps as many as a dozen NGC galaxies or pairs in the sky near which plausible tidal tails still appear prominent and yet where the main galactic bodies either seem suspiciously close or else may already have blended into one. For such imagined mergers (of what we supposed had been not truly unbound galaxies but merely long-period eccentric orbiters) the observations seem to demand a substantial loss of orbital energy in each single fierce encounter. An urgent question thus becomes: Can such rapid loss be justified theoretically for galaxies imagined composed, for simplicity, of only those collisionless mass points known as stars?

The vote is not in yet. However, as the (admittedly contrived) example in Figure 11 illustrates, it is far from obvious that the final answer will be negative. These tentative n-ring calculations by Larry P. Cox and myself simply underscore a point whose vehemence seems previously to have been appreciated (from impulsive-tide idealizations) only by Sastry (1972) and by Alladin *et al.* (1972). It is that just this one head-on penetration of the two systems – and all the ensuing 'splash' from the briefly doubled gravitational force, especially toward the axis – costs these model galaxies roughly one-half the maximum kinetic energy developed in their free fall from infinity. Consequently in Figure 11 at least 80% of the total mass soon tumbles into a single heap.

Acknowledgement

This work was supported in part by the (U.S.) National Science Foundation.

References

Aarseth, S. J.: 1966, *Monthly Notices Roy. Astron. Soc.* **132**, 35.
Alladin, S. M.: 1965, *Astrophys. J.* **141**, 768.
Alladin, S. M., Sastry, K. S., and Ballabh, G. M.: 1972, paper presented at First European Astronomical Meeting, Athens.
Arp, H. C.: 1966, *Atlas of Peculiar Galaxies*, Cal. Tech. Bookstore, Pasadena = *Astrophys. J. Suppl.* **14**, No. 123.
Bertola, F. and d'Odorico, S.: 1973, *Astrophys. Letters* **13**, 161.
Chamberlin, T. C.: 1901, *Astrophys. J.* **14**, 17.
Clutton-Brock, M.: 1972, *Astrophys. Space Sci.* **17**, 292.
Eneev, T. M., Kozlov, N. N., and Sunyaev, R. A.: 1973, *Astron. Astrophys.* **22**, 41.
Holmberg, E.: 1941, *Astrophys. J.* **94**, 385.
Hunter, C. and Toomre, A.: 1969, *Astrophys. J.* **155**, 747.
Kozlov, N. N., Sunyaev, R. A., and Eneev, T. M.: 1972, *Doklady Akad. Nauk SSSR* **204**, 579 = = *Soviet Phys. – Doklady* **17**, 413.
Lauberts, A.: 1973, 'Numerical Models of Gravitationally Interacting Galaxies', Uppsala dissertation.
Ostriker, J. P. and Peebles, P. J. E.: 1973, *Astrophys. J.* **186**, 467.
Pfleiderer, J.: 1963, *Z. Astrophys.* **58**, 12.
Pfleiderer, J. and Siedentopf, H.: 1961, *Z. Astrophys.* **51**, 201.
Sastry, K. S.: 1972, *Astrophys. Space Sci.* **16**, 284.
Stockton, A.: 1974, *Astrophys. J.* **187**, 219.
Tashpulatov, N.: 1969, *Astron. Zh.* **46**, 1236 = *Soviet Astron. – AJ* **13**, 968.
Tashpulatov, N.: 1970, *Astron. Zh.* **47**, 277 = *Soviet Astron. – AJ* **14**, 227.

Toomre, A.: 1970, in W. Becker and G. Contopoulos (eds.), 'The Spiral Structure of Our Galaxy',
    IAU Symp. **38**, 334.
Toomre, A. and Toomre, J.: 1971, *Bull. Amer. Astron. Soc.* **3**, 390.
Toomre, A. and Toomre, J.: 1972, *Astrophys. J.* **178**, 623.
Tully, R. B.: 1972, 'The Kinematics and Dynamics of M51', Univ. of Maryland dissertation.
van den Bergh, S.: 1969, *Astrophys. Letters* **4**, 117.
Vorontsov-Velyaminov, B. A.: 1958, *Astron. Zh.* **35**, 858 = *Soviet Astron. – AJ* **2**, 805.
Vorontsov-Velyaminov, B. A.: 1961, in G. C. McVittie (ed.), 'Problems of Extra-Galactic Research',
    IAU Symp. **15**, 194.
Weliachew, L. and Gottesman, S. T.: 1973, *Astron. Astrophys.* **24**, 59.
Wright, A. E.: 1972, *Monthly Notices Roy. Astron. Soc.* **157**, 309.
Yabushita, S.: 1971, *Monthly Notices Roy. Astron. Soc.* **153**, 97.
Zwicky, F.: 1956, *Ergebnisse der Exakten Naturwissenschaften* **29**, 344.
Zwicky, F.: 1959, *Handbuch der Physik* **53**, 373.

DISCUSSION

E. M. Burbidge: In NGC 4038/9, how do your calculations fit with the observed velocities measured by G. Burbidge and myself (*Astrophys. J.* **145**, 661, 1966) and by Vera Rubin *et al.* (*Astrophys. J.* **160**, 801, 1970)? As I recollect, there is very little velocity difference between the two nuclei, and the largest velocities are negative ones on the far side of one of the components.

Toomre: That small velocity difference between the two nuclei of NGC 4038/39 causes no embarrassment at all to the very symmetric model that my brother and I concocted. More serious, possibly, is that our idealized disks seem to rotate in nearly opposite directions, whereas the real spins seem more closely parallel.

Arp: Would your gravitational perturbation mechanism give the 'antenna-like' tails if NGC 4038/9 started out as a close double, for example as a result of fission of an original simple body?

The question has also to be faced with regard to the Heidmann-Kalloghlian-Markarian pairs that seem to be flying apart. When these pairs were closer together, were they interacting gravitationally? Were they producing tails?

Toomre: I can't possibly swear that some sort of fissioning of galaxies might not give rise to tails like those of 4038/39, or even write ALAR TOOMRE in the sky, but I just think it very unlikely unless we are willing to postulate that the laws of gravity or inertia were different when any two such galaxies were still very close together. Otherwise, the tides then would have been so violent that the two galaxies should have become altogether disrupted and splattered.

Wright: Where does the angular momentum come from to form tails if we have a single body fissioning?

Arp: If the body is fissioning it is coming apart (and probably also rotating) – like an encounter that started at the closest passage. I am asking whether this starting condition will give the same end result as Toomre's usual tail calculations.

Miller: Some years ago when Prendergast and I were working on the large N-body calculations (*Astrophys. J.* **151**, 699, 1968), we did some experiments in which two star clusters fell together. After the collision, there were three blobs, one representing each of the initial clusters, and one at rest at the centre. The central blob was the largest. But this was a pure star case; models with some gas, and with star formation, can behave differently in detail although the gross features are much the same.

With regard to Arp's question, before we got the spiral model we had some models that started from a single gas blob and flew apart. The angular momentum and rate of star formation were both critical parameters governing the behaviour of these models. Most of those that flew apart went into two blobs, but at least one with high angular momentum and a low rate of star formation broke into four parts.

There is a question that I'd like to ask the observers in this audience: granting the observational difficulties involved, what is known (spectroscopically, for example) of the physical properties of these bridges and tails? In particular, if they show O and B stars, and gaseous emission, there may be serious trouble with time-scales. The stars needed to drive the gaseous emission most likely should have formed shortly after the time of closest approach and, according to the time-scales indicated by the Toomres (typically $3-5 \times 10^8$ yr), O and B stars could not live long enough to excite the bridges observed. It is difficult to see how to maintain sufficiently large gas densities in these bridges and tails to support continuing star formation.

Sargent: Dr Searle and I thought that a good test of the hypothesis for forming bridges described by Dr Toomre would be to examine the stellar populations of the bridges and to see if they are the same as those of the parent galaxies.

So far we have looked at the main body and the straight tail of the 'Mice', NGC 4676. We find that the stellar population of the tail is definitely of earlier type than that of the main body. This implies either that stars are drawn selectively from the parent galaxy or that star formation can go on in the drawn out material.

Miller: This has worried me about Toomre's calculations.

Toomre: Who says that the stars are mainly being formed now? I would guess that they formed mostly during the violent part of the tail-making.

Miller: If they were formed then, they would not still be blue and giving emission lines.

Toomre: Perhaps Wal Sargent can remind us whether the observation (Theys *et al.*: *Publ. Astron. Soc. Pacific* **84**, 851, 1972) of λ 3727 emission in one of the tails absolutely requires O and B stars?

Sargent: I think so.

Contopoulos: What is the time-scale involved from the point of closest approach?

Toomre: About 1.2×10^8 yr.

G. de Vaucouleurs: I don't quite see the difficulty, since the main body of the galaxy is bound to be redder than the tail which is essentially a piece of the disk. Perhaps you should put some gas in your models?

Miller: The problem is that the densities get too low for star formation to continue for a long time after the encounter.

Toomre: This problem about star formation is certainly a difficulty but it is hardly a disproof, since we simply don't understand well how stars are formed. Is there anyone here who would like to claim otherwise?

McAdam: Does the success in modelling tails with only stars suggest that magnetic fields with dust or gas cannot be common between galaxies or clusters? Is viscous damping or drag thought to be necessary?

Toomre: Our calculations say very little about magnetic fields or viscosity – except that they seem unlikely to be of any great importance in forming the bridges or tails.

Savedoff: For fast passages, one expects less interaction. Are there any features you would expect to be visible if the 5000 km s^{-1} relative motion in Stephan's Quintet were interpreted as an encounter?

Toomre: If the fly-by is, say, five or more times the escape velocity, the effect is very small. At 5000 km s^{-1} one would expect very little interaction.

Contopoulos: Already at 2000 km s^{-1} we found that nothing happens.

Tifft: On the basis of Zwicky's study of bridges, I have the impression that emission from bridges is rare, and that their most likely makeup is old stars.

Sargent: I only know about that one bridge – or really tail.

Wright: On the question of star formation in tails, I have the distinct impression that tails (at least initially) grow *thinner* with time – quite contrary to intuition. If there is gas in the tail, the consequent density increase is at least *one* of the conditions necessary for star formation. Additionally we have neglected self-gravitation and there is even the possibility of an intergalactic thermal instability leading to squeezing of the tail.

A second point is that we have just finished a survey at Parkes of 44 systems of interacting galaxies in a search for continuum radio emission. Our initial results show that interacting systems are *not* preferentially radio emitters nor are stronger than a comparison sample of non-interacting galaxies.

Allen: We have made similar radio continuum observations with the Westerbork telescope at 21 cm (Allen *et al.*: *Nature* **241**, 260, 1973) of 9 interacting systems, among which are NGC 4038/39 and NGC 4676 A/B. Our sensitivity and angular resolution of 25″ were sufficient to show that in no case could radio emission be detected from the bridges and tails of these systems. Emission was detected from the nuclei of some of the galaxies, and from dusty regions in the main bodies of some others. However, within the small sample which we have, the radio emission seems apparently neither more frequent nor more intense than that of galaxies which are not obviously interacting. (See also Burke, B. F. and Miley, G. K.: *Astron. Astrophys.* **28**, 379, 1973.)

Toomre: I accept that the average interaction does not produce strong radio emission, but there are also objects such as the companion of M51 to be considered.

Lewis: Would you comment on the likely effect of collisions among cluster galaxies in Coma or Virgo for providing substantial quantities of stellar material in the intra-cluster space.

Toomre: As I indicated, tidal damage of all sorts tends to be spectacular only when the encounters are roughly parabolic or even slower. Under such circumstances, I get the impression that the cast-away material amounts to no more than 10–20 %. In any cluster, where most close passages would of necessity be pairwise hyperbolic, all such loss of material would be yet further reduced. It's going to be very difficult to load a cluster with 10 times the present galactic mass by means of debris pulled out from former galaxies.

G. de Vaucouleurs: What we want is 1/2 to 1/3 of the mass of the galaxies.

Toomre: If there are junk-piles of debris to be found, I'm personally inclined to look at cD galaxies or, if mergers occur, even at ellipticals.

Contopoulos: In the case of two merging galaxies, does most of the energy go into random motions of stars or into escaping stars?

Toomre: I am not sure yet. Probably most goes simply into the internal motions or heat. In principle, at least, it is energetically possible for two finite stellar systems to merge and throw away nothing.

E. M. Burbidge: How does the expected number of encounters agree with the space density of galaxies and their velocities? I believe Ambartsumian concluded that the number of doubles found required that they were actually formed as doubles.

Toomre: Indeed so. Any stray (i.e. hyperbolic) encounters seem too unlikely by several orders of magnitude. I therefore strongly concur that most of the present or recent interacting pairs (about 1 in 100 of the NGC objects) must somehow already have been created as bound doubles. As Nelson Limber (*Astrophys. J.* **142**, 1346, 1965) wrote: 'A possibility that may warrant further consideration is that these systems represent old double galaxies, whose semi-major axes and orbital eccentricities are such as to have brought about only now, for the first time, the very close encounters that we observe.' We may indeed be dealing with pairs of galaxies which until recently were still falling together, or which once fell together, missed, and are now coming close for the second or third time.

G. Burbidge: Are you saying that these have been bound systems for 10^{10} years and the close encounter has just happened?

Toomre: Yes. It's of course very improbable but I remind you that, if we see a certain fraction of such encounters now, there must have been many more in the past. Logically, a substantial fraction of all galaxies in the sky would thus have to be remnants of encounters. Would you like to contemplate which single class of galaxies is numerous enough to accommodate such remnants?

Davies: It should be emphasized that the outer parts of spiral galaxies contain a high proportion of neutral hydrogen. These same outer regions are the ones involved in tidal interactions. One might therefore expect that 10 to 50 or even 100 % of the material in the tidal arms might be gaseous hydrogen. This should be investigated optically and at 21 cm.

Abell: You have described several systems, which appear to be so well understood in terms of tidal interaction as to provide strong evidence that tidal interaction does, in fact, occur, and hence that the two galaxies involved must be at about the same distance from us. Am I correct in assuming that among such good examples there is *no* case in which the two members of the pair have very different redshifts (which would suggest non-cosmological components to redshifts)?

Toomre: You are correct. The nearest thing to an exception might be the often-quoted Zwicky triplet that appears as No. 175 in Arp's *Atlas*. One of those galaxies has a redshift perhaps 7000 km s^{-1} less than the other two. But as we discuss on p. 648 of our *Astrophys. J.* **178** article, that situation is probably a superposition, since a so-called bridge there certainly looks to us like a tail.

Yet I don't want Chip Arp to think I always argue against him. If anyone wants to claim, as I suppose I do, that NGC 7603 is merely superposed on its neighbour with that 8000 km s^{-1} excess redshift, let him also remember that there would then still be those wisps (tails?) of that Seyfert-like main galaxy to be accounted for.

Abell: Do you think that the merging of galaxies by two-body collisions in clusters could build cD galaxies; especially those with multiple nuclei, as in NGC 6166?

Toomre: Yes, I *suspect* so. I believe at least Aarseth and Ostriker have guessed likewise – but the problem remains: how can we tell for sure?

THE MAGELLANIC STREAM*

D. S. MATHEWSON and M. N. CLEARY

Mount Stromlo and Siding Spring Observatory, Australian National University, Australia

and

J. D. MURRAY

Division of Radiophysics, CSIRO, Australia

Abstract. A southern sky survey of H I in the velocity range -340 km s^{-1} to $+380$ km s^{-1} has shown that a long filament of H I extends from the Small Magellanic Cloud (SMC) region down to the South Galactic Pole and connects with the long H I filament discovered recently by Wannier and Wrixon (1972) and van Kuilenburg (1972). There is also some evidence that the feature continues on the other side of the Magellanic Clouds and crosses the galactic plane at $l=306°$. The whole filament, which follows very closely a great circle over its entire 180° length, is given the name 'The Magellanic Stream'. It may have been produced by gravitational interaction between the SMC and the Galaxy during a close passage (20 kpc) of the SMC some 5×10^8 yr ago although it is impossible to account for the observed radial velocities along the Stream unless some force other than gravity is invoked to act on the Stream as well.

Recently a southern sky survey has been made for H I in the velocity range -340 km s^{-1} to $+380$ km s^{-1} using the 18-m reflector at Parkes. This survey has revealed (Figure 1a) a long filament of H I which extends from the region of the Small Magellanic Cloud (SMC) to the South Galactic Polar cap and beyond. This is given the name 'The Magellanic Stream'. All H I is plotted in this diagram except the 'zero-velocity' (i.e. local) gas. The average velocity half-width of the H I in the Magellanic Stream is about 30 km s^{-1} which is much broader than the average half-width of 7 km s^{-1} of gas in the local spiral arm.

Figure 1b shows the distribution of H I with high positive velocity (greater than 200 km s^{-1} with respect to the local standard of rest and clearly resolved from spiral arm emission) discovered on the other side of the SMC and extending through the galactic plane at $l=306°$ and up to $b=+30°$ which may also belong to the Magellanic Stream. The two clouds centred on $l=289°$, $b=20°$ and $l=268°$, $b=20°$ were discovered by Wannier *et al.* (1972), although the latter cloud has probably been produced by the expanding shell of the Gum Nebula and should not be included in the Magellanic Stream.

Figure 2 shows the full extent of the Magellanic Stream inked in on an Aitoff projection of the sky in galactic coordinates. The section between $l=320°$, $b=-80°$ and $l=80°$, $b=-65°$ had been surveyed previously by van Kuilenburg (1972); see also Dieter (1965) and Hulsbosch and Raimond (1966). Wannier and Wrixon (1972) used a more sensitive receiver to increase the observed length of this section of the Stream up to $l=90°$, $b=-30°$. The cross-hatched patches are the clouds of high velocity H I discovered much earlier by northern hemisphere observers mostly at positive latitudes in the longitude range 80° to 180°.

A striking feature of the Magellanic Stream is that it follows closely a great circle

* See also *Astrophys. J.* **190**, 291, 1974.

John R. Shakeshaft (ed.), The Formation and Dynamics of Galaxies. 367–374. All Rights Reserved.

over its entire 180° across the sky. If all the H I in the Stream were at the distance of
the SMC (63 kpc), its mass would be about 10^9 $\mathfrak{M}_\odot$ which is equal to the combined
mass of H I in the LMC and SMC. The gas mass of the Inter-Cloud region accounts
for half of this total.

Figure 3 shows the variation of radial velocity (V_{GSR}) with angular distance (θ)
along the Stream using the coordinate system of Wannier and Wrixon (1972) in their
Figure 2. The radial velocities referred to the local standard of rest (LSR) have been
corrected for the galactic rotation at the Sun. These corrected velocities ($V_{GSR} =$
$= V_{LSR} + 225 \sin l \cos b$) are essentially velocities with respect to a non-rotating
Galaxy. The systematic variation of radial velocity found by Wannier and Wrixon
is seen to continue until the Magellanic Clouds are reached but not beyond them,
which produces a nagging doubt as to whether this H I is really part of the Stream.

Many interpretations of the Magellanic Stream have been considered but the most

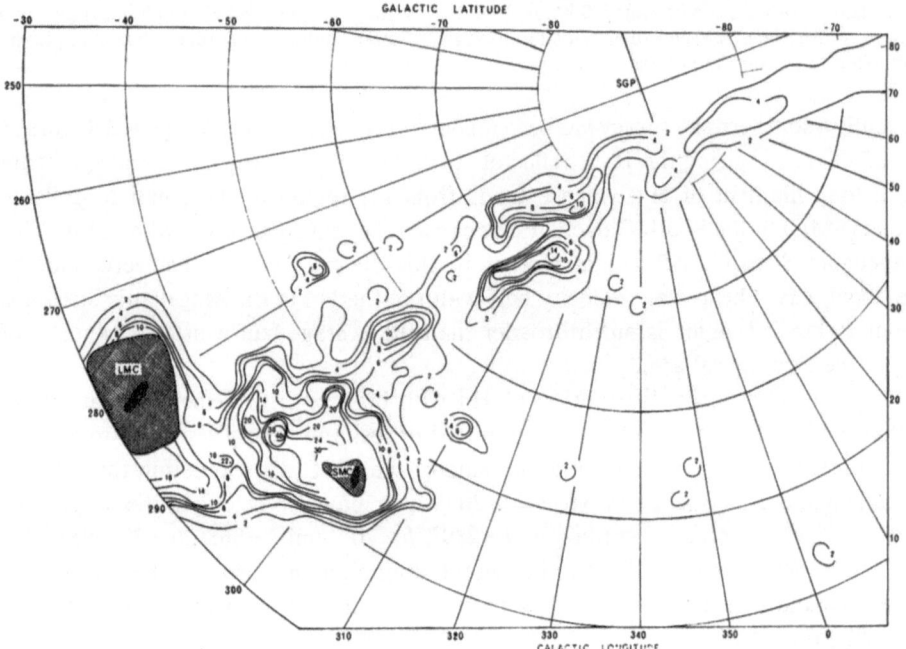

Fig. 1a.

Fig. 1a–b.   The contours give the surface densities of the H I in the Magellanic Stream obtained using
the 18-m reflector at Parkes. The contour unit is 2×10^{19} atom cm^{-2}. The cross-hatched regions
represent the approximate optical extent of the Large Magellanic Cloud (LMC) and Small Magellanic
Cloud (SMC).
(a) The section between the Magellanic Clouds and the South Galactic Polar region; all H I within
the velocity range -340 km s^{-1} to $+380$ km s^{-1} is plotted except the 'zero-velocity' H I, i.e. local
spiral arm gas.
(b) The section between the Magellanic Clouds and $+30°$ galactic latitude; all H I greater than
$+200$ km s^{-1} relative to the local standard of rest is plotted. The two clouds centred on $l = 289°$,
$b = 20°$ and $l = 268°$, $b = 20°$ were taken from Wannier et al. (1972). The latter cloud has probably
been produced by the expanding shell of the Gum Nebula and should not be included in the
Magellanic Stream.

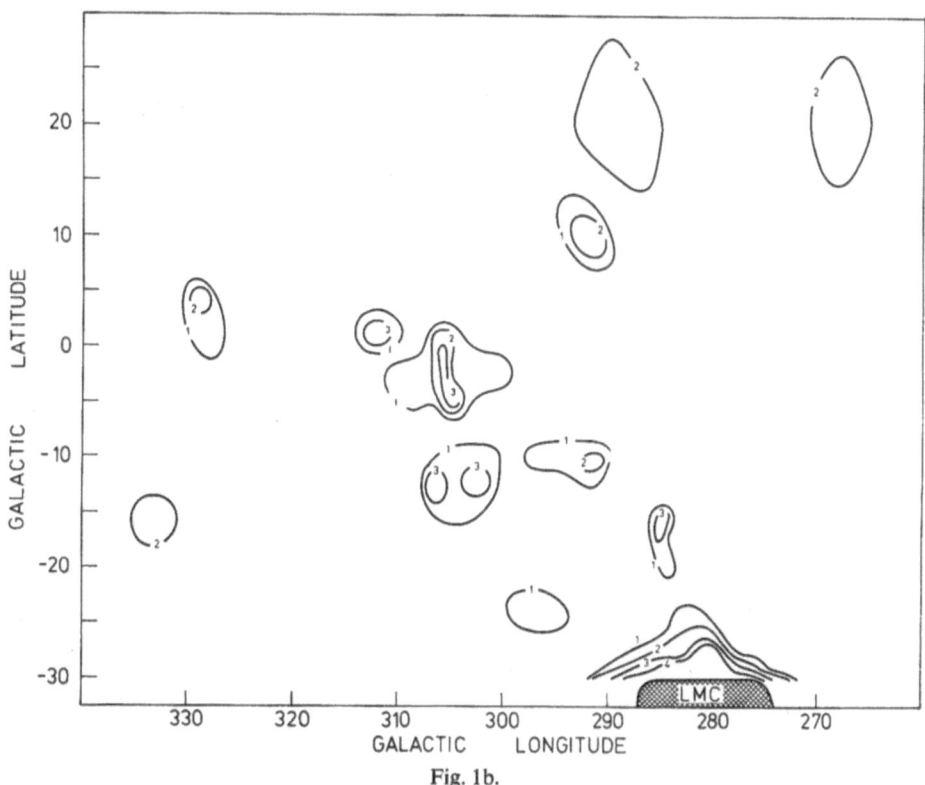

Fig. 1b.

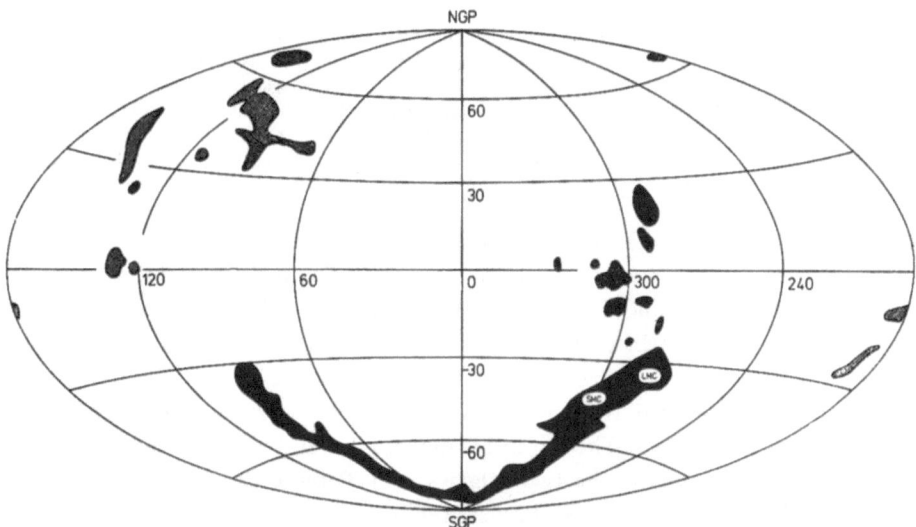

Fig. 2.   The Magellanic Stream is inked in on this Aitoff projection in galactic co-ordinates. The cross-hatched areas are the high velocity H I clouds discovered much earlier by northern hemisphere observers (cf. Hulsbosch, 1972).

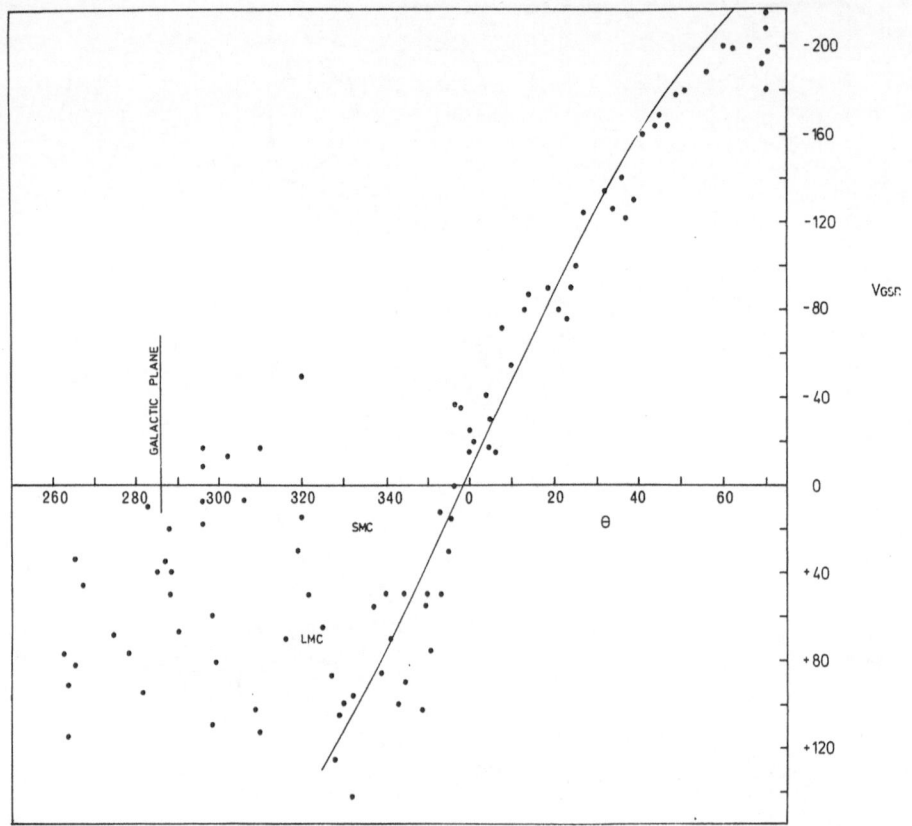

Fig. 3.   Plot of radial velocity in km s^{-1} (V_{GSR}) versus the angular distance (θ) along the Magellanic Stream using the co-ordinate system of Wannier and Wrixon (1972) in their Figure 2. Their radial velocities have been used in the section from $\theta = 70°$ to $\theta = 30°$. The systemic radial velocities of the SMC (20 km s^{-1}) and the LMC (68 km s^{-1}) have also been plotted. The radial velocities (V_{GSR}) refer to a non-rotating galaxy and are obtained from the velocities relative to the local standard of rest by the relation $V_{\text{LSR}} + 225 \sin l \cos b$. The full line represents the relation $V_{\text{GSR}} = -240 \sin (\theta + 2°)$.

favoured one is that gravitational interaction between the SMC and the Galaxy has pulled out the Magellanic Stream from the SMC. Toomre, in unpublished work referred to by Mirabel and Turner (1973), has constructed an orbit for the SMC about the Galaxy in which perigalactic passage occurred some 5×10^8 yr ago at a distance of 20 kpc from the galactic centre. He found that the SMC would have undergone severe disruption, material being drawn out into a bridge and a tail characteristic of such violent tidal events. The tail in his model lay well behind the LMC but projected onto the line joining the two galaxies, and Toomre thought that the Inter-Cloud H I (Hindman *et al.*, 1963; Turner, 1968) may belong to the tail. The bridge lay between the SMC and the Galaxy (see Mirabel and Turner, 1973) but was angled at about 40° to the Magellanic Stream. However, it may be possible to allow the SMC to have a more highly inclined orbit about the Galaxy, which would shift the bridge closer to the South Galactic Pole and more nearly coincident with the Magellanic Stream.

Clutton-Brock (1972) has produced plots to show the tearing of a galaxy by the close passage of a very much larger galaxy. He finds that, under these conditions, the gas and stars are drawn out from the smaller galaxy to form a very prominent bridge and tail. If the orbit of the SMC were highly inclined to the galactic plane then it may be seen from Clutton-Brock's Figures 4g and 4i that in 5×10^8 yr after perigalactic passage, a bridge and tail would be formed by the SMC giving a spatial distribution very similar to that of the Magellanic Stream. Wright (private communication) has made similar computations to those of Clutton-Brock and has reached a similar conclusion.

Although the existence and position of the Magellanic Stream may be explicable in this way, the models do not account for the radial velocities observed along the Stream. In particular the high negative radial velocity ($V_{GSR} = -216$ km s^{-1}) at the tip of the Stream at $l = 90°$, $b = -30°$ is almost an order of magnitude greater than the -30 km s^{-1} predicted by the models of Clutton-Brock and Wright. Oort has suggested (private communication) that this difficulty may be overcome if the Magellanic Stream 'snowploughs' through an intergalactic wind and is thereby braked and blown towards the galactic centre. This increases the component of the velocity of the Stream along our line of sight and also increases the velocity of infall of the Stream. Oort calculates that an intergalactic wind of density 2×10^{-4} atoms cm^{-3} would be sufficient to produce the observed radial velocity at the tip of the Stream, a density of the same order as is necessary to explain the stability of the Local System (Oort, 1970). In addition Oort invokes an intergalactic wind of this density to explain the high velocity H I clouds found by northern hemisphere observers at positive latitudes between longitudes 80° and 180°.

One difficulty with this concept is that denser parts of the Stream, such as that around $l = 300°$, $b = -73°$, would be relatively unaffected by snow-ploughing through the intergalactic wind and would continue along their original orbits. They should then have more positive radial velocities than the nearby, less dense sections whereas the reverse is the case. In addition it is difficult to explain the high positive radial velocities in the Stream near the Magellanic Clouds (e.g. 130 km s^{-1} at $l = 286°$, $b = -47°$ and 90 km s^{-1} at $l = 294°$, $b = -58°$) compared to the systemic radial velocities of 20 km s^{-1} for the SMC and 68 km s^{-1} for the LMC.

It is tempting to speculate that part of the Magellanic Stream may already have hit the galactic plane within 5 kpc of the Sun at $l \approx 100°$ and that its momentum pushed some disk gas to z distances of several kiloparsecs. As the width of the Stream is about 5 to 10 kpc, the disk gas would have been disturbed over a broad range of galactic longitudes around $l = 100°$. At this point the mechanism put forward by Oort (1970) to explain the origin of the high velocity H I clouds (cross-hatched in Figure 2) may operate, so that the gas falls back into the galactic plane under the pressure of the intergalactic wind and the gravitational field of the Galaxy. This action of the Magellanic Stream on the H I in the plane replaces the need for super-explosions postulated by Oort (1970) to replenish the gas in the halo.

The discovery of the Stream suggests many follow-up observations, the most important of which is to search for optical emission. In this regard Bok (1966) commented that a striking feature of the distribution of globular clusters in the Clouds is that many of them are far from the two galaxies, and indeed some cannot be assigned membership specifically to either Cloud (Gascoigne and Lyngå, 1963).

This paper would not be complete without mentioning earlier related work. Nearly twenty years ago de Vaucouleurs (1954) discussed the possibility of a connection between the Magellanic Clouds and the Galaxy and coined the name 'The Magellanic Stream', and Kerr and Sullivan in 1969 considered that the high velocity H I clouds known at that time might be satellites of the Galaxy (perhaps debris scattered around the orbit of the LMC) at distances of the order of 50 kpc. It should also be noted that Hulsbosch (1972), when discussing the long South Polar filament found by van Kuilenburg, and Wannier and Wrixon, suggested that "it may be a tidal arm expelled from the LMC by an encounter with the Galaxy".

Acknowledgements

The authors acknowledge valuable discussions about the interpretation of this feature with Prof. Toomre, Dr Kalnajs and Dr Hulsbosch, and they thank Prof. Oort for communicating the results of his work on the interpretation of the Magellanic Stream prior to publication. The authors also thank David Cooke, Frank Trett and Bob Phelps of the resident engineering staff at Parkes for keeping the receiver operational, and Gary da Costa and Graham White from Mt. Stromlo for assistance with the observations.

References

Bok, B. J.: 1966, *Ann. Rev. Astron. Astrophys.* **4**, 95.
Clutton-Brock, M.: 1972, *Astrophys. Space Sci.* **17**, 292.
de Vaucouleurs, G.: 1954, *Observatory* **74**, 23.
Dieter, N. H.: 1965, *Astron. J.* **70**, 552.
Gascoigne, S. C. B. and Lyngå, G.: 1963, *Observatory* **83**, 38.
Hindman, J. V., Kerr, F. J., and McGee, R. X.: 1963, *Australian J. Phys.* **16**, 570.
Hulsbosch, A. N. M. and Raimond, E.: 1966, *Bull. Astron. Inst. Neth.* **18**, 413.
Hulsbosch, A. N. M.: 1972, *Studies on High-Velocity Clouds*, dissertation, Leiden University.
Kerr, F. J. and Sullivan, W. T.: 1969, *Astrophys. J.* **158**, 115.
Mirabel, I. F. and Turner, K. C.: 1973, *Astron. Astrophys.* **22**, 437.
Oort, J. H.: 1970, *Astron. Astrophys.* **7**, 381.
Turner, K. C.: 1968, Annual Report of the Director, Dept. of Terrestrial Magnetism 1967–68, Carnegie Institution of Washington, Washington, D.C., p. 291.
van Kuilenburg, J.: 1972, *Astron. Astrophys.* **16**, 276.
Wannier, P. and Wrixon, G. T.: 1972, *Astrophys. J. Letters* **173**, L119.
Wannier, P., Wrixon, G. T., and Wilson, R. W.: 1972, *Astron. Astrophys.* **18**, 224.

DISCUSSION

Carrick: What mass was used for the LMC in the tidal calculations?
  Mathewson: $6 \times 10^9 \, M_\odot$.
  Lewis: Does the Magellanic Stream have any connection with the H I companion of NGC 300

found by Shobbrook and Robinson (*Australian J. Phys.* **20**, 131, 1967) in the region of the South Galactic Pole?

Mathewson: The large cloud of H I found by them near NGC 300 is, in fact, part of the Magellanic Stream.

G. de Vaucouleurs: It is not entirely clear to me that the gaseous stream you observe is the same sort of thing as the streams of stars discussed by Dr Toomre.

Toomre: To a first approximation, all the passengers (gas and stars) ride alike so far as gravity is concerned.

G. de Vaucouleurs: What we see in most of the tails is stars. Why do we see gas in this case but no stars?

Mathewson: They may be there, but we've not yet had time to look for them.

Oort: Unless the gas in the stream is very clumpy, its density must be so low that one cannot expect new stars to be formed.

Mathewson: Some preliminary observations we have made with the Parkes 210-ft dish indicate that the stream does indeed have a clumpy structure.

van den Bergh: If the tidal interaction between the Galaxy and the Clouds was able to detach

Model by Wright (see next page).

about one-quarter of the neutral hydrogen in the Clouds then it should also have been able to rip off some of the globular clusters.

Mathewson: I agree, and they should be searched for.

Ekers: Detection of the stars or globular clusters torn out with the gas could provide a definitive test of the theory of intergalactic braking, since these old stars will presumably follow the unbraked trajectory and have radial velocities quite different from the gas.

Wright: I have a simple dynamical model (see previous page) for the Galaxy-LMC interaction that does not completely work! The *positional* features agree well with the Mathewson, Cleary, Murray Stream but the *radial velocities* disagree at small Galactic latitudes. Although we must probably look to some resistive mechanism to produce the high negative radial velocities, I believe the stream is basically an inter-active phenomenon.

Oort: What model did you use for the Magellanic Cloud?

Wright: The mass ratios of our Galaxy to the LMC are not critical within the range 5 to 50, nor is the rotation plane of the LMC within $\pm 45°$. Nor even the perigalactic distance within 15 to 30 kpc. However, the model shown had a mass ratio of 20 to 1, a perigalactic distance of 25 kpc and, before the interaction took place, consisted of a cold, rotating disk of stars in circular orbits. The *present* observed velocities in the LMC are not necessarily in disagreement with the present computed velocities.

Toomre: I would like to bet right here and now that if indeed the Magellanic Stream turns out to have a tidal origin, it will be as some sort of bridge-like debris torn from the *Small* Cloud, rather than from the *Large* Cloud as proposed by Dr Wright.

My reasoning, for what it is worth, rests mainly on the need for a plausible *tail* to complement the claimed near-side tidal damage to whichever Cloud was so mistreated by our Galaxy. I know of nothing as yet that could fairly be described as such beyond the LMC, although the possibly mis-named 21-cm 'bridge' between the Clouds might conceivably be a tidal tail of the SMC (cf. Mirabel, I. F. and Turner, K. C.: *Astron. Astrophys.* **22**, 437, 1973).

Kerr: It is worth noting that the bridge between the Magellanic Clouds, which has been known for 20 yr, is known only in gas, except near the SMC where stars and clusters have been found.

RECENT RADIO STUDIES OF BRIGHT GALAXIES

J. H. OORT

Sterrewacht te Leiden, The Netherlands

Abstract. Some results are discussed of recent – still largely unpublished – high-resolution studies of bright spirals, both in the radio continuum and in the 21-cm line radiation. Special emphasis is given to observations of M51, M81 and NGC 4258.

The most important new data are (1) Estimates of the contrast in gas density between arms and interarm regions (M81 and M101); the contrast appears to be quite strong. It provides an important parameter for the density wave. (2) Evidence for a pronounced decrease of gas density along the arms when these are followed towards the centre. Considerable line radiation is observed from outer regions where the optical intensity of the arms is very low. (3) Extension of rotation curves to larger distances from the centre. (4) Data on the motions in the spiral waves. (5) Determination of the shift between synchrotron and optical arms, providing direct evidence for the formation of stars as a consequence of the passage of the interstellar gas through the spiral density wave. From the shift measured in M51 the formation time is found to be roughly ten million years. If the data for (2) are ascribed to the depletion of the gas by star formation, the average net fraction of the gas consumed in star formation is found to be between 2 % and 3 % per passage through the spiral wave (Table I). (6) Separation of nuclear and disk synchrotron radiation in spirals. (7) Evidence for a recent expulsion of about 10^8 solar masses from the nuclear region of NGC 4258.

A new era has started for the study of spiral galaxies by means of radiowaves. The completion of large synthesis radiotelescopes has for the first time opened the possibility of studying galaxies with sufficient resolution to observe individual spiral arms, and much of the following discussion is based on such observations made by Leiden and Groningen astronomers with the 1.5-km synthesis radiotelescope at Westerbork in The Netherlands.

1. Gas

1.1. GENERAL SPIRAL DISTRIBUTION

For a proper study of spiral structure and its dynamics it is desirable to observe galaxies in which spiral arms can be traced throughout the entire system although, for a confrontation with spiral-wave theory, the arms should not be *very* open. Among the nearer spirals above $+30°$ declination, M81 and M51 appear to be the most suitable. M31, in which better resolution can be obtained, has too high an inclination for the arms to be traced unambiguously throughout, a difficulty which presents itself more seriously still in our own Galaxy. However, in our Galaxy the larger angular scale enables one to study much weaker features. In particular, this circumstance makes it a unique object for investigating the expanding features in the central region, and also the influx of intergalactic gas.

Hydrogen line observations at 21 cm with an angular resolution of $1'.5 \times 3'.0$ have been made of M33 in Cambridge (Wright *et al.*, 1972) and detailed optical measures of the numerous emission nebulae have enabled French observers to study the very asymmetrical motions of the irregular individual arms in this galaxy. For M31 there is a still unpublished investigation by Emerson in Cambridge, while observations with a 70×100 pc beam for a region within about $\frac{1}{2}°$ of the centre have been obtained at Westerbork. At Westerbork extensive high-resolution line observations have also been made for M51, M81, M101 and IC 342, all of which are normal spirals with well-defined spiral arms, M101 and IC 342 being more open and somewhat less regular

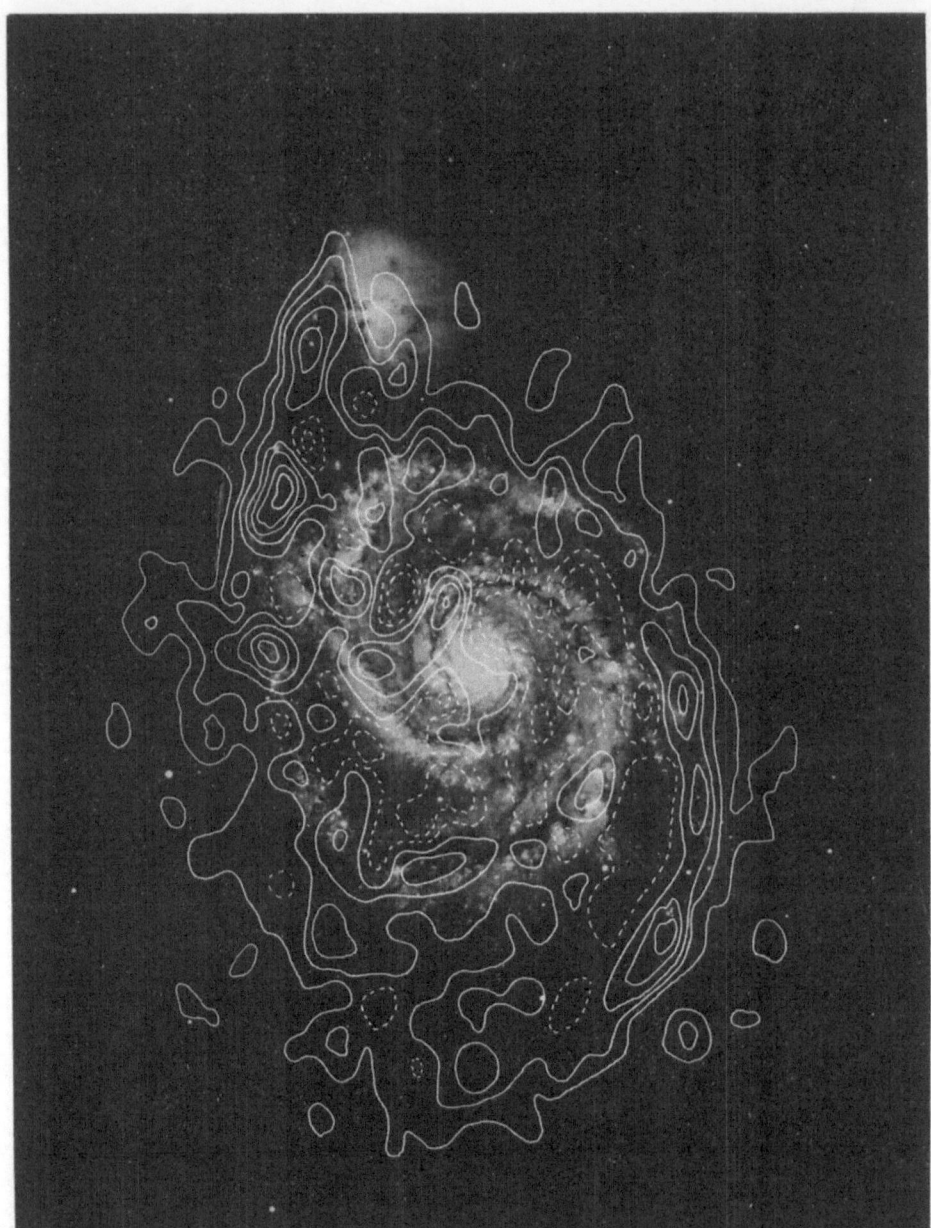

Fig. 1. Neutral hydrogen distribution in M51 measured with the Synthesis Radio Telescope at Westerbork. Half-power beamwidth 24″ in right-ascension and 32″ in declination. Contour interval 3.75×10^{20} atom cm^{-2} (Shane and Bajaja, preliminary results).

than M51 and M81. Equally extensive measurements were made on the explosive spiral NGC 4258 and on the barred spiral NGC 5383.

Figures 1 and 2 show the hydrogen distribution in M51 and M81, as observed in Westerbork with beams of 28″ and 50″ respectively, corresponding to 1.3 and 0.8 kpc. The two galaxies are of different types: M81, which has a strong central bulge, has been classified by Sandage as Sab, and M51 as Sc. The pitch angles of the arms are

M 81

NEUTRAL HYDROGEN
WSRT
contour interval:
3.10^{20} atoms cm^{-2}

Fig. 2.   Neutral hydrogen distribution in M81 measured with the Synthesis Radio Telescope at Westerbork. Half-power beamwidth is shown in the lower right-hand corner. Co-ordinate intervals are 1 min in α and 5′ in δ. (Rots and Shane, in preparation).

not very different, being 16° out to 7–8 kpc in M51 and varying from 11° at $R=4$ kpc to about 17° at $R=9$ kpc in M81. There is a pronounced difference in the frequency of emission nebulae, which are abundant in M51 and rather scarce in M81. The fraction of the mass that is in gaseous form, as judged from the H I emission, is perhaps about 3 times larger in M51. In both galaxies the H I is strikingly concentrated in the spiral arms. It also extends to large distances from the centre, where the optical surface brightness has become small.

For the first time detailed H I observations of a barred spiral have become available, viz. of the giant SBc galaxy NGC 5383. The results, on which Dr Allen has reported (this volume, p. 425), are remarkable. Again the hydrogen extends to very large distances from the centre but, contrary to what is found in most other galaxies for which high-resolution data are available, NGC 5383 has a pronounced concentration of H I in and around the optically very bright nucleus.

It is important for the theory of the spiral waves for us to know the contrast of the gas densities in and between the arms. Unfortunately the observations are not yet quite good enough to provide a trustworthy answer. This is due to three factors: the insufficient sensitivity, the lack of knowledge about the true zero level in M81 (because the zero spacings were still missing), and finally the rapid decrease of the density with decreasing R which I shall discuss in the following. All we can say from the present data in M81 and M51 is that the ratio between the density in the central ridges of the arms and that between arms is probably at least 3. For the outer arms in NGC 4258 it seems to be much higher.

1.2. RATE OF STAR FORMATION AT DIFFERENT DISTANCES FROM THE CENTRE

The estimates of the contrast between arms and interarm regions are difficult because of the superposition of another very striking phenomenon. Along the spiral arms the gas density decreases strongly when passing from the outer to the inner regions. In the arms of M81 the column density decreases from about 1.6×10^{21} atom cm^{-2} at $R=9$ kpc to about 0.6×10^{21} at $R=3.3$ kpc; inside this distance it becomes un-measurable with our present equipment. (The column densities quoted are directly observed values, *not* corrected to a face-on view). This distribution contrasts with that of *light* in the arms (cf. Figure 3). The maximum H I density extends to regions where the arms are almost indistinguishable optically, then outside $R=11$ kpc there is a steep drop in the gas density. Similar changes with R are found in M51 and M101 (for the latter, cf. the report by Allen) and the phenomenon is a very general one. It was found already in 1957 when the first line observations of external galaxies were made (cf. the investigation of M31 by van de Hulst *et al.* (1957)), but Roberts (1971) was the first to draw attention to its common character. He pointed to the frequent existence of outer rings of enhanced 21-cm radiation in spiral galaxies. In M31 the central 'hole' has a radius of about 27′ $(R\sim7.5$ kpc); the broad region of high H I density extends from there to about 18 kpc, beyond which the density falls off steeply. In M31 the region where the H I density first becomes high corresponds with that where the brightest optical arms appear. In M51, on the other hand, the arms are

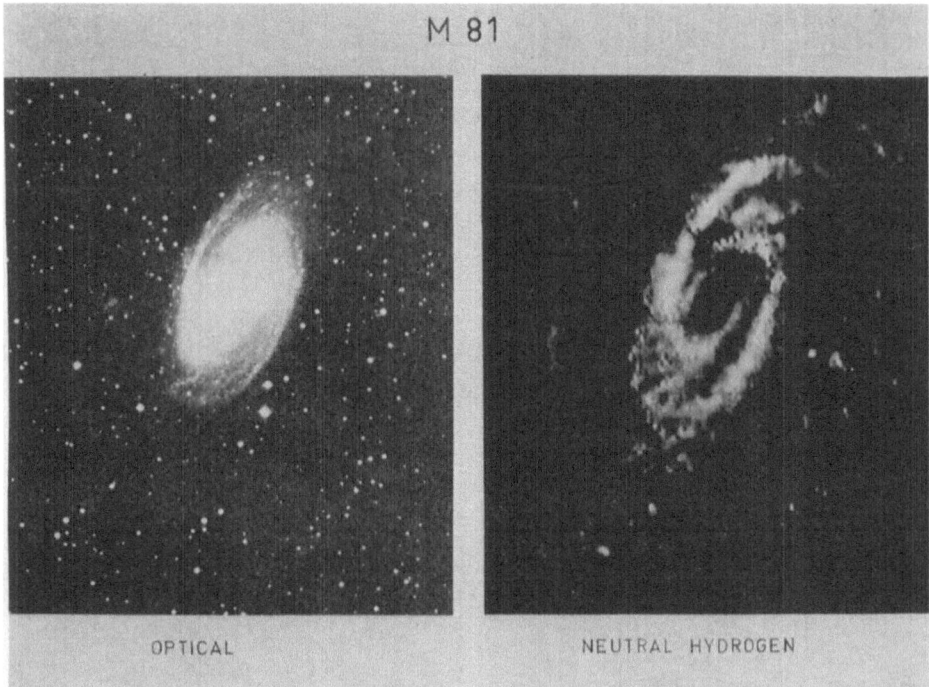

Fig. 3.   Contrast between distribution of optical luminosity (at the left) and neutral hydrogen (at right) in M81 (Rots, unpublished).

optically quite bright in the inner half where the H I density is still low, but the 21-cm arms continue to very large R where the optical arms are barely, if at all, visible. A striking example is furnished by NGC 4258 (cf. Figure 4), where the H I surface density is exceptionally high in the faint outermost arms. A striking example of the contrast between the distribution of light and gas in a very late-type spiral is given by the Cambridge observers for M33 (Figure 5).

No high-resolution 21-cm observation data are available for early-type spirals. NGC 4594 has been observed at Westerbork, but the observations have not yet been reduced. In this case the *optical* data give already fairly convincing evidence that the interstellar density is very low inside the well-known dark ring, which starts at $R = 12$ kpc and extends to 24 kpc. The column density of the dust in the ring is at least an order of magnitude higher than that inside the ring's inner edge. We hope that 21-cm observations will show whether the inner region is equally devoid of *gas*.

The relatively low gas density in the inner parts of spiral galaxies is probably due to a higher rate of star formation. Observations in the radio continuum have added rather conclusive evidence that star formation in spiral galaxies is at present largely initiated by the compression occurring where the interstellar gas moves through the spiral density wave. This has been shown most clearly in M51, as I shall illustrate in the second section of my report.

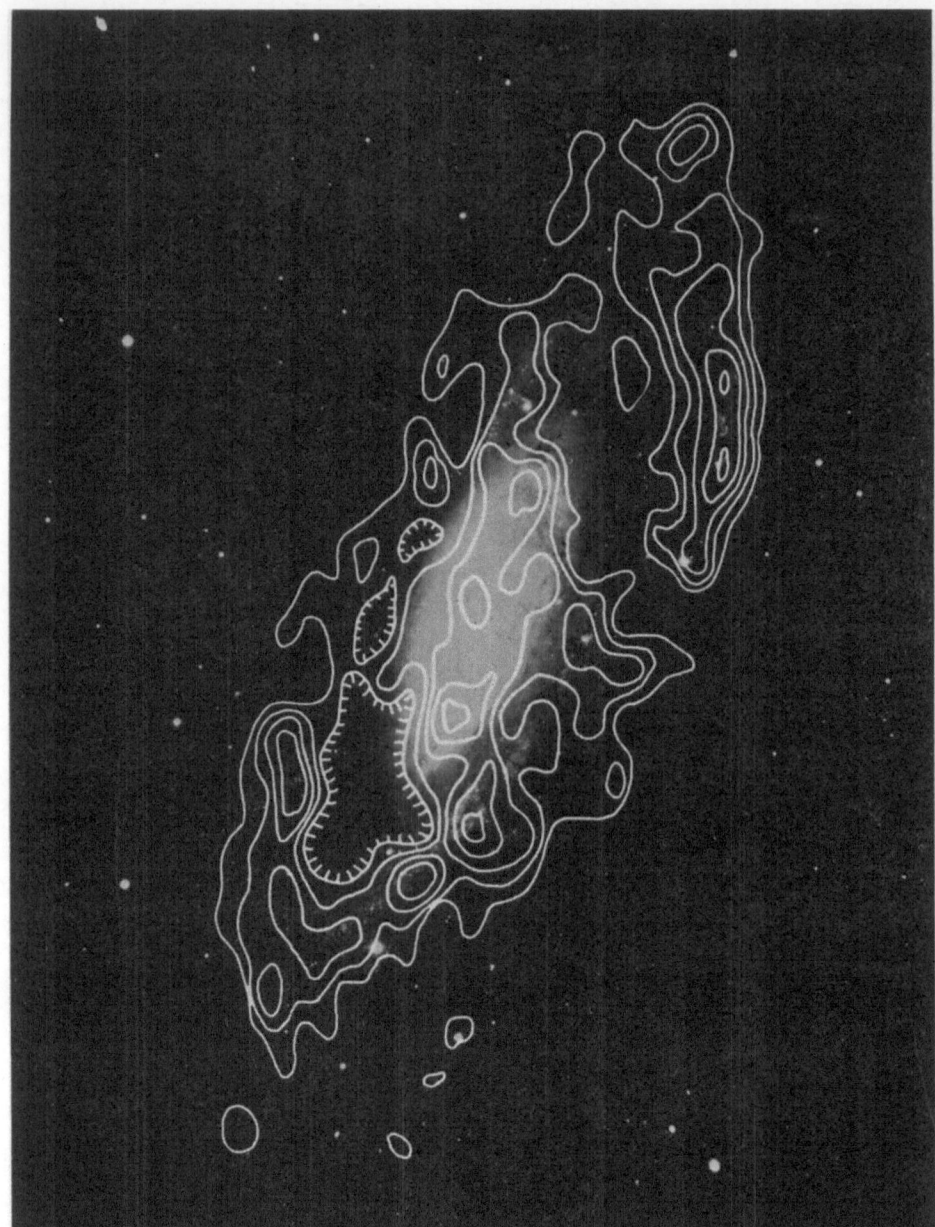

Fig. 4.   Column density of hydrogen in NGC 4258 (Westerbork). Half-power beamwidth 50″ × 70″,
contour interval 4×10^{20} atom cm^{-2} (Shane, preliminary results).

Actually, the observations show only the formation of OB stars and H II regions,
but it is plausible that the formation of less massive stars occurs in the same regions.
If, during the disk stage of spiral galaxies, the bulk of star formation takes place by
this mechanism, then the rate of star formation must evidently be proportional to the

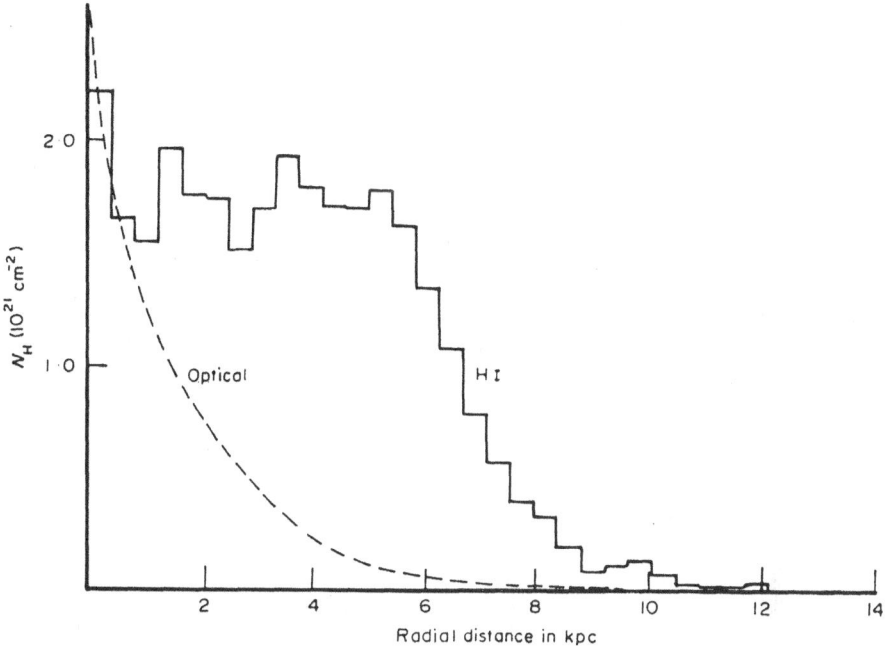

Fig. 5. Distribution of light and neutral hydrogen in M33 (Wright *et al.*, 1972).

frequency at which the interstellar gas passes through the spiral pattern; this, in turn, depends on the difference between the circular angular velocity, ω_c, and the angular velocity of the pattern, Ω. Naturally, the rate will also be proportional to the overall gas density ϱ. Other factors, such as the occurrence of a shock at the passage through the spiral wave, and the strength of this shock, as well as the thickness of the gas layer may play a role; but it is not implausible that the principal determining factors are $(\omega_c - \Omega)$ and ϱ. In the following considerations I therefore confine attention to the latter factors.

Suppose that for each passage through the spiral wave a fraction α of the gas disappears permanently into stars; then, if the initial gas density at the epoch when the stars started to be mainly formed by spiral waves is denoted by ϱ_i, the present gas density will be

$$\varrho = \varrho_i\, e^{-n\alpha},$$

where $n = 10(\omega_c - \Omega)/\pi$ is the number of times the gas has passed through the spiral wave during the 10×10^9 yr of its presumed existence (the unit of time being taken as 10^9 yr).

In practically all spirals investigated, ω_c increases considerably with decreasing distance R from the centre; n therefore *in*creases, and ϱ/ϱ_i will decrease. Qualitatively this provides a natural explanation of the decrease in gas density as we follow the arms inwards, which, as we have seen, is such a striking phenomenon in most spirals.

For a *quantitative* confrontation with the observed ratios of gas column densities we must know the way in which the initial density varied with R. If, in order to obtain an approximate estimate, we assume that this varied as R^{-2}, like the density variation of older disk populations as judged from the general light as well as the mass distribution, we find that we get a good representation of the observed relative gas density in the arms of M81 if $\alpha = 0.022$. This is shown in Table I. It should be noted that, because we consider only ratios of densities, the pattern speed drops out. The last column shows the run of the density that would follow if the fraction of the gas transformed into stars per passage through the wave is taken to be 50% higher.

TABLE I

Observed and computed column densities of the gas in M81

R kpc	ϱ 10^{20} atom cm^{-2}	$\varrho(R)/\varrho(9)$		
		Observed	Computed	
			$\alpha = 0.022$	0.033
9.0	16.5			
4.8	12.0	0.73	0.54	0.22
3.3	6.0	0.36	0.35	0.07
2	3::	0.2::	0.27	0.03

Note: Values in the last line marked :: are very uncertain estimates.

The numbers in this column differ strikingly from the observed density ratios and the difference gives an indication of the margin of uncertainty of α. An additional uncertainty, however, is caused by the inaccuracy of the rotation curve. Also, the period during which fully-developed spiral structure has existed might be appreciably shorter than the assumed age of the galaxy (10^{10} yr), in which case α would be increased proportionally.

We may conclude that the observed run of density can well be explained in this way, and that the average fraction of the gas used up in star formation at each passage through the spiral wave is roughly 2%. The fraction that actually goes into stars is larger, but part of this is returned to the medium as the stars evolve. I estimate that the uncertainty in α is about ± 0.005.

It is of interest to apply the same idea to our own Galaxy. From the known density distribution of the gas we find a similar value of α. If for the vicinity of the Sun we take $\omega = 25$ km s^{-1} kpc^{-1}, and assume with Lin and others that Ω is about 15 km s^{-1} kpc^{-1}, we find $n = 32$. This gives $\varrho/\varrho_i = 0.49$. The actual ratio of the column density of the gas to that of stars plus gas in the vicinity of the Sun is considerably smaller than this, viz. about 0.10. We must therefore conclude that only about one tenth of the stars in the column have been formed by the spiral wave. The rest must either have formed in the original collapse into a disk or at later times independently of the spiral wave.

1.3. ROTATION CURVES

The high neutral density in the outermost regions of the disks of spiral galaxies makes it possible to extend the rotation curves beyond what was accessible by optical methods. Figure 6 (from Roberts and Rots, 1973) shows typical rotation curves for some of the nearer giant galaxies. It presents two interesting features.

Firstly, there is clear evidence for a considerable range in the shapes of these curves: while the Sab and Sb systems M81 and M31 show a pronounced maximum at a

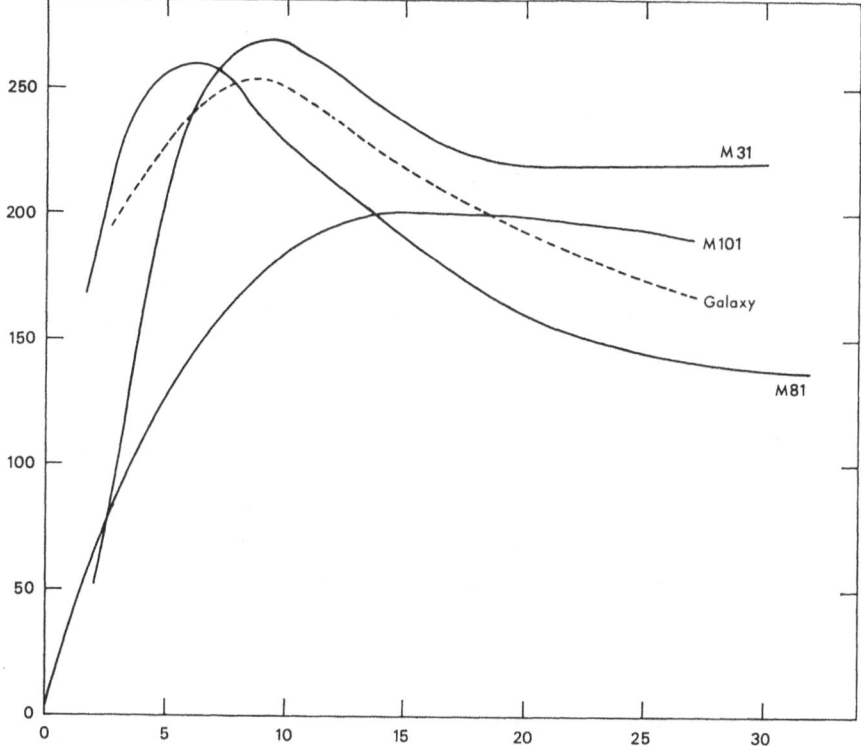

Fig. 6.   Some typical rotation curves (Roberts and Rots, 1973). Ordinates km s^{-1}, abscissae kpc.

distance well within the optically bright parts, the Scd galaxy M101 has a rotation curve which reaches a maximum only in the outermost region of the visible galaxy, beyond which it remains almost constant.

Secondly, the rotation velocity in the outer parts of giant spirals has a tendency to remain practically constant. Judging from present data, this seems to be a general property of giant spirals. Recent Westerbork observations show that the rotation velocity of NGC 4258 remains virtually constant at a value of 220 km s^{-1}, from $R=3$ kpc to the farthest distance from the centre at which it can be measured ($R\sim 20$ kpc). For the giant SB galaxy NGC 5383 it was found constant and high from $R\sim 4$ kpc to the outermost observed points around $R=35$ kpc (Sancisi; cf.

report by Allen, p. 425). A similar behaviour was found in M51 by W. W. Shane and Elly Dekker (unpublished). If the rotation velocity remains constant over a large range in R this indicates that the mass inside R increases approximately proportionally with R. These rotation curves therefore indicate that the masses of giant galaxies may be considerably higher than has previously been assumed.

In NGC 4258 the rotation curve on the northern side shows a sharp maximum within about 800 pc from the centre. This is shown both by the 21-cm observations and by the optical emission lines (Chincarini and Walker, 1967). The rotation in this central part closely resembles that of the nuclear disk in our Galaxy.

1.4. LARGE-SCALE DEVIATIONS FROM CIRCULAR MOTIONS

Because of its three-times-smaller distance, M81 is better than M51 for a study of the systematic streaming motions that should accompany the spiral density waves. The 21-cm observations clearly show the presence of waves, but observations and reductions are not yet sufficiently refined to permit a trustworthy derivation of their amplitudes. The observations will be discussed in the report by Lindblad (this Volume p. 399). For M51, 21-cm observations as well as very extensive Fabry-Pérot Hα observations (Tully, R. B.: 1974, *Astrophys. J. Suppl.* **27**, 415) are available. The former indicate again fairly convincingly that wave motions of the expected nature are present.

The velocity fields shown by the high-resolution observations of the later-type spirals IC 348 and M101 have not yet been analysed sufficiently to permit a full report. In both galaxies there is a large asymmetry in the distribution of the neutral hydrogen. In M101 Rogstad (1971) finds that the north-east half contains 1.5 times as much H I as the south-west half (see also the report by Allen). A general investigation of such asymmetries has been made by the French observers. From a careful discussion of 21-cm observations of the Scd spiral M33 the Cambridge astronomers concluded that the velocities showed no evidence of a regular spiral wave. If there is such a wave its amplitude must be smaller than 3 km s^{-1} (Wright *et al.*, 1972).

For the frequency of large asymmetries see Bottinelli, 1971.

2. Synchrotron Radiation from Spiral Arms, Disks and Nuclei

At 1400 MHz, the frequency at which practically all continuum observations of high resolution have been made, the radiation of all spirals, except possibly a few very-late-type systems, is largely synchrotron in origin. The surface brightness is generally low and observation with unfilled-aperture telescopes is difficult. Only in a small number of galaxies has it so far been possible to study the distribution in sufficient detail to determine its general relationship to the optical arms.

The most illuminating cases are M51 (cf. Figure 7, Mathewson *et al.*, 1972), and M31 (Pooley, 1969; van der Kruit, 1972). In both galaxies the radiation is strongly concentrated near the brightest optical arms. The contrast of the intensity in the arms to that in the interarm regions is considerable.

Fig. 7. Isophotes of 1415 MHz radiation in M51 and NGC 5195 superimposed on a 200-in. plate taken by Humason. The contour unit is 0.8 K brightness temperature. The half-intensity beam is 24″ × 32″ (Mathewson *et al.*, 1972).

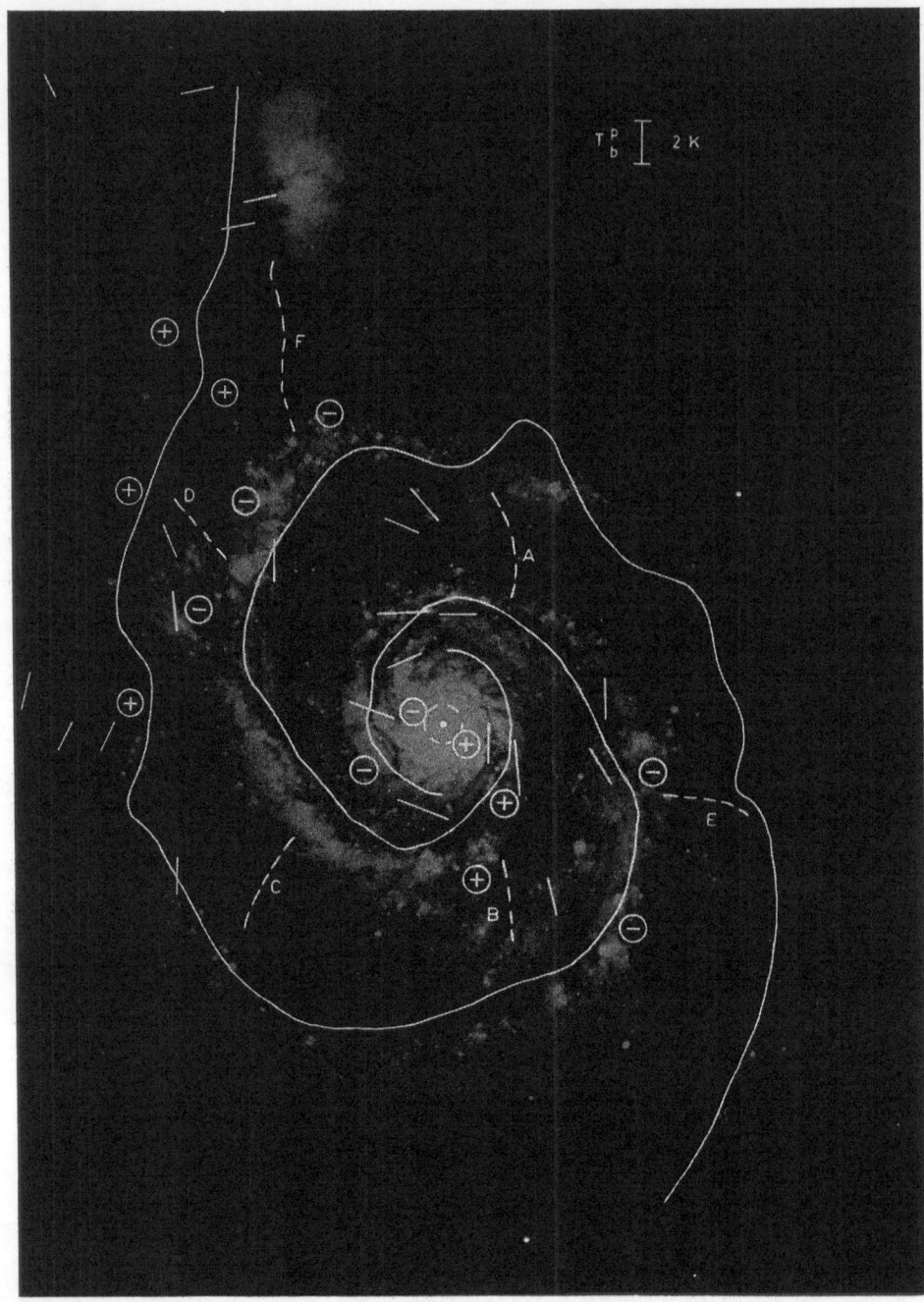

Fig. 8.   The peak lines of the synchrotron emission in M51 (Mathewson *et al.*, 1972).

In the few galaxies which are usable for such a detailed comparison there is a distinct separation between the synchrotron ridges and the ridges of the H II regions. This can best be observed in M51. Around $R=2'$, or 5 kpc, the latter precede the former by about 18° in the direction of the rotation (Figure 8). Furthermore, it is interesting to note that the synchrotron arms coincide closely with the dust concentrations. These phenomena show convincingly that star formation is initiated through the compression of the interstellar medium by the spiral wave, and more particularly in the shock caused by the wave. From the separation between the synchrotron arms and the H II arms, the time between the passage of the shock and

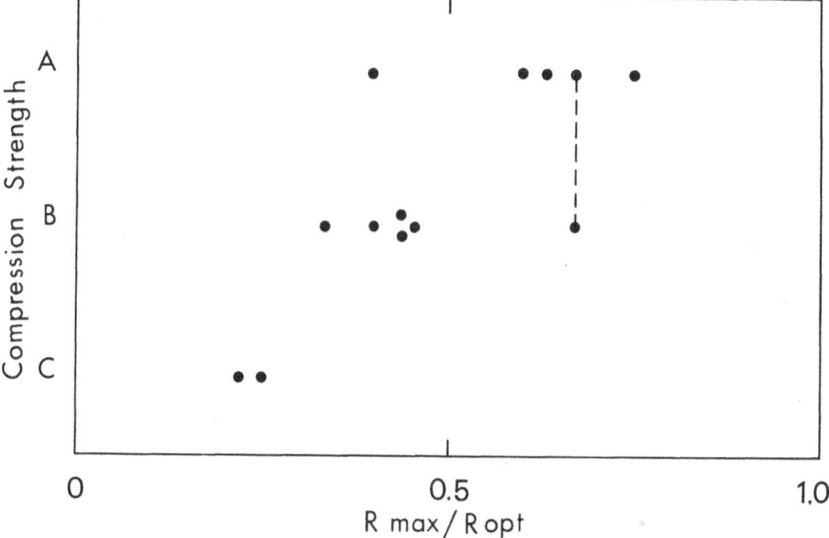

Fig. 9.   'Compression strength' and differential rotation. Abscissae are ratios between the radius where the rotation velocity has a maximum and the optical outer radius of the galaxy; these ratios are used as measures of the amount of differential rotation (van der Kruit, 1973).

the birth of OB stars is found to be about 10 million years. The synchrotron emission provides a particularly sensitive means for measuring the compression, as the emissivity can go up with nearly the third power of the compression.

From a dozen spirals for which the ratio of the intensity in the arms to that in the base disk could be determined with the Westerbork array, van der Kruit (1973) found that this ratio, which he called the 'compression strength', is higher for galaxies with strong differential rotation (cf. Figure 9). It is also correlated with the van den Bergh type and with the value of M/L, possibly because the rate of star formation increases with the compression.

In M51 the synchrotron intensity increases as we follow the arms inward, in contrast with the H I radiation which *decreases* considerably. In many other galaxies, however, such as M31 and probably also M81, there is an extensive hole in the synchrotron radiation, much like that in the hydrogen distribution. However, right in the centre we often see strongly enhanced radiation due to activity in the nucleus. In M31 the hole extends to about 5 kpc. The nucleus will be discussed in the report by Ekers (p. 257).

An interesting feature in M51 is the link with the satellite NGC 5195. There is a bridge of synchrotron emission between the two galaxies; also the radio brightness of the northern disk is about twice as high as that on the southern side opposite the companion.

Aperture synthesis has made it possible to separate the radiation from the nuclear region and the disk component in a considerable number of galaxies. Radio luminosities of the nuclear regions have now been measured in 44 spirals, although for 16 of these only upper limits have been found. There is a very large range in this

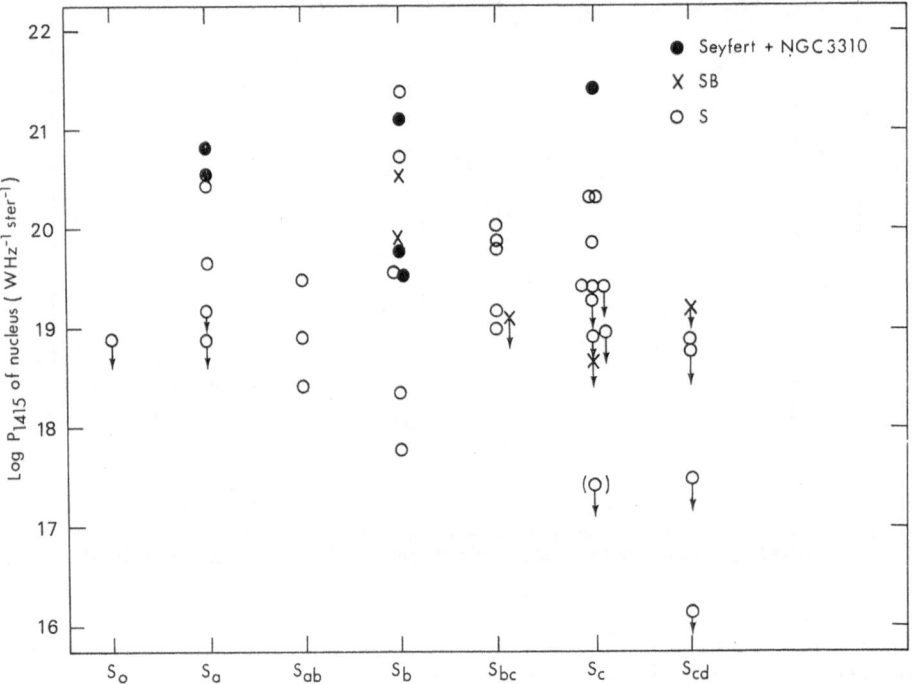

Fig. 10.   Luminosity of the nucleus at 1415 MHz against Hubble type (van der Kruit, 1973).

luminosity, as shown in Figure 10. The brightness of the nucleus is not clearly correlated with the Hubble type, except that the Scd galaxies have all very faint nuclei, mostly unobservable in fact. Within each of the other types the nuclear power varies over a range of about 1000:1. The average brightness of the nuclei of Seyfert galaxies is a factor of about 100 higher than that of normal galaxies. In general the activity of the nucleus appears to have no intimate relation with the general structure of a galaxy. It might be that such activity develops intermittently in all galaxies, on which view Seyfert galaxies may be spirals in which the activity is in an initial, violent stage.

The disk brightness shows a similarly large range, without a very clear correlation with other properties of the spirals, although on average the types earlier than Sbc seem to have lower radio surface brightness than Sbc and Sc galaxies. The spread in the disk brightness is probably connected with a spread in the production of relativistic particles, and may be related either with the frequency of supernovae or with the activity of the nucleus.

There is some evidence for a correlation between disk and nuclear brightness. Of the ten systems in which the brightness temperature of the 'base disk' could be measured by van der Kruit, the three with $T_b \geqslant 2°$ have an average of 19.4 for the logarithm of the power of the nucleus, while in the four with $T_b \leqslant 0°3$ the average $\log P$ is less than 18.

In the case of the ellipticals the occurrence of strong radio emission is highly correlated with the optical luminosity and the mass, and it seems that only the most massive elliptical galaxies can become strong radio sources. There is an indication that the more flattened systems are generally weaker radio emitters. Ellipticals have a very large range in radio brightness and for about 60 or 70% of the bright E's the emission lies below the limit to which observations have been made. All E galaxies with measurable radio emission have a bright radio nucleus, and about half of these have also a radio halo. In a number of cases the nucleus has a small linear size and a flat spectrum.

3. Explosive Events

Because synchrotron radiation is strongly enhanced by compression of the medium, observations of this emission are particularly suited for studying the effects of explosive events. The most interesting case observed is that of the large spiral NGC 4258 in which, superimposed on the ordinary spiral arm radiation, a much stronger pattern of totally different structure was found from observations with the Westerbork telescope. It shows two opposite curved radio ridges running across the optical spiral arms (Figure 11). These ridges have a remarkably smooth structure and extend to the outermost limits of the optical and the H I arms. In their inner parts they coincide precisely with a set of filamentary Hα arms which had been discovered already a dozen years earlier by French observers (Courtès and Cruvellier, 1961), and which were likewise found to be of a novel nature (Figure 12). They are similar in having an unusually smooth texture, without any evident localized emission patches. And they contain no blue stars, being, in fact, invisible on ordinary exposures in blue light. The radio ridges have steep edges on the sides which precede in the direction of the rotation, and are followed by wide plateaus of enhanced radiation. It seems that the Hα filaments are similarly followed by regions of somewhat enhanced Hα radiation. There are indications that both the anomalous Hα arms and the anomalous radio arms lie in the same plane as the ordinary spiral arms. In a face-on picture the outer parts of the radio ridges are straight and roughly perpendicular to the outer optical arms, the various fragments of which have nearly circular shapes (cf. Figure 13).

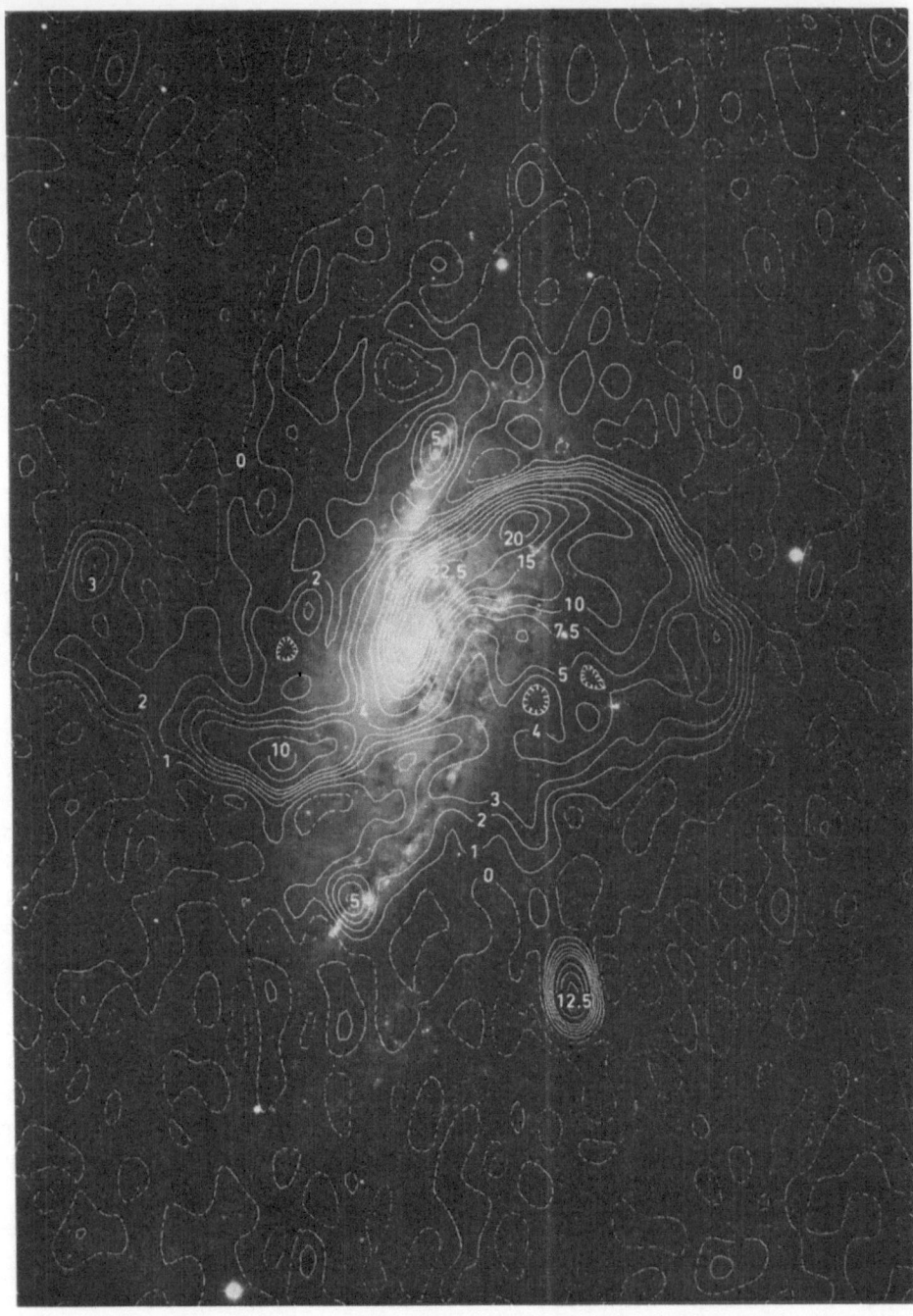

Fig. 11.   NGC 4258. Radio contour map at 1415 MHz superimposed on optical 200-in. photograph
by Sandage. The position of the optical nucleus is indicated by a cross (van der Kruit *et al.*, 1972).

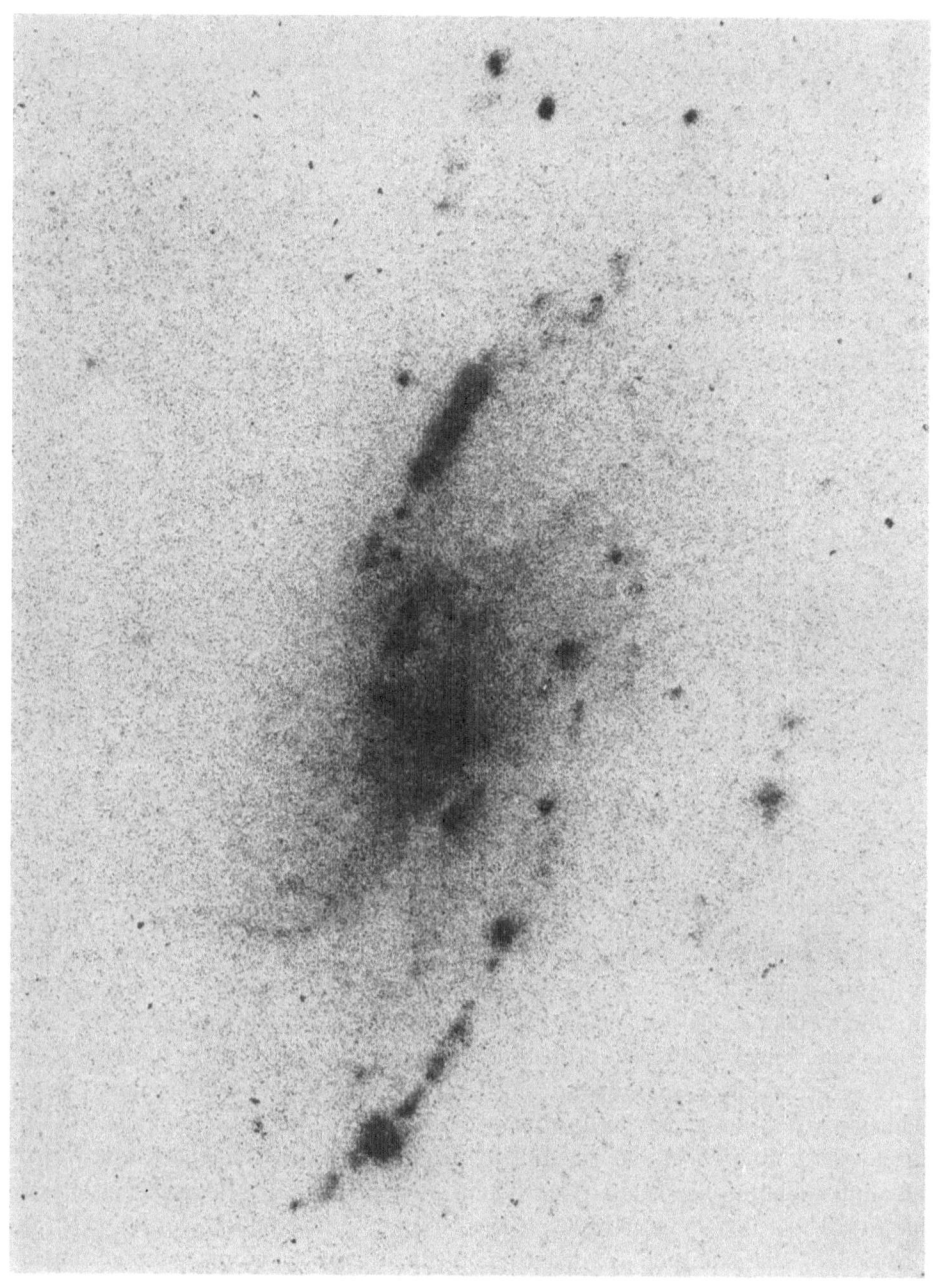

Fig. 12.   NGC 4258, Hα photograph through interference filter of 10 Å half-width
(Deharveng and Pellet, 1970).

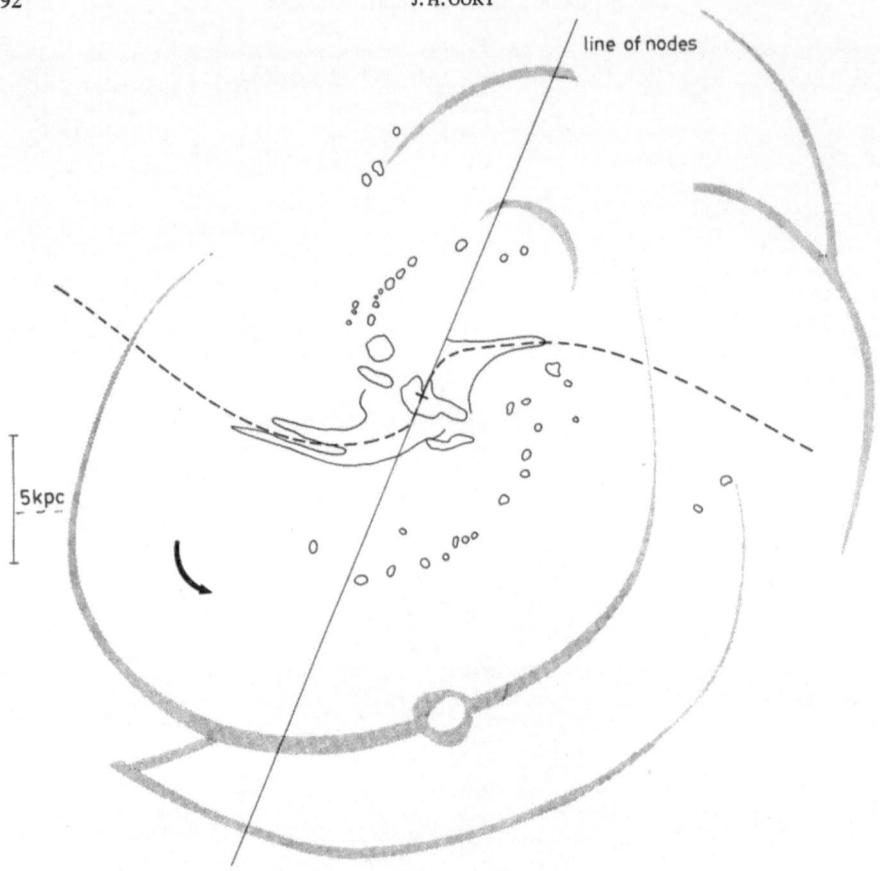

Fig. 13.   NGC 4258. Schematic face-on sketch of anomalous radio ridges (dashed lines) and Hα
filamentary arms (contours), as well as ordinary spiral arms (patches and shaded arms).

It seems possible to explain the anomalous arms and their accompanying 'plateaus'
as the consequence of a vast expulsion of gas from the nucleus into two opposite
directions making relatively small angles with the spiral disk (van der Kruit *et al.*,
1972). The expelled gas would have swept up the gas initially present in the disk and
in doing so would have acquired angular momentum. The radio ridges and Hα
filaments would be the present location of the expelled clouds, the Hα emission being
due to the ionization of the medium by the collision. The enhanced synchrotron
radiation might have come directly from the nuclear explosion. The plateaus may be
due to the fact that the ejection continued over a certain period during which the
direction of the expulsion rotated over an angle corresponding with the angular
extent of the plateaus. In this model the expulsion would have taken place about
18 million years ago, with initial velocities ranging from about 800 to 1600 km s^{-1}.

→

Fig. 14a.   Contour diagram of column density of neutral hydrogen in NGC 4258 having velocities
between + 70 and + 130 km s^{-1} relative to the centre.

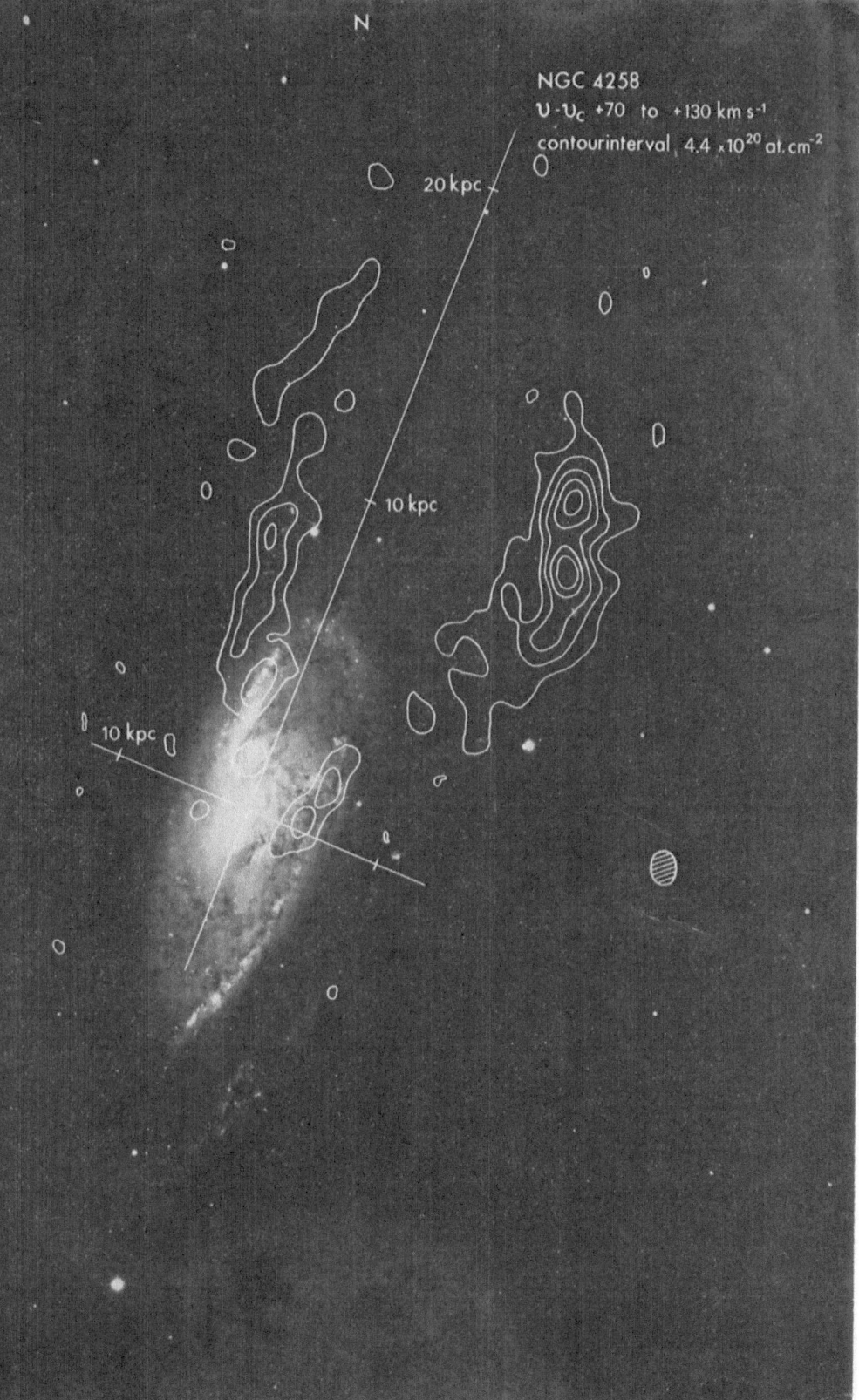

NGC 4258
$\upsilon - \upsilon_c$ +70 to +130 km s^{-1}
contour interval 4.4 $\times 10^{20}$ at. cm^{-2}

N

20 kpc

10 kpc

10 kpc

Fig. 14a.

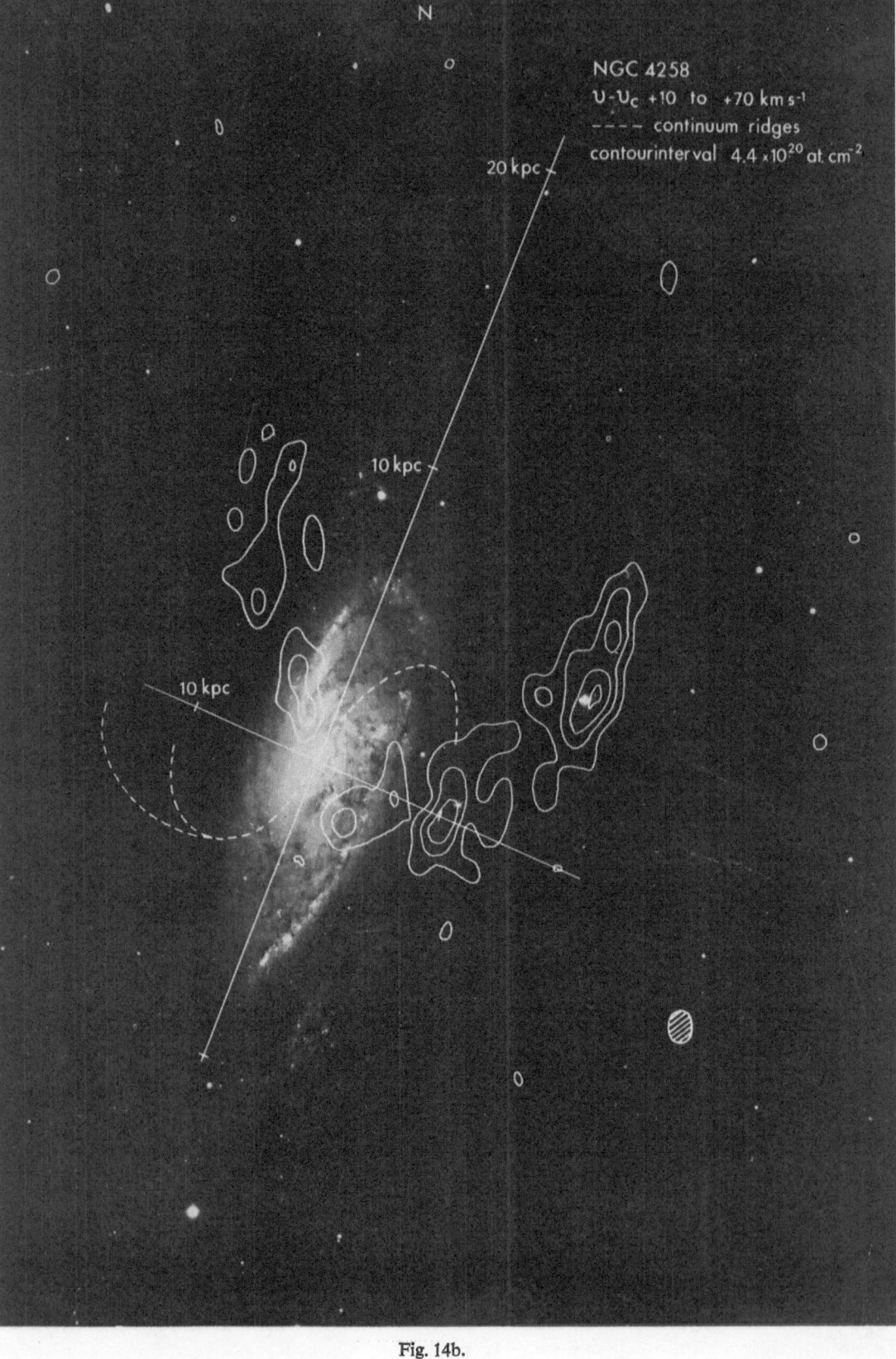

Fig. 14b.

The total ejected mass would have been between 10^7 and 10^8 $M_\odot$. Support for this interpretation is found in the fact that the optical arms show signs of having been swept away in the regions through which the ejected clouds would have moved. An event like this would result in a re-arrangement of a large fraction of the interstellar medium into a very regular spiral pattern.

As NGC 4258 is at present the only case in which a phenomenon of this sort and this magnitude has been observed, it is not possible to say anything about the frequency with which such vast ejections occur in spiral galaxies, nor about their significance for the general problem of spiral structure.

Direct support for the hypothesis of a large-scale expulsion from the nucleus of NGC 4258 has now been found from the 21-cm line observations. These show that there are large masses of gas around the minor axis moving with radial velocities between $+30$ and $+130$ km s^{-1} relative to the centre (cf. Figure 14). On the opposite side of the minor axis there is a considerable amount of hydrogen with high *negative* velocity relative to the centre. If this gas lies in the equatorial plane of the galaxy it would be moving towards the centre. In the model proposed by van der Kruit *et al.* the gas within about 5 kpc from the centre should indeed be falling back and have velocities in the inward direction similar to those observed.

Closer to the nucleus Chincarini and Walker (1967) had likewise found considerable motions in radial directions. These observations have recently been confirmed and extended by van der Kruit on spectra taken with the 200-in. Hale reflector. This gas close to the nucleus may well lie in the plane of the disk and be moving *in*ward.

The total mass of the neutral hydrogen with radial velocities between 60 and 130 km s^{-1} is roughly 6×10^7 $M_\odot$, while that with $+50$ km s^{-1} velocity is 9×10^7 $M_\odot$. Though these data do not suffice to determine the epoch of the hypothetical expulsion it might well have been the same event as that which produced the anomalous synchrotron arms. The amounts of high-velocity hydrogen observed on the minor axis are of the same order as those which had been proposed to explain the anomalous features in the plane of the spiral. The 21-cm observations appear to confirm that an enormous expulsion of gas has occurred in this galaxy. On the other hand we find no very clear signs of a heaping-up of neutral gas near the radio ridges. If our model is correct we must therefore conclude that the interstellar medium that was swept up is still largely ionized.

Except for our own Galaxy this is the first case in which the mass involved in the explosion from a galactic nucleus has been directly measured. The total mass must be higher than the 1.5×10^8 $M_\odot$ given above, because not all of the exploded hydrogen will have a large velocity component in the radial direction, and because the gas with highest velocity may have escaped.

The high density of H I in the inner region of NGC 4258 (cf. Figure 4), which contrasts with the central 'holes' observed in other spiral galaxies, may well be connected

---

←

Fig. 14b.   Same for velocities between $+10$ and $+70$ km s^{-1} (D. van Albada, Oort and Shane, preliminary results).

with the explosion. The *total* amount of H I within 5 kpc is about twice that observed to have high velocities. It may well be, therefore, that practically all the gas found within this distance has been ejected from the nucleus. The kinetic energy involved in the expulsion may be roughly estimated at $\sim 10^{57}$ erg.

The Seyfert galaxies provide interesting information on what may be the initial stages of eruptive periods. So far, however, no neutral hydrogen observations for Seyfert spirals are available which could give an indication of the total mass of gas involved. The only cases of eruption which have been observed in the nearest galaxies, and which are *somewhat* similar, are the expanding arms in the central region of our own Galaxy and the well-known filamentary features around M82. In the former, 21-cm observations have shown the presence outside the galactic disk of clouds with a total mass of about $10^6 \, M_\odot$, moving away from the centre at velocities of the order of 100 km s^{-1}, having probably been expelled about 6 million years ago. If the motions of the 3 kpc arm and the expanding arm at $+135$ km s^{-1} are supposed to have been caused by similar expulsions, there should have been another 'eruption' about 12 million years ago in which a mass of the order of $10^7 \, M_\odot$ was involved.

Acknowledgements

I am indebted to many colleagues for advance information on unpublished data and for valuable discussions on the subject of this report, in particular to D. van Albada, R. J. Allen, E. Bajaja, Elly Dekker, P. C. van der Kruit, A. H. Rots, R. Sancisi and W. W. Shane.

References

Bottinelli, L.: 1971, *Astron. Astrophys.* **10**, 437.
Chincarini, G. and Walker, M. F.: 1967, *Astrophys. J.* **149**, 487.
Courtès, G. and Cruvellier, P.: 1961, *Compt. Rend. Acad. Sci. Paris* **253**, 218.
Deharveng, J. M. and Pellet, A.: 1970, *Astron. Astrophys.* **9**, 181.
Mathewson, D. S., van der Kruit, P. C., and Brouw, W. N.: 1972, *Astron. Astrophys.* **17**, 468.
Pooley, G. G.: 1969, *Monthly Notices Roy. Astron. Soc.* **144**, 101.
Roberts, M. S.: 1971, in D. S. Evans (ed.), 'External Galaxies and Quasi-Stellar Objects', *IAU Symp* **44**, 12.
Roberts, M. S. and Rots, A. H.: 1973, *Astron. Astrophys.* **26**, 483.
Rogstad, D.: 1971, *Astron. Astrophys.* **13**, 108.
van de Hulst, H. C., Raimond, E., and van Woerden, H.: 1957, *Bull. Astron. Inst. Neth.* **14**, 1.
van der Kruit, P. C.: 1972, *Astrophys. Letters* **11**, 173.
van der Kruit, P. C.: 1973, *Astron. Astrophys.* **29**, 263.
van der Kruit, P. C., Oort, J. H., and Mathewson, D. S.: 1972, *Astron. Astrophys.* **21**, 169.
Wright, M. C. H., Warner, P. J., and Baldwin, J. E.: 1972, *Monthly Notices Roy. Astron. Soc.* **155**, 357.

DISCUSSION

van Woerden: In M81 you find that a fraction $\alpha = 0.022$ of the gas condenses into stars, per passage through the spiral shock. This must be a *net* fraction; the fraction condensed originally must be higher, but part of the matter condensed returns into the gas by mass loss from stars and by supernova explosions. Can you estimate the fraction originally condensed?

Oort: The fraction 0.022 represents indeed the *net* amount of gas going into stars and remaining in them. I cannot say off-hand how large a fraction goes into stars temporarily and is returned to the interstellar medium.

van Woerden: In M101 the arm/interarm contrast can be estimated more easily than in M81, since almost any radius vector crosses three spiral arms (or segments). Allen, Goss and I estimate preliminarily that the surface density in the arms is about three times the *average* surface density. Since the arms may cover about one-third or one-quarter of the area of the galaxy, the *interarm* density may be quite low – but we have no direct measures of that yet.

G. Burbidge: How did you measure this rotation of the anomalous radio arms in NGC 4258?

Oort: We did not. What I referred to was the rotation measured in the anomalous Hα arms.

Baldwin: Is the major axis of NGC 4258 well defined dynamically from the H I measurements?

Oort: Yes, and it agrees fairly, though not completely, with the axis of the optical picture.

van der Kruit: At Palomar I have measured optically the velocity field in NGC 4258. These measures show three important features:

(a) The velocities in the normal arms indicate the non-circular motions indicated earlier by the Burbidges and Prendergast (*Astrophys. J.* **138**, 375, 1963). However, if we assume that the line of nodes and the inclination are those of the weak, outer structure (p.a. about 145° instead of 157° and $i \simeq 72°$ instead of 64°) we find that these non-circular motions disappear.

(b) The Hα arms are rotating then more slowly than the rotation curve from the normal arms would indicate. In the model this is explained by the fact that the swept-up gas must have a lower rotation velocity than the circular rotation, because the expelled matter from the nucleus has little or no angular momentum.

(c) Along the minor axis and the region of the 'radio plateau' I find motions which, if in the plane, would indicate that these are contractions. If the plateaus are remnants of earlier phases in the explosion this is expected to be the case because, due to the longer time-scale and possibly lower expulsion velocities, this gas will have been stopped and, due to the lower angular momentum, is now falling back to the nucleus.

Finally, the line-strengths in the Hα arms favour collisional ionization over photo-ionization. This question is being studied at present by Miller and Osterbrock.

Oort: The total kinetic energy involved in the explosion is $\sim 10^{57}$ erg.

Arp: How do you envisage the expulsion of the gas? In particular, how narrowly is it directed, does it spread out as it goes, and what initially directs the expulsion?

Oort: The observations indicate that the expulsion has taken place in a rather wide cone. For the radio ridges one requires gas to be ejected at relatively small angles with the equatorial plane.

I have no suggestion as to what initially directs the expulsion, but would refer you to the report on this subject by Saslaw (p. 305).

Pishmish: You showed us that in some galaxies, for example in M81, there is a lack of neutral hydrogen in the nucleus. Would this also imply that hydrogen gas would not exist in any other form, say in its ionized state? I have in mind in particular the giant H II regions, presumably in an early stage of development, when they are surrounded by a 'cocoon' dust cloud making them optically unobservable. Such regions could, however, be detected by radio recombination line studies.

Oort: In M81 there are relatively few large emission nebulae. Nor do radio observations of the continuum give evidence of the presence of a great number of large H II regions. On the basis of this evidence it seems unlikely that the decrease in neutral hydrogen towards the centre could be compensated by ionized hydrogen.

Allen: Since radio recombination lines always seem to be accompanied by radio continuum emission, the presence of optically-obscured giant H II complexes in the central regions of galaxies would not go undetected in the high-sensitivity radio continuum maps now being produced by aperture synthesis radio telescopes. In the case of M101, for example, Westerbork radio continuum maps at 21-cm wavelength currently being prepared for publication by Goss, Israel and myself do not show a relative excess of such giant H II regions in the central areas.

INTERPRETATION OF OBSERVATIONS OF SPIRAL STRUCTURE
IN TERMS OF THE DENSITY WAVE THEORY*

PER OLOF LINDBLAD

Stockholm Observatory, Sweden

Abstract. A review is given of recent theoretical and observational work on the density wave theory of spiral structure. Emphasis is put on the kinematic picture, and the question whether modern observations reveal the existence of density waves is discussed.

The most obvious phenomenon that might lead one's thoughts to a picture of density waves is a straight bar present in a differentially rotating galaxy. In the 1940's B. Lindblad (ref. Contopoulos, 1970a) developed a theory of self-sustained bar-like density waves in a galaxy with moderate differential rotation. This theory was based on the moments of the Boltzmann equation, i.e. the equation of continuity for the distribution function in six-dimensional phase space valid for a stellar system. Lin and Shu (Lin, 1966; Lin *et al.*, 1969; Shu, 1970a, b), on the basis of the same equation of continuity, proved that a self-sustained *spiral* density wave was a possible solution. Their linear solution was restricted to tightly-wound spirals and small density amplitudes. This theory, as followed up by Lin, Shu, and many others, has attracted much interest because it gives quantitative relations between various structural and kinematic parameters and predicts phenomena which can be compared with observation.

The density wave theory has sometimes been criticized for being 'incomplete'. Perhaps part of this criticism is removed if we call it a density wave *model* of spiral galaxies, a coherent model which is in the process of being developed into a theory.

1. The Theory of Density Waves

In view of the development over the past two years we will approach the density wave theory from the picture of orbits in a galaxy. The foundations for this approach were laid by B. Lindblad (1958, 1964, and references given there) in the 1950's.

An orbit in the galactic plane is governed by two frequencies, both of which are functions of the distance R from the centre of the galaxy:

(1) The angular frequency of circulation around the galactic centre, ω.

(2) The frequency of radial oscillation, the epicyclic frequency, κ.

In general, these frequencies are not commensurable and the orbit is a non-closed rosette. Of fundamental importance for the maintenance of large-scale density waves is the empirical fact that $\omega - \kappa/2$ is nearly constant over the main body of our

---

* The density wave theory of galactic spiral structure has been reviewed several times in recent years, e.g. by Contopoulos (1970a, 1972, 1973b), and Lin (1971), and most recently in two lectures by Shu (1973). An elementary introduction to the subject has been given by Wielen (1971).

John R. Shakeshaft (ed.), The Formation and Dynamics of Galaxies. 399–412. All Rights Reserved.
Copyright © 1974 by the IAU.

Galaxy (Figure 1). This seems to be a general characteristic of galaxies in spite of their rather different rotation curves. The ω vs. κ diagram for M101, for instance, looks very much the same as that in Figure 1.

For any particle or star with orbital frequencies ω_i, κ_i, the orbit may always be considered as a closed ellipse with its centre at the centre of the galaxy, the ellipse rotating with the angular velocity

$$\Omega_i = \omega_i - \frac{\kappa_i}{2}.$$

The particles travel in the forward direction around the orbit with respect to its rotating apsidal line. If $\omega - \kappa/2$ is constant over some part of the system, this means

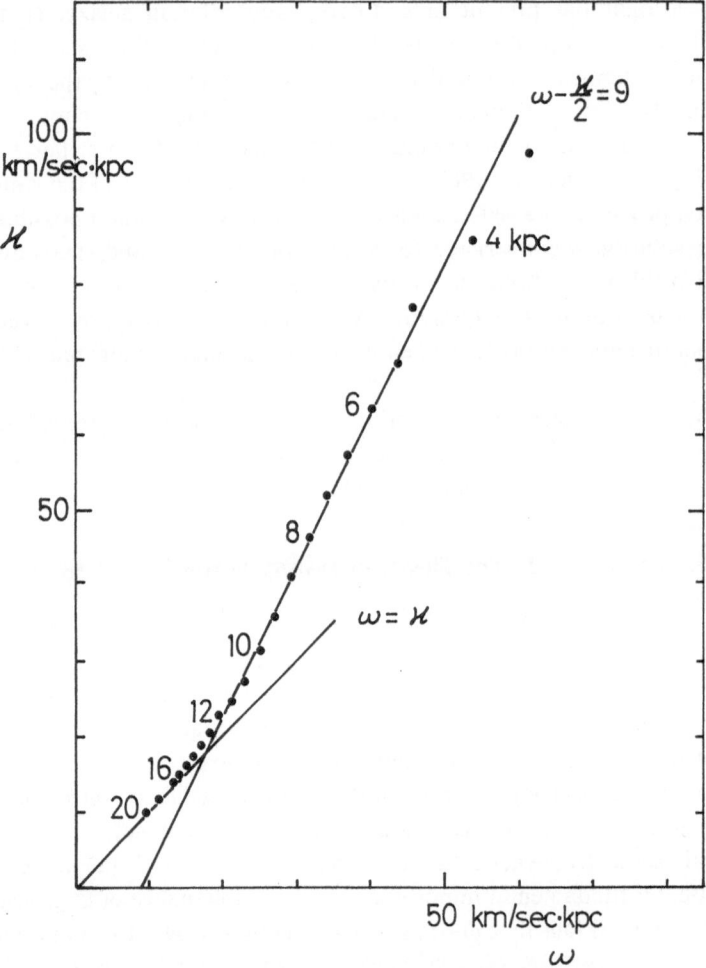

Fig. 1.   Relation between the frequencies ω and κ for the Galaxy. Distances from the centre in kpc are indicated.

that all such orbital ellipses turn with the same speed. This also means that a cloud would disperse along such a closed elliptic orbit. The resulting closed massive ring is generally called a dispersion ring. Another consequence would be that any one-sided density asymmetry will be drawn out into a bi-symmetrical one. This would be the basic reason for the bi-symmetrical character of two-armed spiral and barred type galaxies.

The nearly constant value of $\omega - \kappa/2$ means that, if we arrange a set of orbits into a spiral pattern for instance (Figure 2, due to Kalnajs, 1973), this arrangement will be preserved for some time and will turn with the pattern velocity $\omega - \kappa/2$. The stars in their circulation will move almost tangentially to the arm within the arm

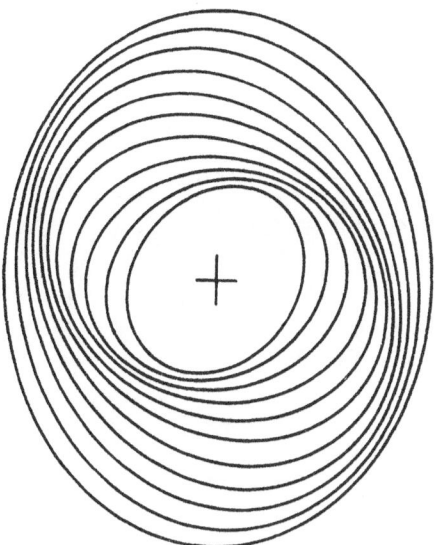

Fig. 2. A set of closed elliptical orbits arranged into a spiral pattern (after Kalnajs, 1973).

pattern and there have their maximum inward motion if the spiral arms are trailing. If mass is added to the orbits, the pattern velocity will be slightly higher than $\omega - \kappa/2$ as given for an unperturbed orbit. What Lin showed is that, if you give mass to these orbits, you can for some time have a self-supporting spiral density wave with a constant pattern velocity.

If the pattern velocity, Ω_p, is larger than some mean value $\langle \omega - \kappa/2 \rangle$, we will in general have three basic resonance regions for different values of R:

(1) The co-rotation region (particle resonance), where $\omega = \Omega_p$.

(2) The inner resonance region, where $\omega - \kappa/2 \doteq \Omega_p$. The unperturbed closed elliptic orbit here turns with the speed of the wave.

(3) The outer resonance region, where $\omega + \kappa/2 = \Omega_p$. An unperturbed closed elliptic orbit with the particles moving in a retrograde direction around the orbit with respect to its apsidal line here turns with the speed of the wave.

We may ask why a galaxy should arrange itself as shown in Figure 2. One reason, suggested by Lynden-Bell and Kalnajs (1972), is that a flat differentially-rotating galaxy with given angular momentum can lower its rotational energy (or increase its entropy) by transporting angular momentum outwards; only in trailing spiral structures do the gravity torques carry angular momentum outwards.

An important contribution to the density wave picture are the computations by Roberts (1969), and more recently by Shu *et al.* (1973), of the gas flow in this spiral gravitational field and the formation of shock waves. In Kalnajs' presentation the shock phenomena arise in the interstellar gas when the dispersion orbits have been turned to such a degree that they intersect (Figure 3). Galactic shocks arise for waves of sufficiently large amplitude, i.e. for sufficiently large eccentricity of the rings.

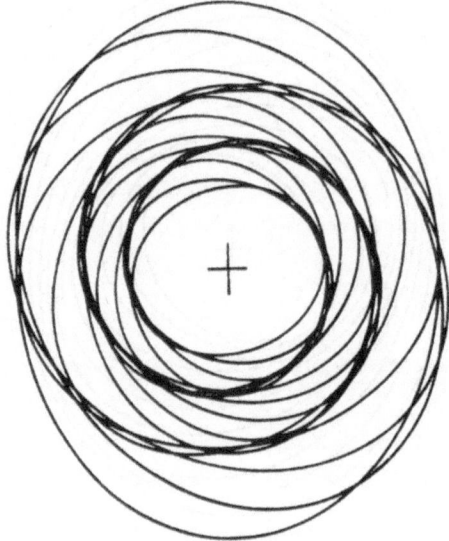

Fig. 3. The amplitude of the elliptical motion increased (when compared to the wavelength of spiral structure) until a galactic shock occurs (after Kalnajs, 1973).

Specifically, Shu *et al.* predict that broad arms with open spiral structure should prevail in galaxies with weak differential rotation, whereas narrow arms and tightly wound spiral structure should prevail in galaxies with strong differential rotation. They also claim that a number of special resonances, in addition to the principal ones mentioned above, can give rise to secondary compressions of the gas.

The galactic shocks would be the location of the most intense star formation. Of great interest in this respect is the work by Shu *et al.* (1972) on galactic shocks in an interstellar medium with two stable phases in pressure equilibrium. The non-linear wave pattern just described is followed by the hot intercloud phase which then compresses the cold clouds to high densities in the shock region.

To be sure, density waves will propagate from co-rotation to the inner resonance region (Toomre, 1969; Shu, 1970b), and galactic shocks will be damped (Kalnajs,

1972; Roberts and Shu, 1972) in a time span which in both cases may be of the order of 10^9 yr. The three principal resonance regions have been considered as sources and sinks for angular momentum and wave energy. Partly for this reason interest has been taken in the last few years in resonance effects, and the resonance regions have been interpreted as key regions for the creation and maintenance of density waves.

Perturbed orbits around the resonance regions have been studied by Contopoulos (1970b) and Vandervoort (1973). As the inner resonance region is approached, they find increasing eccentricity of stable periodic orbits and surrounding tube orbits, which could give the inner region the appearance of a bar-like structure. Near the resonance a second family of elongated stable periodic orbits, roughly perpendicular to the first, appears. If one proceeds inside the resonance, the first set of tube orbits shrinks and eventually disappears, while the periodic orbits of the second family become nearly circular.

Contopoulos (1973a) has further studied resonance effects at co-rotation. He finds four equilibrium points, two of which are stable and situated at spiral density minima, and he argues that large masses could be trapped around these points. Feldman and Lin (1973, see also Feldman 1973) claim that a rigidly-rotating, weakly bar-like structure in the central regions of a galaxy forces a tightly-wound trailing spiral wave at co-rotation. This wave is then supposed to propagate towards the inner resonance region.

Thus, we are given a fairly detailed density-wave model for relatively tightly wound spiral galaxies. The perturbing action is caused by a rigidly-rotating smooth spiral stellar wave with a density contrast of a few per cent and with systematic deviations from circular velocities of a few tens of kilometres per second; it is thus very difficult to observe. On top of these waves rides, like foam on the waves of the ocean, the interstellar gas compressed by shocks and with this the radio and optical spiral tracers. The density of wave energy is comparable to the densities contained locally in turbulent gas motions, starlight, cosmic-ray particles and interstellar magnetic fields (Shu, 1973), thus we can expect that even if the grand design is regular the local spiral structure may be somewhat disordered.

2. Observation of Density Waves

2.1. THE GALAXY

The picture presented above has, of course, evolved through confrontation with observations. For observers it is an appealing working hypothesis and we would like to confront it with the present observing material. Do we see density waves?

The density wave theory has recently been criticized on observational grounds by Piddington (1973a, b). However, one of the main points made by Piddington – that the interstellar gas does not form strings along the spiral arms – now seems to be definitely refuted by the beautiful observations just shown by Professor Oort (p. 375).

Thus, we have good reason to believe that the narrow ridges we find in the velocity distribution of neutral hydrogen as a function of longitude in our Galaxy correspond

to a spiral structure of hydrogen arms. A difficulty when comparing the density wave
theory with observations is that the linear theory does not give a unique value for
the pattern velocity unless we already know the shape of the pattern. To try to fit a
density wave to a spiral pattern found by the assumption of purely circular motions
has its dangers. This is clearly brought out by the interesting attempt to apply density
wave theory, including large scale shock formation, to our Galaxy, made by Roberts
(1972) for a region of the Perseus arm. In Figure 4 we see how a line profile, which
interpreted by an assumption of circular motion would give a complicated density
pattern, can be represented by a single spiral wave, where peaks in the velocity
distribution arise partly from a shock and partly from velocity crowding over intervals
where the radial velocity changes little with distance. Thus, the results based on the
assumption of circular motion may not give even a first approximation to the true
structure and there is the consequential necessity of a complete re-evaluation of
galactic hydrogen observations in terms of the spiral shock theory – a tremendous
but necessary job to be done. Verschuur (1973) has criticized this approach, but I

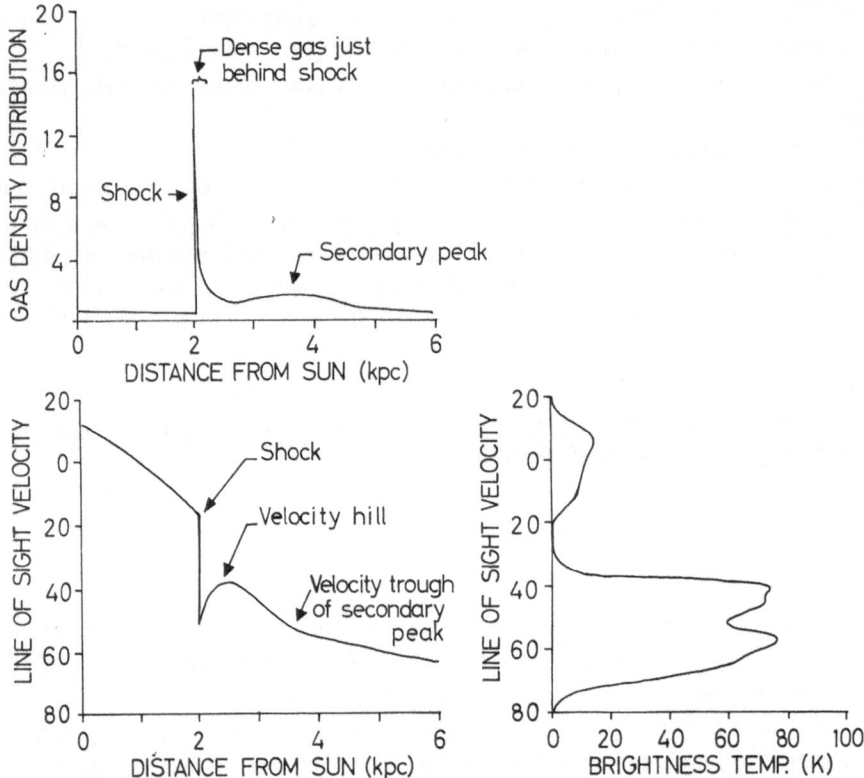

Fig. 4.   Theoretical features of a two-armed spiral shock model in the longitude range 130°–140°.
Upper left: Gas density distribution (unit = average density) as a function of distance from the Sun.
The basic motion of the gas is from left to right. *Lower left*: Line-of-sight velocities versus distance
from the Sun. *Lower right*: Resulting theoretical line profile (after Roberts, 1972).

don't think that his argument solves the problem. Anyhow, confrontation with the observations in the Galaxy, as they are now available in various surveys, emphasizes the need to work out the density-shock wave theory in three dimensions. A first attempt to do this has been made by Tosa (1973).

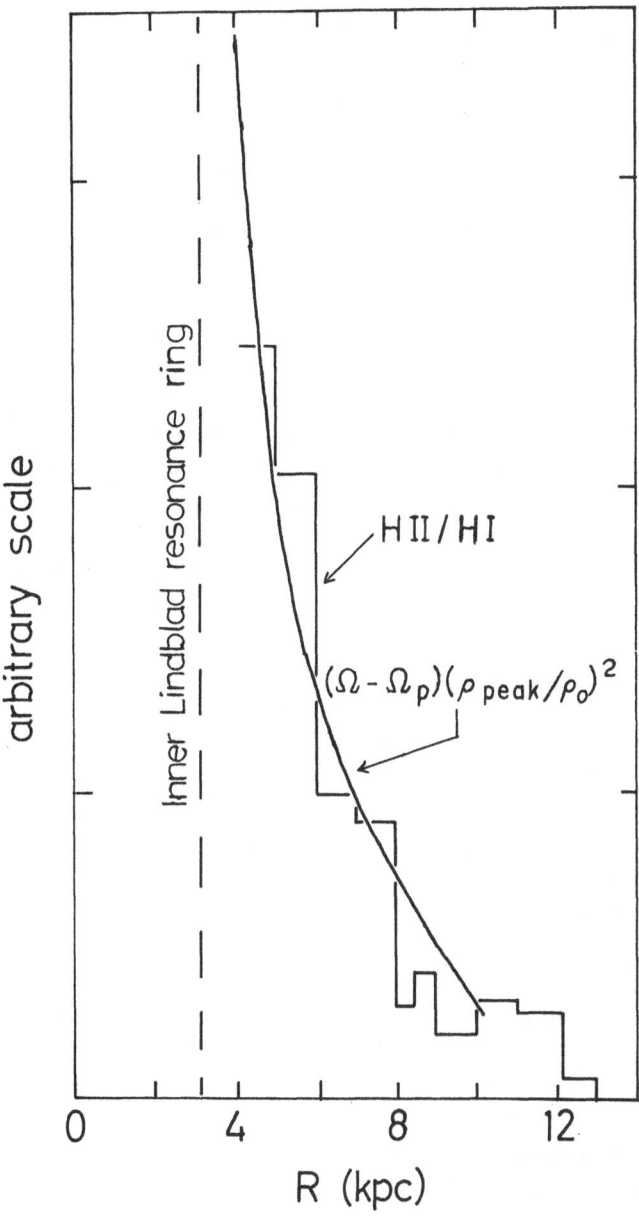

Fig. 5. The abundance of H II relative to that of H I in the Galaxy as compared to a theoretical relation for the frequency of compression times the square of the peak compression strength in a galactic shock (after Shu, 1973). (For $(\Omega - \Omega_p)$ read $(\omega - \Omega_p)$, Editor.)

An important observational phenomenon in our Galaxy, which also seems to have its counterpart in other galaxies, is the discrepancy between the distribution of H II and H I which indicates a difference between the rate of star-formation and average hydrogen density. This has just been discussed by Professor Oort (p. 375). In view of the density wave theory, the rapid increase inward of the ratio H II/H I can be attributed to two causes (Shu, 1973): the increase inward of the frequency $(\omega - \Omega_p)$ at which the interstellar gas is periodically compressed, and the increase inward of the peak compression ϱ_{peak}/ϱ_0. Shu (Figure 5) demonstrates the close resemblance between the run of H II/H I and the quantity $(\omega - \Omega_p)(\varrho_{peak}/\varrho_0)^2$.

The rotation curve of the Galaxy is such that there should exist an inner resonance region for a two-armed spiral wave at about 3–4 kpc from the centre. Attempts to explain the '3-kpc arm' and other apparently expanding features in the vicinity of that region in terms of a dispersion ring have been made by Shane (1972) and by Simonson and Mader (1973).

2.2. EXTERNAL GALAXIES

The external galaxy that has been most thoroughly investigated for comparison with the density wave theory is M51. Well known to everyone are the beautiful results obtained by Mathewson *et al.* (1972), using the Westerbork synthesis telescope to map the 20-cm radio continuum due to synchrotron radiation. A striking feature is the two radio arms. The fact that these radio arms coincide with the dust lanes on the inner side of the optical arms is strong support for the spiral shock-wave theory, the radio emission being due to the compression of the magnetic field along with the gas and the increase in the number density of relativistic electrons.

A very thorough analysis based on optical observations with a Fabry-Pérot interferometer has been carried out by Tully (1974) who finds streaming motions in the region of the inner spiral arms consistent with the density wave theory. From the amplitude of the streaming motions, he obtains the very high value of 15–20% for the ratio of the spiral potential to the axisymmetric potential and, from the inclination of the dust lanes, a rather high pattern velocity which places co-rotation only 2.4 kpc from the centre. An analysis of the non-circular motions inside the inner resonance region shows that they can best be represented by motions of the dispersion-orbit type, which apparently supports Contopoulos' calculations. In accordance with Toomre and Toomre (1972), Tully suggests that the outer spiral structure arises from the tidal effects of NGC 5195, which then would cause the density wave to propagate inwards from co-rotation. Also, Weliachew and Gottesman (1973) find from 21-cm studies that the velocities in the north-eastern part of the galaxy agree with the predictions of the Toomres. In view of the rather elongated orbit of NGC 5195 in the Toomres' model, this would mean that we happen to see M51 in a stage of strong arm formation. Although there are several arguments supporting this view, it still means that we have to apply two different mechanisms to explain spiral arms with shocks delineated by dark lanes in different parts of the galaxy. Van der Kruit (1973), on the other hand, finds that NGC 4321 and M51 show a similar structure in the radio continuum and draws the conclusion that the structure of M51 can only in a minor way be the result of tidal action from NGC 5195.

Thus, it seems very desirable to confirm Tully's analysis with the new 21-cm line measurements which have been carried out with the Westerbork telescope and are being analyzed by Shane and Dekker at Leiden. The total hydrogen map (Figure 1 in the article by Professor Oort, p. 375) shows clearly how the hydrogen follows the

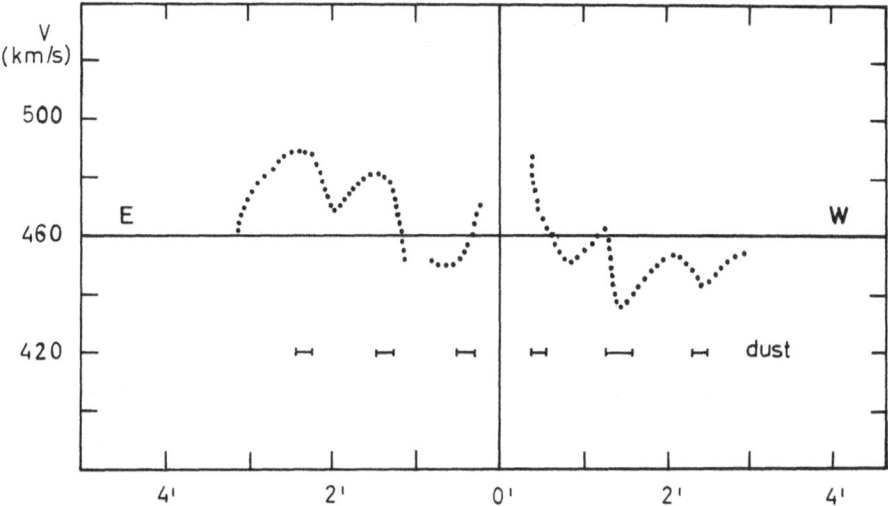

Fig. 6.   Radial motion of neutral hydrogen along the minor axis of M51 as a function of distance from the centre. The position of dust lanes are indicated (courtesy W. W. Shane and E. Dekker).

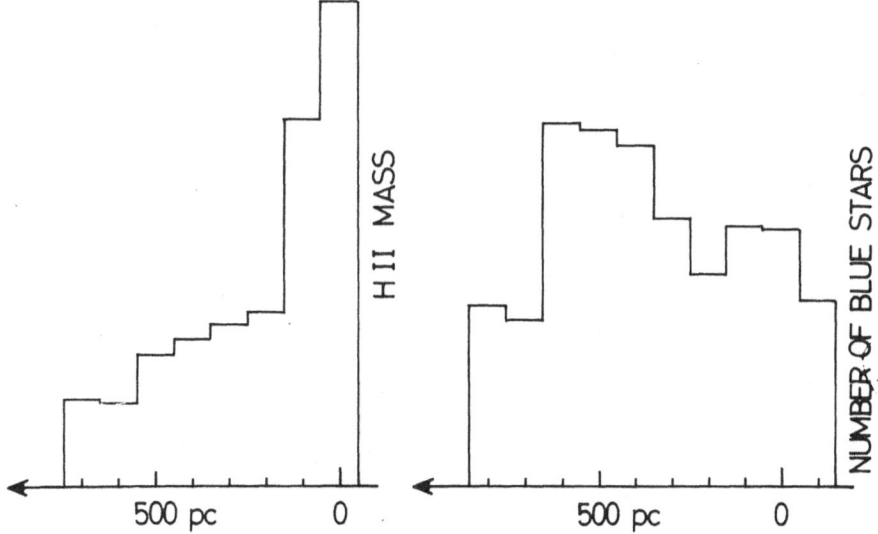

Fig. 7.   Distribution of ionized hydrogen (left) and blue stars (right) as a function of distance from the following edge of a spiral arm in M33 (after Courtès and Dubout-Crillon, 1971).

optical arms. Also, the maximum hydrogen line intensity has a tendency to fall on the dust lanes, where the shock is supposed to be. However, according to Shane (private communication) the hydrogen maximum could very well be shifted with respect to the shock-wave, partly because the clouds in the shock may be optically thick with high self-absorption and partly because of the limited angular resolution which averages over the shock and the following density shoulder. The velocity field derived so far is only preliminary, but it seems that wave-type motions may be present with maximal inward motions coinciding with the dust arms (Figure 6), in agreement with the shock-wave picture.

M81 is an interesting case because of the regular spiral structure and the beautiful hydrogen arms. Here also the hydrogen arms seem to fall right on top of the optical arms (Rots and Shane, 1974; see also p. 378). A very preliminary hydrogen velocity

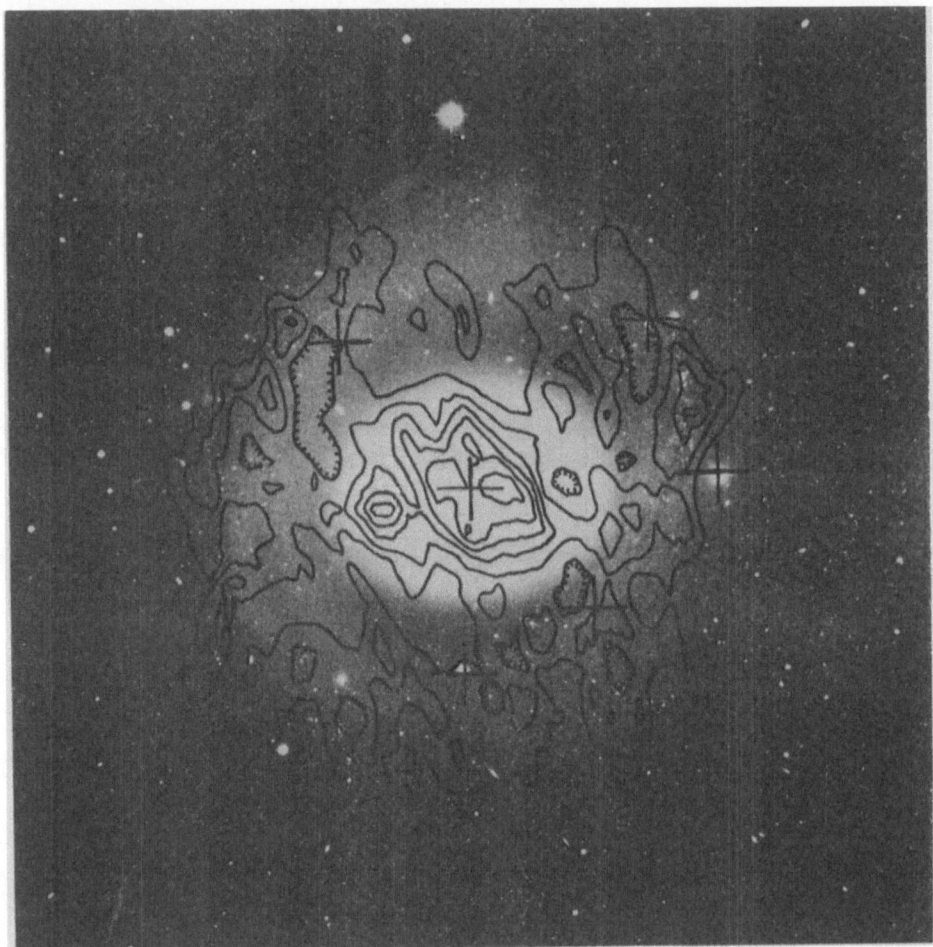

Fig. 8.   Total neutral hydrogen distribution in M94 as observed with the Westerbork telescope
(Bosma *et al.*, 1974)

picture shows indication of waves, although the question of symmetry is not quite clear. A final picture is being produced by Rots at the Kapteyn Laboratory.

M33 has been observed in the 21-cm hydrogen line by Wright *et al.* (1972, Warner *et al.*, 1973) and again the integrated hydrogen is correlated with the optical spiral arms. The authors failed to detect any large-scale density wave with an amplitude greater than 3 km s^{-1}. Courtès and Dubout-Crillon (1971) have studied a part of the southern arm of M33 using interference and coloured filters and have determined the relative distributions of H II regions and bright stars across the spiral arm (Figure 7). If the gravitational force is dominant their result is a strong indication of wave motion. According to Roberts the pattern velocity would then be slower than the circular velocity and the co-rotation point would lie further out than the region studied. Dixon (1971) has investigated the same arm further out by *B, V* photometry. He finds

Fig. 9.   Velocity field of neutral hydrogen in M94 as observed with the Westerbork telescope
(Bosma *et al.*, 1974)

an age progression of stars across the spiral arm opposite to that of Courtès and Dubout-Crillon. In Roberts' picture this would indicate a co-rotation distance somewhere between the radii of the two surveys, i.e. about 3 kpc from the centre, giving a relatively high pattern velocity also in this system.

The case of NGC 4258 (van der Kruit *et al.*, 1972) has been presented by Professor Oort, who suggested that outflow of matter from the nucleus leads to a reformation of the spiral structure. This would not concern the basic density wave in the stellar disk which is, presumably, little perturbed by the expulsion of gas from the nucleus. The outflowing matter could perhaps be collected in the spiral shocks eventually and thus add to the material tracing the spiral structure.

It would be of the outmost importance for our understanding of density waves if we could identify the different resonance regions in the galaxies observed. It is interesting in this connection to look at the galaxy M94 (NGC 4736), where an inner ring structure was discussed in the paper by van der Kruit (this volume, p. 431). This rather tightly wound spiral also has an extensive outer ring which, according to recent observations with the Westerbork telescope (Bosma *et al.*, 1974), is clearly outlined by the neutral hydrogen (Figure 8). The ring seems to be attached at two points to the inner region, but its connection with the inner spiral structure is difficult to establish. The preliminary velocity field given in Figure 9 seems to show a discontinuity across the gap between the outer ring and the main body. It has been suggested (P.O. Lindblad, 1960) that, in the case of M94, this gap corresponds to the co-rotation region. The numerical many-body computations on which this suggestion was based may also to some extent illustrate the theoretical discussions of the co-rotation region by Contopoulos, and by Feldman and Lin, mentioned above. The kinematics of this galaxy seem to be worth a detailed study, and an analysis combined with a theoretical discussion and numerical computations should be rewarding.

In summary, there is strong evidence that we observe density waves in galaxies, although a quantitative confirmation of the density wave theory, or some variant of the theory, is not yet in our hands. This is partly due to the state of the theory, where much work still has to be done before we get a complete picture, and partly to that of the observations, where the really relevant data have only just begun to come in.

With the present vigorous activity and rapid development on both the theoretical and observational side there is promise that we will learn many new interesting facts in the near future.

References

Bosma, A., van de Hulst, J. M., and Sullivan, W. T.: 1974 (in preparation).
Contopoulos, G.: 1970a, in W. Becker and G. Contopoulos (eds.), 'The Spiral Structure of Our Galaxy', *IAU Symp.* **38**, 303.
Contopoulos, G.: 1970b, *Astrophys. J.* **160**, 113.
Contopoulos, G.: 1972, *The Dynamics of Spiral Structure*, Lecture Notes, Astronomy Program, Univ. of Maryland.
Contopoulos, G.: 1973a, *Astrophys. J.* **181**, 657.

Contopoulos, G.: 1973b, *The Density Wave Theory of Spiral Structure*, in Proceedings of the Advanced Study Institute on the Dynamical Structure and Evolution of Stellar Systems, Saas-Fee (Publ. Geneva Obs.)

Courtès, G. and Dubout-Crillon, R.: 1971, *Astron. Astrophys.* **11**, 468.

Dixon, M. E.: 1971, *Astrophys. J.* **164**, 411.

Feldman, S. I.: 1973, Ph.D. Thesis, MIT.

Feldman, S. I. and Lin, C. C.: 1973, *Studies in Appl. Math.* **52**, 1.

Kalnajs, A. J.: 1972, *Astrophys. Letters* **11**, 41.

Kalnajs, A. J.: 1973, *Proc. Astron. Soc. Australia* **2**, 173.

Lin, C. C.: 1966, *J. SIAM Appl. Math.* **14**, 876.

Lin, C. C.: 1971, in C. de Jager (ed.), *Highlights of Astronomy* **2**, 88.

Lin, C. C., Yuan, C., and Shu , F. H.: 1969, *Astrophys. J.* **155**, 721.

Lindblad, B.: 1958, *Stockholm Obs. Ann.* **20**, No. 6.

Lindblad, B.: 1964, *Astrophys. Norv.* **9**, 103 (= *Stockholm Obs. Medd.* No. 145).

Lindblad, P. O.: 1960, *Stockholm Obs. Ann.* **21**, No. 4.

Lynden-Bell, D. and Kalnajs, A. J.: 1972, *Monthly Notices Roy. Astron. Soc.* **157**, 1.

Mathewson, D. S., van der Kruit, P. C., and Brouw, W. N.: 1972, *Astron. Astrophys.* **17**, 468.

Piddington, J. H.: 1973a, *Astrophys. J.* **179**, 755.

Piddington, J. H.: 1973b, *Monthly Notices Roy. Astron. Soc.* **162**, 73.

Roberts, W. W.: 1969, *Astrophys. J.* **158**, 123.

Roberts, W. W.: 1972, *Astrophys. J.* **173**, 259.

Roberts, W. W. and Shu, F. H.: 1972, *Astrophys. Letters* **12**, 49.

Rots, A. H. and Shane, W. W.: 1974, *Astron. Astrophys.* **31**, 245.

Shane, W. W.: 1972, *Astron. Astrophys.* **16**, 118.

Shu, F. H.: 1970a, *Astrophys. J.* **160**, 85.

Shu, F. H.: 1970b, *Astrophys. J.* **160**, 99.

Shu, F. H.: 1973, Lectures delivered at the Advanced Study Institute on the Interstellar Medium, Schliersee, preprint.

Shu, F. H., Milione, V., Gebel, W., Yuan, C., Goldsmith, D. W., and Roberts, W. W.: 1972, *Astrophys. J.* **173**, 557.

Shu, F. H., Milione, V., and Roberts, W. W.: 1973, *Astrophys. J.* **183**, 819.

Simonson, S. C. and Mader, G. L.: 1973, *Astron. Astrophys.* **27**, 337.

Toomre, A.: 1969, *Astrophys. J.* **158**, 899.

Toomre, A. and Toomre, J.: 1972, *Astrophys. J.* **178**, 623.

Tosa, M.: 1973, *Publ. Astron. Soc. Japan*, **25**, 191

Tully, R. B.: 1974, *Astrophys. J. Suppl.* **27**, 415.

van der Kruit, P. C.: 1973, *Astron. Astrophys.* **29**, 249.

van der Kruit, P. C., Oort, J. H., and Mathewson, D. S.: 1972, *Astron. Astrophys.* **21**, 169.

Vandervoort, P. O.: 1973, *Astrophys. J.* **180**, 739.

Verschuur, G. L.: 1973, *Astron. Astrophys.* **24**, 193.

Warner, P. J., Wright, M. C. H., and Baldwin, J. E.: 1973, *Monthly Notices Roy. Astron. Soc.* **163**, 163.

Weliachew, L. and Gottesman, S. T.: 1973, *Astron. Astrophys.* **24**, 59.

Wielen, R.: 1971, *Mitt. Astron. Ges.* **30**, 31.

Wright, M. C. H., Warner, P. J., and Baldwin, J. E.: 1972, *Monthly Notices Roy. Astron. Soc.* **155**, 337.

DISCUSSION

G. de Vaucouleurs: The nature of the outer rings varies a little with the Hubble classification of the inner part; in some galaxies, such as NGC 1068 and NGC 1291, two faint spiral arms emerge from the outer ring and the ring itself is made up of many short spiral arcs.

Lindblad: Such observations should throw light upon how arms are excited at co-rotation.

G. de Vaucouleurs: It is an interesting empirical fact that many of the galaxies which have a central ring-structure, such as M94 (NGC 4736), NGC 4221, 6753 and 7217, also have a large outer ring.

Allen: Some months ago, Piddington (*Astrophys. J.* **179**, 755, 1973) asserted that there was no evidence in the then available data to support the widely-held belief that galaxian spiral arms consist, among other things, of elongated, connected concentrations of neutral hydrogen. Piddington used the radio synthesis data of Rogstad and Shostak (*Astron. Astrophys.* **13**, 99, 1971) to support his claim that the maxima in the H I distribution of M101 did not fall on the obvious optical spiral arms. The higher angular resolution of the Westerbork observations which we have presented (p. 425) (a factor of 9 increase in linear resolution over the observations by Rogstad and Shostak) reveals a very good general correspondence of small-scale features in the H I with the optical spiral structure except in the central regions of the galaxy, where these H I features are almost absent.

I also draw attention to Shu's study (*Astron. Astrophys.*, in press) of the enhancement of synchrotron emission by compression of interstellar gas and magnetic fields. Because of the Parker instability, which causes compressed field to 'pop' out of the galactic plane, Shu finds that the synchrotron emissivity goes as the first power of the density compression, rather than the third power as many have assumed.

Lindblad: Similar results have been derived by Tosa.

Oort: The compression would depend upon how fast the gas cools, and this would depend upon the initial density.

Allen: Yes indeed.

SOME RECENT DEVELOPMENTS IN THE THEORY
OF SPIRAL STRUCTURE

G. CONTOPOULOS

University of Thessaloniki, Greece

1. Particle Resonance

The particle resonance in a galaxy is of special interest because, according to the recent theory of Lynden-Bell and Kalnajs (1972), it provides the excitation of the density waves. In particular it is of interest to know if the density wave reaches the particle resonance, because the dispersion relation derived by Lin and Shu (1966) has a forbidden region if the parameter

$$Q = (\langle \dot{r}^2 \rangle / \langle \dot{r}^2 \rangle_{min})^{1/2},$$

where $\langle \dot{r}^2 \rangle_{min}^{1/2} = (0.2857)^{1/2} \, 2\pi G \sigma_0 / \kappa$ (with σ_0 the surface density and κ the epicyclic frequency), is larger than one. If $Q = 1$, Lin and Shu find two waves going through the resonance but the amplitude tends to infinity there.

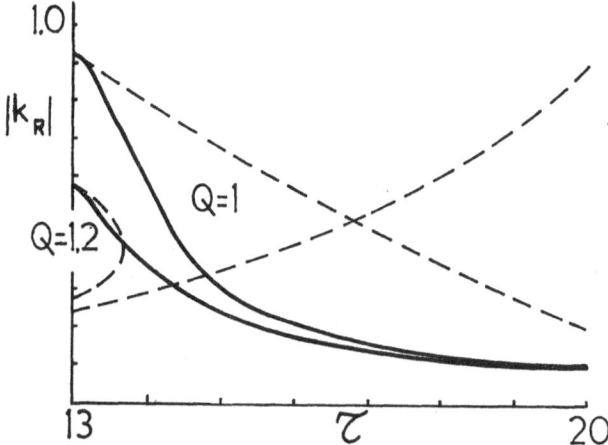

Fig. 1.   The real part of the wave number in Schmidt's model of our Galaxy for $Q = 1$ and $Q = 1.2$ (--- Lin and Shu, —— our preliminary results). The particle resonance is assumed at $r = 16.7$ kpc.

We have derived a dispersion relation applicable near the particle resonance taking into account:
   (1) the variation of the amplitude, considering the wave number k as complex,
   (2) the resonant stars, which give a pole in the response integral, and
   (3) the fact that k is relatively small in this region.
   The preliminary dispersion relations for $Q = 1$ and $Q = 1.2$ are shown in Figure 1.
   We find only one wave going through the resonance, with group velocity directed always inwards. Near the resonance the wave number is small, and the corresponding

John R. Shakeshaft (ed.), The Formation and Dynamics of Galaxies. 413–415. *All Rights Reserved.*

waves are quite open. Finally we find only a small difference between trailing and leading waves.

2. The Density Wave at the Centre of a Galaxy

If there is no inner Lindblad resonance in a galaxy the density wave reaches the centre. A special treatment of this region is necessary because most orbits near the centre cannot be considered as epicycles. Mrs E. Athanassoula-Georgala (Athens) has studied this problem considering both elongated and epicyclic orbits. A dispersion relation valid near the centre is derived without the use of the asymptotic approximation.

3. Trapping of Orbits around Barred Spirals

M. Mihalodimitrakis (Thessaloniki) has extended the work of de Vaucouleurs and Freeman (1972) in studying the orbits around barred spirals. He considered two models of barred spirals: (a) a homogeneous ellipsoid and (b) a homogeneous parallelepiped, while de Vaucouleurs and Freeman considered an inhomogeneous ellipsoid. Many cases were studied with various values of the axis-ratios and of the rotational velocity.

Three families of periodic orbits surrounding the bar were found, two of them direct and one retrograde. One of the two direct families is new (Figure 2). A large set of non-periodic orbits is trapped in rings around the bar.

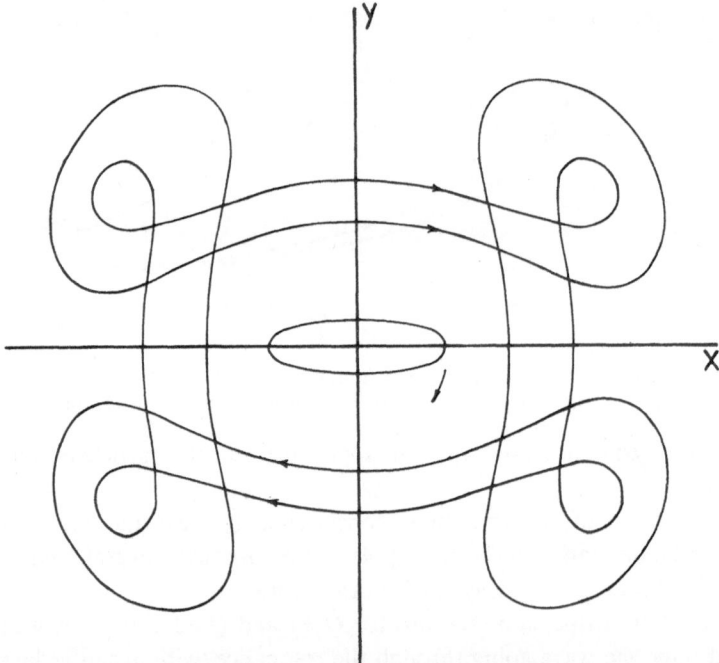

Fig. 2.   A new family of periodic orbits around a barred spiral.

Several periodic orbits were also found around the Lagrangian points L_4, L_5, perpendicular to the axis of the bar, many of them stable.

The Lagrangian points L_1, L_2 on the axis of the bar are always unstable in the case of the ellipsoid but they are stable in some cases of the parallelepiped model.

A theory to describe the trapping of orbits around the (stable) Lagrangian points is developed, similar to the theory of the Lagrangian points L_4, L_5 of a regular spiral galaxy (Contopoulos, 1973).

4. Quasi-Linear Theory

A quasi-linear theory of density waves has been developed by Mrs E. Dekker (Leiden), extending the work of Lynden-Bell and Kalnajs (1972).

The growth coefficient γ is derived from the equation

$$2\gamma = \dot{E}/E,$$

where $\dot{E}$ is the rate of energy change of the wave due to resonant stars, while E is the energy stored in non-resonant stars. One immediate application is that, if there is no inner Lindblad resonance, γ is positive, i.e. the wave grows.

The growth of the wave affects the zero-order distribution function f_0. As the amplitude of the wave increases, f_0 changes until a stationary finite wave is reached. In this research the usual methods of the quasi-linear theory of plasma waves are applied.

This research was supported in part by the Greek National Research Foundation.

References

Contopoulos, G.: 1973, *Astrophys. J.* **181**, 657.
de Vaucouleurs, G. and Freeman, K. C.: 1972, *Vistas in Astronomy* **14**, 163.
Lin, C. C. and Shu, F. H.: 1966, *Proc. Nat. Acad. Sci.* **55**, 229.
Lynden-Bell, D. and Kalnajs, A. J.: 1972, *Monthly Notices Roy. Astron. Soc.* **157**, 1.

DISCUSSION

G. de Vaucouleurs: Does the study of a bar relate to a bar alone or a bar in a disk?

Contopoulos: A bar alone.

Heidmann: With respect to the Lin and Shu theory we have 21-cm line observations of lenticular galaxies which might be of interest. We found that lenticulars have neutral hydrogen components comparable in mass and extent to those of Sa galaxies (*Astron. Astrophys.* **21**, 303, 1973). This then raises the question as to why lenticulars have no spiral arms but Sa's have.

Lindblad: Is anything known about the rotation curves of the lenticular galaxies you mentioned?

Heidmann: Normal rotation curves were obtained, for instance for NGC 3115, by optical astronomers. For our best case, NGC 5102, the width of the 21-cm line is compatible with a normal rotation.

Wilson: What growth times does Mrs Dekker get for resonances with the epicyclic orbits?

Contopoulos: She has no numerical results yet. However, Kalnajs (*IAU Symp.* **38**, 318, 1970) found numerically a mode for a model of M31 with a growth time of a few $\times 10^8$ yr.

Wilson: I have done a similar linear calculation giving instabilities in the galactic gas of 1 kpc wavelength and with a growth time of about 4×10^8 yr.

STUDIES ON THE GALACTIC DENSITY WAVE

JAMES W.-K. MARK

Massachusetts Institute of Technology, Cambridge, Mass., U.S.A.

Abstract. For spiral galaxies, investigations have been made as to the behaviour of density waves at the critical regions of the Lindblad and co-rotation resonances and in particular as to their role in the maintenance and generation of these waves. We shall also discuss the conditions under which an inner resonance ring may be formed. Furthermore, we find that the co-rotation region acts as a wave-amplifier.

1. Inner Resonance and Resonance Rings

Consider a galaxy with Ω and κ denoting its circular and epicyclic frequencies as a function of radial distance from the galactic centre. The inner Lindblad resonance occurs when $(\Omega - \Omega_p) = \kappa/m$, where Ω_p is the constant pattern frequency of a density wave with number of arms m. This resonance is known to be very efficient in absorbing the inward-propagating galactic density waves (Toomre, 1969; Mark, 1971).

For the Schmidt model of our Galaxy, using the pattern speed of $\Omega_p = 13.5 \text{ km s}^{-1}$ kpc^{-1} (Lin *et al.*, 1969), we found (Mark, 1971) that the wave has a peak at about 5 kpc. This correlates well with the peak in the observed ratio of neutral to ionized hydrogen. Lin and Feldman (1970) had first suggested such a correlation as a consequence of the density wave theory but the exact position of the wave peak was not known to them. We must emphasize that our theory predicts that the position of this resonance 'ring' of ionized hydrogen is usually one to two kiloparsecs outwards of r_L, the radius at which the resonance condition is satisfied, since the stars which are in resonant interaction with the wave can have radial excursions due to epicyclic motions even though their average radial distance may be nearly equal to r_L. Figure 1 illustrates this situation.

An interesting property of the wave at inner resonance is its behaviour when the dispersion speed of the stars is varied. For low dispersion speeds, the wave has very short wavelengths in the region of inner resonance and should give the impression of a 'ring' when the thickness of the gaseous arms (in ionized hydrogen) is taken into account. Also the wave amplitude is larger at the ring than in the spiral structure further out. Thus we expect the ratio of ionized to neutral hydrogen to have a peak corresponding to the wave peak. But if the dispersion speeds are large, the spiral structure is more open and the wave is absorbed without ever winding up. We expect here no 'ring' structure in the density of ionized hydrogen and only a smooth decay of the spiral structure inwards through this region. In intermediate cases, the ring may be only partially closed. These expectations should of course be checked by a calculation of the relation of wave amplitude to ionized hydrogen density.

Short-wavelength leading waves convect outwards, with decreasing amplitude in this direction. No sharp peak in wave amplitude can form in the resonance region. Such waves can be excited by mechanisms located near the central region of galaxies.

John R. Shakeshaft (ed.), The Formation and Dynamics of Galaxies. 417–423. All Rights Resrved.
Copyright © 1974 by the IAU.

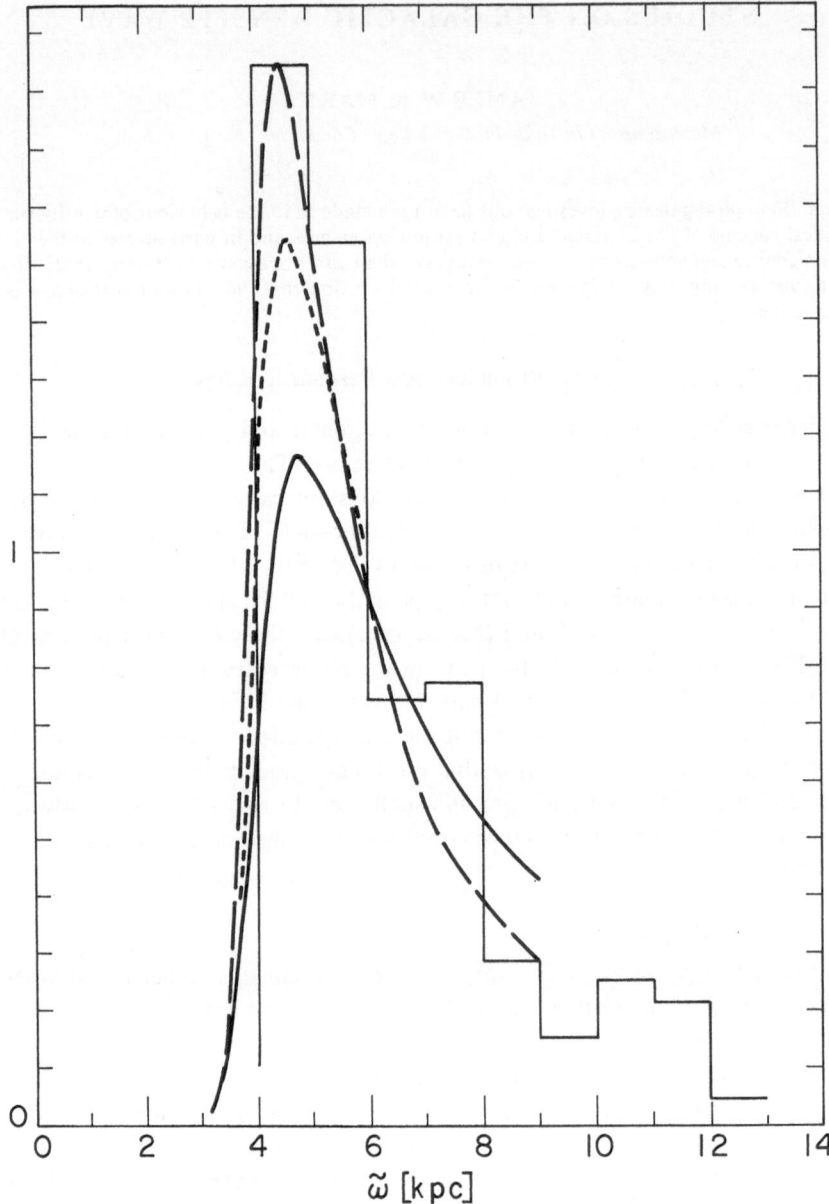

Fig. 1.   The plot of observed H II/H I density ratio versus radial distance is compared with three theoretical curves of wave surface mass density (multiplied by a geometrical factor). Mezger's (1969) number density of H II regions is used for the ionized hydrogen, while the H I density is that obtained by van Woerden (1965). The theoretical curves are obtained using the 1965 Schmidt model and $\Omega_p = 13.5$ km s^{-1} kpc^{-1}, $r_L = 3.2$ kpc. The solid, dotted and dashed curves represent a sequence of decreasing velocity dispersion for the stars.

2. Inner Resonance and the Mechanism of Absorption

We have also found that the details of this absorption of wave angular momentum (action) can be described by a continuity equation which includes source terms (i.e. a wave-action principle). In particular, the wave properties in this region are completely described by the density, flux and source of wave action respectively. For example, the action density is negative for the short-wavelength trailing waves and these waves convect inwards towards the inner resonance. The source density is always positive and thus contributes only to the damping of the wave. As this wave convects inwards, the action density first increases in magnitude because the wave

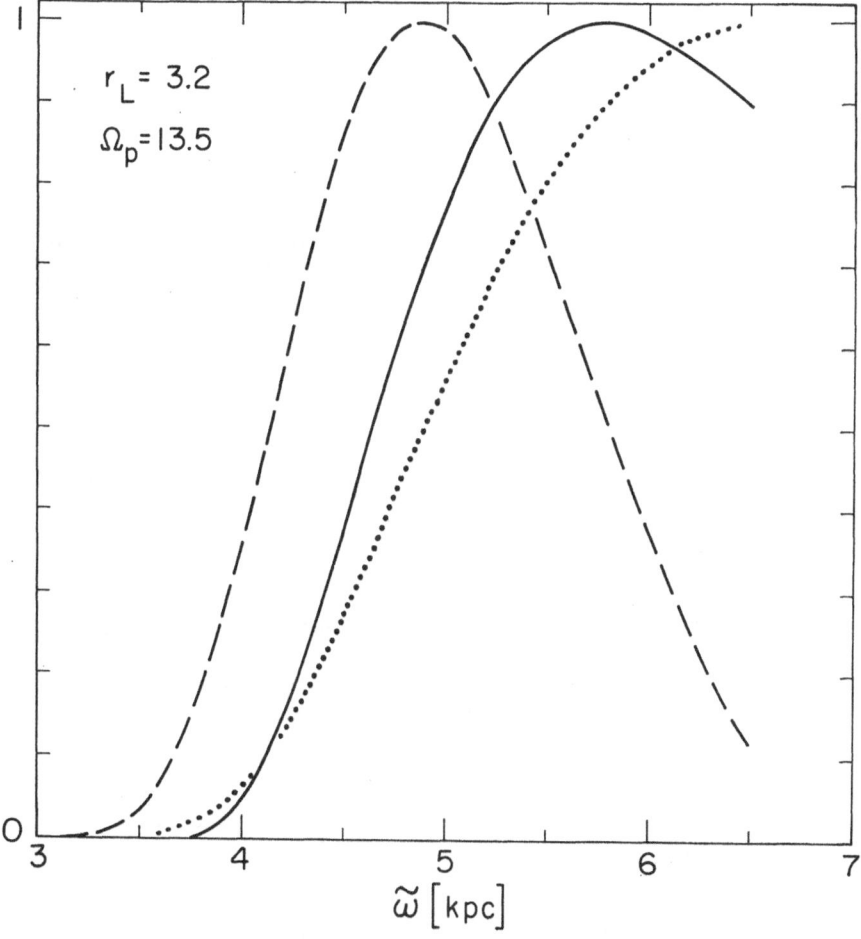

Fig. 2. Plotted against radial distance from the galactic centre are the magnitude of the wave angular momentum density (solid curve), the wave angular momentum flux (dotted curve) and the source density (dashed curve) of angular momentum that damps the wave at inner Lindblad resonance. The model used and pattern speed are the same as in Figure 1, while the vertical scale is in arbitrary units.

flux decreases. But eventually, at about 2–3 kpc outwards of the radius r_L, the magnitude of the action density decreases sharply inwards because of the source. By the time the radius r_L is reached, the wave is completely negligible in amplitude. The behaviour of the relevant physical quantities is illustrated in Figure 2 for this case of short-wavelength trailing waves.

Within the approximations used, the integrated source strength agrees with that supplied by Lynden-Bell and Kalnajs (1972). Thus we have confirmed by detailed calculation that their mechanism for absorption is correct.

3. Preliminary Results on the Possibility of Forming a Bar-Like Density Concentration at the Inner Resonance Region*

Lin (1969) has suggested a mechanism for maintaining the wave which requires the driven response of the inner resonance to possess a bar-like component. Contopoulos (1970) had found a four-fold symmetry in the resonant periodic and tube orbits. He used a wave of nearly constant amplitude. We have recalculated the stellar orbits in this 3–5 kpc region of the galaxy, assuming steady driving by a density wave which takes into account the rapid decay of the wave amplitude inside 5 kpc.

The resonant periodic orbits are individually still roughly elliptical in shape, in agreement with Contopoulos' (1970) calculations for constant wave amplitude. We differ from him in that all these resonant orbits have major axes pointing in the same general direction with deviations up to some 30°. This suggests that a bar-like density concentration has formed. In addition to this, many of the non-periodic orbits tend to suggest the presence also of a density with fourfold symmetry.

However, although the above results are suggestive, it is not easy to determine from them whether the density concentrations are large enough to provide the necessary feedback. A continuum-type calculation is now being pursued which may hopefully give us some more definite numerical estimates on the actual density concentrations.

4. The Co-Rotation Region and an Amplifier of Density Waves

Lynden-Bell and Kalnajs (1972) suggested that the co-rotation resonance may be a source of negative wave action which excites the inward-travelling trailing waves. We studied this resonance for the case where the galactic disk is neutrally stable to Jeans' instability. In order to obtain a consistent solution near this resonance, it turned out to be necessary to consider a process of wave reflection and transmission rather than the simple interaction of one wave with the resonant stars as envisaged by Lynden-Bell and Kalnajs.

At the co-rotation resonance an outward-travelling trailing wave is reflected and transmitted into other trailing waves. The wave generation effect suggested by Lynden-Bell and Kalnajs resulted only in a slight amplification of the reflected wave relative

---

* The work in this section is performed in collaboration with Mr Robert Berman.

to the incident one. An inward-travelling trailing wave is similarly reflected and transmitted at the co-rotation resonance. We used the Schmidt model of our Galaxy.

However, another mechanism is operating at the co-rotation region which acts as an amplifier of wave energy (or angular momentum). This new mechanism depends on the presence of negative energy waves inside co-rotation and positive energy waves outside co-rotation. In the above-mentioned reflection and transmission process, these waves of different intrinsic energy content are coupled across the co-rotation

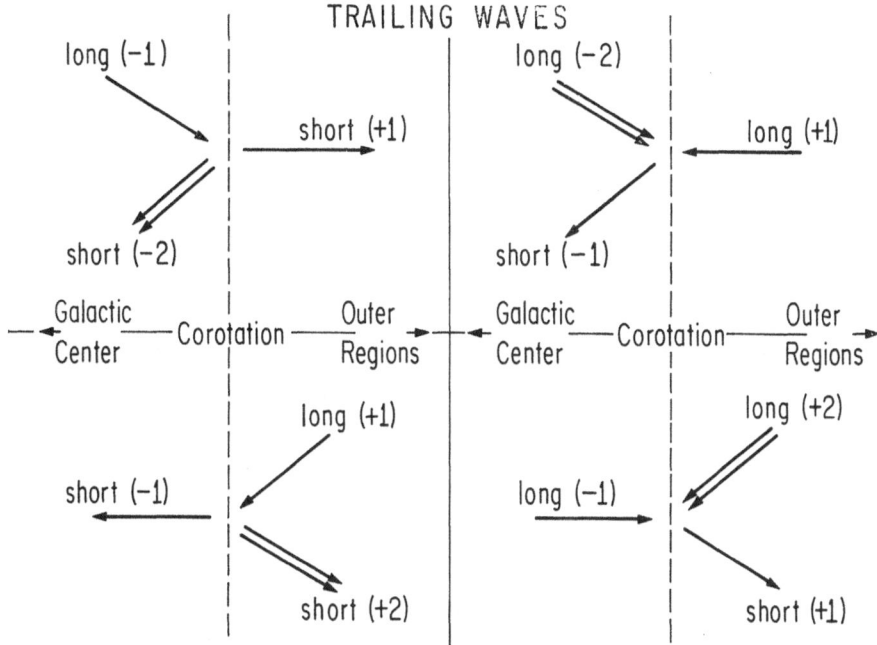

Fig. 3.   Schematic diagram of two wave amplifiers (left frames) and two 'de-amplifiers' (right frames). The co-rotation circle (dashed line) separates central and outer regions of the galaxy. Arrows indicate the direction and strength of the waves. The strength (energy content) is also given in arbitrary units by the signed numbers that follow the words 'long' or 'short' which identify the waves as members of the long or short wavelength branches.

region. Omitting the small contribution of the resonant stars, the reflected wave has twice the energy of the incident one, while energy conservation requires that the transmitted wave carries away an amount of energy equal in magnitude but opposite in sign to that carried by the incident wave. Figure 3 describes in a schematic fashion the four situations that occur for this particular case where the galactic disk has only a stellar component and where this is neutrally stable to axisymmetric Jeans' instability.

This wave amplification process provides one detailed mechanism whereby galaxies convert their abundant source of kinetic energy (and angular momentum) of circular rotation into the energy (and angular momentum) of the spiral wave patterns. The conversion occurs at the co-rotation region where the wave pattern frequency Ω_p is nearly equal to the circular frequency Ω of the local stars. Such a process depends crucially on the fact that the major density waves (i.e., those inside of co-rotation) represent disturbances whose excitation requires the removal of wave energy (in the inertial frame). Amplification occurs due to the above-mentioned coupling with other waves which have the opposite behaviour in that their excitation requires an input of wave energy. Thus we may expect higher amplification to be possible if other 'dissipative' mechanisms occur which satisfy the criterion that they can remove rotational kinetic energy from the galactic disk. A slowly rotating galactic halo can preferentially remove rotational energy from the rapidly rotating disk and further wave amplification is found. Moreover, since the presence of halo stars and of the gas in the disk both increase the effectiveness of wave coupling in the above-described wave amplifier, therefore wave amplification is further enhanced. After much analysis, it is found that the behaviour of the waves at co-rotation can be described by the turning point equation

$$\frac{d^2 w(\zeta)}{d\zeta^2} + [\zeta^2 + m_D + m_H + \varrho(\zeta)] \, w(\zeta) = 0.$$

Here, ζ measures (scaled) distance away from co-rotation; $w(\zeta)$ is the reduced wave function; m_D and m_H are respectively the disk and halo amplification coefficients; and $\varrho(\zeta)$ gives the corrections due to the resonant stars at co-rotation. Figure 1 corresponds to the mathematically simpler case where $m_D = m_H = 0$ and also $\varrho(\zeta)$ is omitted. But under the actual circumstances applicable to galaxies, the values of these coefficients are such that wave amplifications of at least an order of magnitude should be possible.

References

Contopoulos, G.: 1970, *Astrophys. J.* **160**, 113.
Lin, C. C.: 1969, in W. Becker and G. Contopoulos (eds), 'The Spiral Structure of Our Galaxy', *IAU Symp.* **38**, 377.
Lin, C. C. and Feldman, S.: 1970, Ref. C. C. Lin, *Highlights of Astron.* **2**, 88, 1971.
Lin, C. C., Yuan, C., and Shu, F. H.: 1969, *Astrophys. J.* **155**, 721.
Lynden-Bell, D. and Kalnajs, A.: 1972, *Monthly Notices Roy. Astron. Soc.* **157**, 1.
Mark, J. W.-K.: 1971, *Proc. Nat. Acad. Sci.* **68**, 2095.
Mezger, P. G.: 1969, in W. Becker and G. Contopoulos (eds), 'The Spiral Structure of Our Galaxy', *IAU Symp.* **38**, 107.
Toomre, A.: 1969, *Astrophys. J.* **158**, 899.
van Woerden, H.: 1965, Ref. J. H. Oort, *Trans. IAU* **12A**, 789.

DISCUSSION

Contopoulos: The effect of a density wave upon particles originally moving in circular or epicylic orbits will be to change their orbits and, if strong enough, to 'trap' them in non-linear, resonant, quasi-periodic orbits. This 'trapping' process is probably the one normally responsible for absorption of the waves at the inner Lindblad resonance and for the excitation of the waves at the co-rotation distance.

Mark: I agree that trapping is the mechanism causing absorption of the waves at the inner resonance, but I have come to the conclusion that the trapping process at the co-rotation resonance is not effective in *exciting* the wave. My solutions seem to be different from that of Dr Contopoulos. For trailing waves, one solution is that of a long wave propagating outwards, which becomes reflected and transmitted at the co-rotation region into two short trailing waves.

HIGH-RESOLUTION STUDIES OF NEUTRAL HYDROGEN
IN NGC 5383 AND M101

R. J. ALLEN, W. M. GOSS, R. SANCISI, W. T. SULLIVAN III,
and H. VAN WOERDEN

Kapteyn Astronomical Institute, University of Groningen, The Netherlands

Abstract. Maps with 0.'5 resolution are presented of the distribution of neutral hydrogen in two galaxies. The barred spiral NGC 5383 contains much hydrogen, in strong differential rotation, in its outer parts; some H I concentrations with possibly anomalous velocities are observed in the regions of bar and nucleus. In the giant Scd spiral M101, the H I distribution corresponds closely with the optical spiral pattern, except for a lack of hydrogen in the central region.

1. Introduction

The Synthesis Radio Telescope at Westerbork has been operating in the continuum at 21 cm since 1970, at 6 cm since October 1972, and at 50 cm since last month. In addition, a filter spectrometer has been available since November 1971, allowing observations in the 21-cm line with $\approx 0.'5$ angular resolution and 27 km s^{-1} velocity bandwidth. We present here some preliminary results of the observing programmes on the barred spiral galaxy NGC 5383 and the giant Scd galaxy M101.

2. NGC 5383; H I Distribution and Velocity Field

Very little is known about the distribution and motions of H I gas in barred spiral galaxies. The operating principles of our synthesis telescope favour observations at high northern declinations where the choice of suitable barred spirals is limited, and we have chosen to observe NGC 5383, of type SBb, with an optically measured redshift of about 2260 km s^{-1} (distance 47 Mpc with $H \simeq 50$ km s^{-1} Mpc^{-1}) and a well-defined bar about 100″ long (22 kpc) with a dust lane down the middle.

Figure 1 is a preliminary map of the H I distribution in this galaxy as determined with the 25″ × 37″ beam of the Westerbork telescope (linear resolution about 6.5 kpc). This is the first detection of H I in this galaxy as far as we know. The continuum radiation has been subtracted using channels located outside the velocity range of the H I emission. The map shows the angular distribution of the H I intensity integrated over all velocities; the effective noise on this map is such that the outer contour may be somewhat uncertain. At any point in the galaxy the H I is in general spread over several channels, the individual channels being 27 km s^{-1} wide and spaced by 20 km s^{-1}. The channel maps show peak brightness temperatures in the H I of 10 to 15 K, with an rms noise of 2–3 K. Since the signal/noise ratio is small over most of the galaxy, we will limit the present discussion to areas of stronger signals. Some of the features of Figure 1 are the following:

(a) There are several concentrations of H I which lie outside the apparent optical

John R. Shakeshaft (ed.), The Formation and Dynamics of Galaxies. 425–430. All Rights Reserved.
Copyright © 1974 by the IAU.

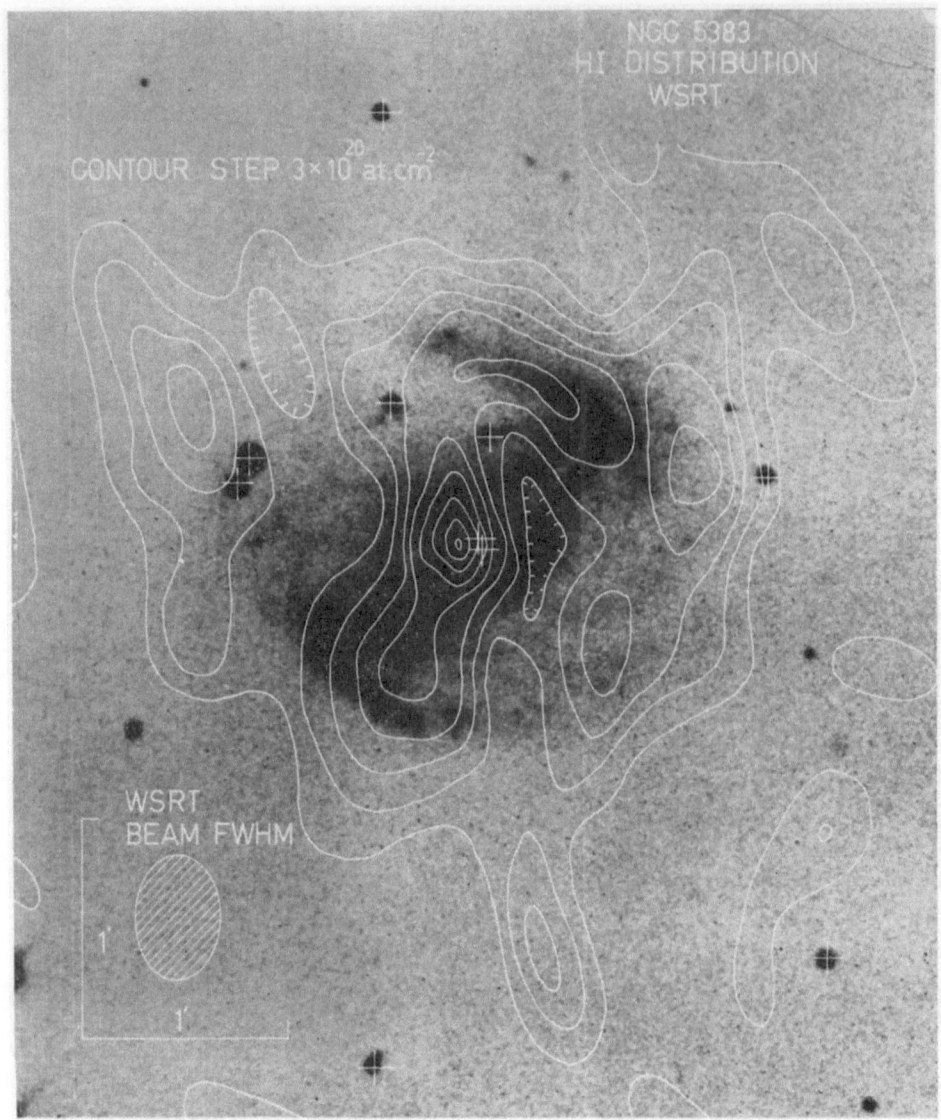

Fig. 1.   The distribution of H I in NGC 5383 as observed with the 25″ × 37″ beam of the Westerbork telescope. The bar of this galaxy is about 100″ long, or 22 kpc at an assumed distance of 47 Mpc. North is at the top, east at the left in all figures.

boundary of the galaxy; one of them lies just beyond the eastern border of Figure 1.

(b) There is H I emission in the region of the nucleus and at the ends of the bar, and evidence for weak emission ($\simeq 5$ K or 2–$3 \times$ rms noise) in the direction of the bar; further observations to improve on the sensitivity are planned.

(c) The major part of the H I distribution is 6′ in diameter, or 80 kpc. The H I mass is then about $8 \times 10^9 \ M_\odot$. There is some evidence for faint extensions in the H I on the southern side of the galaxy, up to perhaps 5′ away.

Some preliminary results on the velocity field of NGC 5383 have been obtained from the H I observations:

(a) The systemic velocity of the galaxy is 2250 km s^{-1} (heliocentric) with an estimated uncertainty of about 15 km s^{-1}. The total radial velocity spread is 350 km s^{-1}.

(b) There are large areas of rather constant radial velocity on both sides of the apparent minor axis of the galaxy. Along the major axis (approximately east-west) the radial velocity remains roughly constant (within 30 km s^{-1}) from about 0'5 up to about 2'5 from the centre. This implies that a substantial fraction of the mass of this galaxy lies outside the bar. The dynamical major axis determined from these observations agrees fairly well with the optical major axis adopted by Burbidge *et al.* (1962).

(c) Outside the region of the bar, the velocity field resembles that expected for a disk-shaped mass distribution in circular differential rotation. Within the region of the bar, there is a strong suggestion of velocity anomalies which may be associated with the bar.

(d) Good agreement between our H I radial velocity measurements and the optical observations along the bar made by Burbidge *et al.* (1962) is obtained in regions where a comparison can be made, namely, near the nucleus and near the ends of the bar.

An estimate of the hydrogen-to-total mass ratio gives a value in good agreement with that commonly obtained for Sb galaxies (Roberts, 1969).

3. M101; Distribution of the Small-Scale Structure in H I

The desire to obtain high-resolution observations of the H I content of this galaxy was one of the many reasons for constructing the multi-channel spectrometer at Westerbork. The observations obtained by Rogstad and Shostak (1971) with a resolution of 4' were insufficient to resolve one spiral arm from another in this galaxy, since these arms have separations of about 3' or less. The Westerbork maps, obtained in 16 half-days of observing, provide an angular resolution of 25" × 30" or about 1 kpc at an assumed distance of 7 Mpc. The data have not yet been corrected for attenuation by the primary beam at the edge of the maps. In addition, the absence of interferometer spacings shorter than 36 m means that structures in the H I on scales of about 10' and larger are not adequately observed; we therefore confine our present remarks to structures on angular scales smaller than 10'.

Figure 2 shows on the left a 'radiograph' representation (Ekers *et al.*, 1973) of the H I distribution in M101 and, to the same scale, an optical photograph taken from Arp (1966). The radiograph contains small crosses at nine fiducial points as well as at the positions of six alignment stars. The mottled appearance of the background arises from noise in the observations; the size of the smallest spots is approximately that of the telescope beam.

The following features can be seen in Figure 2:

(a) The H I tends to appear in long, more-or-less connected features closely coincident with optical spiral arms and H II region complexes. On deeper optical exposures almost all of the hydrogen features have optical counterparts, including the outer H I

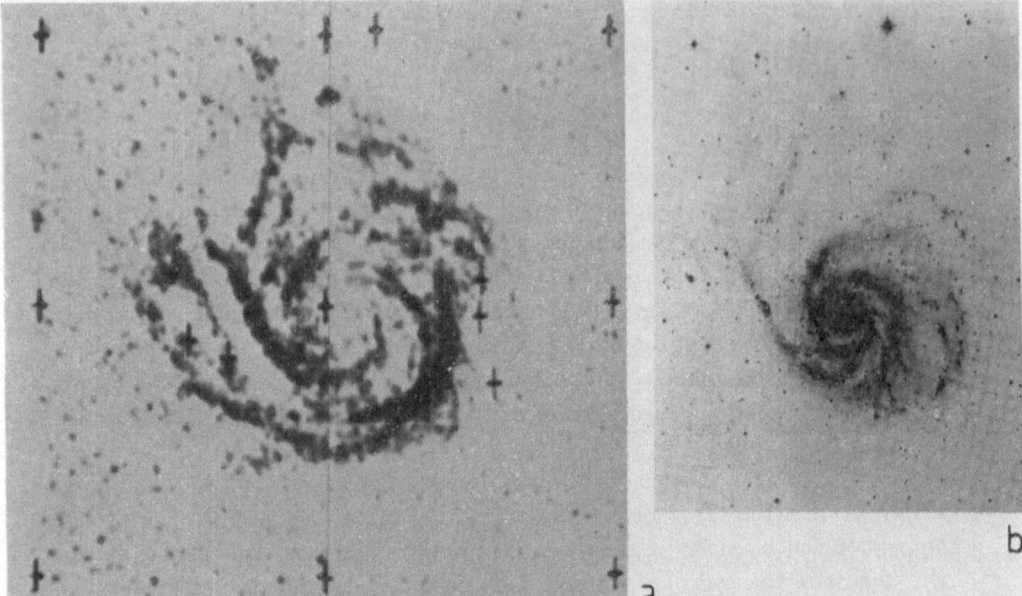

Fig. 2.   (a) Radiograph showing the distribution of neutral hydrogen surface density in M101. Large-scale ($> 10'$) components in the distribution are not recorded. The radiograph contains crosses at nine fiducial points as well as the positions of six stars. The mottled appearance of the background in the radiograph arises from noise in the observations; the size of the smallest spots is about that of the telescope beam.
(b) Blue photograph of M101, from Arp's *Atlas of Peculiar Galaxies*. Scale same as Figure 2(a).

spiral arms to the south-east of the centre. This close correspondence weakens some of the arguments advanced by Piddington (1973) in his criticism of the observational evidence favouring the density-wave theory of spiral structure.

(b) Not every optical feature has an H I counterpart; in particular the central regions of the galaxy seem devoid of small-scale H I features, a fact already noted by Rogstad (1971) and by other observers.

(c) We have detected hydrogen in NGC 5477, an object classed as a small irregular companion of M101. The H I feature is located on the far left of the radiograph about one centimetre above the middle fiducial mark. The optical photograph does not cover this feature.

The radio H I contours are superposed on the optical photograph in Figure 3, illustrating again the detailed correspondence of the narrow, elongated H I features with optical spiral arms. The observed peak-to-average brightness temperature ratio is about 3, but since the H I is in many places not resolved across the spiral arms the true peak-to-average ratio is certainly greater.

Further discussion of these preliminary results on M101 will be published in *Astronomy and Astrophysics* (see **29**, 447, 1973 for Paper I).

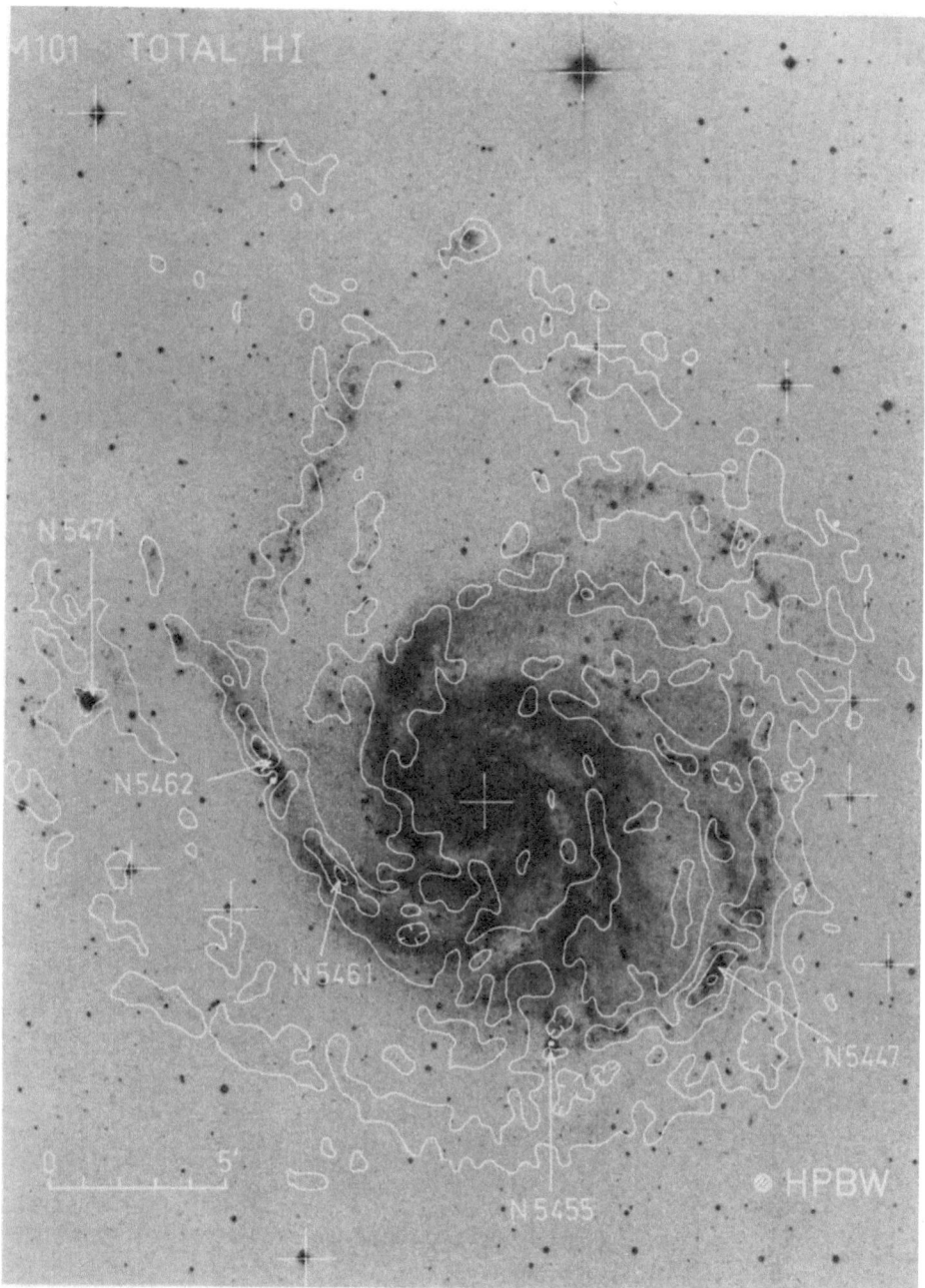

Fig. 3. Overlay of contours of hydrogen surface density (N_H) on the blue photograph. Contour levels are approximately: $(4, 12, 20, 28) \times 10^{20}$ atom cm^{-2}. However, for reasons discussed in the text, the N_H scale increases somewhat away from the centre of the galaxy and the zero level of the N_H contours is uncertain. The angular scale is shown at lower left (1 arc min = 2 kpc), the synthesized beam in the lower right corner.

Note the detailed correspondence of optical and radio features. The brightest H II regions, indicated by their NGC numbers, have all been detected in the continuum. The cross in the nucleus of the galaxy corresponds to: (1) an optical H II region, (2) a weak associated radio source, and (3) the centre of the galaxy as defined by the light distribution on blue photographs. The other crosses are fiducial marks in the hydrogen map corresponding to positions of stars. The three historical supernovae in M101 (SN1909a at 12′ NW, SN1951 at 6′ E,and SN1970g at 6′ S) are indicated by white dots.

Acknowledgements

We thank Professors Oort and van der Laan, D. H. Rogstad, R. D. Ekers, T. S. van Albada, K. W. Weiler and the telescope group, and H. W. van Someren Greve and the reduction group. The Westerbork Radio Observatory is operated by the Netherlands Foundation for Radio Astronomy with financial support from the Netherlands Organisation for the Advancement of Pure Research (ZWO).

References

Arp, H. A.: 1966, *Atlas of Peculiar Galaxies*, California Institute of Technology, Pasadena.
Burbidge, E. M., Burbidge, G. R., and Prendergast, K. H.: 1962, *Astrophys. J.* **136**, 704.
Ekers, R. D., Allen, R. J., and Luyten, J. R.: 1973, *Astron. Astrophys.* **27**, 77.
Piddington, J. H.: 1973, *Astrophys. J.* **179**, 755.
Roberts, M. S.: 1969, *Astron. J.* **74**, 859.
Rogstad, D. H.: 1971, *Astron. Astrophys.* **13**, 108.
Rogstad, D. H. and Shostak, G. S.: 1971, *Astron. Astrophys.* **13**, 99.

DISCUSSION

E. M. Burbidge: You are presumably getting a high angular momentum out of these observations, and this is interesting in that the meagre other data on barred spirals also suggest high angular momenta for barred spirals (cf. Burbidge, E. M., Burbidge, G. R., and Prendergast, K. H.: *Astrophys. J.* **136**, 704, 1962).

THE MOTIONS IN THE CENTRAL REGION OF NGC 4736*

P. C. VAN DER KRUIT

Hale Observatories, Carnegie Institution of Washington, California Institute of Technology,
Pasadena, Calif., U.S.A.

Abstract. Spectroscopic observations of the central part of NGC 4736 show strong noncircular motions connected with the triple radio source. It is also found that a central ring of H II regions is probably expanding at about 30 km s⁻¹.

1. Introduction

The Sab galaxy NGC 4736 (M94) has a number of interesting features. One of these is the faint external ring that is visible on deep photographs and which will be discussed in more detail elsewhere in this volume (p. 410). I will concentrate here on the central region. Figure 1a shows the galaxy's spiral structure and the central triple

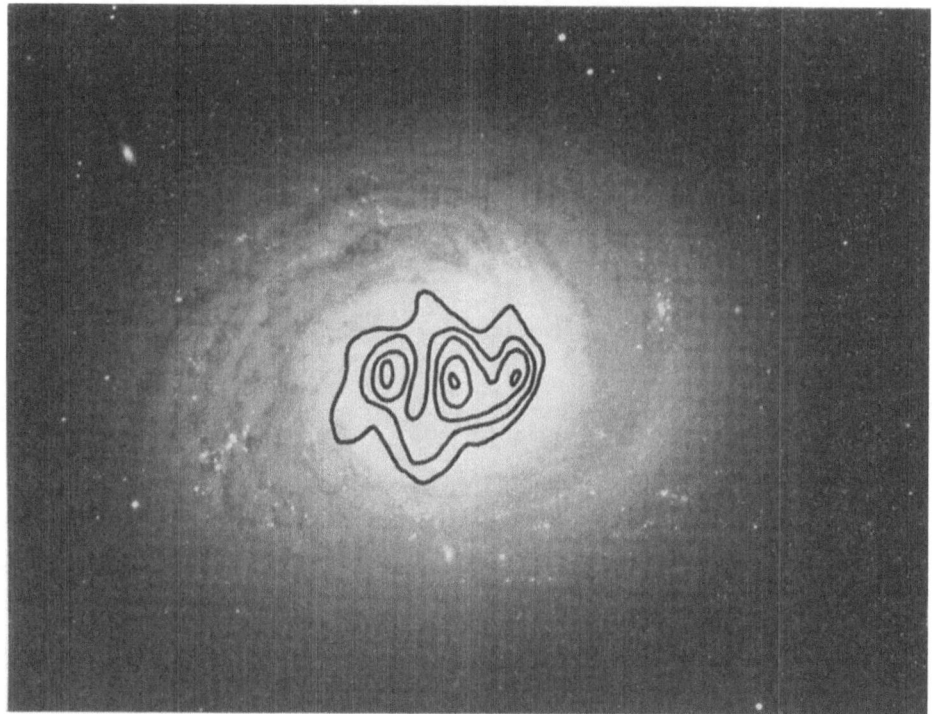

Fig. 1a.

Fig. 1a–b. The various aspects of NGC 4736: (a) The optical disk with spiral structure has an apparent major axis in position angle 107°. The triple radio source is aligned almost east-west. (b) The central ring of H II regions has a major axis in position angle 123° (Hα-interference plate taken with a 200-in. prime-focus image tube; bandwidth from −1600 to +2000 km s⁻¹ and exposure time 20 min). The radio source components fall just inside the bright H II regions. In both pictures north is at top and east at left. The scales are 4″12 and 1″44 mm⁻¹, respectively.

* See also: *Astrophys. J.* **188**, 3, 1974.

John R. Shakeshaft (ed.), The Formation and Dynamics of Galaxies. 431–437. *All Rights Reserved.*

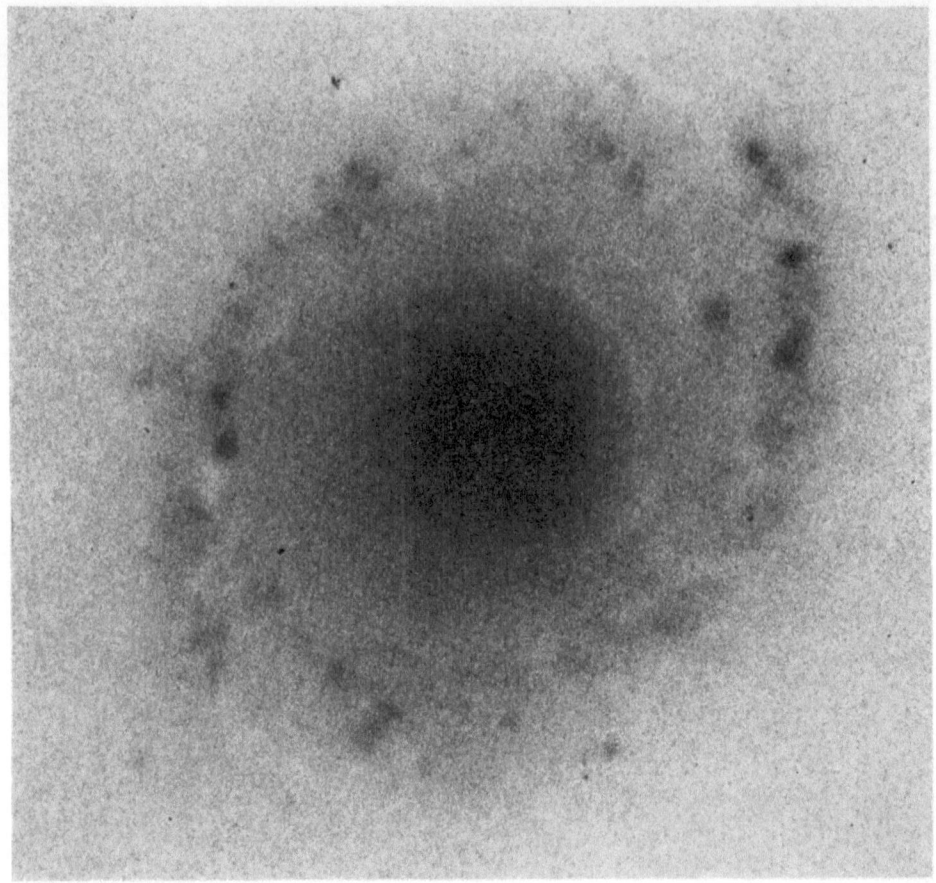

Fig. 1b.

radio source (van der Kruit, 1971b) as measured with the Westerbork synthesis radio telescope at 1415 MHz. The radio source extends nearly east-west. The line of nodes for the plane of the galaxy is in position angle 107° (Burbidge and Burbidge, 1962, $[B^2]$) as measured from the outline of the optical image of Figure 1a. At a radius of about 50″ from the nucleus there is a 'knotty ring' of H II regions, described by B^2. This is shown in the Hα plate taken at the prime focus of the 200-in. (508-cm) Hale telescope and reproduced in Figure 1b. The 'ring' has an elliptical shape; the position angle of the major axis is 123° (according to B^2) and is markedly different from that of the region in Figure 1a. Just beyond the H II regions there is a very sharp drop in (optical) brightness, described by Sandage (1961), which reinforces the ring-like character of this region. The outer components of the radio complex fall just inside the ring of H II regions. An extensive spectroscopic survey has been made of this inner region in order to determine whether there are effects on the disk of NGC 4736 from the nuclear activity, which is evident from the radio complex.

2. Observations and Velocity Field

The image-tube spectrographs at the Cassegrain foci of the 60-in. (152-cm) and the Hale 200-in. (508-cm) telescopes at Palomar Mountain were used to obtain 22 spectra in 12 position angles on the sky. The spectra were all taken with red gratings; the lines recorded and measured are Hα, [N II] λλ6548, 6583 and [S II] λλ6716, 6731. The emission lines are seen in the 'knotty ring' exclusively and, in agreement with Duflot (1962) and Chincarini and Walker (1967), this region was found to exhibit very small changes in velocity as a function of distance from the centre. This makes it possible to characterize each position angle from the nucleus with one velocity only. A discussion of the errors indicates that the mean error in this velocity ranged from 14 km s^{-1}, if it was derived from one spectrum only, to 8 km s^{-1} if three spectra were available.

On the assumption that the velocity field is symmetric around the nucleus (as Chincarini and Walker show from absorption-line measurements), the systemic velocity of NGC 4736 is +304±4 km s^{-1}, which agrees well with the value of 307 km s^{-1} found from measurements of the 21-cm line emission at Westerbork (Allen, private communication).

The measured velocities as a function of position angle from the nucleus are shown

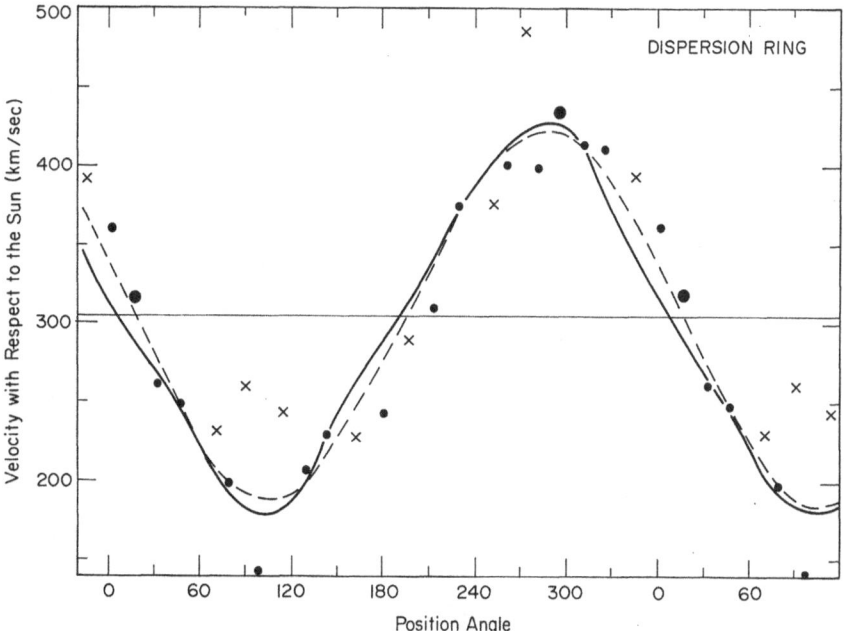

Fig. 2.   Observed velocities as a function of position angle from the nucleus and the predicted velocities from the dispersion ring model. The small dots are points derived from one spectrum, crosses from two spectra, and large dots from three. The dashed line is the basic rotation and the full line the total motion in the elliptical dispersion ring. Note that the range of position angle from 0° to 120° has been repeated.

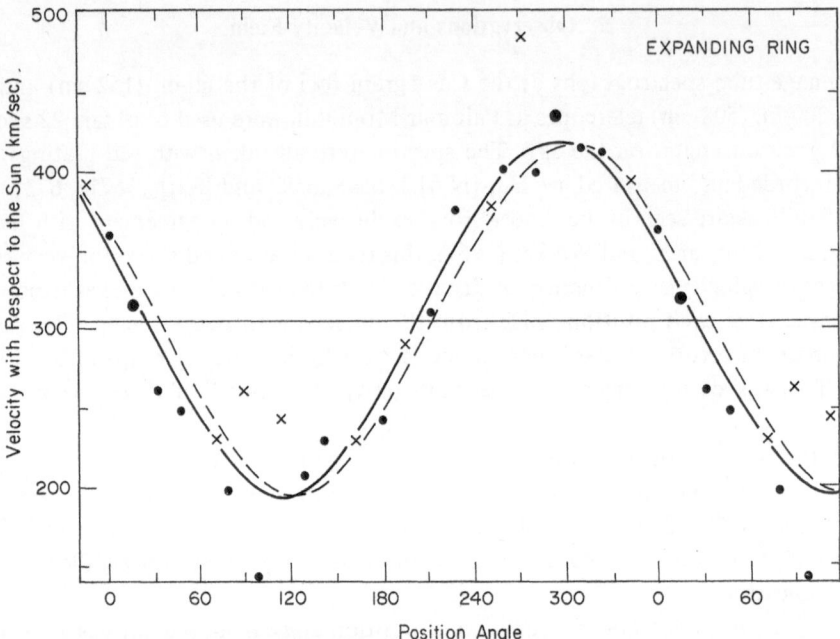

Fig. 3. The same data points with the expanding ring model. The dashed curve is the rotation component only (153 km s⁻¹); the expansion component is 28 km s⁻¹.

in Figures 2 and 3, together with two possible solutions. As can be seen from the figures, the observed points follow the general pattern expected for a rotating ring, except for those in position angles 90°, 100°, 114° and 270°. Since this is the direction of the triple radio source it may be concluded that we are seeing here direct effects of the nuclear explosion on the motions in the disk.

Excluding the measurements at these positions, the best fit to the observed run of velocity with position angle by a field of pure and uniform rotation gives a dynamical major axis in position angle $113° \pm 2°$. This was also found by Chincarini and Walker from spectra in only three position angles. With the present accuracy, however, it is now established that this is significantly different from the line of nodes determined from the shape of the outer regions (107°) and from that determined from the shape of the knotty ring and the accompanying sharp drop in brightness (123°). Also the scatter of the points around the best-fitting sinusoid is not more than is expected from the errors.

3. Interpretations and Discussion

A possible model for NGC 4736 is one in which the line of nodes is in position angle 113° and in which the various apparent major axes are due to perturbations in the structure of the disk. However, a model in which there is such a complicated geometry without noncircular motions of appreciable magnitude is difficult to envisage. Also, the nuclear activity (apparent from the radio sources) and its effects on the disk (at least in the velocity field) suggest the possibility of a causal relation between the ring and the nucleus. I will therefore treat this region in a way similar to that in which the central region of our Galaxy has been explored, namely, in terms of expanding features or of stable, elliptic orbits.

The latter idea was suggested from stellar dynamics by B. Lindblad and has been referred to as a Lindblad resonance or dispersion ring. It has been applied to the central region of our Galaxy by Simonson and Mader (1973), following Shane (1972), and to M51 by Tully (1974). In these cases they fitted the geometry to the observed velocities and indicated a possible solution. Here I will compare the observed velocities with those to be expected from the observed geometry.

The northern side of NGC 4736 is the nearer. This follows from the facts (a) that dust is seen in that side and (b) that the eastern part is approaching, together with the assumption that the spiral structure is trailing. Now if we assume that the line of nodes is in position angle 107° (as indicated by the outer structure) and that the inclination is 40° (from the same outline), the deprojected shape and orientation of the ring can be found and also the predicted velocity field. This is illustrated in Figure 2 where the dashed line indicates the basic rotation and the full line the predicted total velocities. One can see that the 'epicyclic' velocities have components in the line of sight *with the wrong sign*, so that there is clear disagreement between model and observations. It also means that such a model can be made to agree with the observed shape of the ring *and* the observed velocities only if the line of nodes is in a position angle *larger* than 123° or if the southern side were the near side. Neither of these possibilities is very realistic.

Another interpretation is that the ring is close to circular and has a field of radial motions. This means that the line of nodes is in position angle 123° and the inclination is 46°. Figure 3 shows such a fit to the data, where the (uniform) rotation velocity is 153 km s^{-1} and the (uniform) *expansion* velocity is 28 km s^{-1}. The scatter around the line (forgetting about the discrepant velocities in p.a. 90°, 100°, 114°, 270°) is less than in the model with uniform rotation and I conclude that the best model is a circular ring expanding at about 30 km s^{-1} (at least along the apparent minor axis).

I propose that the ring is the result of activity in the nucleus of NGC 4736. This is reinforced by the fact that the outer components of the triple radio source are found close to the brighter H II regions and that strongly deviating velocities are found at these positions. The model for this is expected to be similar to that proposed for the central region of our Galaxy (van der Kruit, 1971a), although there is for NGC 4736 no indication of whether the expulsion took place *in* the plane or at an angle to it. It must also be assumed that the direction of expulsion has been rotating, presumably along with the nucleus.

It is at present not possible to separate the observed velocities into tangential, radial, and perpendicular components, and hence no detailed model for the expulsion or its properties can be given. However, the dimensions and the velocities are comparable to those of the 3 kpc arm in our Galaxy, and consequently most properties, such as the energy involved, expulsion velocities, etc., will be of the same order of magnitude.

The drop in brightness just outside the bright H II regions might be the contact surface between the expanding gas and the quiescent-gas layer. Both the continuum and the emission lines drop in brightness very sharply there and it could easily be a shock region. As in the density wave theory of spiral structure, such a shock might have triggered star formation and given rise to the ring of bright H II regions. It is hoped that subsequent observations will enable a more detailed model to be constructed.

Finally, it is of interest to compare the fit of the dispersion ring model to those in our Galaxy and M51. One can easily check that in both these cases the orientation of the ellipse is precisely that in which the predicted pattern of the velocities in the line of sight is similar to that for an expanding ring. This, and the fact that on the minor axis we always see expansions, make the dispersion ring solution for our Galaxy and M51 rather suspect. It becomes even more so now that NGC 4736 exhibits a case where the observed geometry and velocities are incompatible with that solution.

Acknowledgements

I am grateful to the Carnegie Institution of Washington for a research fellowship and to Dr H. W. Babcock for his hospitality at the Hale Observatories. I also thank the Carnegie Institution for the privilege of receiving the Dr Knut Lundmark Award to permit my travel to Australia.

References

Burbidge, E. M. and Burbidge, G. R.: 1962, *Astrophys. J.* **135**, 366.
Chincarini, G. and Walker, M. F.: 1967, *Astrophys. J.* **147**, 407.
Duflot, R.: 1963, *Compt. Rend. Acad. Sci. Paris* **255**, 2714.
Sandage, A. R.: 1961, *The Hubble Atlas of Galaxies*, Carnegie Institution of Washington, Washington, D.C.
Simonson, S. C. and Mader, G. L.: 1973, *Astron. Astrophys.* **27**, 337.
Tully, R. B.: 1974, *Astrophys. J. Suppl.* **27**, 415
van der Kruit, P. C.: 1971a, *Astron. Astrophys.* **13**, 405.
van der Kruit, P. C.: 1971b, *Astron. Astrophys.* **15**, 110.

DISCUSSION

Miller: Your last slide seems to indicate that the radial velocity data could be explained by a shift in phase that would correspond to a reassignment of the line of nodes. Is your inference of non-circular motion based on data other than the velocity field?

van der Kruit: This shift corresponds to non-zero velocities in the line of sight along the apparent minor axis, which are therefore radial motions. It is exactly this shift which leads to the invocation of expanding motions, because an epicyclic approximation for a dispersion ring would give a different run of the curve *and* a shift in phase in the wrong direction. The inference is thus drawn from a *combination* of data inferred from the velocity field and the observed appearance, and the differences between these two.

Mark: Does the ring connect with the spiral structure further out?

van der Kruit: The spiral structure of NGC 4736 is multi-armed. An inspection of all photographs indicates that part of the 'knotty' structure of the ring might be due to crossing dust lanes, but it is not possible to follow the spiral structure clearly through the ring.

Mark: The density-wave theory can account for such a ring, at the inner Lindblad resonance, without the necessity for an explosion. There is a better theoretical basis for the dispersion ring model, and in addition one finds the peak of ionized hydrogen at 4 to 5 kpc radius in our Galaxy, with a natural explanation in terms of a shock associated with the density wave model.

van der Kruit: In our Galaxy and M 51 it is very difficult to decide between the dispersion ring model and an expansion model, because the geometries are such that the line-of-sight velocities predicted by each are indistinguishable. In NGC 4736, on the other hand, the observed velocities are consistent with an expansion model but not with a dispersion ring model. It is true that this argument depends upon an epicyclic approximation which may not be valid, but nor is there any direct observational evidence that a density wave operates in the multi-armed NGC 4736.

ON THE STRENGTH OF THE GALACTIC SHOCK WAVE
AND THE DEGREE OF DEVELOPMENT OF SPIRAL STRUCTURE

WILLIAM W. ROBERTS, JR.

University of Virginia, Charlottesville, Va., U.S.A.

MORTON S. ROBERTS

National Radio Astronomy Observatory, Charlottesville, Va., U.S.A.*

and

FRANK H. SHU

University of California at Berkeley, Calif., U.S.A.

Abstract. The luminosity of a spiral arm is believed to originate primarily in the very young, newly forming stars; and the spiral arm itself to be a spiral wave which is capable of triggering the formation of the young stars selectively along the wave crest. A semi-empirical study of the density wave patterns predicted in the density wave models of twenty-five external galaxies has been made and one result of this study is presented here. It is found that those galaxies of the sample whose models predict the possibility of strong shock waves are also the galaxies which exhibit long, well-developed spiral arms; and those galaxies whose models predict weak shock waves are also the galaxies which exhibit less-developed spiral structure. This trend is seen through a correlation between $w_{\perp 0}$, the velocity component of basic rotation normal to a spiral arm, which is an important parameter in determining the shock strength on the one hand, and luminosity class, which is a measure of the degree of development of spiral structure on the other.

The spiral structure often observed in disk-shaped galaxies is commonly thought to be associated with wave phenomena. The luminosity of a spiral arm originates primarily in the very young, newly-forming stars, and the spiral arm is believed to be a spiral wave which is capable of triggering from the gas the formation of young stars selectively along the wave crest. The wave itself has been seen from two different viewpoints: first, as a density wave (Lindblad and Langebartel, 1953; Lindblad, 1963; Lin and Shu, 1964, 1966; Lin *et al.*, 1969; Lin, 1971; Roberts and Yuan, 1970) in which gravitational forces are dominant, with magnetic forces playing only a minor role; and second, as a hydromagnetic wave (Piddington 1967a, b; 1970; 1973a, b) in which magnetic fields are dominant.

Of these two, only the density wave model has been developed sufficiently to provide quantitative predictions regarding spiral structure. In this model, galactic shock waves form in the gaseous component of the galactic disk as a necessary consequence of the theory of waves for sufficiently large amplitudes (Shu *et al.*, 1973). As the nonlinear counterpart of the small-amplitude linear density wave, the galactic shock is visualized as a possible triggering mechanism for the gravitational collapse of gas clouds, leading to star formation along a spiral arm (Roberts 1969, 1972).

One might wonder how important such waves are in influencing the evolution

* Operated by Associated Universities, Inc., under contract with the National Science Foundation.

of a galaxy and in governing the generation and appearance of the spiral structure. Furthermore, one could question the entire basis of the wave interpretation and ask: are those galaxies in which strong shock waves are predicted also the galaxies which exhibit long, well-developed spiral arms?

In recent work (Roberts, Roberts and Shu, in preparation; also Shu *et al.*, 1971), a semi-empirical study has been made of the density wave patterns predicted for twenty-five galaxies. Included in this study are all galaxies with observed rotation curves, except the Magellanic Clouds, which satisfy the following two criteria:

(i) a luminosity classification has been assigned by van den Bergh (1960a, b),

(ii) the velocity data prescribing the rotation curve are complete over a significant fraction of the photometric radius of the galaxy.

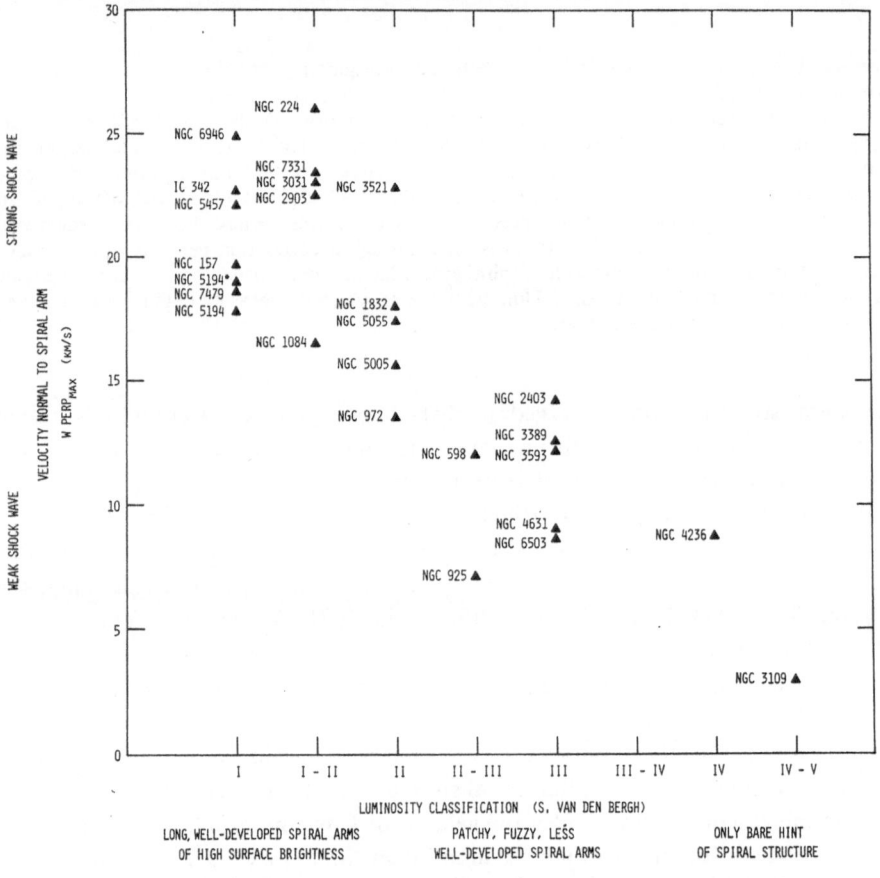

Fig. 1. A trend for a sample of twenty-five galaxies indicative of a correlation between $w_{\perp 0}$ – the velocity component of basic rotation normal to a spiral arm – and shock strength on the one hand, and luminosity classification and degree of development of spiral arms on the other. Those galaxies in which strong shock waves are predicted are found to exhibit long, well-developed spiral arms; and those galaxies in which weak shock waves are predicted are found to exhibit less-developed spiral structure.

By means of a standardized least squares fit to the rotation velocity data, an overall mass model for each galaxy is constructed, consisting of a mass model (Toomre, 1963) to cover the disk, together with one or two spheroidal mass models to cover the nuclear bulge, where necessary. On the assumption that the co-rotation radius is that of the easily visible disk and the outermost H II region, the density wave pattern for each galaxy is determined.

Figure 1 illustrates one result of this investigation. Plotted on the vertical axis is the maximum value reached by $w_{\perp 0}$, the velocity component of basic rotation normal to a spiral arm. The strength of the shock varies roughly as the square of $w_{\perp 0}$, and therefore the galactic shock is expected to be very strong in those galaxies with high values of $w_{\perp 0}$ and very weak in those with low values of $w_{\perp 0}$. Plotted on the horizontal axis is the luminosity classification by van den Bergh. Galaxies with long, well-developed arms of relatively high surface brightness are classified in category I; those with short, patchy, less-developed arms in categories II and III; and those with only a hint of spiral structure in categories IV and V.

Apparent in Figure 1 is a trend indicative of a possible correlation between $w_{\perp 0}$ and shock strength on the one hand, and degree of development of spiral structure on the other. Those galaxies whose models predict the possibility of strong shocks are found to be the galaxies which exhibit long, well-developed spiral arms; and those whose models predict weak shocks are found to be the galaxies which exhibit less-developed spiral structure.

The strengths of the compression regions along the spiral arms of eleven of these galaxies have been determined by van der Kruit (1973) from studies with the Westerbork Synthesis Radio Telescope, and his subsample shows a trend in general agreement with that for the larger sample of twenty-five galaxies in Figure 1 here.

References

Lin, C. C.: 1971, in C. de Jager (ed.), *Highlights of Astronomy* **2**, 88.
Lin, C. C. and Shu, F. H.: 1964, *Astrophys. J.* **140**, 646.
Lin, C. C. and Shu, F. H.: 1966, *Proc. Nat. Acad. Sci.* **55**, 229.
Lin, C. C., Yuan, C., and Shu, F. H.: 1969, *Astrophys. J.* **155**, 721.
Lindblad, B.: 1963, *Stockholm Obs. Ann.* **22**, 3.
Lindblad, B. and Langebartel, R. G.: 1953, *Stockholm Obs. Ann.* **17**, 6.
Piddington, J. H.: 1967a, *Monthly Notices Roy. Astron. Soc.* **136**, 165.
Piddington, J. H.: 1967b, *Planetary Space Sci.* **15**, 1625.
Piddington, J. H.: 1970, *Australian J. Phys.* **23**, 731.
Piddington, J. H.: 1973a, *Astrophys. J.* **179**, 755.
Piddington, J. H.: 1973b, *Monthly Notices Roy. Astron. Soc.* **162**, 73.
Roberts, W. W.: 1969, *Astrophys. J.* **158**, 123.
Roberts, W. W.: 1972, *Astrophys. J.* **173**, 259.
Roberts, W. W. and Yuan, C.: 1970, *Astrophys. J.* **161**, 877.
Shu, F. H., Milione, V., and Roberts, W. W.: 1973, *Astrophys. J.* **183**, 819.
Shu, F. H., Stachnik, R. V., and Yost, J. C.: 1971, *Astrophys. J.* **166**, 465.
Toomre, A.: 1963, *Astrophys. J.* **138**, 385.
van den Bergh, S.: 1960a, *Astrophys. J.* **131**, 215.
van den Bergh, S.: 1960b, *Astrophys. J.* **131**, 558.
van der Kruit, P. C.: 1973, *Astron. Astrophys.* **29**, 263.